Ecology of Estuaries: Anthropogenic Effects

Author
Michael J. Kennish, Ph.D.
Institute of Marine and Coastal Sciences
Fisheries and Aquaculture TEX Center
Cook College, Rutgers University
New Brunswick, New Jersey

CRC Press
Boca Raton Ann Arbor London

Library of Congress Cataloging-in-Publication Data

Kennish, Michael J.
  Ecology of estuaries : anthropogenic effects / Michael J. Kennish.
    p.   cm.
  Includes bibliographical references and index.
  ISBN 0-8493-8041-3
  1. Estuarine ecology.   2. Estuarine pollution—Environmental
aspects.   3. Man—Influence on nature.   I. Title.
QH541.5.E8K47   1991
574.5′26365—dc20
                                            91-29291
                                               CIP

Direct all inquiries to CRC Press, Inc., 2000 Corporate Blvd., N.W., Boca Raton, Florida, 33431.

© 1992 by CRC Press, Inc.

International Standard Book Number 0-8493-8041-3

Library of Congress Card Number 91-29291
Printed in the United States
0 1 2 3 4 5 6 7 8 9

THE AUTHOR

Michael J. Kennish, Ph.D., is a Supervising Research Scientist in the Institute of Marine and Coastal Sciences at Cook College, Rutgers University, New Brunswick, New Jersey.

He graduated in 1972 from Rutgers University, Camden, New Jersey, with a B.A. degree in geology and obtained his M.S. and Ph.D. degrees in the same discipline from Rutgers University, New Brunswick, New Jersey, in 1974 and 1977, respectively.

Dr. Kennish's professional affiliations include the American Fisheries Society (Mid-Atlantic Chapter), American Geophysical Union, American Institute of Physics, Estuarine and Coastal Sciences Association, Estuarine Research Federation, New England Estuarine Research Society, Atlantic Estuarine Research Society, Southeastern Estuarine Research Society, Gulf Estuarine Research Society, Pacific Estuarine Research Society, National Shellfisheries Association, New Jersey Academy of Science, and Sigma Xi. He is also a member of the Advisory Board of the Fisheries and Aquaculture TEX Center of Rutgers University, overseeing the development of fisheries and shellfisheries in estuarine and marine waters of New Jersey.

Although maintaining research interests in broad areas of marine ecology and marine geology, Dr. Kennish has been most actively involved with investigations of anthropogenic effects on estuarine ecosystems and with studies of deep-sea hydrothermal vent environments. He is the author of *Ecology of Estuaries* (Volumes 1 and 2) published by CRC Press, the editor of *Practical Handbook of Marine Science,* published by CRC Press, and the co-editor of *Ecology of Barnegat Bay, New Jersey,* published by Springer-Verlag. In addition to these four books, Dr. Kennish has published articles in scientific journals and presented papers at numerous conferences. Currently, he is the co-editor of the journal, *Reviews in Aquatic Sciences,* the marine science editor of the journal, *Bulletin of the New Jersey Academy of Science,* and Series Editor of Marine Science Books for CRC Press. His biographical profile appears in Who's Who in Frontiers of Science and Technology and Who's Who Among Rising Young Americans.

DEDICATION

TO

The faculty and staff of the Institute of Marine and Coastal
Sciences of Rutgers University

PREFACE

The rapid development of estuarine ecology as a field of scientific inquiry reflects a growing awareness of the immense societal importance of these coastal ecosystems. The many, varied, and increasingly complex problems arising from anthropogenic use of estuaries have focused attention on the pressing need to protect estuarine resources and to assess each system's physical, chemical, and biological conditions. Unequivocally, the study of estuaries involves both ecological and societal issues. As a result of the heightened public concern related to the water quality of coastal regions in the U.S., for example, numerous detailed and quantitative research programs were undertaken in the 1960s, 1970s, and 1980s, which provided a basic understanding of the processes operating in many estuaries. The information derived from these programs has also aided decision-making bodies at all levels — that plan for the utilization of natural resources — to make sound decisions dealing with these dynamic waters.

Estuaries and nearshore oceanic waters are susceptible to a multitude of human wastes from a burgeoning population in the coastal zone. These highly sensitive ecosystems serve as repositories for dredge spoils, sewage sludge, and industrial and municipal effluents. Areas with the highest levels of contaminants border metropolitan centers. The literature on estuarine ecology is replete with references to impacts on biotic populations attributable to organochlorine insecticides, PCBs, organotin compounds, heavy metals, toxic nonmetals, oil spills, and nutrient loadings. Considering the pervasive and enigmatic anthropogenic problems plaguing these coastal ecotones and the tremendous economic importance of estuaries to man, it behooves aquatic scientists from all disciplines to make a concerted effort to determine the overall impact of human activity on them.

The principal objective of this book is to examine anthropogenic effects on estuaries. The volume has been designed as a text for undergraduate and graduate students as well as a reference for scientists conducting research on estuarine systems. However, administrators, managers, decision makers, and other professionals involved in some way with investigations of estuaries can also find value in the publication. I have attempted to assemble citations of the major research articles and books treating anthropogenic effects on estuaries and to present the subject matter in an organized framework that should facilitate its use.

I am grateful to the reviewers of this work. In addition, I am indebted to the scientists, colleagues, family members, and friends who have provided encouragement and inspiration during the production of the book. In Rutgers Univesity, I express deep appreciation to K. W. Able, J. F. Grassle, J. N. Kraeuter, R. E. Loveland, R. A. Lutz, and N. P. Psuty for innumerable exchanges of ideas on estuaries and coastal marine waters. Special thanks are given to R. A. Lutz for his unending help over the years. The Editorial Department of CRC Press is acknowledged for its efficiency and guidance during the publication process. Finally, I am most appreciative of my wife, Jo-Ann, and sons, Shawn and Michael, for their love and understanding during the preparation of the manuscript.

TABLE OF CONTENTS

Introduction
Types of Pollutants 1
Waste Disposal Strategies 3
Biological Effects of Waste Disposal 6
    Bioaccumulation 6
    Biotransformation and Metabolism of Pollutants
    in Estuarine Organisms 10
        Biotransformation 10
        Elimination of Xenobiotic Compounds 11
        Toxicity 14
Plan of This Volume 19
References 26

1 Organic Loading
Introduction 33
Eutrophication Problems 34
    Sewage 34
        Methods of Disposal 34
        Sewage Sludge Composition 36
    Eutrophication and Organic Carbon Enrichment 38
        Eutrophication of Chesapeake Bay 39
            Scale of Eutrophication 41
    Nutrients 43
    Estuarine Susceptibility to Nutrient Loading 47
    Dispersion Models of Release Effluent 49
        Puff Model 51
        Particle Model 52
    Nutrient Loading, Red Tides, and Paralytic
    Shellfish Poisoning 53
Summary and Conclusions 55
References 57

2 Oil Pollution
Introduction 63
Sources of Oil Pollution 64
Composition of Oil 67
    Classes or Series 69

Alkanes 69
Cycloalkanes 69
Alkenes 70
Alkynes 70
Aromatic Hydrocarbons 70
Other Compounds 71
Biogenous Hydrocarbons 71
Toxicity of Crude Oil 71
Fate of Polluting Oil 72
Abiotic and Biotic Factors 72
Spreading 74
Evaporation 76
Photochemical Oxidation 77
Dissolution 78
Emulsification 79
Sedimentation 79
Microbial Degradation 81
Effects of Polluting Oil on Estuarine Habitats and Organisms 86
General 86
Effects of Oil on Estuarine Habitats 88
Salt Marches 88
Mudflats and Sandflats 91
Mangroves 91
Seagrasses 93
Temperate Subtidal Estuarine Habitats 93
Effects of Oil on Estuarine Organisms 94
General 94
Types of Organisms 95
Phytoplankton 95
Zooplankton 96
Benthos 97
Fish 105
Birds 109
Mammals 114
Case Studies 115
The *Florida* Oil Spill 115
The *Arrow* Oil Spill 116
The *Exxon Valdez* Oil Spill 117

Summary and Conclusions 119
References 121

3 Polynuclear Aromatic Hydrocarbons
Introduction 133
Distribution 137
    PAHs in Sediments 137
    PAHs in Biota 150
        Accumulation, Depuration, and Toxicity 150
        NOAA Mussel Watch Project 160
    PAHs in Water 163
PAH Transformation 166
    Photodegradation 166
    Chemical Oxidation 166
    Biotic Transformation 166
PAHs in Estuarine Systems 167
    Examples 167
        Chesapeake Bay 167
        Narragansett Bay 169
Environmental Cycle of PAH Compounds 170
Summary and Conclusions 172
References 175

4 Chlorinated Hydrocarbons
Introduction 183
Organochlorine Pesticides 185
    DDT 185
        DDT in Bivalves 191
        DDT in Fish 191
    Cyclodiene Pesticides 192
        Aldrin 192
        Dieldrin 193
        Endrin 193
        Lindane 194
        Chlordane 194
    Other Pesticides 195
        Kepone 195
        Mirex 195
        Toxaphene 196

Chlorinated Benzenes and Phenols 198
    Hexachlorobenzene 198
    Pentachlorophenol 198
Polychlorinated Biphenyls 199
    Introduction 199
    Physical-Chemical Properties 200
    Environmental Effects 203
        PCB Concentrations in Water 207
        PCB Concentrations in Sediments 209
        PCBs in Aquatic Organisms 211
            Factors Affecting PCB Concentrations in
            Aquatic Biota 214
            Uptake of PCBs by Estuarine Organisms:
            National Surveys 215
            Effects of PCBs on Biotic Groups 219
    Polychlorinated Terphenyls 233
Case Studies 234
    Chesapeake Bay 234
    San Francisco Bay 236
        Organochlorines in Sediments 236
        Organochlorines in Biota 239
    Puget Sound 243
Summary and Conclusions 246
References 248

5 Heavy Metals
Introduction 263
Sources of Heavy Metals 265
    Estuarine Systems 265
        River Input 265
        Atmospheric Input 271
        Anthropogenic Input 271
            Antifouling Paints 273
            Waste Disposal 276
            Smelting 282
Bioaccumulation of Heavy Metals 283
Heavy Metals in Estuarine Systems 289
    Case Studies 289

Chesapeake Bay 289
San Francisco Bay 295
Summary and Conclusions 302
References 303

6 Radioactivity

Introduction 317
Radioactivity and Radiation 318
  Radioactivity Defined 318
  Common Types of Ionizing Radiation 321
    α Particle 321
    β Particle 322
    Neutron 322
    γ Rays 323
  Sources of Radiation 323
    Natural Sources 323
      Cosmic Radiation 323
      Primordial Radionuclides 325
    Anthropogenic Sources 326
      Nuclear Fuel Cycle 326
      Nuclear Explosions 328
  Types of Radioactive Wastes 332
Radioactivity and Estuarine Organisms 334
  General 334
  Impact of Radioactivity on Estuarine Organisms 339
    Bioaccumulation 339
    Effects of Radiation on Organisms 339
      Somatic Effects 339
      Genetic Effects 344
Radioactive Waste Disposal 345
  General 345
  Storage or Disposal Systems 346
  Waste Management Strategies 349
Summary and Conclusions 350
References 352

7 Dredging and Dredged-Spoil Disposal

Introduction 357
Dredging Devices 359

Environmental Effects of Dredging and
Dredged-Spoil Disposal 363
  Adverse Effects 363
    Primary Impacts on Habitat and Organisms 363
    Effects on Water Quality 369
      Turbidity 369
      Nutrients 380
  Beneficial Effects 380
    Increased Circulation 380
    Sediment Supply and Habitat Restoration 382
    Nutrients 383
Case Studies 383
  Chesapeake Bay 383
  Alberni Inlet, British Columbia 386
Regulation of Dredged Material Disposal 388
Summary and Conclusions 390
References 392

8 Effects of Electric Generating Stations
Introduction 399
Historical Development 401
Effects of Power Plant Operation 402
  Thermal Discharges 402
    Water Quality 405
    Estuarine Organisms 406
  Waste Heat Utilization 417
  Biocides 419
  Impingement and Entrainment 423
    Impingement 425
    Entrainment 433
      Thermal Stresses 434
      Mechanical Effects 435
      Biocidal Effects 435
Case Studies 437
  Oyster Creek Nuclear Generating Station 437
    Impact on Aquatic Biota 438
      Construction Effects 438
      Cooling System Effects 439

Pilgrim Nuclear Power Station 443
   Impact on Aquatic Biota 444
      Thermal Discharges 444
      Impingement 444
      Entrainment 446
Summary and Conclusions 446
References 450

Index 461

Introduction

TYPES OF POLLUTANTS

Estuarine and marine environments have been used as major repositories of anthropogenic wastes for decades. With their increasing usage as reservoirs for a multitude of wastes during the 20th century, shallow estuarine and coastal marine ecosystems were gradually subjected to significant impacts not only in sensitive habitat areas, but also in aquatic communities inhabiting them. Growing public awareness of pollution effects in the oceanic realm in the 1960s and 1970s led to a reassessment of the value of these critical regions. Their high biotic productivity, rivaling the most intensively cultivated farmlands, was emphasized. Even though they comprise only 8% of the total area of the ocean, estuaries and nearshore oceanic areas accounted for 50% of the world fisheries harvest. Only the highly productive upwelling areas, essentially responsible for the remaining 50% of the world fisheries harvest, yielded comparable statistics.[1] Furthermore, the untold aesthetic and recreational worth of these coastal ecosystems could not be set into an economic perspective for many individuals.

Contaminants enter estuarine and marine waters via several key pathways, specifically direct pipeline discharges from coastal communities, discharges and dumping from ships, riverine input, atamospheric deposition, and nonpoint source runoff from land.[2] The most common anthropogenic wastes disposed in the coastal zone are industrial and municipal wastes, sewage sludge, and dredged materia.[1] Pollutants typically associated with these wastes include heavy metals, synthetic organic compounds (xenobiotics), organic carbon, nutrient elements, and pathogens. The New York Bight apex provides an example of an area that has received wastes

1

from all of these sources. The Hudson River plume and dredged material, containing domestic and industrial wastes released to the Hudson-Raritan estuary by inhabitants of the New York-New Jersey metropolitan complex in addition to materials transported into the estuary from upstream and from nonpoint surface runoff, contributed most of the pollutants to the apex in past years. Of secondary importance as a pollutant source in the bight is the ocean disposal of sewage sludge and industrial wastes. Direct discharges from coastal communities and atmospheric deposition supply the smallest amount of pollutants.[3]

In addition to the planned disposal of wastes at sea, accidental spills are responsible for occasional, yet substantial, quantities of pollutants in estuarine and marine waters. Large oil spills at times have eradicated entire aquatic communities. Intertidal and subtidal biotopes incur much of the damage when the oil washes ashore, as exemplified by the *Torrey Canyon* disaster in 1967.[4] Petroleum hydrocarbons associated with oil spills, as well as oil dispersant materials applied during clean-up operations, are toxic to a wide spectrum of marine organisms. Some aromatic compounds, in particular high molecular weight polynuclear aromatic hydrocarbons (PAHs), constitute potent mutagens and carcinogens. A great variety of compounds may be released during any oil spill, and their toxicity to floral and faunal populations depends on a host of factors (e.g., age of the organism, life history stage, and season of the year).[5]

Most of the oil that enters the sea does not originate from oil spillage. Approximately 45% of all the oil found in estuarine and marine waters derives from urban runoff, municipal wastes, effluents from nonpetroleum industries, and polluted rivers. About 35% originates from oil transportation activities, with only approximately 25% of this total attributable to major spillages and accidents. Natural oil seeps are responsible for the remaining 20% of the oil in the sea.[6]

Marine pollution monitoring and survey programs worldwide have detected substantive changes in the biota of impacted study sites. Environmental disasters, such as mercury poisoning in Minamata Bay, Japan,[5] instilled widespread fear that the contamination of estuarine and marine organisms could be routed back to human populations and pose a serious health threat. As evidence of progressive and, in some cases, irreparable degradation of coastal ecosystems mounted, societal concerns precipitated a series of regulatory, political, and legal actions designed to ameliorate the surge of anthropogenic disturbances of ecological systems. Two federal statutes were promulgated in the early 1970s to better control waste disposal in the sea: the Federal Water Pollution Control Act (FWPCA) of 1972, also known as the Clean Water Act, and the Marine Protection, Research, and Sanctuaries Act (MPRSA) of 1972.[7-9] The FWPCA regulates the discharge of effluents into freshwater and marine waters, and the MPRSA, the release of sewage sludge, dredged material, and industrial wastes at sea. Incineration of wastes at sea is also controlled by the MPRSA.

A reevaluation of ocean disposal by the U.S. Environmental Protection Agency (EPA) in the 1970s concluded with the adoption of a policy for the phasing out of the dumping of sewage sludge and most other domestic and industrial wastes in the ocean. Muir[10] has shown that this policy resulted in the discontinuation of sewage sludge disposal in the New York Bight by the City of Philadelphia in 1980. Nearshore ocean dumping in the New York Bight apex has been essentially terminated, with dumping relegated to deeper waters offshore at the 106-mi site along the outer continental shelf and slope.[11] Recent observations from submersibles indicate that the impact of dumping on the benthos as the 106-mi disposal site may be mollified by dilution and dispersal of the wastes.[12] The main objective has been to remove waste disposal from more sensitive, shallow water areas, where impacts on biotic systems may be most devastating.

WASTE DISPOSAL STRATEGIES

Capuzzo et al.[13] scrutinized the strategies currently in place for dealing with ocean disposal of wastes. Two primary options generally are considered for disposal of most wastes in the ocean: containment and dispersal; however, for those contaminants not easily controlled by containment or dispersal strategies, (e.g., xenobiotic compounds and pathogens), alternate measures, such as point source control and recycling, may be most effective. According to Capuzzo et al.,[13] the disposal of most wastes in the ocean cannot be handled via a containment strategy. Notable exceptions are dredged material and high-level radioactive wastes; in the case of dredged material, a submarine pit and cap can successfully confine the wastes, whereas careful containment of the high-level radioactive wastes prior to disposal should be a mandatory consideration. Recommendations of xenobiotic compounds, whose long-term effects and persistence in the environment often are unknown, incorporate as a pirority item the minimization of discharge, with the amount released being contingent upon the existing toxicity and bioaccumulation data available. For pathogens, better identification and enumeration of their presence must be achieved, together with a more effective means of disinfecting sewage.[13]

The efficacy of disposal options arises from both environmental and public health concerns related to waste disposal. These concerns focus on the accumulation of chemical contaminants as well as pathogenic organisms in marine resources destined for human consumption. In addition, the direct impact of substances on estuarine and marine ecosystems, for example, the buildup of degradable forms of organic matter and nutrients which when left uncontrolled promotes localized eutrophication, organic enrichment, and oxygen depletion, poses a problem for mankind. Finally, the toxicity of chemical contaminants potentially lowers the reproductive success and survival of aquatic organisms, thereby adversely affecting not only biotic communities but also the functioning of the ecosystem. Waste management strategies, therefore, must invoke plans that minimize the risk of contaminant exposure of biota, and by doing so will optimize conditions for a low impact on ecological systems. The execution of these plans will likewise allay fears of the transference of contaminants from the aquatic systems back to man.

Criteria have been formulated for the selection of waste disposal sites at sea by the Joint Group of Experts on the Scientific Aspects of Marine Pollution (Table 1).[14] In the past, site selection has been strongly based on proximity to the shoreline, for example, the 12-mi (apex site),.65-mi (continental shelf site), and 106-mi (deepwater site) disposal sites in the New York Bight.[3] From the standpoint of waste dispersal, however, a number of physical factors and processes should be evaluated prior to site selection. Investigations

TABLE 1
Site Selection Considerations for Waste Disposal in the Ocean

Physical	Sedimentary	Biological
Ocean flow	Physical and chemical	Fishing grounds
Surface waves	properties of wastes	Aquaculture sites
Wind-driven surface	and sediment	Breeding and
currents	Sorption capacity	nursery grounds
Interior circulation	Distribution	Migration routes
Turbulent diffusion	coefficients	Productivity
Shear diffusion	Sedimentation	Recreational
Vertical mixing	Sediment disperson	areas
Modeling advection	Bioturbation	
and diffusion	Sediment stability	

From GESAMP (IMCO/FAO/UNESCO/WMO/WHO/IAEA/UN/UNEP Joint Group of Experts on Scientific Aspects of Marine Pollution), Scientific Criteria for the Selection of Waste Disposal Sites at Sea, Reports and Studies No. 16, Inter-Governmental Maritime Consultative Organization, London, 1982. With permission.

to establish a maximum dispersal strategy must assess the density stratification of the water column, water depth, surface wave action, turbulence, and prevailing long-term mean and transient short-term currents.[1] Of these factors, water column stratification may be most important since it controls transport and mixing processes as well as the hydrodynamic stability of flow that governs dispersal of wastes.[13,14] Duedall et al.[15] detailed how sewage sludge accumulated in the New York Bight apex at the depth of the pycnocline within a seasonally stratified water column; they noted that advection of surface waters enhanced horizontal transport of the material. Lateral transport of the waste decreases and its deposition increases as stratification of the water column breaks down.[16] Bottom currents, bioturbation activities by benthic macrofauna, and storms that roil seafloor sediments, all foster remobilization of waste products and contaminants.[1]

Criteria that should be met in a maximum dispersal strategy are the following (Capuzzo et al., p. 498):[13] "...(1) dispersal of particles in areas of high resuspension and ventilation should ensure that enrichment of benthic habitats from organic carbon and nutrients will be avoided; and (2) discharge and dispersal of metals should be maintained at a level such that the deposition rate of these contaminants in marine sediments is within the same order of magnitude as natural input." In essence, the dispersive characteristics of the receiving waters dictate the dispersal strategies implemented at a given locale. It is necessary to formulate disposal strategies that maximize the dispersive characteristics of an area.

BIOLOGICAL EFFECTS OF WASTE DISPOSAL

The effects of waste input to estuaries can be manifested on several biological levels of organization. Aside from effects seen at the community level, responses of estuarine and marine organisms to anthropogenic wastes occur at the population, organismic, cellular, and subcellular levels of organization.[17] Assessment of these effects hinges, in part, on data acquisition on the bioaccumulation and toxicity of chemical contaminants comprising the waste. The ability of organisms to biotransform contaminants also is a key factor in regulating the accumulation of the compounds in tissues.

BIOACCUMULATION

Most marine species have the capability of accumulating pollutants to concentrations orders of magnitude higher than those in seawater.[18,19] Pollutants that are lipophilic and have low ater solubility tend to have great biomagnification potential.[19] One may express the ability of an organism to concentrate pollutants by means of a bioconcentration factor (BCF), defined by Fowler (p. 2)[18] "...as the ratio of the amount of substance per unit fresh weight of tissue to that in an equal weight of seawater." These concentration factors do not take into consideration the relative significance of the different routes of pollutant uptake by the organism. BCF values determined from field observations of animals in polluted waters appear to agree reasonably well with those derived from laboratory investigations.[6]

The bioaccumulation of pollutants in the tissues of aquatic organisms is particularly important because it has been widely used to delineate the degree of contamination of estuarine waters. Toxicologists often employ the terms "bioconcentration" and "biomagnification" when describing bioaccumulation. Bioconcentration refers to an organism's ability to accumulate a contaminant significantly in excess of that in the ambient water. Biomagnification, in turn, is the concentration of a pollutant up the food chain such that relatively low levels accumulate in organisms at the base of the chain and higher levels, possibly reaching harmful or lethal doses, in organisms at the to of the chain.[20] While certain substances may consistently exhibit a bioconcentration effect in biota, they do not necessarily biomagnify up the food chain. Mance[21] has shown this to be the case among heavy metals.

The biomagnification of certain toxins, for example, chlorinated hydrocarbon compounds (DDT, PCBs, etc.), poses a threat to higher trophic level organisms, including humans. Although it is widely believed that many contaminants pass through succeeding trophic levels, this phenomenon has not been unequivocally established for a large number of them.[6] More research most be performed on uptake rates, distribution in the body, and release rates of pollutants in test animals. The attainment of a "critical" concentration of lipophilic pollutants in the tissues of aquatic organisms, for instance, often triggers pathological symptoms. Evaluation of uptake and release kinetics of biota with regard to these pollutants is most important.[19]

The bioavailability of contaminants in estuarine and marine waters depends on biological, chemical, and physical processes that also modulate their concentrations. The bioaccumulation of pollutants is strongly coupled to bioavailability.[17] Biological processes capable of influencing the bioavailability of contaminants include microbial degradation and bioturbation. Among chemical processes, dissolution, redox reactions, and sorption-desorption of compounds can alter the chemical form of contaminants. Currents and tidal exchange are physical forces that may also affect their bioavailability.

When the chemical uptake by an organism surpasses the elimination rate, bioaccumulation can take place.[22] Adsorption of chemicals onto body surfaces, ingestion of detritus and food, and exchange of water at feeding and respiratory surfaces represent major routes of entry of contaminants into aquatic organisms. They tend to be balanced by active and passive removal mechanisms, such as the excretion of metabolized by-products and the generation of particulate products (e.g., eggs, feces, and molts).[17] Direct chemical uptake has been demonstrated for algae, zooplankton, annelids, arthropods, mollusks, and fish.[22] Phytoplankton, zooplankton, and macroalgae, for example, rapidly assimilate heavy metals and radionuclides from water, food, or sediments.[23-30] In the case of phytoplankton and zooplankton, the large surface area to volume ratio plays a prominent role in the high accumulation rates. Faunal groups other than zooplankton which accumulate heavy metals and radionuclides from estuarine waters are macroinvertebrates[30-36] and fishes.[37-39] Several factors influence the uptake of these substances by the aforementioned groups of organisms, especially the physical and chemical form of the element and organismal size.

More provocative targets of bioaccumulation studies are organo-chlorine and petroleum hydrocarbon comounds. DDT and PCBs, two organochlorine compounds gaining national attention in the 1960s and 1970s, accumulate to relatively high levels in some organisms. Phytoplankton, zooplankton, fish, waterfowl, and other biota can concentrate these compounds directly from water and sediments; the faunal populations likewise extract them from the foods they consume. Because they are lipophilic, these organic pollutants tend to partition out into the fatty tissues of the organ-isms.[40-50] By direct lipid/water partitioning, bacteria accumulate certain chlorinated hydrocarbon compounds to levels 10^2 to more than 10^4 times their concentrations in water.[51] Phytoplankton and zooplankton take up DDT and PCBS from water via adsorption and absorption processes, although assimilation from food may be more important for zooplankton, leading to elevated organo-chlorine body burdens. Marine zooplankton reportedly have con-centration factors as high as 10^5 for DDT and PCBs.[52,53] Fish accu-mulate organochlorines from water principally by absorption across the gill membrane.[18,54] A similar absorption process occurs in the uptake of other pollutants (e.g., PAHs).[55] As in zooplankton, however, the dietary pathway of chlorinated hydrocarbon accumu-lation may be equally or even more important in larger inverte-brates and fish. Clearly, future work must assess the relative sig-nificance of dietary vs. water pathways in organchlorine contami-nation of estuarine organisms.

Petroleum hydrocarbons are likewise assimilated quite rapidly from water by zooplankton and benthic invertebrates.[56] Bivalve mollusks, in particular, have only a limited capability of metabolizing petroleum hydrocarbon compounds; hence, they can accumulate large quantities of these contaminants.[57,58] The early life history stages of estuarine and marine animals — eggs, larvae, and juvenile forms — usually are more sensitive to these compounds than adults. When subjected to crude oil or refined petroleum products, they may experience physiological, behavioral, or developmental abnormalities. The low molecular weight compounds, which are volatile, evaporate relatively quickly to the atmosphere and, consequently, usually have less of an impact on biota. High molecular weight tars remain less toxic than middle molecular weight constituents. Aliphatics are less toxic than aromtic compounds. The sensitivity of organisms to oil pollution depends on multiple factors, such as the age, size, and maturity of the individual, time of the year, type of oil, and other factors.[5]

In addition to accumulating petroleum hydrocarbon compounds from water, estuarine fauna also assimilate them from food. Experimental evidence on zooplankton substantiates the significance of food-chain transfer of petroleum hydrocarbons.[59] Some data have been published suggesting that the dietary pathway may be less important among detritivores.[18] Petroleum hydrocarbon uptake in fish derives from drinking, feeding, and absorption through the gills.[18]

Other chemicals easily accumulated (i.e., bioconcentrated) by estuarine organisms are PAHs. While oil spillages constitute a substantial source of PAHs in localized areas, alternate (principally anthropogenic) routes exist for their entry into estuarine and marine environments: automobile and industrial emissions, sewage effluents, forest and grass fires, urban runoff, and harbor dredged spoils. More than 95% of the total PAHs in aquatic environments originate from petroleum spillage and atmospheric deposition.[60] The combustion of fossil fuels concomitant with atmospheric transport appears to be a primary source of the compounds.

Many estuarine organisms conspicuously concentrate PAH compounds. The bioconcentration of these substances is manifested most clearly in those organisms incapable of metabolizing them (e.g., algae and mollusks). As the molecular weight of the PAH increases, a tendency exists for bioconcentration factors to rise as well. Bivalves, most obviously, accumulate large quantities of PAH compounds and thus have been utilized as sentinel organisms to assess temporal and spatial trends of the contaminants in estuarine and coastal marine waters.[61] The ability or inability or organisms to metabolize PAHs influences the biological transport, bioaccumulation, disposition, and toxicity of the pollutants in aquatic habitats.[62]

BIOTRANSFORMATION AND METABOLISM OF POLLUTANTS IN ESTUARINE ORGANISMS
Biotransformation

As xenobiotic compounds accumulate in estuarine and marine fauna, metabolism or biotransformation of the foreign substances acts to ameliorate their toxicological effects. Lech and Vodicnik (p. 526)[63] define biotransformation as "...the biologically catalyzed conversion of one chemical into another", such that the metabolites of the foreign compounds are more easily disposed of by the organism. For example, processes of oxidation, reduction, hydrolysis, or conjugation convert lipophilic organic contaminants to more wate-soluble products that the animal eliminates.[19] The rate of elimination of the substances from an organism effects their toxicity as well as their bioaccumulation potential. The more rapidly a toxic chemical is eliminated; the less likely tissue damage will ensue.[22] In connection with this, water-soluble (polar) compounds tend to be more easily excreted than lipid-soluble (nonpolar) types.

The metabolic pathways that enable estuarine and marine fauna to eliminate potentially dangerous chemical compounds are referred to as toxicogenic/detoxification systems.[64] These pathways most often involve reactions carried out under the influence of enzymes occurring in the soluble, mitochondrial, or microsomal fractions of the liver, or, in some species, in the intestine, kidney, and lung.[63] For instance, the oxidation of xenobiotic compounds in estuarine fish, mammals, and some invertebrates via hydroxylation, O-dealkylation, N-dealkylation, or epoxidation takes place by the activation of a group of hepatic microsomal enzymes known as the cytochrome P-450 mixed function oxygenase (MFO) system.[9,55] Mammalian species generally biotransform a wide range of substrates. In contrast, the ability of invertebrates to metabolize xenobiotic compounds is not uniform, as exhibited by mollusks, crustaceans, and polychaetes. The hard clam, *Mercenaria mercenaria,* as well as most mollusks, apparently lack the MFO cytochrome P-450 system and therefore, cannot metabolize certain contaminants, such as PAH compounds. Brown shrimp, *Penaeus aztecus,* in addition to other crustacean populations (e.g., the copepod, *Calanus plumchrus*), have greater PAH-metabolizing abilities. Hence, these crustaceans release PAHs much more rapidly than the molluscan species. Pesch[64] observed that the polychaete, *Neanthes arenaceodentata,* can metabolize benzo(a)pyrene, cyclophosphamide, and dimethylnitrosamine. Polychaetes generally possess an MFO system,[65,66] enabling them to metabolize a variety of mutagenic and carcinogenic compounds. They are deemed to be valuable test organisms for assessing the bioavailability and toxicity of contaminants in sediments.[64]

Giam et al.[19] and Lee and Quattrochi[67] describe the MFO systems in fish and invertebrates. NADPH-cytochrome *c* reductase, cytochrome P-450, and phospholipids comprise the primary components of the MFO systems of these faunal groups. In fish, MFO activity peaks in microsomes from hepatic tissues. Crabs show the greatest MFO activity in the microsomes of the green gland and stomach, although activity can be detected in most tissues. Low activity has been discerned in the blood, gill, reproductive tissues, eyestalk, cardiac muscle, and hepatopancreas. The highest MFO activity in polychaetes occurs in the intestine.

Factors known to influence MFO systems in mammals include age, sex, and diet. Among fish and crustaceans, gonadal status, molting, and season of the year affect MFO activity. Hormonal factors possibly regulate MFO of marine systems.

Elimination of Xenobiotic Compounds

Estuarine and marine organisms gradually lose their pollutant burden by eliminating contaminants in either soluble or particulate form. The release of the soluble pollutant fraction may arise via metabolic excretion processes or simple desorption and ion exchange. The loss of the contaminated particulate fraction is coupled to the voiding of feces, molting, mucus secretion, and release of reproductive products.[18]

Estuarine biota tend to lose pollutants more slowly than they accumulate them. Results of depuration experiments on estuarine fauna indicate an initially rapid decline in pollutant concentration when individuals are first transferred from contaminated to uncontaminated waters, owing to the release of unmetabolized compounds, desorption of loosely bound pollutants, and voiding of unassimilated materials. Subsequently, however, the rate of contaminant loss diminishes, reflecting the more difficult release of the pollutants that are tightly bound to tissues.[18]

The biological half-time is a term used to quantify the elimination process. Defined by Fowler (p. 20)[18] as "the time for one half of the pollutant to be lost from an organism or a pool within the organism," the biological half-time of pollutants differs significantly among estuarine organisms. Fowler[18] examined the biological half-times of various inorganic and organic pollutants in marine organisms. Half-times of heavy metals and radionuclides in phytoplankton based on experimental studies are on the order of several hours to a few days, suggesting rapid elimination of these contaminants due largely to their desorption from extracellular binding sites. The depuration rate of a pollutant in zooplankton depends on the exposure time of the animal to the contaminant in spite of the rapid metabolic turnover rates of the group. Work on heavy metal and radionuclide excretion kinetics in marine zooplankton reveals that depuration proceeds from both fast and slow exchanging pools over the long term. Hence, elements with low assimilation efficiencies (e.g., Ce, Cs, and Pu) have biological half-times of a few days or less in zooplankton, whereas substances tightly bound within the organism (e.g., methylmercury) exhibit biological half-times of several months.

The clearance rates for chlorinated or petroleum hydrocarbons in zooplankton vary among species and are contingent upon the duration of exposure as well as the size of the organism and excretion processes. Molting of planktonic crustaceans, while enhancing the elimination of organochlorine compounds, may be less important in the loss of petroleum hydrocarbons. Smaller euphausiids that molt more frequently than larger forms release DDT at greater rates. The excretion of fecal pellets rich in chlorinated hydrocarbons represents a second, and perhaps more important, route for the loss of organochlorine body burdens in zooplankton. It also is a significant pathway for the elimination of water-soluble hydrocarbons and particulate oil fractions from individuals. Nevertheless, petroleum hydrocarbons tend to persist in zooplankton even after extensive depuration. For example, some petroleum hydrocarbon residues have been detected in zooplankton after several weeks of depuration, despite biological half-times for the contaminants of only 2 to 3 d.[59]

Benthic macroinvertebrates differ from zooplankton in that they eliminate heavy metals and radionuclides at much slower rates, commensurate with their lower metabolic rates. In general, biological half-times for these pollutants range from several days to more than 1 year, and they are a function of the type of element and the process of bioaccumulation. Mollusks, in particular, sequester these substances; reported biological half-times of certain metals and radionuclides in bivalves commonly exceed 1 year. Crustaceans release inorganic pollutants more rapidly than mollusks through molting and defecation.

As in the case of heavy metals and radionuclides, chlorinated hydrocarbons appear to be relatively persistent in benthic invertebrates, being eliminated more slowly than they are accumulated. The biological half-times for PCBs in bivalves, polychaetes, and shrimp seem to be on the order of weeks. Losses of pesticides, including dieldrin and endrin, probably proceed much more rapidly in these groups. Biological half-times for organochlorine pesticides of <1 week have been delineated in various species.

Petroleum hydrocarbons can be quickly lost from contaminated benthos. Depuration rates of petroleum hydrocarbons in bivalves are often quite high, with half-times of <1 week registered for these compounds when the organisms are contaminated with them for only a few hours. Thus, Fowler[18] and Lee[68] documented biological half-times of 2 to 7 d for bivalves subjected to short-term exposures of various petroleum hydrocarbons. When exposed to chrone contamination, however, bivalves exhibit slower depuration rates and longer half-times, approaching 60 d in some cases.[68,69] Apart from the duration of the exposure periods, the concentrations of petroleum hydrocarbons strongly influence the elimination rates for these compounds.

Crustaceans, polychaetes, and sipunculids metabolize petroleum hydrocarbons and eliminate them relatively rapidly. The primary pathway for the elimination of these contaminants from crustaceans is via defecation; however, molting also leads to reduced wholebody petroleum hydrocarbon levels. Similar to crustaceans,

polychaetes and sipunculids lose petroleum hydrocarbons principally by fecal excretion. Essentially all of these compounds are
released from the taxa within 2 weeks of exposure.

Fish have highly variable biological half-times for metals and
radionuclides, as well as for organochlorine and petroleum hydrocarbon compounds. The loss of heavy metals and radionuclides by
fish commonly follows a double rate function.[18] Fish can swiftly
eliminate elements with low assimilation efficiencies (e.g., ^{121}Pu),
but those elements assimilated and bound to internal tissue are lost
much more slowly. Hence, biological half- times of greater than
100 d have been ascertained for methylmercury lost from *Pleuronectes platessa.*

Biological half-times of organochlorine compounds in fish are
species specific, with clearance times ranging from a few hours to
weeks. Estuarine and marine species can metabolize DDT and
other organochlorine pesticides; however, they are less effective at
metabolizing PCBs. In general, fish have slower elimination rates
for organochlorine compounds than accumulation rates.

Estuarine and marine fish also eliminate petroleum hydrocarbons more slowly than they accumulate them. However, they can
metabolize the contaminants and excrete them as water-soluble
metabolites. Marine fish lose PAH compounds at different rates;
for instance, benzo(a)pyrene is lost more slowly than napthalenes.
Elimination rates for petroleum hydrocarbons tend to diminish
with time, as evidenced by long-term elimination rates of fish.[18]

Toxicity

Pollutant toxicity to estuarine and coastal marine organisms is a
subject of enormous interest today among aquatic ecologists and
toxicologists because of the continued release of large volumes of
potentially harmful contaminants in the coastal zone and the threat
that this practice poses to food chains involving man. Due to the
vulnerability of aquatic communities to the impact of a multitude
of chemicals and other xenobiotics, the need exists for the acquisition and assessment of toxicological data in the investigation of
water pollution and in the formulation of water quality standards.
Aquatic toxicology is a multidisciplinary science requiring the
input and integration of knowledge from the basic sciences, with
key areas of study often being aquatic ecology, behavior, biochem-

istry, histology, and physiology. These coastal ecosystems involve complex interactions of chemical (e.g., hydrolysis, photolysis, and oxidation/reduction), biological (bioaccumulation, biotransformation, and biodegradation), and physical (e.g., molecular structure, solubility, volatility, and sorption), factors that influence the environmental concentrations of contaminants and their potential impact on aquatic biota.[70]

Rand and Petrocelli[70] define toxicity as a relative property of a chemical that concerns its potential harmful effect on a living organism. As stated by these workers, the toxicity of a chemical to an organism is a function of its concentration and the duration of exposure. The toxic action of a chemical may be manifested in the death of the organism (i.e., lethal toxicity), in effects on the organism other than its death (i.e., sublethal toxicity), in effects that manifest themselves quickly — by convention, within a period of a few days (i.e., acute toxicity), and in effects that manifest themselves over a longer period measurable in weeks or months rather than days (i.e., chronic toxicity).[6]

Experimental conditions in toxicity tests typically expose a sample of test organisms to a certain concentration of toxin, and observations are made of the time of death of the animals. Mortality of the organisms is not simultaneous. It usually follows a sigmoid relationship to the period of exposure, when the cumulative percentage of mortality is plotted against elapsed time.[5] Customarily, the mortality data are transformed to probits by plotting on a probability scale; the elapsed times, in turn, are transformed to log times. By drawing best fit lines through the observed data points, a set of probit lines or time-mortality curves result (Figure 1). The time of death of 50% of the test organisms, the median lethal time, LT_{50}, can be read off the probit lines.

An alternative, and more widely used, method of plotting the toxicity data is to graph the percentage mortality (transformed to probits) against the concentration of the toxin (transformed to log concentration). This method yields a set of concentration-mortality curves from which the lethal concentration of LC_{50} can be read for each observation time. The median lethal concentration, sometimes referred to as the median tolerance limit (TL_M), defines the concentration of a toxin which causes 50% of the organisms to die within a specific period of time. This time is commonly 24, 48, 72,

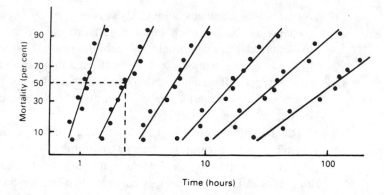

FIGURE 1. Plots of cumulative percent mortality against time showing probit lines on logarithmic-probability graph paper. Dotted lines reveal how median lethal times are determined from plot. Lines on the left are for higher toxin concentrations. (From Abel, P. D., *Water Pollution Biology,* Ellis Horwood, Chichester, U.K., 1989. With permission.)

and 96 h, with the median lethal concentration recorded as 24 h LC_{50}, 48 h LC_{50}, 72 h LC_{50}, or 96 h LC_{50} (Figure 2).

The 96-h median lethal concentration (96 h LC_{50}) is the most frequently used measure of acute toxicity when testing fish and macroinvertebrates. In the case of fish, the lack of movement, particularly gill movement, and lack of reaction to prodding usually signify death. Death in some invertebrates is more difficult to determine, and for these organisms, a median effective concentration (EC_{50}), rather than an LC_{50}, is estimated. The effect used for estimating EC_{50} with some invertebrates (e.g., daphnids and midge larvae) includes immobilization, and others (e.g., crabs, crayfish, and shrimp), immobilization and loss of equilibrium. Table 2 lists the most commonly used fish and invertebrates species in acute toxicity tests.[71]

Societal concern for the amelioration of anthropogenic impacts and preservation of estuarine and coastal marine environments has fostered the rapid expansion of aquatic toxicology as a discipline of study. The development of toxicity test methods for saltwater organisms (both invertebrates and fishes) intensified during the past 10 to 15 years.[71] Earlier aquatic toxicity tests, especially prior to the 1960s, concentrated on freshwater organisms, particularly fishes.

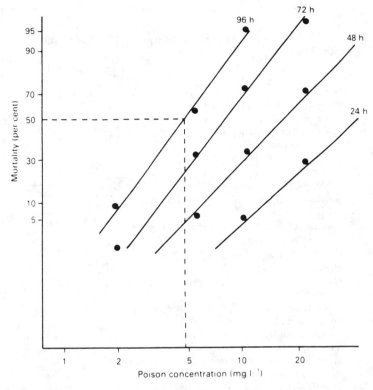

FIGURE 2. Plots of cumulative percent mortality at specified observation times (probability scale) against toxin concentration (log scale) showing resultant probit lines. Dotted lines reveal the 96 h LC_{50} value. (From Abel, P. D., *Water Pollution Biology*, Ellis Horwood, Chichester, U.K., 1989. With permission.)

While laboratory toxicity tests incorporate unnatural conditions, they nevertheless provide a conservative estimate of the possible effects of contaminants in the field. Large numbers of chemicals can be screened via acute toxicity tests, and the sensitivity of organisms to the chemicals evaluated. The major advantage of acute toxicity tests is that they yield a quick, relatively inexpensive, and reproducible estimate of the toxic effects of a contaminant. Disadvantages include the following: (1) the tests do not supply substantive information concerning the sublethal or cumulative effects of a contaminant and (2) potential chronic toxicity cannot be predicted from the tests.[71] The chronic or full life cycle toxicity

TABLE 2
Fish and Invertebrate Species Commonly Used for Acute Toxicity Tests

Freshwater
  Vertebrates
    Rainbow trout, *Salmo gairdneri*
    Brook trout, *Salvelinus fontinalis*
    Fathead minnow, *Pimephales promelas*
    Channel catfish, *Ictalurus punctatus*
    Bluegill, *Lepomis macrochirus*
  Invertebrates
    Daphnids, *Daphnia magna, D. pulex, D. pulicaria*
    Amphipods, *Gammarus lacustris, G. fasciatus,*
      G. pseudolimnaeus
    Crayfish, *Orconectes* sp., *Cambarus* sp., *Procambarus*
      sp., or *Pacifastacus ieniusculus*
    Midges, *Chironomus* sp.
    Snails, *Physa integra*
Saltwater
  Vertebrates
    Sheepshead minnow, *Cypinodon variegatus*
    Mummichog, *Fundulus heteroclitus*
    Longnose killifish, *Fundulus similis*
    Silverside, *Menidia* sp.
    Threespine stickleback, *Gasterosteus aculeatus*
    Pinfish, *Lagodon rhomboides*
    Spot, *Leiostomus xanthurus*
    Sand dab, *Citharichthys stigmaeus*
  Invertebrates
    Copepods, *Acartia tonsa, A. clausi*
    Shrimp, *Penaeus setiferus, P. duorarum, P. aztecus*
    Grass shrimp, *Palaemonetes pugio, P. vulgaris*
    Sand shrimp, *Crangon septemspinosa*
    Mysid shrimp, *Mysidopsis bahia*
    Blue crab, *Callinectes sapidus*
    Fidler crab, *Uca* sp.
    Oyster, *Crassotrea virginica, C. gigas*
    Polychaetes, *Capitella capitata, Neanthes* sp.

From Parrish, P. R., in *Fundamentals of Aquatic Toxicology: Methods and Applications,* Rand, G. M. and Petrocelli, S. M., Eds., Hemisphere Publishing, New York, 1985, 31. With permission.

test, according to Petrocelli (p. 96),[72] "...is designed to expose all life stages of the test animal — viable gametes, newly fertilized ova, early stages of developing embryos, or newly hatched larvae

— to a range of chemical concentrations estimated (from acute toxicity test exposures) to bracket the threshold for significant deleterious effects." For a comprehensive treatment of chronic toxicity testing, readers should examine the work of Petrocelli.[72]

PLAN OF THIS VOLUME

Wilson[20] recognized five main categories of contaminants commonly discharged into estuarine and coastal marine environments. They are organic matter, oil, heavy metals, organochlorines, and radioactivity. These contaminants and their basic properties which make them a cause of environmental concern are illustrated in Figure 3. In sum, the impact of the pollutants on estuarine and coastal marine systems depends on their concentrations, their toxicity, and their persistence.

In this volume, Wilson's five main categories of contaminants in estuaries are assessed, together with PAHs, dredging and dredged-spoil disposal, and the effects of thermal discharges, impingement, entrainment, and toxic chemicals from electric generating stations. Chapter 1 addresses organic loading in estuaries and the problems (e.g., eutrophication) resulting from the enrichment of these waters due to the addition of large quantities of nutrient elements. The

POLLUTANT

organic matter

oil

heavy metals

organochlorines

radioactivity

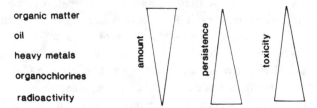

FIGURE 3. Major categories of contaminants commonly discharged into estuarine and coastal marine waters, together with the properties which make them an environmental concern. (From Wilson, J. G., *The Biology of Estuarine Management,* Croom Helm, London, 1988. With permission.)

principal sources of organic pollutants in estuaries and coastal marine waters include municipal sewage and domestic wastes and, secondarily, industrial discharges such as those related to food processing, textile, and paper manufacture. However, the largest contribution of organic and inorganic nutrients often originates from nonpoint source runoff. The susceptibility of U.S. estuaries to nutrient loading is investigated. Shallow coastal bays characterized by poor circulation appear to be the most susceptible estuaries to the effects of nutrient loading, although conditions of hypoxia or anoxia are becoming more perceptible in larger systems (e.g., Chesapeake Bay and Long Island Sound) as well. Several physical and mathematical models have been applied to the study of pollution and eutrophication in estuaries.

Chapter 2 provides an account of petroleum pollution in estuaries — both ascribable to major oil spillages and chronic additions. While major oil spillages have attracted much greater public attention, the effects of chronic oil pollution are probably much more significant worldwide in terms of the total area impacted and the number of aquatic communities stressed. Spectacular accidents involving tankers or offshore oil platforms do not represent the major source of petroleum hydrocarbons in estuarine or marine waters. Approximately 45% of the total enters the sea from urban runoff, polluted rivers, nonpetroleum industries, and municipal wastes and effluents. Another 20% is released from natural oil seeps. Accidents and major spillages contribute roughly 25% of the oil at sea.

The effects of oil spills in estuaries are considerably longer lived than in well oxygenated, exposed coastal areas, because circulation in these systems can entrap oil in bottom currents, thereby depositing oil among the benthos over protracted periods of time. Furthermore, these low energy environments may frequently experience hypoxic or anoxic conditions that inhibit degradation of petroleum hydrocarbon compounds. Hence, oil spills in these habitats can have a drastic and long-term impact on biota and must be prevented at all costs, if possible.

When oil spills do occur in the coastal zone, cleaning operations are typically intense. Clean-up measures generally undertaken focus on the application of one or more of the following: (1) emulsifier dispersants, (2) gelling agents, (3) dense oleophilic material to sink the oil, (4) burning the oil, and (5) physical removal techniques. In any event, clean-up operations are usually costly, time consuming, and difficult to successfully complete.

Chapter 3 describes PAHs in estuaries, chemicals that include some of the most carcinogenic, mutagenic, and toxic compounds found in aquatic environments. PAHs are compounds consisting of hydrogen and carbon arranged in the form of two or more fused benzene rings in linear, angular, or cluster arrangements with substituted groups possibly attached to one or more rings. The incomplete combustion of fossil fuels, industrial and sewage effluents, accidental spillage or natural seepage of oil, and grass and forest fires supply most of these chemicals to estuaries. They enter these coastal ecotones principally via riverborne influx, nonpoint source runoff, and atmospheric deposition. Once there, they disperse into the water column, become incorporated in bottom sediments, undergo chemical oxidation and biodegradation, and concentrate in the biota. PAHs have a strong affinity for particulate manner; consequently, only about one third of all PAHs occur dissolved in the water column.

The toxic effects of PAHs on estuarine organisms are well documented. PAHs tend to accumulate in bottom sediments where benthic organisms bioaccumulate, biotransform, and biodegrade them. Unsubstituted lower molecular weight PAH compounds containing two or three rings, exhibit significant acute toxicity to organisms, but are noncarcinogenic. The higher molecular weight PAHs, containing four to seven rings, are much less toxic, but many of them remain carcinogenic, mutagenic, or teratogenic to a large number of invertebrates, fishes, birds, and mammals. Certain biota (e.g., bivalves) lack MFO (enzyme) activity and, therefore, concentrate PAHs. Some of these organisms (e.g., mussels) have proven to be ideal sentinels for monitoring the contaminants. The National Oceanic and Atmospheric Administration (NOAA) directs a nationwide survey of PAHs in estuaries and coastal waters utilizing sentinel organisms, specifically mussels and oysters. Surveys of PAHs in the Chesapeake Bay system are examined.

Chapter 4 recounts the accumulation and fate of chlorinated hydrocarbon compounds in estuaries. Also referred to as organochlorine compounds, these xenobiotics are perceived to be a threat to aquatic communities, affecting organismal physiology and reproduction and ultimately leading to the death of susceptible species. This chapter covers the following organochlorine pesticides in estuaries: DDT and its metabolites; the cyclodiene pesticide group (i.e., aldrin, dieldrin, endrin, lindane, and chlordane); and keypone, mirex, and toxaphene. In addition, the chlorinated benzenes and phenols, hexachlorobenzene and pentachlorophenol, and polychlorinated biphenyl (PCB) compounds are detailed.

As a group, the chlorinated hydrocarbon compounds are highly refractory organic substances whose persistence and toxic nature has caused readily identifiable ecological damage in some localized areas. Because they accumulate in the fatty tissues of organisms, these compounds characteristically undergo biomagnification, with upper trophic level consumers having the highest concentrations of the contaminants. Thus, the most conspicuous biotic impacts of chlorinated hydrocarbon compounds often appear among fish and mammal populations that accumulate the pollutants to harmful levels. Some of these organochlorines (e.g., PCBs) have also been coupled to both acute and chronic health effects in humans (i.e., cancer, liver disease, skin lesions, and reproductive disorders). Consequently, the National Status and Trends program of NOAA and the nationwide Mussel Watch program have been designed in part to assess organochlorine contamination in aquatic biota, targeting nationwide patterns of organochlorine accumulation in fish and shellfish.

Chapter 5 deals with heavy metals in the estuarine environment. Some of these metals (e.g., copper and zinc), although essential for proper metabolism of biota at lower concentrations, are potentially toxic to estuarine organisms above a threshold availability. Moreover, metals such as cadmium, lead, and mercury, which serve no biological function, can be hazardous even when small quantities are added via anthropogenic activity. Trace elements leading to severe contamination problems in some coastal habitats include aresenic, cadmium, chromium, copper, lead, mercury, selenium, and silver; these are commonly of greatest concern in estuaries and coastal marine waters. Anthropogenic sources of heavy metals in estuaries are numerous, the more obvious being smelting operations, ash disposal, municipal and industrial wastewaters, sewage-sludge disposal, dredged-spoil dumping, and the burning of fossil fuels. Boston Harbor, Newark Bay (New Jersey), and Puget Sound (Commencement Bay and Elliott Bay, Washington) are representative estuarine systems subjected to localized contamination of heavy metals.

One important aspect of heavy metal toxicity concerns the potential for metal bioaccumulation, as demonstrated by bivalve mollusks (e.g., mussels, oysters, and clams). Although estuarine organisms frequently accumulate these elements, little biomagnification results. The highest concentrations of heavy metals are reported in biota inhabiting locally polluted habitats rather than in higher trophic level organisms attributable to a biomagnification effect.

Heavy metals may also form organometal complexes that can be particularly hazardous to estuarine and marine communities. Tributyl tin, a component of antifouling paints, is an organometal compound that poses a danger to aquatic life in estuaries and harbors throughout the world. It has been implicated in severe ecological damage of a number of estuarine systems.

The input of radioactivity in estuaries and oceans is treated in Chapter 6. Radionuclides found in estuaries originate from naturally occurring terrigenous and cosmic sources. Nuclear explosions, power and propulsion reactors, as well as wastes from hospitals, research laboratories, and industries, represent sources of artificial radioactivity that have entered estuaries. Since 1944, a significant quantity of radioactive wastes accumulating in estuarine and marine systems have been derived from nuclear weapons testing and power productions. Fallout from nuclear weapons testing spread low concentrations of radioactivity across the surface waters of the marine biosphere between 1946 and 1980. The amounts and types of radioactive wastes generated in power productions are primarily a function of the nuclear fuel cycle employed. A typical nuclear fuel cycle consists of uranium mining, milling, conversion, uranium enrichment, nuclear fuel fabrication, irradiation of fuel elements in a nuclear reactor, and the reprocessing or disposal of spent fuel. Accidents, such as the loss of nuclear submarines (e.g., the *USS Tresher* in 1963 and the *USS Scorpion* in 1968) and airborne nuclear weapons contribute to anthropogenic releases of radionuclides to the marine environment. One of the primary concerns of radioactivity in estuarine and marine environments, similar to heavy metals, is their uptake by organisms and transference through food chains to man.

Four classes of radionuclides have been discerned in estuarine and marine environments: (1) naturally occurring nuclides with short half-lives continually produced by cosmic interaction with either atmospheric constituents or extraterrestrial materials; (2) primordial nuclides of elements that also have stable elements (e.g., ^{40}K and ^{87}Rb); (3) primordial parent nuclides of the three natural radioactive decay series, ^{238}U, ^{235}U, and ^{232}Th, and the families of shorter-lived daughter nuclides; and (4) artificially produced nuclides released from the aforementioned sources. Several radioactive waste management strategies related to estuarine and marine environments have been devised, the two most acceptable being dilution and dispersion and isolation and containment. Isotopic dilution with staple isotopes and ejection of radioactive wastes to outer space have been proposed as possible future remedies to the waste disposal problem in the biosphere. The most likely disposal sites for high-level radioactive wastes, however, are geologic formations in stable repositories on land or subseabed locations in mid-plate, mid-gyre regions of the oceans.

Chapter 7 treats the subject of dredging and dredged-spoil disposal in estuaries. In order to maintain navigational waterways and port facilities for commercial (primarily) and recreational activities, substantial volumes of sediment are removed and relocated in estuarine and coastal marine environments. Along with maintenance dredging, dredging of sediment may be required for new construction. In either case, the dredged spoils generated account for large volumes of waste substances placed in coastal waters.

Although dredging enhances commercial and recreational uses of the estuary, it also modifies estuarine water quality and impacts aquatic communities. Dredging and dredged-spoil disposal affects, at least temporarily, plankton and nekton by increasing turbidity, nutrients, pollutants (e.g., heavy metals), and suspended solids, while decreasing dissolved oxygen levels. Apart from these physical and chemical effects, direct operation of the dredge destroys the benthic habitat and disrupts the benthos, which get transported along with sediment to spoil-disposal sites. Mortality of the benthos occurs from mechanical damage by the dredge itself, as well as by smothering of sediment when the organisms are picked up or deposited. Moreover, removal of intact sediments affects advective processes that control the deposition and erosion of sedimentary grains.

Potential impacts of dredged-material disposal at a dumpside invariably relate to changes in the grain size or sediment type of the habitat. Another major concern is sediment contamination by chemicals from industrial and domestic wastes. Both acute and chronic toxicity of biota have been reported from contaminated sediment disposal sites.

Construction and operation of electric generating stations in the coastal zone likewise impact aquatic communities in estuaries and coastal marine waters, as expressed in Chapter 8. During construction, the bathymetry, bottom composition, circulation patterns, and water quality in proximity to the facilities are often greatly altered owing to dredging operations required to build intake and discharge canals, docks, bulkheads, and other structures. A dramatic increase in suspended sediment usually takes place, creating a number of deleterious effects, notably a diminution of primary production both in the water column and benthos. Lowered primary production, in turn, potentially disrupts food chains and has a direct effect on primary consumer organisms.

Possible impacts of operation of electric generating stations on aquatic biota fall under four main categories: (1) thermal loading or calefaction; (2) entrainment; (3) impingement; and (4) release of toxic substances. Open-cycle, once-through cooling systems generate substantial amounts of waste heat that enter receiving waters. A multitude of thermal-induced stresses frequently arises, often manifested in the interference of normal physiological processes and/or behavioral responses of the biota. Heightened vulnerability to disease, as well as stresses imposed by changing gaseous solubilities (e.g., dissolved oxygen) and reaction rates of toxicants, exacerbates community impacts.

Impingement of organisms against intake screens and entrainment of smaller forms in the cooling water system of power plants account for greater overall mortality of biota than thermal loading. Adult invertebrates and fishes too large to pass through intake screens become impinged on them, resulting in the death and injury of many individuals. Eggs, larvae, and adult planktonic forms entrained in the facilities with cooling water are subjected to thermal, chemical, and physical stresses. Immense numbers of various life history stages of invertebrates and fishes can be killed by the entrainment process. However, it is generally very difficult to quantify the entrainment impact on adult populations in an estuary. The same holds true for impingement effects.

Toxic chemicals released by electric generating stations during routine operation include chlorine, heavy metals, and in the case of nuclear reactors, radionuclides. Chlorine is used in these facilities to control biofouling on heat exchanger surfaces. Trace metals (e.g., copper) originate from the dissolution of piping in the condenser cooling system. Low levels of radioactive waste are discharged into receiving waters. All of these toxic materials have the potential to adversely affect aquatic communities and, therefore, must be carefully regulated to minimize impacts on organisms in estuarine environments.

REFERENCES

1. Capuzzo, J. M., Burt, W. V., Duedall, I. W., Park, P. K., and Kester, D. R., The impact of waste disposal in nearshore environments, in *Wastes in the Ocean,* Vol. 6, *Nearshore Waste Disposal,* Ketchum, B. H., Capuzzo, J. M., Burt, W. V., Duedall, I. W., Park, P. K., and Kester, D. R., Eds., John Wiley & Sons, New York, 1985, 3.
2. Bourdeau, P. and Barth, H., Estuarine, coastal and ocean pollution: EEC policy and research, in *Estuarine and Coastal Pollution: Detection, Research, and Control,* Vol. 18, Moulder, D. S. and Williamson, P., Eds., Pergamon Press, Oxford, 1986, 1.
3. Swanson, R. L., Champ, M. A., O'Connor, T., Park, P. K., O'Connor, J., Mayer, G. F., Stanford, H. M., and Erdheim, E., Sewage-sludge dumping in the New York Bight apex: a comparison with other proposed ocean dumpsites, in *Wastes in the Ocean,* Vol. 6, *Nearshore Waste Disposal,* Ketchum, B. H., Capuzzo, J. M., Burt, W. V., Duedall, I. W., Park, P. K., and Kester, D. R., Eds., John Wiley & Sons, New York, 1985, 461.
4. Smith, J. E., Ed., *"Torrey Canyon" Pollution and Marine Life,* Cambridge University Press, London, 1968.
5. Clark, R. B., *Marine Pollution,* Clarendon Press, Oxford, 1986.

6. Abel, P. D., *Water Pollution Biology,* Ellis Horwood, Chichester, U.K., 1989.

7. U.S. Congress, The Federal Water Pollution Control Act Amendments of 1972, Public Law 92-532, 86 Stat. 1052, Washington, D.C., 1972.

8. U.S. Congress, Marine Protection, Research, and Sanctuaries Act of 1972, Public Law 92-500, 86 Stat. 1060, Washington, D.C., 1972.

9. U.S. Congress, An Act to Amend the Marine Protection, Research, and Sanctuaries Act of 1972, Public Law 95-153, 92 Stat. 1255, Washington, D.C., 1977.

10. Muir, W. C. History of ocean disposal in the mid-Atlantic Bight, in *Wastes in the Ocean,* Vol. 1, *Industrial and Sewage Wastes in the Ocean,* Duedall, I. W., Ketchum, B. H., Park, P. K., and Kester, D. R., Eds., John Wiley & Sons, New York, 1983, 273.

11. Runyon, R., New Jersey Department of Environmental Protection, personal communication, 1989.

12. Grassle, F., Rutgers University, personal communication, 1990.

13. Capuzzo, J. M., Burt, W. V., Duedall, I. W., Kester, D. R., and Park, P. K., Future strategies for nearshore waste disposal, in *Wastes in the Ocean,* Vol. 6, *Nearshore Waste Disposal,* Ketchum, B. H., Capuzzo, J. M., Burt, W. V., Duedall, I. W., Park, P. K., and Kester, D. R., Eds., John Wiley & Sons, New York, 1985, 491.

14. GESAMP (IMCO/FAO/UNESCO/WMO/WHO/IAEA/UN/UNEP Joint Group of Experts on Scientific Aspects of Marine Pollution), Scientific Criteria for the Selection of Waste Disposal Sites at Sea, Reports and Studies No. 16, Inter-Governmental Maritime Consultative Organization, London, 1982.

15. Duedall, I. W., O'Connors, H. B., Parker, J. H., Wilson, R. W., and Robbins, A. S., The abundances, distribution, and flux of nutrients and chlorophyll *a* in the New York Bight apex, *Est. Coastal Mar. Sci.,* 5, 81, 1977.

16. Hatcher, P. G., Berberian, G. A., Cantillo, A. Y., McGillivary, P. A., Hanson, P., and West, R. H., Chemical and physical processes in a dispersing sewage sludge plume, in *Ocean Dumping of Industrial Wastes,* Ketchum, B. H., Kester, D. R., and Park, P. K., Eds., Plenum Press, New York, 1981, 347.

17. Capuzzo, J. M. and Kester, D. R., Biological effects of waste disposal: experimental results and predictive assessments, in *Oceanic Processes in Marine Pollution,* Vol. 1, *Biological Processes and Wastes in the Ocean,* Capuzzo, J. M. and Kester, D. R., Eds., Robert E. Krieger Publishing, Malabar, FL, 1987, 3.

18. Fowler, S. W., Biological transfer and transport processes, in *Pollutant Transfer and Transport in the Sea,* Vol. 2, Kullenberg, G., Ed., CRC Press, Boca Raton, FL, 1982, 1.

19. Giam, C. S., Ray, L. E., Anderson, R. S., Fries, C. R., Lee, R., Neff, J. M., Stegeman, J. J., Thomas, P., and Tripp, M. R., Pollutant responses in marine animals: the program, in *Pollutant Studies in Marine Animals,* Giam, C. S. and Ray, L. E., Eds., CRC Press, Boca Raton, FL, 1987, 1.

20. Wilson, J. G., The Biology of Estuarine Management, Croom Helm, London, 1988.
21. Mance, G., Pollution Threat of Heavy Metals in the Aquatic Environment, Elsevier, London, 1987.
22. Spacie, A. and Hamelink, J. L., Bioaccumulation, in *Fundamentals of Aquatic Toxicology: Methods and Application,* Rand, G. M. and Petrocelli, S. R., Eds., Hemisphere Publishing, New York, 1985, 495.
23. Fowler, S. W., Heyraud, M., and Cherry, R. D., Accumulation and retention of plutonium by marine zooplankton, in *Activities of the International Laboratory of Marine Radioactivity,* 1976 Report, International Atomic Energy Agency, Vienna, 1976, 42.
24. Davies, A. G., Pollution studies with marine plankton. II. Heavy metals, *Adv. Mar. Biol.,* 15, 381, 1978.
25. Topcuoglu, S. and Fowler, S. W., Factors affecting the biokinetics of technetium (^{95m}Tc) in marine macroalgae, *Mar. Environ. Res.,* 12, 25, 1984.
26. Guimaraes, J. R. D. and Penna-Franca, E., ^{137}Cs, ^{60}Co, and ^{125}I bioaccumulation by seaweeds from the Angra dos Reis nuclear power plant region, *Mar. Environ. Res.,* 16, 77, 1985.
27. Sanders, J. B. and Cibik, S. J., Reduction of growth rate and resting spore formation in a marine diatom exposed to low levels of cadmium, *Mar. Environ. Res.,* 16, 165, 1985.
28. Graneli, E., Persson, H., and Edler, L., Connection between trace metals, chelators, and red tide blooms in the Laholm Bay, SE Kattegat — an experimental approach, *Mar. Environ. Res.,* 18, 61, 1986.
29. Toudal, K. and Riisgaard, H. U., Acute and sublethal effects of cadmium on ingestion, egg production, and life-cycle development in the copepod *Acartia tonsa, Mar. Ecol. Prog. Ser.,* 37, 141, 1987.
30. Balogh, K. V., Comparison of mussels and crustacean plankton to monitor heavy metal pollution, *Water Air Soil Pollut.,* 37, 281, 1988.
31. Devi, V. U., Heavy metal toxicity to fiddler crabs, *Uca annulipes* Latreille and *Uca triangularis* (Milne Edwards): tolerance to copper, mercury, cadmium, and zinc, *Bull. Environ. Contam. Toxicol.,* 39, 1020, 1987.
32. Bruegmann, L. and Lange, D., Trace metal studies on the starfish *Asteriasrubens* L. from the western Baltic Sea, *Chem. Ecol.,* 3, 295, 1988.
33. Langston, W. J., *A Survey of Trace Metals in Biota from the Mersey Estuary,* Marine Biology Association of the U.K., Plymouth, England, 1988.
34. Sinex, S. A. and Wright, D. A., Distribution of trace metals in the sediments and biota of Chesapeake Bay, *Mar. Pollut. Bull.,* 19, 425, 1988.
35. Prasad, P. N., Neelakantan, B., and Shanmukhappa, H., Trace metal metabolism in marine crustaceans, *Environ. Ecol.,* 7, 279, 1989.
36. Evans, R. D., Andrews, D., and Cornett, R. J., Chemical fractionation and bioavailability of cobalt-60 to benthic deposit-feeders, *Can. J. Fish., Aquat. Sci.,* 45, 228, 1988.

37. Jensen, A. and Cheng, Z., Statistical analysis of trend monitoring data of heavy metals in flounder *(Platichthys flesus)*, *Mar. Pollut. Bull.*, 18, 230, 1987.

38. Vas, P., Observations of trace metal concentrations in a carcharhinid shark, *Galeorhinus galeus*, from Liverpool Bay, *Mar. Pollut. Bull.*, 18, 193, 1987.

39. Harding, L. and Goyette, D., Metals in northeast Pacific coastal sediments and fish, shrimp, and prawn tissues, *Mar. Pollut. Bull.*, 20, 187, 1989.

40. Bazulic, D., Najdek, M. Pavoni, B., and Orio, A. A., PCB effects on production of carbohydrates, lipids, and proteins in the marine diatom *Phaeodactylum tricornutum, Comp. Biochem. Physiol.*, 91C, 409, 1988.

41. Knickmeyer, R. and Steinhart, H., Cyclic organochlorines in plankton from the North Sea in spring, *Est. Coastal Shelf Sci.*, 28, 117, 1989.

42. Tavares, T. M., Rocha, V. C., Porte, C., Barcelo, D., and Albaiges, J., Application of the mussel watch concept in studies of hydrocarbons, PCBs, and DDT in the Brazilian Bay of Todos os Santos (Bahia), *Mar. Pollut. Bull.*, 19, 575, 1988.

43. Flores Baez, B. P. and Bect, M. S. G., DDT in *Mytilus edulis:* statistical considerations and inherent variability, *Mar. Pollut. Bull.*, 20, 496, 1989.

44. Murty, A. S., *Toxicity of Pesticides to Fish*, Vol. 1, CRC Press, Boca Raton, FL, 1986.

45. Greig, R. A. and Sennefelder, G., PCB concentrations in winter flounder from Long Island Sound, 1984-1986, *Bull. Environ. Contam. Toxicol.*, 39, 863, 1987.

46. Morris, R. J., Law, R. J., Allchin, C. R., Kelly, C. A., and Fileman, C. F., Metals and organochlorines in dolphins and porpoises of Cardigan Bay, West Wales, *Mar. Pollut. Bull.*, 20, 512, 1989.

47. Borlakoglu, J. T., Wilkins, J. P. G., and Walker, C. H., Polychlorinated biphenyls in fish-eating sea birds — molecular features and metabolic interpretations, *Mar. Environ. Res.*, 24, 15, 1988.

48. Ohlendorf, H. M. and Fleming, W. J., Birds and environmental contaminants in San Francisco and Chesapeake Bays, *Mar. Pollut. Bull.*, 19, 487, 1988.

49. Phillips, D. J. H. and Spies, R. B., Chlorinated hydrocarbons in the San Francisco estuarine ecosystem, *Mar. Pollut. Bull.*, 19, 445, 1988.

50. Boon, J. P., Eijgenraam, F., Everaarts, J. M., and Duinker, J. C., A structure-activity relationship (SAR) approach towards metabolism of PCBs in marine animals from different trophic levels, *Mar. Environ. Res.*, 27, 159, 1989.

51. Grimes, D. J. and Morrison, S. M., Bacterial bioconcentration of chlorinated hydrocarbon insecticides from aqueous systems, *Microb. Ecol.*, 2, 43, 1975.

52. Cox, J. L., Uptake, assimilation, and loss of DDT residues by *Euphausia pacifica*, a euphausiid shrimp, *Fish. Bull., U.S.*, 69, 627, 1971.

53. Ware, D. M. and Addison, R. F., PCB residues in plankton from the Gulf of St. Lawrence, *Nature*, 246, 519, 1973.

54. Addison, R. F., Organochlorine compounds in aquatic organisms: their distribution, transport, and physiological significance, in *Effects of Pollutants on Aquatic Organisms,* Lockwood, A. P. M., Ed., Cambridge University Press, London, 1976, 127.

55. Neff, J. M., Polycyclic aromatic hydrocarbons, in *The Aquatic Environment: Sources, Fates, and Biological Effects,* Applied Science, London, 1979.

56. Jordan, R. E. and Payne, J. R., *Fate and Weathering of Petroleum Spills in the Marine Environment: A Literature Review and Synopsis,* Ann Arbor Science, Ann Arbor, MI, 1980.

57. Strömgren, T., Effect of oil and dispersants on the growth of mussels, *Mar. Environ. Res.,* 21, 239, 1987.

58. Berthou, F., Balouët, G., Bodennec, G., and Marchang, M., The occurrence of hydrocarbons and histopathological abnormalities in oysters for seven years following the wreck of the *Amoco Cadiz* in Brittany (France), *Mar. Environ. Res.,* 23, 103, 1987.

59. Harris, R. P. Berdugo, V., Corner, E. D. S., Kilvington, C. C., and O'Hara, S. C. M., Factors affecting the retention of a petroleum hydrocarbon by marine planktonic copepods, in *Fate and Effects of Petroleum Hydrocarbons in Marine Organisms and Ecosystems,* Wolfe, D. A., Ed., Pergamon Press, Oxford, 1977, chap. 30.

60. Eisler, R., Polycyclic Aromatic Hydrocarbon Hazards to Fish, Wildlife, and Invertebrates: A Synoptic Review, Biol. Rep. 85(1.11), U.S. Fish and Wildlife Service, Washington, D.C., 1987.

61. National Oceanic and Atmospheric Administration. A Summary of Data on Tissue Contamination from the First Three Years (1986–1988) of the Mussel Watch Project, NOAA Tech. Mem. NOS OMA 49, National Oceanic and Atmospheric Administration, Rockville, MD, 1989.

62. Varanasi, U., Ed., *Metabolism of Polycyclic Aromatic Hydrocarbons in the Aquatic Environment,* CRC Press, Boca Raton, FL, 1989.

63. Lech, J. J. and Vodicnik, M. J., Biotransformation, in *Fundamentals of Aquatic Toxicology: Methods and Applications,* Rand, G. M. and Petrocelli, S. R., Eds., Hemisphere Publishing, New York, 1985, 526.

64. Pesch, G. G., Sister chromatid exchange and genotoxicity measurements using polychaete worms, *Rev. Aquat. Sci.,* 2, 19, 1990.

65. Lee, R. F. and Singer, S. C., Detoxifying enzymes system in marine polychaetes: increases in activity after exposure to aromatic hydrocarbons, *Rapp. P.-v. Reun. Cons. Int. Explor. Mer.,* 179, 29, 1980.

66. Lee, R. F., Singer, S. C., Tenore, K. R., Gardner, W. S., and Philpot, R. M., Detoxification system in polychaete worms: importance in the degradation of sediment hydrocarbons, in *Marine Pollution Functional Responses,* Vernberg, W. B., Calabrese, A., Thurberg, F. P., and Vernberg, F. J., Eds., Academic Press, New York, 1979, 23.

67. Lee, R. and Quattrochi, L., Cytochrome P-450 and mixed-function oxygenase systems in marine invertebrates, in *Pollutant Studies in Marine Animals,* Giam, C. S. and Ray, L. E., Eds., CRC Press, Boca Raton, FL, 1987, 51.

68. Lee, R. F., Accumulation and turnover of petroleum hydrocarbons in marine organisms, in *Fate and Effects of Petroleum Hydrocarbons in Marine Organisms and Ecosystems,* Wolfe, D. A., Ed., Pergamon Press, Oxford, 1977, chap. 6.

69. Boehm, P. D. and Quinn, J. G., The persistence of chronically accumulated hydrocarbons in the hard shell clam *Mercenaria mercenaria, Mar. Biol.,* 44, 227, 1977.

70. Rand, G. M. and Petrocelli, S. R., Introduction, in *Fundamentals of Aquatic Toxicology: Methods and Applications,* Rand, G. M. and Petrocelli, S. M., Eds., Hemisphere Publishing, New York, 1985, 1.

71. Parrish, P. R., Acute toxicity tests, in *Fundamentals of Aquatic Toxicology: Methods and Applications,* Rand, G. M. and Petrocelli, S. M., Eds., Hemisphere Publishing, New York, 1985, 31.

72. Petrocelli, S. R., Chronic toxicity tests, in *Fundamentals of Aquatic Toxicology: Methods and Applications,* Rand, G. M. and Petrocelli, S. M., Eds., Hemisphere Publishing, New York, 1985, 96.

1 Organic Loading

INTRODUCTION

The significance of nutrient and carbon inputs (i.e., organic loading) to estuarine and marine ecosystems has been the subject of detailed investigations for decades.[1-5] While eutrophication resulting from continued enrichment of organic and inorganic nutrients has unequivocally compromised the water quality of many hectares of estuaries worldwide, the total impact attributable solely to sewage wastes is difficult to determine because other contaminants usually accompany high organic loads and may obfuscate their effects.[6] Moreover, the complexity of circulation in estuaries and nearshore oceanic regions has far-reaching implications for the dilution, transport, and distribution of pollutants.[7,8] The concentration, chemical form, and availability of contaminants are dependent on the biological, chemical, and physical processes occurring in the environment and are, consequently, estuary specific.[9] The assimilative capacity of a particular system for nutrient enrichment remains finite, and although small additions of nitrogen and phosphorus compounds commonly enhance its productivity, excessive amounts can alter the species composition, diversity, and dynamics of the biotic communities.

EUTROPHICATION PROBLEMS

SEWAGE
Methods of Disposal

Sewage wastes and associated nutrients enter estuaries via three principal pathways: (1) piped outfalls, (2) riverborne discharges, and (3) sewage-sludge dumping.[10] Santa Monica Bay, California (U.S.) receives treated wastewater from a major sewage treatment plant, Hyperion, by means of a submarine outfall system. Approximately 1.6×10^6 m^3/d of mixed primary-secondary effluent diffuses into the bay.[11] Over the past 40 years, the length of the discharge pipes has increased from 1.6 km offshore with the outfall at 13.7 m depth to 8 km offshore and 70 m depth for effluent and 11 km offshore and about 100 m depth for the processed solids. The monitoring of total coliform bacteria in the vicinity of the outfalls has revealed high water quality despite increased flows and waste loads in recent years.

In spite of the lower initial sewage concentrations derived from riverborne inputs than from direct pipeline discharges, increased nutrient levels and dissolved and particulate organic constituents in riverine flows ascribable to sewage enrichment can greatly influence water quality conditions in estuaries and nearshore oceanic areas, especially in proximity to large metropolitan centers. Another important difference between indirect riverborne inputs and direct pipeline discharges regards the state of the nitrogen, phosphorus, and other nutrient elements in the sewage effluent. In riverine systems, the nitrogen from sewage may be as plant biomass or as nutrients released by microbial decomposition of organic products. Because autotrophs assimilate some of these nutrients while being transported downstream, only a fraction of the nutrients may ultimately enter estuaries and coastal waters by this pathway. In pipeline discharges, however, the nitrogen primarily occurs in combined, unoxidized, and partially oxidized forms.[10]

Two examples of riverborne inputs include the River Mersey, which flows into Liverpool Bay (England), and the Raritan River, New Jersey, which empties into Raritan Bay (U.S.). The discharges of the River Mersey contain a combination of domestic sewage, industrial wastes, and agricultural runoff. The Raritan River likewise carries a wide range of contaminants from untreated and treated domestic sewage to discharges from various upriver agricultural, industrial, and manufacturing sites.[12] The Hudson-Raritan estuary complex represents a large source of nitrogen, phosphorus, and carbon for the coastal waters of the New York Bight. Only a small amount of nitrogen entering this estuary (approximately 10%) fuels primary production within the system;[13-16] most of the nitrogen flushes into surface waters of the New York Bight apex.[17] Up to 80% of the available nitrogen in the apex is used by phytoplankton within 20 km of the mouth of the Hudson-Raritan estuary.[9,18] Sewage and industrial effluents released from the Hudson and Raritan Rivers possibly contribute to episodic blooms (i.e., 100,000 cells per milliliter) of "red tide" flagellates in the apex, and in mid-summer, periods of dissolved oxygen depression and nearly anaerobic water occasionally develop in the lower estuary and apex by the decomposition of organic matter following this excess plant growth. Anoxia in the benthos of the apex may arise from seasonal and annual variations in productivity and stratification of the water column.[17]

Sewage sludge dumping directly affects the seabed and the benthic community. Additionally, water quality over sewage sludge dumping grounds is generally enriched in nitrogen and phosphorus which can lead to eutrophic conditions. Topping[10] alluded to four well-documented sewage sludge dumping grounds, the outer Thames, New York Bight, Liverpool Bay, and Garroch Head (near Glasgow). The impact of sludge-derived nutrients on the water quality at each locale has been coupled to turbulence and water exchange. According to Topping (p. 325),[10]

"...In the Outer Thames, nutrients are rapidly dispersed into water that is already substantially enriched and a very large area of enhancement results masking any local effects. In New York Bight, there is better dispersal into clean renewing water and, therefore, much less nutrient build-up, but the pattern of various highly polluted inputs creates highly variable water quality situations which, however, seldom achieve major plankton anomalies because of the rapidity of water renewal. In Liverpool Bay, dispersal is only moderate and highly variable but somewhat more persistent water quality anomalies are common. Since these are associated with some toxic or unfavorable conditions, phytoplankton development is poor and favors the growth of *Phaeocystis*, improving towards modest diatom growth when more offshore water types are present. At Garroch Head, dispersal is poor but the water is deep and

the build up of nutrients in the water column is, therefore, much less.
The low level of near-bottom turbulence retains much of the nutrients
in the sludge to leak away slowly. Moderately enhanced phytoplankton
and rich zooplankton result."

Sewage Sludge
Composition

In the past, sewage was commonly discharged to estuarine or
coastal marine waters untreated either by comminution or chlorination.[19] Prior to being released to these habitats, the sewage may
have been subjected to several processes to improve the effluent.
Perkins[19] and Klein[20] tabulate the following stages in sewage treatment:

1. Screening removes larger solids.
2. A comminutor macerates solids into fine particles.
3. Sedimentation eliminates settleable solids.
4. The settled sewage is chlorinated.
5. Aerobic treatment is applied to the settled sewage.
6. The sludge is disposed.

With greater public awareness of eutrophication problems in
estuaries during the last 2 decades, more cities and towns in the
coastal zone have been pressured into constructing secondary and
tertiary treatment facilities. In the 1960s and 1970s, it was customary for these communities to do nothing or to improve sewage
effluent only through stage 2 (above) before discharging it to the
environment.[19] Secondary treatment, although still polluting in
respect to eutrophication, is much more economical than tertiary
treatment, and therefore, has been adapted by many communities.

Capuzzo et al.[9] and Duedall et al.[21] have described the composition of sewage sludge. Consisting of a mixture of liquids and
solids (up to 10% by weight), sewage sludge contains the remains
of solid waste along with bacteria that have actively degraded
organic products from domestic waste as well as degradable industrial organic chemicals. Since some contaminants such as PCBs,
heavy metals, and chlorinated pesticides are not easily degraded
and sorb readily to bacterial and organic particle surfaces, the
sludge ultimately concentrates potentially toxic chemicals, some of

which originate from nonpoint sources (i.e., combined sewer systems, atmospheric fallout, and rainwater).

Pathogenic microorganisms, contaminants of biological concern, enter sludge during the processing of human and animal wastes. Four groups of pathogens — bacteria, viruses, protozoa, and helminths — occur in sewage effluents and sludge (Table 1).

TABLE 1
Pathogens Likely to be Associated with Sewage Sludge

Bacteria	Viruses	Protozoa	Helminths
Salmonella spp.	Poliovirus	*Entamoeba histolytica*	*Echinococcus granulosus*
Shigella spp.	Coxsackie A and B	*Acanthamoeba* spp.	*Hymenolepis nana*
Vibrio spp.	Echovirus	*Giardia* spp.	*Taenia saginata*
Mycobacterium spp.	Adenovirus		*Fasciola hepatica*
Bacillus anthracis	Reovirus		*Ascaris lumbricoides*
Clostridium perfringens	Parvovirus		*Enterobius vermicularis*
Yersinia spp.	Rotavirus		*Strongyloides* spp.
Campylobacter spp.	Hepatitis A		*Trichuris trichiura*
Pseudomonas spp.	Norwalk and related gastroenteric viruses		*Toxocara canis* *Trichostrongylus* spp.
Leptospira spp.			
Listeria monocytogenes			
Escherichia coli			
Clostridium botulinum			

From Alderslade, R., The problems of assessing possible hazards to the public health associated with the disposal of sewage sludge to land, in *Recent Experience in the United Kingdom: Characterization, Treatment, and Use of Sewage Sludge,* Hermite, P. L. and Ott, H., Eds., Reidel Publishing, Dordrecht, 1981, 372. With permission.

They pose significant health threats to humans who either ingest contaminated shellfish or swim in contaminated water. Among bacteria, the genus *Salmonella* includes the organism responsible for typhoid, *Shigella* spp. cause dysentery, and some species of the *Clostridia* produce exotoxins pathogenic to man.[19] Certain viruses in contaminated areas give rise to infectious hepatitis. Sewage treatment, as stated by Capuzzo et al.,[9] can lower the number of pathogens in wastewater effluent, specifically during the sludge-forming process. Thermophilic digestion and other sewage treatment techniques have the potential to further reduce the population size of the pathogenic organisms.

In the New York Bight, barge dumping of sewage sludge takes place at a site 170 km off the New Jersey coastline. Another sewage sludge dumpsite located in shallower waters (approximately 25 m deep) 19 km offshore of the New Jersey coastline in the New York Bight apex was closed in 1987. As mentioned previously, sewage sludge is discharged to the ocean in the Southern California Bight through pipelines rather than by barge dumping.

EUTROPHICATION AND ORGANIC CARBON ENRICHMENT

Many estuaries worldwide receive large amounts of nutrients from wastewaters of sewage treatment facilities and unsewered domestic sources. Compared to municipal wastewaters, industrial sources of nutrients and carbon remain insignificant. In most cases, it is extremely difficult to delineate the impact of nutrient enrichment via point sources of sewage wastewater discharges from that of nonpoint sources, such as agricultural and atmospheric additions, as well as urban and rural runoff. The contribution of nitrogen and phosphorus compounds from waterfowl alone may overshadow that of other sources during certain seasons in some estuaries. In general, the effect of sewage discharges on estuarine and coastal marine systems is manifested most conspicuously among phytoplankton and benthic communities. Because runoff plays a major role in the eutrophication of these coastal habitats, the entire drainage basin, and not just the body of water, must be considered when evaluating nutrient enrichment processes.

Eutrophication refers to a progressive enrichment of estuarine waters due to the addition of large amounts of inorganic nutrients.[22] Whereas enrichment with inorganic nutrients in fact maintains rich and productive biotic communities in some systems, excess nitrogen input fosters phytoplankton blooms that have led to oxygen deficits in bottom waters of many areas and ultimately the death of much aquatic life. The most sensitive systems appear to be those characterized by poor circulation where oxygen-depleted waters cannot be effectively oxygenated. These affected bodies often are shallow coastal bays having low freshwater inflow and restricted tidal ranges. However, even larger estuaries (e.g., Chesapeake Bay and Long Island Sound) have fallen victim to hypoxia and anoxia, resulting in a series of undesirable events and culminating in multiple fishkills and the loss of valuable shellfish beds.[23,24] Although anoxia has been documented in Chesapeake Bay since the 1930s,[25] in recent years it has become more prolonged and widespread.[26,27] Its ecological effects can be devastating.[28]

Eutrophication of Chesapeake Bay

Day et al.[22] examined nutrient enrichment problems in Chesapeake Bay. They observed that eutrophic conditions are common in the tributaries of the bay, for instance, the Patuxent River where nitrate and phosphorus concentrations have risen dramatically together with phytoplankton abundances. The decomposition of phytoplankton and organic matter associated with sewage inputs and other nutrient sources (e.g., industrial outfalls) causes a decrease in dissolved oxygen in bottom waters. Oxygen depletion zones may be exacerbated by seasonal increases in density stratification in the estuary attributable to the spring freshet which increases water column stability and minimizes advective transport of oxygen from the surface to the deep layer. Hence, during the period from June to September, anoxic conditions purportedly spread along the bottom of the bay's channel in waters >10 m deep.[29] Clearly, areas of oxygen depletion in Chesapeake Bay are coupled to biological, chemical, and physical factors.

Two types of oxygen depletion zones can be distinguished in estuaries: (1) anoxic zones (bottom waters with <0.1 mg/l dissolved oxygen) and (2) hypoxic zones (bottom waters with <2.0 mg/l dissolved oxygen).[30] While hypoxia imposes severe stress on many benthic and demersal organisms, anoxia threatens the existence of benthic populations and communities. The literature is replete with references to hypoxia and anoxia.[31-34] In a partially mixed estuary characterized by thermohaline stratification nearly year-round (e.g., Chesapeake Bay)[35] benthic oxygen demand (typically ranging from 1 to 4 g $O_2/m^2/d$ in estuaries)[27,36] is the primary sink for oxygen, and vertical transport across the halocline is the primary source. Another potentially important sink for oxygen is the oxidation of organic matter above the seafloor; an additional source of oxygen replenishment is longitudinal advective transport. Gravitational circulation accounts for most of the advective transport in a partially mixed estuary. Based on the work of Kuo and Neilson (p. 281),[27] the oxygen budget in the deep waters of the Virginia estuaries that are tributaries to the Chesapeake Bay may be written as

$$V\frac{d}{c} = -BA - kMV + EA\frac{c_1 - c}{d} + Q_1c_1 - Q_2c - Q_vc \qquad (1)$$

where c and c_1 are the mean DO concentrations in the hypoxic and overlying water masses, respectively, V and A are the volume and areal extent of the hypoxic water mass, t is time, B is the benthic oxygen demand per unit area per unit time, M is the total oxygen demand of oxidizable material (or biochemical oxygen demand) per unit volume, k is the oxidation rate, E is the vertical diffusion coefficient, d is the mean distance between the centers of the upper and lower water masses, and Q_1, Q_2, and Q_v are the flow rates due to gravitational circulation. Conservation of water mass requires that $Q_1 = Q_2 + Q_v$; when this is accounted for and all terms are divided by V, Equation 1 may be given as

$$\frac{dc}{dt} = -\frac{B}{h} - kM + \frac{E}{h}\frac{c_1 - c}{d} + \frac{U}{L}(c_1 - c) \qquad (2)$$

where h and L are the mean depth and length of the lower water mass, and U is the average net upestuary velocity in the lower layer.

For Chesapeake Bay, as well as other major estuaries in Virginia, the frequency, duration, and severity of hypoxic conditions depends on the strength of the gravitational circulation, which is strongly controlled by density distribution. Clearly, salinity outweighs temperature as a modulator of the density patterns. It is the longitudinal salinity gradient that acts as the forcing function for the gravitational circulation.[27] Freshwater inflow, in turn, principally influences rates of salinity change with depth and distance.[29]

Taft et al.[29] recorded increased water column stability in the estuary between February and May concomitant with higher riverine flow. From summer to early winter, however, reduced riverine input lowered this stability. Their data suggest that a quantitative cause and effect relationship exists between river flow and oxygen depletion rate, although this relationship has yet to be unequivocally established.

In waters below 10 m depth, respiration in the water column is the main factor contributing to oxygen depletion. Benthic respiration remains secondary in importance. Organic matter accumulating from the previous year (summer and fall) and settling into the deep water during winter provides most of the oxygen demand. Some new primary production in spring, despite being less of a driving force in oxygen dissipation in the bay, sinks and contributes to the summer deep-water oxygen depletion.[29]

Scale of Eutrophication

Jaworski[37] compared the scale of eutrophication and nutrient loadings of Chesapeake Bay and its tributary estuaries (Tables 2 and 3). While the data show no excessive eutrophication problems in Chesapeake Bay, some of its tributary estuaries exhibit eutrophic conditions. Based on the level of eutrophication, Jaworski[37] described three types of estuaries:

1. Hypereutrophic systems having very excessive nuisance conditions, anoxia, and "undesirable" biological communities.
2. Eutrophic systems having excessive nuisance conditions, low dissolved concentrations, and "undesirable" biological communities.
3. Noneutrophic systems having biologically healthy and productive components with "desirable" biological communities.

Jaworski[37] considered the Potomac estuary, with considerably greater external loading than the Chesapeake, to be hypereutrophic during the period 1969 to 1971 and eutrophic during 1977 and 1978. The Patuxent River he deemed to be eutrophic during these same periods. In contrast, the Rappahannock and York Rivers were noneutrophic. More recently, water quality in the upper Potomac

TABLE 2
Scale of Eutrophication in Chesapeake Bay and its Tributary Estuaries

Ecosystem		Ecological description	Surface area (m^2) (10^6)	Vol. (m^3) (10^6)	Avg. depth (m)	Retention time (years)
Patuxent	1963	Noneutrophic	137	660	4.8	1.70
	1969-71	Eutrophic	137	660	4.8	1.70
	1978	Eutrophic	137	660	4.8	1.70
York	1969-71	Noneutrophic	210	910	4.3	0.72
Rappahannock	1969-71	Noneutrophic	400	1,780	4.5	1.27
James	1969-71	Eutrophic	600	2,400	3.6	0.39
Potomac	1913	Noneutrophic	1,250	7,150	5.8	1.07
	1954	Eutrophic	1,250	7,150	5.8	1.07
	1969-71	Hypereutrophic	1,250	7,150	5.8	1.07
	1977-78	Eutrophic	1,250	7,150	5.8	1.07
Chesapeake Bay	1969-71	Localized				
(Including Tribs)		eutrophic	11,500	74,000	6.5	1.16
(Excluding Tribs)		conditions	6,500	52,000	8.4	1.32

From Jaworski, N. A., in *Estuaries and Nutrients,* Neilson, B. J. and Cronin, L. E., Eds., Humana Press, Clifton, NJ, 1981, 37. With permission.

TABLE 3
External Nutrient Loadings for Chesapeake Bay and its Tributary Estuaries

		External phosphorus			External nitrogen			At. N:P ratio of load
		(g/yr) (10^6)	(g/m^2/ yr)	(g/m^3/ yr)	(g/yr) (10^6)	(g/m^2/ yr)	(g/m^3/ yr)	
Patuxent	1963	170	1.24	0.26	930	6.7	1.4	12
	1969-71	250	1.82	0.38	1,110	8.1	1.7	10
	1978	420	3.06	0.64	1,500	11.4	2.4	8
York	1969-71	160	0.76	0.18	1,190	5.6	1.3	17
Rappahannock	1969-71	180	0.45	0.10	1,500	3.8	0.8	19
James	1969-71	1,780	2.70	0.70	10,300	15.6	4.2	13
Potomac	1913	910	0.73	0.13	18,600	14.8	2.6	46
	1954	2,000	1.63	0.28	22,600	18.1	3.1	26
	1969-71	5,380	4.30	0.80	25,200	20.2	3.5	11
	1977-78	2,520	2.01	0.35	32,800	26.2	4.6	30
Chesapeake Bay	1969-71							
(Including Tribs)		15,000	1.30	0.20	109,100	9.5	1.5	16
(Excluding Tribs)		7,350	1.10	0.10	70,160	10.8	1.3	22

From Jaworski, N. A., in *Estuaries and Nutrients,* Neilson, B. J. and Cronin, L. E., Eds., Humana Press, Clifton, NJ, 1981, 37. With permission.

has improved dramatically due to a massive wastewater management effort costing more than \$1 billion. Gone are the low dissolved oxygen levels, nuisance algal conditions, and high coliform densities in the upper estuary that occurred in the 1950s and 1960s.[34]

Assessing the external loadings of phosphorus and nitrogen, Jaworski[37] obtained highest readings for the 1969 to 1971 time frame in the James, Patuxent, and Potomac systems. These three estuaries appeared to be nitrogen limited based on nitrogen to phosphorus ratios (an N/P value <16 presumably is nitrogen limited). However, water quality data of the upper Potomac estuary in 1977–1978 revealed N/P ratios of 30 to 40, reflecting a change to a phosphorus-limited system. The York, Rappahannock, and Chesapeake systems with N/P ratios of 16 or above seemed to be phosphorus limited. They were also noneutrophic. Based on calculations of nutrient loadings of more than 10 estuaries on the East Coast of the U.S. and their perceived trophic state, excessive eutrophic conditions probably can be prevented if phosphorus loading is <1.0 g/m²/year. Jaworski[37] suggests a "permissible" phosphorus loading of 0.75 g/m²/year or less.

NUTRIENTS

The macronutrients nitrogen, phosphorus, and silicon are the key elements with respect to autotrophic growth in estuaries. While nitrogen is generally the chief limiting element to primary production in estuaries and coastal marine waters,[38-42] phosphorus may be limiting to plant growth during certain seasons of the year in some systems. Boynton et al.[36] illustrated that phosphorus concentrations consistently exceeded those of nitrogen during episodes of peak productivity in a wide variety of estuarine ecosystems. Silicon is chiefly required by floral groups that secrete siliceous skeletons, but may be required by other aquatic plants as well.[43] Estuarine plants likewise require major elements (e.g., calcium, carbon, magnesium, oxygen, and potassium), minor and trace elements (e.g., cobalt, copper, iron, molybdenum, vanadium, and zinc), and organic nutrients (e.g., the vitamins biotin, thiamin, and vitamin B_{12});[44] however, the availability of these elements and nutrients usually does not limit plant growth.

The Redfield ratios of atomic weights of the elements C:N:Si:P equal 106:16:15:1. The dissolved inorganic N:P ratios in estuaries can be highly variable, ranging from <1.0 to >200. The ratio typically undergoes a seasonal cycle with lowest values of <16:1 during the season of peak phytoplankton production.

The nitrogen and phosphorus concentrations in 28 estuarine systems are listed in Table 4. Nitrogen values range from <1 to 60 μg-at/l. It is evident from these measurements that phytoplankton production is partially constrained by the need for nitrogen and phosphorus in atomic proportions of 16:1.

Most nutrient elements enter estuaries in riverine discharges; a minor fraction is usually supplied by seawater and precipitation. Sewage disposal yields substantial amounts of nitrogen in some systems, particularly those impacted by large metropolitan centers. For example, Valiela[41] showed that ammonia concentrations in the western end of Long Island Sound near New York City equalled 45 to 100 μg-at/l, but only 0 to 5 μg-at/l away from the city. Similarly, nitrate readings in proximity to the city were 8 to 20 μg-at/l compared to 0.5 to 8 μg-at/l for the remainder of the sound. Often an unknown but significant quantity of nitrogen and phosphorus is derived from diffuse or nonpoint (land) sources, including overland runoff during storms and ground water discharges. In some estuaries, such as Chesapeake Bay, atmospheric deposition (wet and dry) contributes substantial concentrations of nitrogen.[24] The relative input of nutrients from the aforementioned sources changes seasonally and yearly. The amount of nutrients entering estuaries from sewage treatment discharges is more constant than that from the other sources, especially land runoff, riverine discharges, and precipitation, which are strongly coupled to the seasonal hydrological cycle.

Responses of estuaries to nutrient addition largely depend on whether the systems are mixed or stratified. The two-layered circulation pattern characteristic of stratified estuaries creates a nutrient trap in which nutrients build to greatest concentrations toward the head of the estuary. Higher velocities of surface flow and increased turbulence and mixing mollify the nutrient accumulation, whereas accelerated rates of production of organic matter enhance the nutrient accumulation.[45] Poor circulation in upstream bottom water fosters anoxic conditions as organic matter accumulates. Organic carbon, as well as organic nitrogen and phosphorus, requires oxygen for complete mineralization, which reduces oxygen

TABLE 4
Concentrations of Nitrogen and Phosphorus in 28 Estuarine Ecosystems[a]

	Nutrient concentration	
Estuary	Nitrogen	Phosphorus
River dominated		
Pamlico River, North Carolina	1.5	8.0
Narragansett Bay, Rhode Island	0.6	1.6
Western Wadden Sea, Netherlands	3.0	2.0
Eastern Wadden Sea, Netherlands	4.0	2.5
Mid-Patuxent River, Maryland	4.2	2.3
Upper Patuxent River, Maryland	10.0	2.0
Long Island Sound, New York	1.5	0.5
Lower San Francisco Bay, California	20.6	3.8
Upper San Francisco Bay, California	11.5	2.0
Barataria Bay, Louisiana	4.6	0.8
Victoria Harbor, British Columbia	11.5	2.0
Mid-Chesapeake Bay, Maryland	4.5	0.6
Upper Chesapeake Bay, Maryland	5.0	6.0
Duwamish River, Washington	60.0	3.0
Hudson River, New York	5.0	0.16
Apalachicola Bay, Florida	10.0	0.1
Embayments		
Roskeeda Bay, Ireland	0.4	2.2
Bedford Basin, Nova Scotia	0.6	0.5
Central Kaneohe Bay, Hawaii	0.8	0.3
S. E. Kaneohe Bay, Hawaii	1.0	0.5
St. Margarets Bay, Nova Scotia	1.1	0.5
Vostok Bay, Russia	1.0	0.05
Lagoons		
Beaufort Sound, North Carolina	0.5	0.5
Chincoteague Bay, Maryland	3.2	2.5
Peconic Bay, New York	1.9	1.3
High Venice Lagoon, Italy	2.4	0.05
Fjords		
Baltic Sea	1.3	0.1
Loch Etive, Scotland	1.1	0.06

[a] Nutrient concentrations in µg-at/l.

From Boynton, W. R., Kemp, W. M., and Keefe, C. W., in *Estuarine Comparison,* Kennedy, V. S., Ed., Academic Press, New York, 1982, 69. With permission.

concentrations in the bottom water. When the velocity of the deeper water is sufficiently great, an oxygen deficit leading to anoxia can be averted because of the rate of replacement of the deep water. In mixed estuaries, the distribution of nutrients is much more uniform throughout the water column. Moreover, oxygen transport offsets the oxygen demands of the decomposition process. An estuary, being a highly dynamic body of water in space and time, may exist primarily in a mixed or stratified state, or exhibit characteristics of both concurrently in different sections of the estuary.[46] This complexity in physical states can obscure the assessment of nutrient enrichment. Ultimately, it is the relative rates of loading, physical and metabolic cycling, mixing, and flushing that must be evaluated in cases of oxygen depletion and anoxia of these coastal ecosystems.

Turner et al.[47] contend that the intensity of water column stratification is the principal factor determining the spatial and temporal depletion of oxygen in Alabama estuarine and adjacent continental shelf waters. Vertical stratification increases, as does the occurrence of hypoxic events, during periods of low wind velocity. Vertical diffusion rates tend to decline with greater water column stratification. Where oxygen-depleted water spreads across broad expanses, lateral advective transport presumably is less significant than vertical diffusive transport in determining the oxygen balance in coastal and continental shelf waters.

The overall health of an estuary subjected to nutrient loading can be described in terms of various physical (i.e., hydrological, geological, and chemical), biological, and ecosystem elements (Table 5). A number of measures have been applied in past investigations to assess the integrity of estuaries subjected to nutrient enrichment (Table 6). For example, biological measures include descriptions of species abundance, diversity, evenness, equitability, and richness; they aid in understanding changes at the population and community levels. At the ecosystem level, indices involving ratios between indicators of healthy and unhealthy conditions have proven to be most useful.[46] Examples of these measures are the ratio of oxygen availability to oxygen demand, the ratio of actual to critical levels of dissolved nutrients, the ratio of aerobic to anaerobic species in bottom habitats, and the indices of trophic state. Because of the diverse array of physical, chemical, and

---

TABLE 5
Elements of Estuarine Health

Hydrological	Adequate flushing, especially by freshwater inflow
Geological	High diversity of benthic environments
	An adequate mix of erosional and depositional types
	An adequate mix of sediment particle-range types
Chemical	High water quality (in top and bottom layers)
	Nutrient levels within safe ranges
	Oxygen levels above safe minimum levels
	High quality of bottom surface layer
	Organic content not excessive
	Surface layer aerobic
Biological	Typical species within normal ranges of abundance (for the time of year)
	Nuisance species within acceptable limits
	Basic biological processes neither unduly elevated nor depressed
Ecosystem	Spatial structural diversity high
	Nutrient cycling functioning properly
	Temporal variability of the above factors within normal range

From Darnell, R. M. and Soniat, T. M., in *Estuaries and Nutrients,* Neilson, B. J. and Cronin, L. E., Humana Press, Clifton, NJ, 1981, 225. With permission.

---

biological properties from estuary to estuary, however, no single measure can be used to adequately ascertain the biological integrity of a specific estuarine site and problem.[48]

ESTUARINE SUSCEPTIBILITY TO NUTRIENT LOADING

Biggs et al.[49] adapted the Vollenweider approach to describe estuarine susceptibility to pollutants and eutrophication. While Vollenweider's work[50] dealt with the relationship between nutrient loading and planktonic algal trophic response in lakes, Biggs et al.[49] extended this concept to estuaries by developing a classification scheme based on physical and hydrologic data and pairing it with a watershed classification based on anthropogenic activity and physical characteristics of the watershed. The integration of these classifications is designed to evaluate the potential ability of an estuary to flush pollutants generated by man's activity in the watershed.

The accommodative capacity of an estuary for pollutants is contingent upon its hydraulic loading which reads as follows

TABLE 6
Measures of Ecosystem Health

Hydrological	Flushing rate vs. oxygen demand (for the most vulnerable portion of the estuary)
Geological	Ratio of areas covered by erosional vs. depositional environments
Chemical	Ratio of actual to critical levels of dissolved nutrients (especially P and N)
	Ratio of oxygen availability to oxygen demand
Biological	Chlorophyll content of water column
	Plankton species composition and abundance
	Biofouling species succession and total production
	Benthic species composition, diversity, etc. (by habitat type)
	Ratio of aerobic to anaerobic species in the benthos
	Total respiration of unit area benthic samples
Ecosystem	Some measure of areal structural heterogeneity of the system
	Some measure of areal functional heterogeneity of the system
	Some measure of temporal variability of the system
	Some index of trophic state that combines several of the above measures in a meaningful way

From Darnell, R. M. and Soniat, T. M., in *Estuaries and Nutrients,* Neilson, B. J. and Cronin, L. E., Eds., Humana Press, Clifton, NJ, 1981, 225. With permission.

$$Q_s = \frac{\overline{Z}}{\tau} \tag{3}$$

where Q_s = hydraulic loading; Z = mean depth; and τ = hydraulic residence time; and

$$\tau = \frac{V}{Q + V_T} \tag{4}$$

where V = low tide estuary volume; Q = annual freshwater inflow; and V_T = annual intertidal volume. Unequivocally, freshwater and tidal inflow and mean depth of the estuary modulate the accommodative capacity of a given system. A large mean depth or short hydraulic residence time translates into a higher hydraulic loading value. In essence, an estuary with a large hydraulic loading number must have either a large volume for dilution or a high flushing rate to rapidly remove a pollutant.

Of the 78 estuaries in the continental U.S. examined by Biggs et al.,[49] seven of the ten systems with the lowest hydraulic loading values occur in the Gulf region of Texas. They characteristically are shallow bodies of water with restricted tidal ranges and low freshwater inflow. All have a greater potential for being impacted by anthropogenically produced loadings. In terms of watershed impact, activities of man were discovered to be most intense in the Atchafalaya and Vermillion Bays, Chesapeake Bay, Columbia River, Galveston Bay, Hudson River/Raritan Bay, Potomac River, and San Francisco Bay. Table 7 lists the least and most susceptible estuaries to the general population, heavy industry, and agricultural activities.

DISPERSION MODELS OF RELEASED EFFLUENT

Transport and dispersion problems in estuaries have been scrutinized by employing physical and mathematical models.[51] Gade[51] explicated the difficulties of applying physical modeling to pollution studies in estuaries, emphasizing scaling problems of the generally distorted physical models. The most effective and frequently used scale models are Froude models of estuaries where tidal action predominates. Sugimoto[52] and McClimans and Gjerp[53]

TABLE 7
Summary of Estuarine Pollution Susceptibility

Most susceptible systems

General population	Heavy industry
Brazos River**	Brazos River**
Ten Thousand Islands	North/South Santee Rivers**
San Pedro Bay**	Galveston Bay**
North/South Santee Rivers**	Sabine Lake**
Galveston Bay**	San Pedro Bay**
Suisun Bay**	Connecticut River*
Sabine Lake**	Calcasieu Lake
St. Johns River*	Hudson River/Raritan Bay
Apalachicola Bay	Charleston Harbor
San Antonio Bay**	Perdido Bay
Connecticut River*	Potomac River
Great South Bay*	San Antonio Bay**
Merrimack River	Mobile Bay
Atchafalaya/Vermillion Bays*	Suisun Bay**
Matagorda Bay*	Great South Bay*

TABLE 7 (CONTINUED)
Summary of Estuarine Pollution Susceptibility

Most susceptible systems

Baffin Bay   San Pedro Bay**
Chesapeake Bay   Sabine Lake**
Agricultural activities   Corpus Christi Bay
  Brazos River**   Galveston Bay**
  Suisun Bay**   San Antonio Bay**
  North/South Santee Rivers**   Winyah Bay
  St. Johns River*   Albemarle Sound
  Matagorda Bay*   Neuse River
  Atchafalaya/Vermillion Bays*   Laguna Madre

Least susceptible systems

General population   Boston Bay
  Willapa Bay**   Coos Bay**
  St. Catherines/Sapelo Sound **   Broad River**
  Penobscot Bay**   Willapa Bay**
  Humboldt Bay**   Bogue Sound**
  Broad River**   Puget Sound*
  Hood Canal**   Narragansett Bay
  Coos Bay**   Santa Monica Bay
  Casco Bay**   Saco Bay
  Grays Harbor   St. Andrew/St. Simons Sound*
  Chincoteague Bay*   Agricultural workers
  Bogue Sound**   Humboldt Bay**
  St. Andrew/St. Simons Sound*   Hood Canal**
  Sheepscot Bay*   Penobscot Bay**
  Apalachee Bay   Coos Bay**
  Rappahannock River   St. Catherines/Sapelo Sound**
  St. Helena Sound   Chincoteague Bay*
  Puget Sound*   Bogue Sound**
Heavy industry   Long Island Sound
  St. Catherines/Sapelo Sound**   Casco Bay**
  Hood Canal**   Willapa Bay**
  Penobscot Bay**   Broad River**
  Casco Bay**   Sheepscot Bay*
  Humboldt Bay**   Klamath River
  Buzzards Bay

Note: Systems that are present in all three categories are marked with two asterisks; systems present in two categories are marked by one asterisk.

From Biggs, R. B., DeMoss, T. B., Carter, M. M., and Beasley, E. L., *Rev. Aquat. Sci.,* 1, 189, 1989. With permission.

compiled information on Froude models. Despite the fact that only tidal currents are simulated by these models and they give lower dispersion than observed in nature, good agreement has been attained between the models and field observations.

Mathematical models utilizing simple box models, diffusion/advection models, or more advanced dynamic models have been applied more successfully to transport and dispersion problems of estuaries than scale models. Most numerical models simulating estuarine conditions deal mainly with tidal currents and related dispersion. They are barotropic.[51]

Puff Model

Nielsen[54] employed two mathematical models, puff and particle, to explain the spreading and dilution of effluents in coastal ecosystems. The puff and particle models are dynamic types, with the puff model based on the superposition principle and the particle model based on a Monte Carlo simulation technique. The puff model simulates the spreading and dilution of effluents by considering the effluent emission from an outlet site as a series of puffs. As iterated by Nielsen,[54] each puff yields a given amount of effluent, M_{puff}, determined by the rate of effluent release, Q, and the time interval, Δt, between each puff release

$$M_{puff} = Q\Delta t \qquad (5)$$

The horizontal movement of each puff equals the product of the Δt and current velocity

$$x_{t+\Delta t} = x_t + u\Delta t \qquad (6)$$

$$y_{t+\Delta t} = y_t + v\Delta t \qquad (7)$$

where x and y are the coordinates for the position of the center of the puff at time t, and u and v are the horizontal velocities in a Cartesian coordinate system.

One assumes that the effluent within a puff fits a two-dimensional Gaussian distribution with variances increasing with time and the growth rate determined by turbulent diffusion. Down to a given depth, d, the concentration is considered constant. Within each puff, the effluent concentration, C_{puff}, can be expressed as

$$C_{puff}(x,y) = M_{puff} \exp\left(-(x-x_0)^2 / 2\sigma_x^2\right)$$
$$\times \exp\left(-(y-y_0)^2 / 2\sigma_y^2\right) d^{-1} \qquad (8)$$

where (x_0, y_0) is the position of the center of the puff and (σ_x^2, σ_y^2) are the variances in the two-dimensional Gaussian distribution. By summing the contributions from all puffs, the effluent concentration at a given location is determined. In regard to effluent spreading and dilution, the amount of effluent in each puff decreases after each Δt according to the decay rate, k

$$\Delta M_{puff} = -k\Delta t \qquad (9)$$

The puff model is a deterministic model that has been most successfully applied to simulate effluent spreading and dilution from a continuous point source. Figure 1 depicts a simplified flow chart for the puff model.

Particle Model

Similar to puff models, particle models have proven to be useful in simulating the spread and dilution of released effluents under time-varying and inhomogeneous conditions. The stochastic particle model presented by Nielsen[54] treats pollution problems when only one or a few impacted locations are of concern. This situation differs from that of the puff model which has been applied to simulations of effluent spreading when the contamination of multiple locations is to be tracked. Because of the interplay of many factors in pollution assessment at a given location, stochastic descriptions take on added significance in particle modeling. Nielsen (p. 36)[54] summarizes this condition:

"... To cope with stochastic fluctuations, it seems appropriate to characterize the environmental states by probability distributions. To perform a simulation, an environmental state is constructed by random sampling (Monte Carlo method) from the various distributions. Then a particle is released from the outfall, and its motion is tracked during a period of time until it is apparent whether it reaches the given location. When a large number of simulations has been performed, the percentage of particles that have reached the location of concern equals the fraction of effluent contributing to contamination at the location."

Figure 2 displays a simplified flow chart for the particle model.

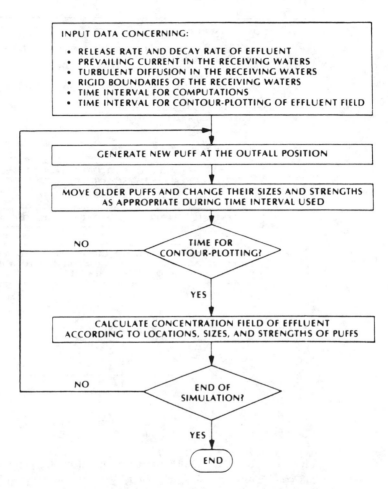

FIGURE 1. Simplified flow chart for the puff model. (From Nielsen, P. B., in *Oceanic Processes in Marine Pollution,* Vol. 2, *Physicochemical Processes and Wastes in the Ocean,* O'Connor, T. P., Burt, W. V., and Duedall, I. W., Eds., Robert E. Krieger Publishing, Malabar, FL, 1987, 33. With permission.)

NUTRIENT LOADING, RED TIDES, AND PARALYTIC SHELLFISH POISONING

Elevated nutrient concentrations associated with sewage disposal in estuarine and coastal marine waters may contribute to intense blooms of dinoflagellates termed "red tides" which can inflict heavy losses on faunal populations, especially in benthic communities. Characterized by reddish, brownish, or yellowish

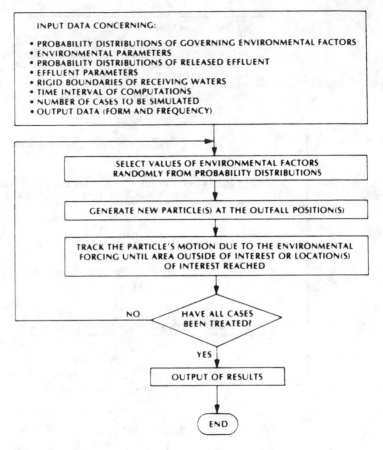

FIGURE 2. Simplified flow chart for the particle model. (From Nielsen, P. B., in *Oceanic Processes in Marine Pollution,* Vol. 2, *Physicochemical Processes and Wastes in the Ocean,* O'Connor, T. P., Burt, W. V., and Duedall, I. W., Eds., Robert E. Krieger Publishing, Malabar, FL, 1987, 33. With permission.)

colorations of the water, red tides result from blooms of dinoflagellates containing the red pigment, peridinin.[55,56] Even though much has been written regarding the potential transmission of pathogens to humans via sewage effluents released to estuarine waters (Table 1), the general populace is less aware of the possible dangers created by certain red tides. The neurotoxin, saxitoxin, produced by dinoflagellates such as *Protogonyaulax catenella* (= *Gonyaulax catenella*), *P. tamarensis* (= *G. excavata*), and *Pyrodinium bahamense* var. *compressa* concentrates in edible shellfish (e.g., clams, mussels, and oysters) which filter the plankters from the water column. When ingested by humans, the toxin-bearing shellfish often cause paralytic shellfish poisoning (PSP), an acute illness typified by numbness of the lips, tongue, and fingertips.[55] PSP can be fatal in severe cases. In spite of the globally widespread occurrences of PSP, the frequency of outbreaks is not great. Between 1793 and 1966, for instance, only 600 cases were reported worldwide.[57]

Walker[58] cited an outbreak of PSP attributable to a massive bloom of *Protogonyaulax tamarensis* in the fall of 1972 which affected Maine, New Hampshire, and Massachusetts. Shellfish were contaminated as far south as Cape Cod, and many hectares of beds were closed. Anderson et al.[59] has warned of potentially significant PSP problems in the estuarine waters of Connecticut and New York.

Maranda and Shimizu[60] discussed a 2-year survey for diarrhetic shellfish poisoning (DSP) in Narragansett Bay. DSP is a transitory gastrointestinal disorder experienced by people who consume shellfish contaminated by lipid-soluble toxins.[61] Members of the dinoflagellate genus, *Dinophysis* have been implicated as the causative organisms of DSP. Examples are *D. fortii* and *D. acuminata* in Japanese waters and possibly *D. acuta* and *D. norvegica* in European waters.[60] Outbreaks of DSP have been documented in Norway[62] and Sweden,[63] and new techniques now available to detect DSP toxins promise to increase the number of reported DSP cases worldwide.[64,65]

SUMMARY AND CONCLUSIONS

The enrichment of estuaries with organic and inorganic nutrients has contributed to eutrophication problems in numerous systems. In many cases, domestic and municipal sewage wastes, as well as nonpoint source runoff, have been the principal sources of nutrients to estuaries. The three main pathways by which sewage wastes and associated nutrients enter estuaries include: (1) piped outfalls, (2) riverborne discharges, and (3) sewage-sludge dumping. Santa Monica Bay, California receives treated wastewater discharges via a piped outfall. Riverborne inputs of sewage wastes are exemplified by influent systems of Liverpool Bay (England) and Raritan Bay (U.S.). Sewage-sludge dumping grounds exist in the outer Thames estuary, New York Bight, Liverpool Bay, Garroch Head (west of Glasgow), and many other estuarine systems.

Sewage sludge consists of a mixture of liquids and solids (up to 10% by weight), with the sludge containing remains of solid wastes plus bacteria. Contaminants (e.g., PCBs, heavy metals, and chlorinated pesticides) sorbed to particle surfaces of the sludge comprise potentially toxic wastes within the sludge. Pathogens, such as bacteria, viruses, protozoa, and helminths, proliferating in sludge and sewage effluents may pose health hazards to humans.

Nutrient enrichment from nonpoint sources, for example, urban and rural runoff, can overshadow nutrient additions from anthropogenic point sources. Nonpoint-source nutrient enrichment often promotes phytoplankton blooms and leads to eutrophication problems. These events result in hypoxia or anoxia in susceptible water bodies culminating in the death of aquatic life, including recreationally and commercially important fin- and shellfish.

An assessment is made of eutrophication problems in the Chesapeake Bay ecosystem. Here, tributary systems of Chesapeake Bay (e.g., Patuxent River estuary) periodically exhibit serious eutrophic conditions. Furthermore, the deeper waters of Chesapeake Bay (below 10 m) have become more frequently plagued by eutrophication in recent years. Oxygen-depletion zones have been exacerbated by seasonal increases in density stratification in the estuary due, in particular, to the spring freshet.

Physical and mathematical models are being utilized more often in the study of pollution and eutrophication problems in estuaries. Difficulties of scaling, however, have limited the usefulness of physical models in pollution investigations. The most effective and frequently employed scale models are Froude models of estuaries. Mathematical models, rather than scale models, have proven to be powerful tools in evaluating transport and dispersion problems, with the application of simple box models, diffusion/advection models, and more advanced dynamic models being especially valuable. Mathematical models gaining favor in the assessment of the spreading and dilution of effluents released to estuarine and coastal marine waters are puff and particle models.

REFERENCES

1. Mayer, G. F., Ed., *Ecological Stress and the New York Bight: Science and Management,* Estuarine Research Federation, Columbia, SC, 1982.
2. Duedall, I. W., Ketchum, B. H., Park, P. K., and Kester, D. R., Eds., *Wastes in the Ocean,* Vol. 1, *Industrial and Sewage Wastes in the Ocean,* John Wiley & Sons, New York, 1983.
3. Ketchum, B. H., Capuzzo, J. M., Burt, W. V., Duedall, I. W., Park, P. K., and Kester, D. R., Eds., *Wastes in the Ocean,* Vol. 6, *Nearshore Waste Disposal,* John Wiley & Sons, New York, 1985.
4. Capuzzo, J. M. and Kester, D. R., Eds., *Oceanic Processes in Marine Pollution,* Vol. 1, *Biological Processes and Wastes in the Ocean,* Robert E. Krieger Publishing, Malabar, FL, 1987.
5. Wolfe, D. A. and O'Connor, T. P., Eds., *Oceanic Processes in Marine Pollution,* Vol. 5, *Urban Wastes in Coastal Marine Environments,* Robert E. Krieger Publishing, Malabar, FL, 1988.
6. Carriker, M. R., Anderson, J. W., Davis, W. P., Franz, D. R., Mayer, G. F., Pearce, J. B., Sawyer, T. K., Tietjen, J. H., Timoney, J. F., and Young, D. R., Effects of pollutants on benthos, in *Ecological Stress and the New York Bight: Science and Management,* Mayer, G. F., Ed., Estuarine Research Federation, Columbia, SC, 1982, 3.
7. van de Kreeke, J., Ed., *Physics of Shallow Estuaries and Bays,* Springer-Verlag, New York, 1986.
8. Nielsen, P. B., Deterministic and stochastic models of dispersion of released effluents into coastal areas, in *Oceanic Processes in Marine Pollution,* Vol. 2, *Physicochemical Processes and Wastes in the Ocean,* O'Connor, T. P., Burt, W. V., and Duedall, I. W., Eds., Robert E. Krieger Publishing, Malabar, FL, 1987, 33.
9. Capuzzo, J. M., Burt, W. V., Duedall, I. W., Park, P. K., and Kester, D. R., The impact of waste disposal in nearshore environments, in *Wastes in the Ocean,* Vol. 6, *Nearshore Waste Disposal,* Ketchum, B. H., Capuzzo, J. M., Burt, W. V., Duedall, I. W., Park, P. K., and Kester, D. R., Eds., John Wiley & Sons, New York, 1985, 3.
10. Topping, G., Sewage and the sea, in *Marine Pollution,* Johnston, R., Ed., Academic Press, London, 1976, 303.
11. Garber, W. F. and Wada, F. F., Water quality in Santa Monica Bay, as indicated by measurements of total coliforms, in *Oceanic Processes in Marine Pollution,* Vol. 5, *Urban Wastes in Coastal Marine Environments,* Wolfe, D. A. and O'Connor, T. P., Eds., Robert E. Krieger Publishing, Malabar, FL, 1988, 49.
12. Multer, H. G., Stainken, D. M., McCormick, J. M., and Berger, K. J., Sediments in the Raritan Bay — lower New York Bay complex, *Bull. N. J. Acad. Sci.,* 29, 79, 1984.

13. Garside, C., Malone, T. C., Roels, O. A., and Sharfstein, B. A., An evaluation of sewage-derived nutrients and their influence on the Hudson River estuary and New York Bight, *Est. Coastal Mar. Sci.*, 4, 281, 1976.
14. Duedall, I. W., O'Connors, H. B., Parker, J. H., Wilson, R. W., and Robbins, A. S., The abundances, distribution, and flux of nutrients and chlorophyll a in the New York Bight apex, *Est. Coastal Mar. Sci.*, 5, 81, 1977.
15. Duedall, I. W., O'Connors, H. B., Wilson, R. E., and Parker, J. H., The Lower Bay Complex, MESA New York Bight Atlas Monogr. No. 29, New York Sea Grant Institute, Albany, 1979.
16. McLaughlin, J. J. A., Kleppel, G. S., Brown, M. P., Ingram, R. J., and Samuels, W. B., The importance of nutrients to phytoplankton production in New York Harbor, in *Ecological Stress and the New York Bight: Science and Management,* Mayer, G. F., Ed., Estuarine Research Federation, Columbia, SC, 1982, 469.
17. Mearns, A. J., Haines, E., Kleppel, G. S., McGrath, R. A., McLaughlin, J. J. A., Segar, D. A., Sharp, J. H., Walsh, J. J., Word, J. Q., Young, D. K., and Young, M. W., Effects of nutrients and carbon loadings on communities and ecosystems, in *Ecological Stress and the New York Bight: Science and Management,* Mayer, G. F., Ed., Estuarine Research Federation, Columbia, SC, 1982, 53.
18. Malone, T. C., Factors influencing the fate of sewage-derived nutrients in the lower Hudson estuary and New York Bight, in *Ecological Stress and the New York Bight: Science and Management,* Mayer, G. F., Ed., Estuarine Research Federation, Columbia, SC, 1982, 389.
19. Perkins, E. J., *The Biology of Estuaries and Coastal Waters,* Academic Press, London, 1974.
20. Klein, L., *River Pollution,* Vol. 3, *Control,* Butterworths, London, 1966.
21. Duedall, I. W., Ketchum, B. H., Park, P. K., and Kester, D. R., Global inputs, characteristics, and fates of ocean-dumped industrial and sewage wastes: an overview, in *Wastes in the Ocean,* Vol. 1, *Industrial and Sewage Wastes in the Ocean,* Duedall, I. W., Ketchum, B. H., Park, P. K., and Kester, D. R., Eds., John Wiley & Sons, New York, 1983, 3.
22. Day, J. W., Jr., Hall, C. A. S., Kemp, W. M., and Yàñez-Arancibia, A., *Estuarine Ecology,* John Wiley & Sons, New York, 1989.
23. Officer, C. B., Biggs, R. B., Taft, J. L., Cronin, L. E., Tyler, M. A., and Boynton, W. R., Chesapeake Bay anoxia: origin development and significance, *Science,* 223, 22, 1984.
24. Correll, D. L., Nutrients in Chesapeake Bay, in *Contaminant Problems and Management of Living Chesapeake Bay Resources,* Majumdar, S. K., Hall, L. W., Jr., and Austin, H. M., Eds., Pennsylvania Academy of Science, Philadelphia, 1987, 298.
25. Newcombe, C. L. and Horne, W. A., Oxygen poor waters of the Chesapeake Bay, *Science,* 88, 80, 1938.

26. Flemer, D. A., Mackierman, G. B., Nehlsen, W., and Tippie, V. K., Chesapeake Bay: A Profile of Environmental Change, Tech. Rep., U. S. Environmental Protection Agency, Washington, D.C., 1983.

27. Kuo, A. Y. and Neilson, B. J., Hypoxia and salinity in Virginia estuaries, *Estuaries,* 10, 277, 1987.

28. Seliger, H. H., Boggs, J. A., and Biggley, J. A., Catastrophic anoxia in the Chesapeake Bay in 1984, *Science,* 228, 70, 1985.

29. Taft, J. L., Taylor, W. R., Hartwig, E. O., and Loftus, R., Seasonal oxygen depletion in Chesapeake Bay, *Estuaries,* 3, 242, 1980.

30. Pokryfki, L. and Randall, R. E., Nearshore hypoxia in the bottom water of the northwestern Gulf of Mexico from 1981 to 1984, *Mar. Environ. Res.,* 22, 75, 1987.

31. Leming, T. D. and Stuntz, W. E., Zones of coastal hypoxia revealed by satellite scanning have implications for strategic fishing, *Nature,* 310, 136, 1984.

32. Gaston, G. R., Effects of hypoxia on macrobenthos of the inner shelf of Cameron, Louisiana, *Est. Coastal Shelf Sci.,* 20, 603, 1985.

33. Renaud, M. L., Annotated Bibliography on Hypoxia and Its Effects on Marine Life, with Emphasis on the Gulf of Mexico, NOAA Tech. Rep. No. 21, National Marine Fisheries Service, Washington, D.C., 1985.

34. Jaworski, N. A., Retrospective study of the water quality issues of the upper Potomac estuary, *Rev. Aquat. Sci.,* 2, 163, 1990.

35. Pritchard, D. W., Chemical and physical oceanography of the bay, in Proc. Governors Conf. on Chesapeake Bay, Wye Institute, Wye, MD, 1968, 49.

36. Boynton, W. R., Kemp, W. M., and Osborne, C. G., Nutrient fluxes across the sediment-water interface in the turbid zone of a coastal plain estuary, in *Estuarine Perspectives,* Kennedy, V. S., Ed., Academic Press, New York, 1980, 93.

37. Jaworski, N. A., Sources of nutrients and the scale of eutrophication problems in estuaries, in *Estuaries and Nutrients,* Neilson, B. J. and Cronin, L. E., Eds., Humana Press, Clifton, NJ, 1981, 83.

38. Sharp, J. H., Review of carbon, nitrogen, and phosphorus geochemistry, *Rev. Geophys. (Supplement), Contrib. Oceanogr.,* 1991, 648.

39. Boynton, W. R., Kemp, W. M., and Keefe, C. W., A comparative analysis of nutrients and other factors influencing estuarine phytoplankton production, in *Estuarine Comparisons,* Kennedy, V. S., Ed., Academic Press, New York, 1982, 69.

40. Carpenter, E. J. and Capone, D. G., Eds., *Nitrogen in the Marine Environment,* Academic Press, New York, 1983.

41. Valiela, I., *Marine Ecological Processes,* Springer-Verlag, New York, 1984.

42. Kennish, M. J., *Ecology of Estuaries,* Vol. 2, *Biological Aspects,* CRC Press, Boca Raton, FL, 1990.

43. Raymont, J. E. G., *Plankton and Productivity in the Oceans,* Vol. 1, *Phytoplankton,* 2nd ed., Pergamon Press, Oxford, 1980.
44. Levinton, J. S., *Marine Ecology,* Prentice-Hall, Englewood Cliffs, NJ, 1982.
45. Redfield, A. C., Ketchum, B. H., and Richards, F. A., The influence of organisms on the composition of seawater, in *The Sea,* Vol. 2, Hill, M. N., Ed., John Wiley & Sons, New York, 1963, 26.
46. Darnell, R. M. and Soniat, T. M., Nutrient enrichment and estuarine health, in *Estuaries and Nutrients,* Neilson, B. J. and Cronin, L. E., Humana Press, Clifton, NJ, 1981, 225.
47. Turner, R. E., Schroeder, W. W., and Wiseman, W. J., Jr., The role of stratification in the deoxygenation of Mobile Bay and adjacent shelf bottom waters, *Estuaries,* 10, 13, 1987.
48. Cairns, J., Jr., Quantification of biological integrity, in The Integrity of Water, Ballentine, R. K. and Guarraia, L. J., Eds., U. S. Environmental Protection Agency, Washington, D.C., 1977, 171.
49. Biggs, R. B., DeMoss, T. B., Carter, M. M., and Beasley, E. L., Susceptibility of U. S. estuaries to pollution, *Rev. Aquat. Sci.,* 1, 189, 1989.
50. Vollenweider, R. A., Scientific Fundamentals of the Eutrophication of Lakes and Flowing Waters, with Particular Reference to Nitrogen and Phosphorus as Factors in Eutrophication, Tech. Rep. DAS/CSE/68.27, Organization for Economic Cooperation and Development, Paris, 1968, 250.
51. Gade, H. G., Estuaries and fjords, in *Pollutant Transfer and Transport in the Sea,* Vol. 2, Kullenberg, G., CRC Press, Boca Raton, FL, 1982, 141.
52. Sugimoto, T., Similitude of the hydraulic model experiment for tidal mixing, *J. Oceanogr. Soc. Jpn.,* 30, 260, 1974.
53. McClimans, T. A. and Gjerp, S. A., Numerical Study of Distortion in a Froude Model, Am. Soc. Civil Eng. Proc. 16th Int. Conf. Coastal Engineering, Vol. 3, 1978, 2887.
54. Nielsen, P. B., Deterministic and stochastic models of dispersion of released effluents into coastal areas, in *Oceanic Processes in Marine Pollution,* Vol. 2, *Physicochemical Processes and Wastes in the Ocean,* O'Connor, T. P., Burt, W. V., and Duedall, I. W., Robert E. Krieger Publishing, Malabar, FL, 1987, 33.
55. Steidinger, K. A., A re-evaluation of toxic dinoflagellate biology and ecology, in *Progress in Physiological Research,* Vol. 1, Round, C. R. and Chapman, D. J., Eds., Elsevier, Amsterdam, 1983, 147.
56. Mee, L. D., Espinosa, M., and Diaz, G., Paralytic shellfish poisoning with a *Gymnodinium catenatum* red tide on the Pacific Coast of Mexico, *Mar. Environ. Res.,* 19, 77, 1986.
57. Kao, C. Y., Tetrodoxin, saxitoxin, and their significance in the study of excitation phenomena, *Pharmacol. Rev.,* 18, 997, 1966.
58. Walker, L. M., Life histories, dispersal, and survival in marine planktonic dinoflagellates, in *Marine Planktonic Life Cycle Strategies,* Steidinger, K. A. and Walker, L. M., Eds., CRC Press, Boca Raton, FL, 1984, 19.

59. Anderson, D. M., Kulis, D. M., Orphanos, J. A., and Ceurvels, A. R., Distribution of the toxic dinoflagellate *Gonyaulax tamarensis* in the southern New England region, *Est. Coastal Shelf Sci.*, 14, 447, 1982.

60. Maranda, L. and Shimizu, Y., Diarrhetic shellfish poisoning in Narragansett Bay, *Estuaries*, 10, 298, 1987.

61. Murata, M., Shimatani, M., Sugitani, H., Oshima, Y., and Yasumoto, T., Isolation and structural elucidation of the causative toxin of the diarrhetic shellfish poisoning, *Bull. Jpn. Soc. Sci. Fish.*, 48, 549, 1982.

62. Dahl, E. and Yndestad, M., Diarrhetic shellfish poisoning (DSP) in Norway in the autumn of 1984 related to the occurrence of *Dinophysis* spp., in *Toxic Dinoflagellates*, Anderson, D. M., White, A. W., and Baden, D. G., Eds., Elsevier, New York, 1985, 495.

63. Krogh, P., Edler, L., Graneli, E., and Nyman, U., Outbreak of diarrhetic shellfish poisoning on the west coast of Sweden, in *Toxic Dinoflagellates*, Anderson, D. M., White, A. W., and Baden, D. G., Eds., Elsevier, New York, 1985, 501.

64. Yasumoto, T., Recent progress in the chemistry of dinoflagellate toxins, in *Toxic Dinoflagellates*, Anderson, D. M., White, A. W., and Baden, D. G., Eds., Elsevier, New York, 1985, 259.

65. Sullivan, J. J., Methods of analysis for DSP and PSP toxins in shellfish — a review, *J. Shellfish Res.*, 7, 587, 1988.

2 Oil Pollution

INTRODUCTION

Estuaries rank among the most productive areas on earth, with gross primary productivity amounting to approximately 10 kcal/m²/ year of organic matter compared to 0.5 to 3 kcal/m²/year and <1 kcal/m²/year for continental shelf waters and the open ocean, respectively.[1] In addition, they serve as critical reproductive and nursery grounds for a wide variety of fishes and important habitats for numerous benthic and planktonic organisms. Because of the high densities of floral and faunal populations occupying these coastal ecotones, environmental consequences of anthropogenic impacts tend to be more severe than in other habitats. Among the most serious pollution threats to the biota inhabiting estuaries is oil, which not only kills organisms directly, but also destroys or damages sensitive habitats that support aquatic communities. Indeed, estuaries and contiguous habitats such as salt marshes, mangroves, mudflats, and coastal marine waters — all having particular ecological significance — are permanently at risk from accidents involving oil.

Public perception of oil pollution in the sea generally centers on spectacular accidents involving supertankers or offshore oil platforms. However, most of the oil found in estuaries is not derived from such events at all. Still, oil tanker accidents account for some of the most devastating impacts on coastal and estuarine ecosystems. Examples of major oil spills affecting coastal regions during the last 25 years are the following:

1. *Torrey Canyon* — In March 1967, 117,000 t of Kuwait crude oil spilled into the sea, ultimately depositing 40,000 t of oil along the beaches of Cornwall (Wales) and Brittany (France).

2. *Florida* — In September 1969, 630 t of light refined oil released from the barge damaged West Falmouth Harbor in Buzzards Bay, Massachusetts.
3. *Arrow* — In February 1970, 10,000 t of heavy refined oil spilled from the tanker into Chedabucto Bay, Nova Scotia, Canada.
4. *Argo Merchant* — In December 1976, 26,000 t of No. 6 fuel oil drained into Nantucket Shoals, about 50 km southeast of Nantucket Island, off Massachusetts.
5. *Amoco Cadiz* — In March 1978, 223,000 t of Arabian and Kuwait light crude oil spread into the English Channel, eventually polluting a 300-km stretch of the Brittany coast.
6. *Exxon Valdez* — In March 1989, more than 240,000 barrels (37.9 million l) of oil from the tanker entered Prince William Sound, Alaska, creating the worst oil spill ever in U.S. waters.

As reported by Hall et al.,[2] approximately 75% of the accidental oil spills in the U.S occur in coastal waters, principally in estuaries, enclosed bays, and wetlands.

This chapter examines the impact of oil pollution on estuarine and coastal marine waters. Focus is placed on the sources and fate of oil in estuaries, the magnitude of its input, and the methods used to mitigate its effects. Assessment of the hazards posed by the oil includes an investigation of the toxicity of the chemical compounds comprising petroleum products in the sea to specific biological ecosystem components. In addition, details of the physical, chemical, and biological factors influencing the fate of oil in the estuarine environment are presented.

SOURCES OF OIL POLLUTION

The total input of oil to estuaries is difficult to estimate because most statistics available do not separate the oil entering estuarine areas from that released to other marine waters.[3] Carlberg[3] estimates that more than 33% of the entire input of oil to the marine hydrosphere may enter estuaries. Abel[4] notes that river runoff, urban runoff, municipal wastes, and effluents from nonpetroleum industries are responsible for approximately 45% of the oil reaching the sea. Another 33% of the oil found there results from activities related to oil transportation, with only about 25% of this amount attributable to accidents and major spillages, and the remainder ascribed to normal operational losses. Roughly 20 to 25% of the oil escapes from the seafloor via natural oil seeps.

While major oil spills and accidents are perceived by much of the public as a principal danger to estuarine and marine environments, it is the chronic oil pollution associated with routine operations of coastal oil refineries and oil installations, as well as discharges of industrial and municipal wastes, that actually affect a greater area.[4] Hall et al.[2] assign a "sizeable proportion" of the oil introduced to waters in the coastal zone to oil-related facilities (i.e., refineries, offshore oil drilling rigs, oil pipelines, onshore storage and distribution facilities, tankers, and tanker unloading facilities in harbors and in offshore deep water). Some of this oil is released purposely and routinely. For example, the discharge of ballast water containing oil contaminants from tankers exacerbates pollution effects. Moreover, wastewaters released from refineries and offshore oil rigs contribute contaminants to offshore and coastal waters. These routine losses can greatly exceed losses from major accidents.

Table 1 records the primary sources of petroleum hydrocarbons in the sea. As evident from these data, the input of hydrocarbons from phytoplankton is much higher than the input from fossil hydrocarbons. The biosynthesis of hydrocarbons equals 26,000 mta compared to 4.94 mta of petroleum hydrocarbons derived from transportation, fixed installations, and other anthropogenic sources. Most of the input to estuaries results from river runoff (1.40 mta), urban runoff (0.40 mta), and municipal and industrial waste (0.45 mta). Tanker accidents and routine transportation activities may supply more than 0.50 mta to these coastal habitats as well.[5]

Based on the data compiled on petroleum intentionally or unintentionally released to estuarine and marine waters, oil inputs originating from the users of petroleum products far surpass those from the oil extraction and transport industries (Table 1). According to Clark,[5] these industries are responsible for little more than 25% of the total input of oil to the sea. Estimates by the National Research Council[6] on the global input of petroleum to the marine environment, however, reveal that marine transportation represents the

TABLE 1
Estimated World Input of Petroleum Hydrocarbons to the Sea[a]

	Oil industry	Other	Total
Transportation			
Tanker operations	0.60		
Tanker accidents	0.30		
Dry docking	0.25		
Other shipping operations		0.12	
Other shipping accidents		0.10	
	1.15	0.22	1.37
Fixed installations			
Offshore oil production	0.06		
Coastal oil refineries	0.06		
Terminal loading	0.001		
	0.12		0.12
Other sources			
Industrial waste		0.15	
Municipal waste		0.30	
Urban runoff		0.40	
River runoff		1.40	
Atmospheric fallout		0.60	
Natural seeps		0.60	
		3.45	3.45
	1.27	3.67	4.94
Biosynthesis of hydrocarbons			
Production by marine phytoplankton		26,000	
Atmospheric fallout		100–4,000	

[a] Millions of tons per year.

From Clark, R. B., *Marine Pollution, 2nd ed.,* Clarendon Press, Oxford, 1989. With permission.

largest single source of petroleum input, accounting for more than 40% of the oil on the basis of a best estimate and 30 to 50% on the basis of a probable range of values (Table 2). Most of this input is from routine operational discharges and not accidents, which gener-

TABLE 2
Input of Petroleum Hydrocarbons Into the Marine Environment[a]

Source	Probable range	Best estimate
Transporation		
Operational	4.4–15.0	7.2
Accidents	2.2–3.0	2.9
Subtotal	6.6–18.0	10.1
Municipal wastes and runoff	4.0–21.5	8.1
Atmosphere	0.3–3.4	2.1
Natural sources	0.2–17.3	1.7
Offshore production	0.3–0.4	0.3
Total	11.4–60.6	22.3

[a]   Million barrels per year. One barrel equals 42 U.S. gal or about 35 Imperial gal. Table converted from NRC (1985) estimates based on metric tons per year.

From National Research Council, *Using Oil Spill Dispersants on the Sea*, National Academy Press, Washington, D.C., 1989. With permission.

ate only about 12% of the oil entering the marine hydrosphere each year.

COMPOSITION OF OIL

Crude petroleum (equivalent to crude oil for discussions in this chapter) consists of a complex mixture of many thousands of gaseous, liquid, and solid organic compounds. Hydrocarbons, with 4 to 26 or more carbon atoms in the molecule, are most abundant (Figure 1).[5] These compounds comprise more than 75% of the weight of the oil.[7] A variety of hydrocarbon structures is discernible, with the hydrogen and carbon atoms arranged in straight, branched, or cyclic chains, including aromatic compounds (Figure 1). Nonhydrocarbons constitute as much as 25% of the oil, i.e., compounds containing oxygen, sulfur, nitrogen, or metals (e.g., copper, iron, nickel, and vanadium).

A number of chemical classes can be differentiated depending on

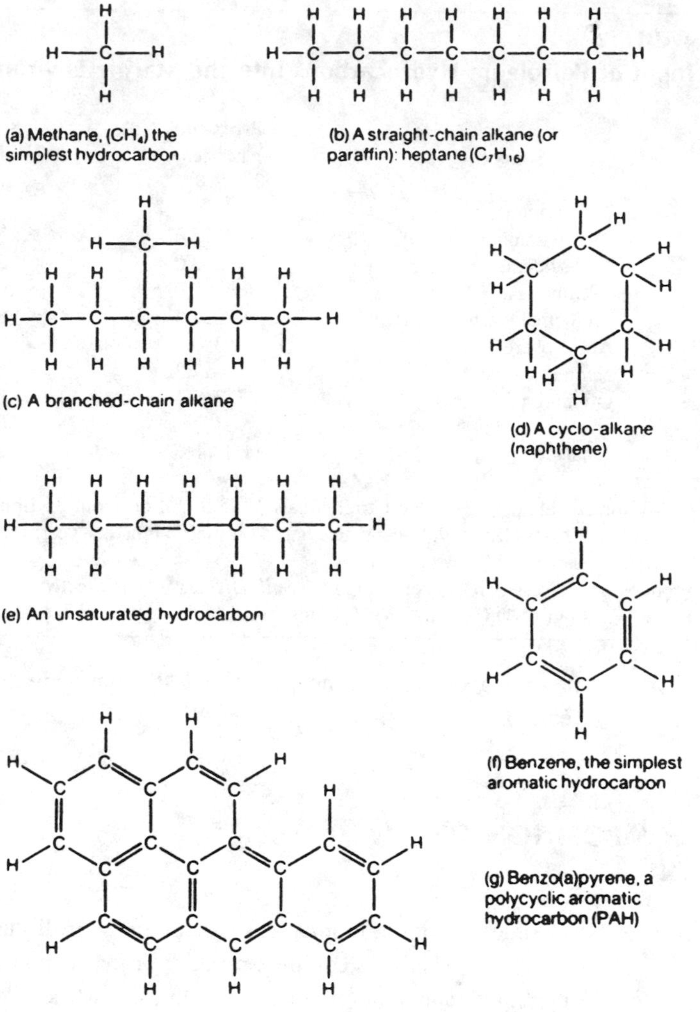

(a) Methane, (CH$_4$) the simplest hydrocarbon

(b) A straight-chain alkane (or paraffin): heptane (C$_7$H$_{16}$)

(c) A branched-chain alkane

(d) A cyclo-alkane (naphthene)

(e) An unsaturated hydrocarbon

(f) Benzene, the simplest aromatic hydrocarbon

(g) Benzo(a)pyrene, a polycyclic aromatic hydrocarbon (PAH)

FIGURE 1.   The structure of some hydrocarbons. (From Clark, R. B., *Marine Pollution, 2nd ed.,* Clarendon Press, Oxford, 1989. With permission.)

how the atoms bind together in the molecules.[3] They are alkanes (paraffins), cycloalkanes (cycloparaffins, naphthenes), alkenes (olefins), aromatic compounds, and other compounds in oil. It is important to note that the relative concentrations of the different hydrocarbon types vary greatly in petroleums from different sources.[7] Hence, the significance of each chemical class may likewise change dramatically from one crude oil to the next. Much of the following description on chemical classes is based on the work of Carlberg.[3]

CLASSES OR SERIES
Alkanes

These compounds, alternatively classified as paraffins, consist of a series of saturated aliphatic hydrocarbons having the empirical formula:

$$C_n H_{2n+2} \tag{1}$$

The molecules are principally straight chained, but branched varieties also occur in the class. Methane (CH_4), or marsh gas, represents the smallest and lightest member of the series of compounds comprising the alkanes or paraffins.[8] The largest molecules of the series can be quite complex, with more than 60 carbon atoms. The straight-chained molecules are commonly referred to as n-alkanes and the branched molecules as i-alkanes. The alkanes, along with the alkenes and alkynes, have been collectively termed aliphatic hydrocarbons.

Cycloalkanes

This group of compounds, also known as cycloparaffins and naphthenes, has the general formula

$$C_n H_{2n} \tag{2}$$

The molecules are shaped into rings. Five or six carbon atoms exist in the ring structure of most members of this group.

Both the cycloalkanes and the alkanes are saturated hydrocarbons; i.e., each carbon atom has attached to it all the hydrogen atoms that it can hold.[8] In the molecules of these two groups, only single bonds join the carbon atoms. However, in alkene, alkyne, and aromatic molecules, multiple bonds join at least one pair of carbon atoms. Because these three hydrocarbon groups have one or more pairs of carbon atoms joined by multiple bonds, they are called unsaturated hydrocarbons.

Alkenes

The alkenes or olefins exhibit the same general formula (2) as the cycloalkanes. They differ, however, in that the molecules of the group are either straight chained or branched and not shaped into rings. The unsaturated structure of the molecules accounts for this difference. The unsaturated aliphatic hydrocarbons contain molecules with one or more carbon-to-carbon double bonds, thereby differing from those of the alkynes which contain at least one carbon-to-carbon triple bond.

While alkenes are common constituents of petroleum products, they typically do not comprise a significant fraction of crude oils. The same is true of the alkynes. Some refining processes result in their addition to petroleum products. In contrast, alkenes may be a significant component of the biogenous hydrocarbons.

Alkynes

As noted above, the alkynes are unsaturated hydrocarbon compounds containing carbon-to-carbon triple bonds. A pair of carbon atoms in alkyne molecules behaves differently from any carbon atoms in alkane and alkene molecules. The first member of the alkyne series is ethyne (acetylene) with the following formula

$$C_2 H_2 \tag{3}$$

The triple bond structure of ethyne is represented as

$$H - C \equiv C - H \tag{4}$$

The general formula of the alkynes is written as

$$C_n H_{2n-2} \tag{5}$$

Aromatic Hydrocarbons

The molecules of these compounds are characterized by the presence of at least one benzene ring containing a group of six carbon atoms linked together in a hexagonal arrangement. The benzene ring has three double bonds. In benzene, the simplest member of the group, a hydrogen atom joins to each carbon atom. A methyl group replaces one of the hydrogen atoms in toluene, the next member of the series.

In addition to benzene and the methylbenzenes (e.g., toluene and xylene), this group includes polynuclear aromatic hydrocarbons (PAHs) having several benzene rings. The polynuclear aromatics can be subdivided into the fused-ring compounds (e.g., alkyl naphthalenes, benz(a)pyrene, and phenanthrene) and the linked rings (e.g., the biphenyls). The aromatic content of crude oils typically varies substantially, but rarely exceeds 40%.

Other Compounds

The heterocyclic compounds in crude oil contain most of the oxygen, nitrogen, and sulfur found in the oil. These are aromatics or naphtheno-aromatics in which oxygen, nitrogen, or sulfur replace one or more carbon atoms in the molecule. Crude oil may have up to 2% oxygen, 0.05 to 0.8% nitrogen, and up to 5% by weight of sulfur. The oxygen occurs principally in phenols and carboxylic acids, nitrogen mainly in pyridine and quinolines, and sulfur in elemental form as well as in hydrogen sulfide, mercaptans, and aliphatic and cyclic sulfides.

Biogenous Hydrocarbons

The largest input of hydrocarbons, as mentioned above, is via biosynthesis (Table 1). While crude oil hydrocarbons consist of thousands of different molecules, ranging from 16 to more than 20,000 in molecular weight, biogenic hydrocarbons have a more limited suite of compounds within a relatively narrow range of molecular weights. Medium-resolution gas chromatography and mass spectrometry can be used to resolve biogenic hydrocarbons (up to at least C_{22}) into individual compounds.[9] Tables 3 and 4 detail the major differences between petroleum and biogenic hydrocarbons.

Toxicity of Crude Oil

A wide range of toxic substances have been identified in crude oil. For instance, benzene, toluene, xylene and other low molecular weight aromatics can be highly toxic. Other toxic constituents are acids (e.g., carboxylic acids) and phenols, as well as sulfur compounds (e.g., sulfides, thiols, and thiophenes) in the oil. PAHs (e.g., 1,2-benzanthracenes, 3,4-benz(a)pyrene, 1,2-benzphenanthrene, diphenylmethane, fluorene, and phenanthrene) may also be toxic to a multitude of aquatic organisms.[3]

The toxicity of the various classes of petroleum hydrocarbons tends to increase along the series from the alkanes, cycloalkanes,

TABLE 3
Major Differences between Petroleum and Biogenic Hydrocarbons

Hydrocarbon	Petroleum	Biogenic
n-Alkanes	Odd and even number of C atoms nearly equally abundant	Odd number of C atoms much more abundant than even numbers
	Adjacent members of series usually vary little in concentration	
Aromatics	Many and complex (polybenzenes and naphthalenes, polynuclear aromatics with multiple alkyl substitutions, naphthalene-aromatics)	Simple, with one or two alkyl substitutions at most
Cycloalkanes (Naphthenes)	Heterogeneous; those with substituted rings more abundant than their parent compounds	Uncommon; one- to three-chain rings
Alkenes (Olefins)	Usually absent in crude oil; may occur in refined petroleum	Major portion of biologic hydrocarbons

From Sanders, H. L., Grassle, J. F., Hampson, G. R., Morse, L. S., Garner-Price, S., and Jones, C. C., *J. Mar. Res.*, 38, 265, 1980. With permission.

and alkenes to the aromatics.[5] In general, the smaller molecules in each series of hydrocarbons are more toxic than the larger ones. The evaporative loss of volatile hydrocarbons from crude oil during the first 24 to 48 h of a spill is the most important process which removes the toxic, lower molecular weight components from the oil.[6] The volatile hydrocarbons, in fact, are an important, albeit ephemeral source of toxic materials present in crude oils, and their rapid loss reduces the potential impact of oil spills on estuarine and marine organisms.

FATE OF POLLUTING OIL

ABIOTIC AND BIOTIC FACTORS

Oil that enters the estuarine environment is affected by a number of abiotic factors as well as by the physical-chemical properties of the oil itself. When released to estuaries in river runoff and domestic and industrial effluents, the oil may be present as dissolved sub-

TABLE 4
Guidelines for the Differentiation of Petroleum and Biogenous Hydrocarbons

Petroleum contains a much more complex mixture of hydrocarbons with much greater ranges of molecular structure and molecular weight.

Petroleum contains several homologous series, with adjacent members usually present in nearly the same concentration. The approximate unit ratio for even and odd numbered alkanes is an example, as is the homologous series of C_{12}-C_{22} isoprenoid alkanes. Marine organisms have a strong predominance of odd-numbered C_{15} through C_{21} alkanes.

Petroleum contains more kinds of cycloalkanes and aromatic hydrocarbons. Also, the numerous alkyl-substituted ring compounds have not been reported in organisms. Examples are the series of mono-, di-, tri-, and tetramethyl benzenes and the mono-, di-, tri-, and tetramethyl naphthalenes.

Petroleum contains numerous naphtheno-aromatic hydrocarbons that have not been reported in organisms. Petroleums also contain numerous heterocompounds containing sulfur, nitrogen, oxygen, and metals, and the heavy asphaltic compounds.

From *Petroleum in the Marine Environment*, National Academy of Sciences, Washington, D.C., 1975. With permission.

stances, emulsions, or particles.[3] Spills of oil from tankers in estuaries and marine waters spread rapidly over the water surface to form a slick that is largely immiscible with the water. The thickness of the slick generally varies from several millimeters down to 1 μm, depending mainly on the nature of the oil, the seawater temperature, and the area available for spreading.[6] For the first 6 to 10 h after a spill, the spreading of oil probably is the most significant process.[10] Cormack[11] summarizes three distinct spreading phases associated with an oil spill: (1) an initial phase in which gravitational and inertial resistance forces control the spreading; (2) an intermediate phase in which gravitational and viscous drag forces predominate; and (3) a final phase in which surface tension and viscous drag forces play a key role.

Within several hours of the initial spill, the composition of the oil changes markedly due to evaporation of low molecular weight components, dissolution of water soluble constituents in the water column, and the emulsification of immiscible components as small droplets.[5] The evaporation of the volatile components of the oil yields a residue of greater viscosity and lower toxicity on the sea surface. The dissolution process likewise involves the lighter, more toxic molecules, but rates of evaporation exceed those of dissolution by about two orders of magnitude. Wave action and water turbulence are important agents in the physical dispersion of the continuous surface layer as small oil droplets into the water column. Hence, the oil slick, minus its volatile and soluble constituents, gradually transforms into dispersed droplets in the sea.[11]

The loss of volatiles causes the density of the oil to approach that of seawater; consequently, it may sink. Under certain conditions, a water-in-oil emulsion forms a viscid mass containing 70 to 80% water. Because of its appearance, this mass has been termed "chocolate mousse". The formation of chocolate mousse results in pancakelike masses on the water surface.

The composition of the oil continues to change through time owing to the effects of photochemical oxidation, which is a function of sunlight intensity and biodegradation by organisms, particularly bacteria, that preferentially attack the low molecular weight molecules. The composition of the oil, water temperature, and nutrient availability (e.g., nitrogen and phosphorus) largely regulate the rate of biodegradation of the oil.[11] Ultimately, the cumulative effects of evaporation, dissolution, photooxidation, and microbial degradation lead to the formation of tarballs or heavy residues ranging in size from <1 mm to 10 to 20 cm in diameter.[5,10] Figure 2 illustrates the time-course of factors affecting an oil spill at sea. The following discussion of factors is drawn heavily from the work of Carlberg,[3] the National Research Council,[6] Jordan and Payne,[10] Cormack,[11] and Murray.[12]

Spreading

The interaction of oil droplets with diffusive and current shear processes in near-surface currents provides the primary mechanism of oil spreading at sea.[13] Gravitational forces, surface tension, inertial forces, and frictional forces are the main forces influencing the lateral spreading of oil on a calm sea.[10] The relative significance of these forces during the early to late phases of spreading are stipulated in the factors section above. In effect, the inertia of the oil body and the oil/water frictional forces retard the spreading forces ascribable to gravity and surface tension. Inertia is the dominant antispreading force for the first several hours subsequent to a spill, being gradually supplanted by the frictional retardation force as the principal antispreading force owing to the increase in the viscous water/ oil mixture and the decrease in the thickness of the oil slick. A mathematical model predicting the spreading of oil on a calm sea has been devised by Fay.[14] This model is deficient in a number of respects;[15,16] notably, it does not consider the effects of oil viscosity, evaporation, and emulsification. Moreover, it assumes a constant volume spill of a single component petroleum that spreads radially, with a monotonically decreasing thickness.[12] Field observations reveal, however, that oil does not spread uniformly, but separates into thin (<0.1 mm) and thick (>0.5 mm) areas with variable levels of spreading pressure.[12,16] The interaction of interfacial tension, gravity, and viscosity in spreading processes, the accumulation of oil at downwelling convergence zones generated by water movement, and the formation of chocolate mousse produces nonuniformity in oil thickness.[6]

The distortion and distribution of a spreading slick are greatly influenced by the advection of waves and currents, with current and wind vectors directing the slick. Wind, waves, and surface currents primarily control the drift or movement of the center of mass of an oil slick.[10] Murray[17] deemed dispersive processes to be far more important than spreading in determining the outcome of a continuous spill. The seasonal permanent current and transient wind-drift current are the two commonly used components in oil spill advection calculations. Typically the effect of wind and wind waves on an oil slick is to break it up into small patches of oil that ultimately

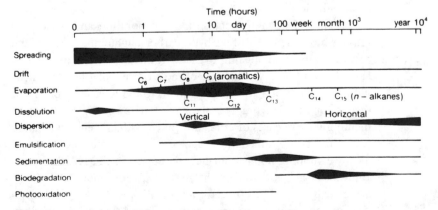

FIGURE 2. Time-course of factors affecting an oil spill at sea. (From Clark, R. B., *Marine Pollution, 2nd ed.*, Clarendon Press, Oxford, 1989. With permission.)

move away from the spill site by the average current and diffuse via turbulent eddies. The oil patches move slowly apart by horizontal eddy diffusion. Wind waves and vertical diffusion mix the oil at the sea surface, producing subsurface advection to depths of about 10 m. This process enables subsurface currents to transport the oil away from the spill site as well.[18] Unfortunately, the downward mixing of oil components below the sea surface brings hydrocarbons into contact with estuarine and marine organisms that would not normally be impacted by the pollutants. The susceptibility of these organisms to the toxic effect of a hydrocarbon compound is contingent upon many factors, especially the concentration of the compound, its aqueous solubility, the type of organism and its life history stage, and the length of time of organismal exposure to the contaminant.[6]

Evaporation

The process by which low to medium molecular weight components of low boiling point in oil are volatized into the atmosphere, termed evaporation,[12] usually accounts for the largest initial change in composition of a polluting oil spill at sea.[18] The rate of evaporative loss is a function of the vapor pressure of the oil at ambient temperature, but may be accelerated by the atmospheric and physical conditions of the sea surface. For example, high velocity winds and rough seas foster aerosol formation that enhances evaporation.[11] Other factors that also influence evaporative loss are the composition and surface area of the oil, air and seawater temperatures, and solar radiation.[6]

Jordan and Payne[10] disclose that evaporation and dissolution remove more than 90% of the hydrocarbons lighter than n-C_{10} within several hours of an oil spill. In a study of natural oil seeps, Sivadier and Mikolaj[19] observed the loss of most of the volatile components within a period of 1 to 2 h subsequent to the release of oil. These results compared favorably to those of Harrison et al.,[20] who, experimenting with releases of crude oil spiked with cumene on a natural water surface, discerned the disappearance of the cumene, as well as all lower boiling point aromatics, within 90 min. McAuliffe[21] and Johnson et al.[22] ascertained the loss of all low molecular weight aromatic hydrocarbons from a crude oil slick within 8 h. The loss of benzene and toluene occurred within 1 h and that of dimethylbenzenes and trimethylbenzenes within 5 and 8 h, respectively. The rapid evaporative loss of the volatile components is biologically advantageous because of their toxicity to estuarine and marine organisms.[20,23,24]

Since low molecular weight hydrocarbons from C_4 to C_{12} may comprise nearly 50% of an average crude oil,[3,25] evaporation facilitates substantial weathering of a polluting oil, manifested in the large decrease in slick volume. Owing to evaporation, pentadecane ($n\text{-}C_{15}$) is the lowest normal alkane commonly recorded in weathered oils, and hydrocarbons lower than $n\text{-}C_{12}$ are rarely recovered in seawater extracts.[10] As a slick volume declines, together with the volatile components, the viscosity and specific gravity of a polluting oil rise contributing to the gradual transformation of the oil.

Photochemical Oxidation

The heavier, less volatile components of a polluting oil at sea, although not very soluble in seawater, can yield more soluble compounds when they interact with oxygen under the influence of sunlight. The photochemical oxidation of oil takes place slowly in seawater, the concentration of oxidation products being low the first few days after an oil spill.[6] Thin oil slicks, approximately 10^{-8} cm thick, appear to be affected more efficiently by photochemical oxidation; in these slicks, photooxidation becomes important 1 or 2 d subsequent to a spill after the volatile components of the oil have been lost.[18]

Most of the photooxidative reactions that transform petroleum hydrocarbons into compounds with chemical and biological activities occur in the surface water layer and in the surface oil film because of the limited penetration of light into seawater.[3] Photooxidation of petroleum hydrocarbons in marine and estuarine waters involves an autocatalytic free-radical chain reaction, culminating in the formation of aldehydes, ketones, hydroxy compounds, and ultimately, low molecular weight carboxylic acids. However, polymerization or condensation reactions of aldehydes and keytones with phenols, as well as esterification between alcohols and carboxylic acids, are responsible for the formation of higher molecular weight intermediates.[26]

Parker et al.[26] demonstrated that the rate of photooxidation of petroleum hydrocarbons at sea is partly dependent on the wavelength of incident light. When thin films of oil were irradiated with light of wavelengths between 300 to 350 nm, peak photooxidation reactions occurred. In essence, at wavelengths of light <300 nm, photolytic reactions were rapid. Slower photooxidation reactions resulted when only wavelengths present in sunlight were used. Hansen[27] also found more rapid photooxidation processes at wavelengths shorter than 300 nm. He discovered that the photolytic processes increased by at least two orders of magnitude when the oil was irradiated by a wider spectral range of light from 200 to 300 nm.

Similar to the effects of evaporation, photooxidation increases the viscosity of the remaining part of a crude oil. In addition, it increases the density of the polluting oil. Sedimentation may also rise due to the formation of solid or semisolid particles.[3] Changes in the spreading properties of the oil attributable to photooxidation may likewise develop, as is evident from the decreased or increased spreading coefficient values recorded in laboratory experiments.[28]

Dissolution

A small fraction of crude oil passes into solution in seawater; although the concentration of the dissolved oil is not great, it poses a threat to estuarine or marine organisms that ingest the soluble compounds. The most soluble compounds of a crude oil are also the most volatile, lower molecular weight constituents. The rate of dissolution of a crude oil depends not only on its chemical composition, including the molecular structure and relative abundance of the components, but also on the physical-chemical conditions of the environment (e.g., salinity, temperature, wind, wave action, turbulence, etc.).[10] In regard to the solubility of different hydrocarbon groups, solubility decreases in the order aromatics, naphthenes, i-alkanes, n-alkanes. The solubility is also inversely related to the molecular weight of the hydrocarbons within each group.[3]

In estuarine and marine environments, dissolved organic matter (DOM) promotes the solubility of petroleum hydrocarbons. The surface-active nature of DOM is responsible for this effect. Humic substances in marine waters enhance hydrocarbon solubilization into a semicolloidal or micellar state.[10]

Thick, highly viscous crude oils tend to be modified less by dissolution than light, thin oils containing a higher concentration of volatile components. The thickness of most viscous oils retards dissolution because of the greater diffusion distances components must traverse to reach the oil-water interface. As weathering processes proceed and the more volatile components are lost, dissolution of the oil gradually diminishes.[10]

Emulsification

Polluting oil can form two types of emulsions: oil-in-water and water-in-oil. Water is the continuous phase in an oil-in-water emulsion and provides for conditions much more conducive to microbial degradation of oil droplets than the case of a water-in-oil emulsion.[12] The formation of water-in-oil emulsions exacerbates the impact of a polluting oil; moreover, the size or volume of an oil slick may actually be enlarged by this process, as was apparent after the *Torrey Canyon* spill. When stable water-in-oil emulsions develop, they float on seawater and agglomerate into viscous masses with a chocolate-like constituency resembling chocolate mousse. Mousse formation enables oil to potentially persist for months at sea, where it can be advected many miles from the original spill site. Agitated sea surfaces promote the formation of stable water-in-oil emulsions.

The consistency of a water-in-oil emulsion is a function of the water content. Hence, a >80% water-in-oil emulsion yields a viscous, chocolate-like constituency (i.e., mousse). A 50 to 80% water-in-oil emulsion has a grease-like constituency, whereas a 30 to 50% water-in-oil emulsion is fluid, resembling the parent oil.[10,29]

The stability of water-in-oil emulsions is related to the chemical components in nonvolatile residues, namely the simultaneous presence of asphaltenes and paraffins.[6,30,31] Experiments by Bridie et al.[30] and Payne and Phillips[31] show that no mousse forms when asphaltenes and waxes are removed from oil samples; however, when readded to the basic oil, mousses can be easily formed. Clearly, the occurrence of surface-active materials, such as the asphaltenes, plays a vital role in the formation of water-in-oil mousse.

Sedimentation

Several processes act on crude oil to enhance its settlement to the seafloor. These processes cause increases in oil density leading to settlement and include: (1) evaporation and dissolution of lower molecular weight compounds; (2) photooxidation and subsequent dissolution of oxygenated materials; (3) formation of viscous and higher density water-in-oil emulsions; (4) incorporation of particulates and the agglomeration of oil particulate mixtures; and (5) incorporation of higher density microorganisms and even macroorganisms (e.g., barnacles).[10] The density of the parent oil, which can vary appreciably, has much to do with its capacity to sink. Number 6 fuel oil, for example, has greater density than most other oil and, therefore, tends to sink more rapidly, even without much degradation.[32] In general, however, an extensive amount of weathering occurs prior to settlement of oil products.

A major factor promoting the sedimentation of oil is the sorption of hydrocarbons to particulate matter (e.g., clay, silt, sand, shell fragments, and organic material) in the water column. Because of the large sediment loads in many river plumes, estuaries, and coastal marine waters, this process often predominates in these areas. Agitated seas seem to hasten the process of the adsorption of oil onto suspended particulate material, and when calm conditions return, the higher specific gravity mixture, which substantially exceeds that of seawater alone (1.025 g/cm^3), sinks the oil.[12] The specific gravity of detrital mineral and particulate organic material commonly ranges from 2.5 to 3.0 g/cm^3; consequently, this material can rapidly raise the specific gravity above the critical level for the deposition of petroleum. The fractionation of the oil prior to its sorption onto particulates and subsequent settlement to the seafloor can also be significant.

Galt[33] recounted how large concentrations of suspended sediment from the Ventura River in California contributed to the sinking of large quantities of oil during the Santa Barbara Channel spill in 1969. By comparison, little oil sank during the *Argo Merchant* oil spill in 1976. As a result, microbial degradation was much less effective in the former case when the oil became trapped in seafloor sediments.

Not only does the composition of the crude oil influence sedimentation, but the composition of the suspended particulates plays a factor as well. For instance, in laboratory studies, the adsorption efficiency for oil has been shown to vary among clay minerals, with the values increasing in the order of montmorillonite, illite, kaolinite, bentonite. Clays interact with oils to form a spontaneous flocculation of colloids or colloidal electrolytes in seawater.[10,34] Thus, the sorption of oil onto clay surfaces and the subsequent particle flocculation probably is responsible for much sedimentation of oil in high turbidity zones of riverine, estuarine, and coastal marine waters.

A potentially important biological pathway for the sedimentation of oil involves the ingestion and voiding of hydrocarbons in the feces of zooplankton. Following the spill of Bunker C fuel oil from the *Arrow* in Chedabucto Bay, Nova Scotia, in 1970, substantial quantities of oil droplets were consumed by zooplankton. Furthermore, the zooplankton feces contained up to 7% oil by weight, and because the density of the fecal matter surpassed that of seawater, it quickly settled to the seafloor. Jordan and Payne[10] have estimated that a population of 2000 individuals per cubic meter of the copepod, *Calanus finmarchicus,* covering an area of 1 km^2 to a depth of 10 m would remove as much as 3 mta of oil daily if the oil concentration equaled or exceeded 1.5 ppb. Since this oil would be bound to fecal matter having a density greater than that of seawater, it would ultimately settle to the seafloor where it would be subjected to further reworking and microbial degradation.

Microbial Degradation

Biodegradation of petroleum hydrocarbons by bacteria, as well as by a variety of fungi and yeasts, becomes increasingly more important in the alteration of oil as its residence time lengthens in the water. Bacteria, in particular, are highly capable of degrading crude oil components. Previous research and observation have shown that more than 200 species of bacteria, yeasts, and filamentous fungi can metabolize one or more hydrocarbon compounds.[10,35] This action occurs over a rather protracted period of time commencing with the termination of evaporative losses and continuing for 1 week to 1 year or more, depending on temperature and other environmental conditions.[6] The microbial breakdown of oil is most rapid in shallow, nearshore environments where a rich nutrient supply enables the proliferation of dense microbial populations. As expected, the higher water temperatures in tropical and subtropical waters accelerate this process, while the lower water temperatures in polar seas hinder it. Despite the limitations imposed on microbial activity by low water temperatures, petroleum degrading microbes are widely distributed and quite abundant in some polar regions, such as the Beaufort Sea and Arctic coastal waters.[36,37] The concentrations of these predominantly psychrophilic (obligate and facultative) microbes can approach those of temperate systems.[36] Microbial action on petroleum hydrocarbons declines substantially in the deep sea at depths below 1000 m.[18]

Microorganisms utilize polluting oil as sources of carbon and energy.[38] Ultimately, the oil is degraded to carbon dioxide and water. Representative groups of microorganisms capable of oxidizing petroleum hydrocarbons and/or their derivatives in estuarine and marine environments include *Acinetobacter, Achromobacter, Aeromonas, Arthrobacter, Bacillus, Bacterium, Brevibacterium, Cornybacterium, Enterobacter, Micrococcus, Mycobacterium, Penicillium, Pseudomonas, Sarcina,* and *Serratia. Pseudomonas* sp., along with the filamentous fungus *Cladosporium resinae,* often are dominant hydrocarbon-utilizing organisms.[10] Microbial degradation principally involves the paraffinic and aromatic components,[6] but Rontani et al.[39] contend that asphaltenes undergo some biodegradation as well. Polar fractions, or nitrogen-, sulfur-, and oxygen-containing compounds, exhibit no evidence of biodegradation.[40]

Among the various components of crude oil, aromatic hydrocarbons are oxidized less quickly than aliphatics by microbes. Resistance to microbial breakdown generally increases through the following series: n-alkanes, i-alkanes, cycloalkanes, and aromatics.[3] Thus, the composition of the oil affects the degradation process.

Laboratory experiments have demonstrated that many intermediate and end products can be obtained by microbial metabolism of oil. For example, acids, alcohols, aldehydes, ketones, peroxides, and sulfoxides have been derived from microbial degradation of oil in the laboratory. Some of these products (e.g., acids, aliphatic alcohols, and equivalent aromatic derivatives) are possibly disruptive to marine life processes.[3] Certain compounds potentially interfere with the chemotaxis of organisms.[41] In addition, oxidized intermediate products formed by the biodegrading organisms may be more toxic than the parent compound.

Carlberg[3] conveyed that biodegradation rates of crude oils, lubricating oils, cutting oils, and oil wastes range from 0.02 to 2.0 g/m^2/d at 24 to 30°C. However, a host of factors influence degradation rates, some of the most important being nutrient availability, temperature, salinity, oxygen, and type of oil. Jordan and Payne[10] have examined these factors. In regard to nutrients, nitrogen and phosphorus represent limiting factors to the rate and magnitude of petroleum hydrocarbon degradation. Temperature, in turn, directly affects metabolic rates of the microbes. As connoted above, low temperatures reduce biodegradation rates of hydrocarbon compounds by suppressing microbial growth rates and metabolic rates; moreover, they allow the toxic components to be retained in the oil, causing a further inhibition of microbial growth. Psychrophilic bacteria, which includes most marine forms, display optimal growth rates between 15 and 20°C. Most marine bacteria have an optimal salinity range of 25 to 35‰, but many of them can be found in waters beyond this range. Nevertheless, these microbes do not reproduce well in salinities from 15 to 20‰.[35] The abundance of bacteria typically is high in estuaries because of the greater concentration of nutrients accumulating there.[42] The estuarine microbes have adapted to a rather wide variation in salinities. Finally, dissolved and free oxygen are important for the complete microbial oxidation of hydrocarbon compounds. Because the breakdown of these hydrocarbons partially depends on the nature of the compounds, the type of parent oil is a major factor in the degradation process.

The application of chemical dispersants (surfactants) to polluting oil spills also influences biodegradation of the oil. While the majority of dispersants appear to enhance biodegradation by emulsification of the oil, some can ameliorate biodegradation rates because the dispersants, along with the dispersed oil, at times prove to be temporarily toxic or inhibitory to natural microbial populations. Furthermore, the dispersants produce new bacterial substrates and the microbes may preferentially attack them before the oil.[6] However, the generation of new surface area by the use of dispersants is the key factor in the biodegradation process since bacterial growth rates increase with interfacial area.[10] The dispersed oil droplets resulting from the application of a chemical dispersant remain in the water column instead of undergoing beaching or sedimentation.[43] In sum, dispersants generally augment the environmental conditions necessary for suitable microbial growth by increasing the surface-to-volume ratios of oil that lead to greater degradation rates at the oil-water interface (Figure 3).[6] They likewise lessen the tendency of oil to form tar balls and mousse which mollify biodegradation rates.[44,45] More data need to be collected on dispersed oil degradation by microbes in the field.[46] To date, much of this research has been confined to laboratory settings.[6]

Microbes are much less successful at degrading oil once it accumulates in bottom sediments. Carlberg[3] presents several reasons for this effect. First, a film or layer of oil covering seafloor sediments has a much smaller effective surface area for microbial degradation than oil dispersed in water. Second, bioturbating organisms intensely rework the upper layer of sediments which can translocate the oil to deeper layers, possibly even into anaerobic layers where the microbial degradation rate of the oil decreases markedly. If the amount of sedimenting oil exceeds the degradation rate of microbes to remove it, oil will accumulate and possibly persist for several years. Table 5 lists the concentration of petroleum hydrocarbons in sediments of estuaries and oceanic areas. Figure 4 illustrates the sedimentation of oil, as well as other abiotic and biotic processes acting on petroleum hydrocarbons, in estuaries and marine systems. The weathering and biodegradation of petroleum hydrocarbons in these environments continue to receive considerable attention.[47-49]

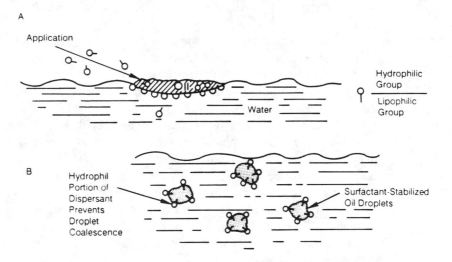

FIGURE 3.   Application of a chemical dispersant to an oil spill. (A) Surfactant concentrates at the oil-water interface. (B) Micelles or surfactant-stabilized droplets form from the oil slick. (From Canevari, G. P., in *Oil on the Sea,* Hoult, D. P., Ed., Plenum Press, New York, 1969, 29. With permission.)

TABLE 5
Hydrocarbon Concentrations in Selected Estuarine and Oceanic Waters

Location	Sample depth	Water depth (m)	Concentration (ppm dry wt)	No. of samples
Buzzard's Bay, MA Wild Harbor River (over 2.5 year period)	Top 10 cm, sand and clay	<3	250–1,650	~180
Station 37, subtidal unpolluted	Top 10 cm	11	38–70	10–20
Silver Beach (over 2.5 year period)	Top 10 cm, sand and clay	<2	500–12,000	~60
Mississippi coastal bog	—	0–1.5	350	1
Narragansett Bay, RI, mouth of Providence River (head of Bay)	Top 8—10 cm	?	820–3,560	2
Middle of Bay	Top 8–10 cm	?	350–440 cm	2
	Top 8–10 cm	?	50–60	2
Vineyard Sound, MA	Top 7 cm	6–9	1.7 n-alkanes only	1
Chedabucto Bay, Nova Scotia (over 2-year period)	Top 5 cm	3	34–420	7
	Top 5 cm	12	11–1,240	7
Coast of France, Le Havre vicinity	"Surface"	Subtidal	450	1
Seine estuary	"Surface"	Subtidal	33	1
Coast of France, Le Havre vicinity (different than that above)	"Surface"	Subtidal	920	1
Bay of Veys	"Surface"	Subtidal	38	1
Port Valdez, Alaska	"Surface"	Subtidal	0.5–2.5 (C_{16}-C_{28} only)	?
Gulf of Batabano, Cuba	?	?	15–85	10
Orinoco Delta, Venezuela	?	?	27–110	10
Gulf of Mexico, open ocean	?	?	12–63	10
Mediterranean, open ocean	?	?	29	1
Cariaco Trench	?	?	56–352	16
Southeast Bermuda to base of rise near Hudson Canyon	Top 5 cm of sediment	>3,000	1–4	5

From *Petroleum in the Marine Environment*, National Academy of Sciences, Washington, D.C., 1975, 57. With permission.

EFFECTS OF POLLUTING OIL
ON ESTUARINE HABITATS
AND ORGANISMS
GENERAL

Estuarine and marine biota are impacted by polluting oil either indirectly via the degradation of critical habitat areas or directly via the toxic effects of water-soluble components of crude oils and refined products on individuals in populations. In addition, the application of chemical dispersants, solvents, and agents that reduce surface tension and facilitate the removal of oil slicks from the water surface,[6] may also be toxic to estuarine and marine life. While some workers implicate the toxicity of oil dispersants as the primary culprit in ecological damage associated with oil spills,[4] others point to laboratory findings that reveal lower acute lethal toxicities of dispersants currently in use than crude oils and their refined products.[6]

The effects of polluting oil on a community of estuarine organisms can vary greatly because the diverse number of chemical compounds present in various proportions typically have different degrees of toxicity to the flora and fauna exposed to them. Moreover, the susceptibility of the organisms to the damage of polluting oil depends on the life history stages exposed to the compounds and the length of time of exposure. Both short- and long-term effects have been documented on impacted biota. Short-term effects may be manifested in high mortality of individuals immediately subsequent to the release or discharge of a polluting oil, and long-term effects, such as lower population abundance through time, may develop from sublethal doses of toxins to egg, larval, and juvenile life stages.

Once a hydrocarbon compound is taken up by an estuarine organism, it follows one of three general pathways: the compound may be metabolized, stored with possible elimination at a later time, or excreted unchanged. The hydrocarbons enter the organisms in several ways: by active uptake of dissolved or dispersed substances; by ingestion of petroleum-sorbed particles, including live or dead organic matter; and by the drinking or gulping of water containing the chemicals, as in the case of fish.[50] The effects and fates of these compounds on biota may be assessed by several different types of chemical and biological investigations. Carlberg,[3] synthesizing the work of Anderson et al.,[51] categorizes them into four general types of laboratory and field studies:

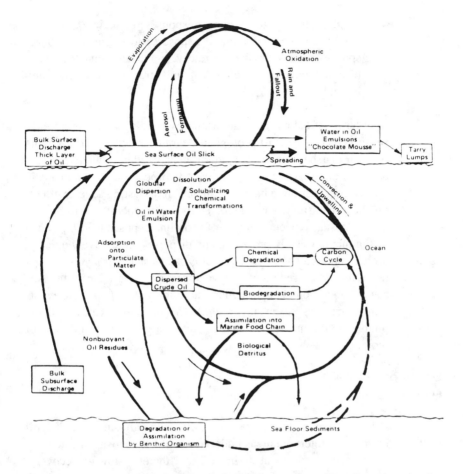

FIGURE 4. Fate and weathering of polluting oil in estuarine and marine waters showing various abiotic and biotic processes that act to alter the oil. (From Burwood, R. and Speers, G. C., *Est. Coastal Mar. Sci.*, 2, 117, 1974. With permission.)

1. Short–term toxicity studies — to determine the range of tolerance to the pollutant, to set exposure levels for sublethal studies, and to evaluate the significance of environmental concentrations.

2. Organism–environment transfer studies — to determine rates of accumulation and release of pollutant by the organism.

3. Physiological studies — to determine the extent and nature of modification of metabolic parameters in response to exposure to sublethal concentrations of the pollutant.

4.   Field studies — to compare responses of marine animals to natural and perhaps chronic exposures to pollutants with responses observed in the laboratory, and to determine the effects of pollutants on the metabolism and structure of marine communities.

In terms of habitat damage, estuaries rank high among the most sensitive ecosystems to ecological degradative effects of polluting oil. Salt marshes, mangroves, mudflats, and other low-energy areas tend to trap oil. The accumulation of hydrocarbon contaminants in the sediments of these habitats can have long-lasting adverse effects on the organisms living there. Unlike high-energy environments such as instance open stretches of shoreline exposed to strong wave action that physically breaks up and disperses oil relatively quickly, aforementioned sheltered areas often sequester petroleum components in bottom sediments for years, re-releasing them to the surrounding environment particularly at times of major episodic events (e.g., hurricanes, storm surges, etc.). This outcome creates an inhospitable environment for organisms that may persist for a decade or more.

EFFECTS OF OIL ON
ESTUARINE HABITATS
Salt Marshes

Salt marsh biotopes are especially vulnerable to polluting oil because, as intertidal areas underlain by thick deposits of fine sediments capable of sequestering oil and thereby slowing recovery, they often serve as final repositories of petroleum hydrocarbons washed ashore by physical processes. As mentioned above, once oil penetrates into bottom sediments, it commonly persists for years or even decades, arresting the development of benthic communities. In addition, the oil does not necessarily remain restricted to a particular location but can move about within the sediments, as exemplified by the mobility of oil in sediments of Wild Harbor River and estuary, Massachusetts, after Number 2 fuel oil spilled from the barge *Florida* on September 16, 1969. In this case, the greatest concentrations of oil occurred in intertidal and subtidal zones of the river and declined away from the shore.[9] However, oiled sediments moved about, coming to rest after a lengthy period of time on deeper, softer sediments.[52] A similar pattern of sediment-oil dispersal was evident subsequent to the Santa Barbara Channel oil spill in 1969.

The release of oil trapped in bottom sediments over a period of time creates a long-term problem for the stability and health of bottom communities. This process may operate for years following an oil spill. For example, workers monitored oil reentering the overlying water from contaminated sediments for at least 5 years after the *Arrow* spilled oil into Chedabucto Bay, Nova Scotia.[53]

Cormack[11] suggested that salt marshes can absorb any damage incurred from a single oil spill, asserting they recover well owing to their rapid renewal of plant growth. However, Hall et al.[2] offer two cases, the spill of Bunker C oil in Chedabucto Bay, Nova Scotia, and Number 2 fuel oil in salt marshes near West Falmouth, MA, where habitat and community damage from a single spill of oil was extensive. According to these scientists, the susceptibility of a salt marsh to damage from oiling appears to be related to the ease of penetration of the hydrocarbons into bottom sediments. The rate of sediment deposition, in turn, may play a primary role in the resiliency of the salt marsh to oiling. The rate of recovery of the marsh depends on how deeply the oil sinks into the substrate. If penetration of petroleum hydrocarbons extends below the level of the roots and rhizomes of the vascular plants, significant degradation of the biotope can ensue. The marsh seems to recover quite quickly when polluting oil does not penetrate to the roots and rhizomes of the vegetation. Bioturbating infauna facilitate the downward movement of oil in salt marsh sediments. Burrowing by fiddler crabs in salt marshes along the East Coast of the U.S. probably is significant in this regard.[2]

Clark[5] encapsulates the range of observed effects of oil pollution on annual and perennial plants inhabiting salt marshes. As related by Clark (p. 52),[5] the kind of annual plant response to oiling depends on the season. He iterates, "...if the plants are in bud, flowering is inhibited; if the flowers are oiled, they rarely produce seeds; if the seeds are oiled, germination is impaired." The annuals generally respond unfavorably to oil pollution. Most of these individuals die from petroleum hydrocarbon contamination, and recovery of the populations on a salt marsh may require two or three seasons.

Perennial plants tend to be more resilient to polluting oil, but large numbers of individuals of certain species can be lost after a single oiling event. The most susceptible forms to oil contamination are shallow-rooted species (e.g., *Salicornia* spp. and *Sueda maritima*) with limited or no food reserves. Perennials having large food reserves, notably those with taproots, exhibit much greater survivorship. They can even tolerate repeated oilings.

Even after thorough contamination of a salt marsh surface with oil and the loss of the marsh grasses, recolonization may be rapid when the sedimentation rate is high. Rapid sedimentation will quickly bury a contaminated surface such that the roots of the colonizing plants do not reach the oiled layers. Stebbings[54] described this phenomenon in a salt marsh on the coast of France where 20 cm of new, uncontaminated sediments deposited in <2 years after an oil spill provided ideal conditions for the recolonization of the marsh habitat.

Although it is apparent that some salt marshes can recover well from single oil events, as discussed by Stebbings,[54] repeated oilings reduce the general resilience of the salt marsh biotope. Observations of both accidental and experimental contaminations of salt marshes by successive spillages of oil indicate greater severity of damage and longer periods of recovery of the floral and faunal communities. The oil firmly adheres to the stems and leaves of the plants, with little being washed off by infiltrating tides. Ultimately, the plants, together with the benthic fauna inhabiting the marsh surface, experience heavy mortality due to continual dosages of petroleum hydrocarbons. Chronic oil pollution of salt marshes either from large, successive spills that preclude complete recovery of the system or from the continuous discharge of small quantities of oil in effluents discharged from refinery outfalls, ballast water treatment plants, and similar sources can have severe, long-term consequences.[55]

Attempts to clean oiled salt marshes using dispersants have not been very effective, and, in fact, may have accelerated habitat damage. For instance, Baker et al.[56] documented the short-term loss of *Salicornia* spp. and long-term (1 to 2 years) reduction in density of *Spartina anglica* ascribable to both oil and dispersed oil. Delaune et al.[57] discerned reduced biomass of salt marsh plants in Louisiana (U.S.) when dispersant was applied prior to oiling. They also found little mortality among the meiofauna because of the oil or dispersant. In studies of the salt marshes of the Atlantic Coast, Lane et al.,[58] reported less sensitivity of high marsh vegetation than mid- and low marsh vegetation to weathered crude oil and Corexit 9527 dispersant. The creek-edge and mid-marsh plants showed the greatest sensitivity to the oil plus dispersant. Oil alone had the smallest impact on the plant communities.

Mudflats and Sandflats

The surfaces of these habitats usually remain wet for extended periods of time as a result of tidal action. Thus, any oil carried onto the surfaces typically is separated from the sediments by a water layer. Consequently, the ensuing tide tends to lift the oil and transport it up or down the intertidal flat. Cormack[11] emphasizes that the use of chemical dispersants should be avoided on mudflats and sandflats because of their proximity to commercially or recreationally important fishing areas and shellfish beds.

The same response is applicable to the salt marsh biotope. While the waterlogged muds and sands of this habitat resist the penetration of crude oil, the application of dispersants promotes the deep penetration of emulsified oil into sediments, creating hazardous conditions for the infauna. For example, heavy mortalities of polychaete worms and bivalve mollusks (i.e., *Cardium edule* and *Mytilus edulis*) were detailed by George[59] after the exposure of oiled mudflats to emulsifier. Cowell[55] discussed the deep penetration of emulsified oil into sediments of Cornish beaches subsequent to the *Torrey Canyon* spill. Dekker and van Moorsel,[60] experimenting with artificial intertidal mudflats treated with Forties crude oil and Finasol OSR-5 dispersant over a period of 10 months, concluded that the effects of crude oil alone on biota were less severe than that of dispersant plus oil. Clams (*Macoma balthica*), cockles (*Cardium edule*), and polychaetes (*Arenicola marina*) experienced high mortality from the dispersant plus oil. The National Research Council[6] has stressed that the application of dispersants on oiled inter- and subtidal environments in general, ranging from salt marshes to mudflats and seagrass beds, may enhance oil penetration into the sediments, thereby exacerbating ecological damage of the habitats.

Mangroves

Cormack[11] favors similar oil spill responses for mangroves and salt marshes. Oil contamination of these habitats should be avoided at all costs, but when it does occur, cleaning efforts must employ the most gentle methods available, such as gentle flushing of the affected areas with water followed by mechanical recovery of the oil. Crude oil infiltrating mangrove swamps can destroy many hectares of forested habitat. More than 20 years of time may be required for the regrowth of the trees.

Polluting oil can be particularly threatening to mangroves since it may clog the lenticels that facilitate oxygen transport in the aerial roots of the trees. When the lenticels become clogged with oil, the oxygen level in the root air spaces often drops to 1 to 2% of normal within 2 d, as demonstrated in *Avicennia* and *Rhizophora*.[5] The acute decrease in the transport of oxygen to the submerged part of the tree no doubt has caused the demise of many hectares of mangroves. Nonetheless, other factors are probably important as well because a number of mangroves exist that have been severely oiled yet have experienced little vegetation loss.

Experimental investigations of red mangroves tainted with oil and dispersant in Panama have uncovered changes in growth, respiration, and transpiration.[61] In these investigations, mature red mangroves were exposed to both oil and oil mixed with dispersant at different sites. Inspection of the sites showed differences between vegetation exposed to untreated oil alone and vegetation exposed to oil premixed with dispersant. The untreated oil spread far into the mangrove forest due to wave and tidal action, contaminating sediments, prop roots, algal mats, and wrack. Interior areas of the impacted forest continued to be heavily oiled after 1 week, whereas the outer fringe of the forest lost much of its oil after several tidal cycles. In contrast, oil premixed with dispersant did not adhere well to sediments or algal mats and coated prop roots only lightly; most of this dispersed oil was removed by tidal action after about 1 week. As a result, very little defoliation (5%) occurred at the forested site exposed to oil premixed with dispersant, but significant defoliation (60 to 70%) took place at the mangrove site subjected only to untreated oil.

The use of dispersants on mangroves already contaminated with oil has not been highly successful. Despite treating oil-coated mangrove trees after 1 d with high-pressure seawater or a nonionic water-based dispersant, Teas et al.[62] observed poor survival. Many of the trees died over a 30-month period even though they had been washed with dispersant or high-pressure seawater. Based on this study, probably the best approach in dealing with oil contamination of mangroves is to avoid pollution of the biotope in the first place.

Seagrasses

In a review of the literature on the effects of oil and dispersants on seagrass beds, the American Society for Testing and Materials (ASTM)[63] advised that dispersants should not be used to treat oil over seagrass beds in shallow lagoons or areas of restricted flushing, including many estuaries. Consideration should be given to the application of dispersants only when a greater impact may be incurred by allowing the oil to move ashore and affect other sensitive habitats onshore. In any case, it is much more sensible to apply a dispersant to a polluting oil in offshore areas, where it poses less of a hazard, than to sensitive habitats near- or onshore.

A number of recent laboratory studies have assessed the effects of crude oil and chemically dispersed oil on seagrass. In experiments on *Thalassia testudinum* using Prudhoe Bay crude oil and Corexit 9527 dispersant, Baca and Getter[64] reported similar toxicity of the oil and dispersed oil on the seagrass. Thorhaug and Marcus[65-67] and Thorhaug et al.,[68] testing the impact of Murban and Louisiana crude oils and seven dispersants on three subtropical and tropical seagrass species (i.e., *Thalassia testudinum*, *Halodule wrightii*, *Syringodium filiforme*), discovered lower survival and growth in the seagrasses treated with chemically dispersed oil than with crude oil alone. *Thalassia testudinum* appeared to be more resistant to contaminant exposure than *H. wrightii* and *S. filiforme*. In conclusion, dispersants contributed to increased mortality and decreased growth of the seagrasses.

Temperate Subtidal
Estuarine Habitats

The concentration of a polluting oil and the length of exposure of aquatic biota to oil components seem to be the primary determinants of ecological damage in these habitats.[6] The benthic organisms of estuaries are highly susceptible to the toxic components of oil and may be coated and smothered by sedimenting oil. Moreover, suspension and deposit feeders become contaminated by oiled organic and inorganic particles ingested on the seafloor. Coelenterates, crustaceans, echinoderms, mollusks, and polychaetes frequently suffer heavy mortalities subsequent to large oil spills that impact estuaries. The persistence of the oil in bottom sediments commonly hinders the recovery process. The gradual leaching of the contaminants back to the water column from the sediments creates an unfavorable environment for recolonization.

The grounding of the barge *Florida* reported in an earlier section exemplifies these effects. The high toxicity of the oil released from the barge, containing 41% aromatics, and the incorporation of oil constituents into bottom sediments after the spill produced a hazardous environment for benthic as well as nektonic organisms. Scallops, clams, oysters, crabs, shrimp, and lobsters succumbed in large numbers. Contamination of the benthic habitat culminated in the long-term closure of commercially viable shellfish beds, as the release of oil from the sediments continued over a protracted period of time. Finfish also died, some immediately after the spill, due to the toxic components of the oil. Finfish mortality was especially high in sheltered, shallow creeks and embayments that harbored juveniles of commercially important species (e.g., bluefish and flounders). The benthic community did not completely recover from the spill in heavily polluted areas.[5,9]

EFFECTS OF OIL ON
ESTUARINE ORGANISMS
General

The lethal and sublethal effects of crude and refined petroleum products on estuarine and marine organisms have been the subject of considerable research over the years. Lethal effects on biota involve the interference of the oil with cellular and subcellular processes leading to an organism's death. The organism can die directly by smothering and suffocation under the oil, or soon thereafter by the impairment of movements to obtain food or escape from predators after being coated by the hydrocarbon contaminants. Sublethal effects generally are more subtle; for example, aberrant physiological or behavioral function resulting from the pollutants, while not causing immediate mortality, may predispose an organism to a greater long-term risk of death. The oil commonly adversely affects reproduction, growth, distribution, and behavior such that gradual shifts in species composition, abundance, and diversity eventually take place.[69] Together, lethal and sublethal effects of oil contamination are manifested as acute and chronic biological effects occurring immediately after or much later than the actual polluting event.

Nearly all estuarine and marine life forms have purportedly been impacted by oil spills — temporary and permanent plankton, fishes, benthos, seabirds, and mammals — with the severity of the impact related to a multitude of factors. Chief among these are (1) the composition of the oil; (2) amount of oil; (3) form of the oil (i.e., fresh, weathered, or emulsified); (4) occurrence of the oil (i.e., in solution, suspension, dispersion, or adsorbed onto particulate matter); (5) duration of exposure; (6) involvement of neuston, plankton, nekton, or benthos in the spill or release; (7) juvenile or adult forms involved; (8) previous history of pollutant exposure of the biota; (9) season of the year; (10) natural environmental stresses associated with fluctuations in temperature, salinity, and other variables; (11) type of habitat affected; and (12) clean up operations (e.g., physical methods of oil recovery and the use of chemical dispersants).[70] Estuarine organisms respond in a multitude of ways to oil pollution as discussed below.

Types of Organisms
Phytoplankton

Because of their proximity to a polluting oil floating on the estuarine surface and their general sensitivity to the toxic components of the oil, plankton may be at greater risk than other biotic groups physically removed from the oil. For those minute organisms floating or swimming in the top few centimeters of the estuary or on a surface film of the water (i.e., neuston), the impact may be greatest of all.[5] In general, laboratory experiments evaluating the effects of oil on whole plankton communities have yielded somewhat conflicting results with those of field observations following oil spills. Most importantly, the laboratory studies reflect greater impacts on the communities than observed in nature.

Although phytoplankton populations are adversely affected by oil, their patchy distribution and high rate of proliferation can mollify the overall impact.[71] Typically, therefore, phytoplankton recover quickly from a surface slick of oil, returning to population levels existing prior to a spill. Depending on petroleum hydrocarbon concentrations, photosynthesis by algae is either enhanced or depressed, as evidenced by work on algal cultures. For instance, petroleum hydrocarbon concentrations below 50 ng/g enhance photosynthesis and those above 50 ng/g depress photosynthesis by algal populations in culture.[5] The concentration of oil necessary to cause death of phytoplankton ranges from about 1 to 10^{-4} ml/l.[69] Cell division ceases or diminishes at concentrations of 10^{-1} to 10^{-5} ml/l, depending on the species.

Several laboratory investigations have dealt with the toxicity of dispersed and undispersed oil on phytoplankton. Tokuda,[72] assessing the affect of dispersed oil, undispersed oil, and dispersant on the diatom *Skeletonema costatum*, said that growth of the phytoplankter was the same under the influence of crude oil alone or dispersed oil. The least growth occurred in individuals subjected to dispersant alone. The toxicities of undispersed and dispersed oil on the phytoplankton were similar. In a related study, Fabregas et al.[73] ascertained comparable toxicities of dispersed oil and dispersant on *Tetraselmis suecica*. Hsaio et al.[74] and Chan and Chiu[75] found undispersed oil to be less toxic to phytoplankton than dispersed oil, but these studies may have been flawed.[6]

Zooplankton

Among the holoplankton of estuaries, copepods have been the target of much research on the effects of oil contaminants. Juvenile *Acartia hudsonica* and *Oithona nana*, when immersed in seawater containing 10 µl/l of oil, were observed to die after 3 to 4 d.[69] Adults of these species experienced higher mortality rates after longer exposures to 10 µl/l of oil in seawater. Members of the genera *Acartia* and *Calanus* died within 24 h when immersed in seawater containing oil in concentrations of 100 µl/l. While *Oithona* also exhibited adverse responses to this level of oil contamination, it appeared to be less sensitive than *Acartia* and *Calanus*.[69]

That zooplankton accumulate significant amounts of petroleum hydrocarbons via ingestion of oil is clearly evident from observations of zooplankters following major oil spills at sea. Subsequent to the *Argo Merchant* spill on Nantucket Shoals in December 1969, for instance, oil-ingesting copepods exhibited high levels of hydrocarbon contamination. Oil-laden fecal pellets of these plankters served as a pathway for the sedimentation of oil contaminants to the seafloor.

Falk-Peterson et al.[76] stated that even low concentrations of oil and dispersants may adversely affect the fertilization and embryonic development of copepods. The feeding of copepods can also be altered by polluting oil. Spooner and Corkett[77] registered changes in the feeding rates of four species of copepods exposed to 10 ppm of Kuwait crude oil. Foy,[78] however, deemed *Calanus hyperboreus* to be quite resistant to Prudhoe Bay crude.

Benthos

Benthic Macroflora: Various impacts of oil contamination of macrobenthic algae have been chronicled in the literature. During the Santa Barbara oil spill in 1969, the green algae *Chaetomorpha aerea, Enteromorpha intestinalis,* and *Ulva angusta* were only slightly damaged in mid- and high intertidal zones.[79] Bunker C oil, which spilled from the *Arrow* into Chedabucto Bay, Nova Scotia, caused much more extensive damage to *Fucus spiralis*; this macroalgal species disappeared from the rocky shores of the bay after the spill.[80] The oil was also responsible for the elimination of the cordgrass, *Spartina alterniflora*, from heavily oiled areas of salt marshes.

Based on laboratory studies, Hsiao et al.,[74] Ganning and Billing,[81] and Thelin[82] presented findings on the toxicity of dispersant-oil mixtures and crude oil alone on *Laminaria saccharina, Phyllophora truncata, Fucus vesiculosus,* and *F. serratus,* respectively. They also examined the responses of these species in terms of the effects on zygotes and seedlings,[78] oxygen uptake,[81] and *in situ* primary production. They noted that crude oil alone was less toxic to the plants than dispersant-oil mixtures.

Mucins secreted by many benthic macroalgae provide a protective covering that prevents oil from adhering to them.[5] The large brown seaweeds in temperate intertidal zones, for instance, have a mucilaginous slime covering their outer surface; this slime layer serves as a barrier to the penetration of oil.[69] However, when oil does adhere to the fronds, the increased weight of the plant can be detrimental, with the seaweeds being torn from their stipes under conditions of strong winds and waves.[5] Nevertheless, the damaged plants seem to retain a considerable resiliency to the pollutants. Even healthy seaweeds unaffected by oil contamination in intertidal, exposed temperate shores undergo large-scale changes in biomass over an annual cycle. In these plants, new growth is added near the base during the warmer (growth) season, but the distal parts of the plant are lost in winter.[69]

Results of investigations on the impact of oil pollutants on vascular plants (i.e., salt marsh plants, mangroves, and seagrasses) are presented above.

Benthic Macrofauna: Both the inter- and subtidal habitats of estuaries are highly susceptible to oil pollution. In many coastal oil spills, the intertidal zone becomes the final depot for petroleum products as winds and currents drive oil slicks ashore. The subtidal zone of estuaries also receives oil contaminants via the weathering of the oil at the sea surface, the sorption of the hydrocarbons onto sediment, and the settlement of the oil-laden particles together with tarry weathered products to the seafloor. In both habitats, the benthic epi- and infauna will often suffer high mortality due to the smothering action of the oil and/or the toxic effect of the oil in conjunction with a surfactant.[83] In addition to the immediate damage to benthic communities, commonly manifested as increased mortality rates of the constituent populations, long-term impacts may develop from the aberrant growth, reproductive output, and altered distribution patterns of the fauna. In essence, therefore, the petroleum hydrocarbons account for a loss of ecological fitness among the benthos.[84]

Most research on oil degradation in estuarine and marine environments has assessed processes operating under subaerial exposure or in the water column. Consequently, much less information has been compiled on the fate of oil incorporated into bottom sediments.[85] Some notable work has been reported in the literature on this topic, however,[86,87] with interesting findings being published on degradation rates of oil in aerobic vs. anaerobic sediments.[88,89] Little data exist on the changes in animal-sediment interactions attributable to oil accumulation in these zones. A benthic community comprised of actively bioturbating infaunal populations fundamentally influences the geochemistry of seafloor sediments.[90] Rhoads and Boyer[91] and Aller[92] attest to the effects of benthic fauna on the physical and chemical properties, respectively, of marine sediments. Processes of irrigation, pelletization, tube construction, mucous secretion (which affect water flow and erodibility of surface sediments), and burrowing and feeding activities of the biota continually modify the biological benthic boundary layer. These animal-sediment interactions can be radically disturbed by buried oil, thereby impacting sediment transport and geochemical properties of the seafloor. A reduction or cessation of particle and fluid bioturbation due to the oil may curtail or eliminate particle transport and the exchange of sediment pore water and overlying water. The recolonization of these disturbed habitats, most conspicuously by pioneering species, ultimately leads to infaunal reworking of the substrate, causing oil-laden sediments to be resuspended in the water column. In this way, the benthic boundary layer remains a potentially chronic, contaminated zone to the bottom-dwelling organisms.

Because of their relatively limited mobility, benthic invertebrates are, at times, killed in large numbers by even small spills of oil. Mulligan et al.,[93] for example, documented a large kill of benthic fauna from a small oil spill in the Great Bay estuary, New Hampshire. The Tamano oil spill in Casco Bay, Maine, likewise was responsible for a heavy kill of benthic invertebrates. Finally, the West Falmouth oil spill resulting from the grounding of the barge *Florida* had a devastating impact on the benthos, with a massive kill of populations rendering some areas azoic.[3,9] Soon after the spill, the opportunistic polychaete worm, *Capitella capitata*, reinvaded the defaunated habitat, attaining rather high densities. Within 1 year, other benthic species began to repopulate the affected areas as well.[3] More recently, investigations have shifted to oil pollution impacts on the benthic environment of coastal marine regions (e.g., Georges Bank.)[94]

Among the benthic invertebrates impacted by oil pollution, bivalve mollusks have received much attention. Within this group are species of value as sentinel organisms in monitoring environmental contaminants. Mussels and oysters, for instance, have proven to be extremely valuable in monitoring temporal trends and spatial distributions of a wide range of pollutants, such as PAHs.[95] As referred to above, George[59] documented widespread mortality of *Cardium edule* and *Mytilus edulis* inhabiting oiled mudflats. Lowe and Pipe,[83] making observations on the responses of *M. edulis* to chronic low-level diesel oil hydrocarbons in a mesocosm system, ascertained a reduction in levels of storage reserves as well as gamete atresia (degeneration) and resorption in experimental animals exposed to the hydrocarbons. Based on studies employing microencapsulated oil, Strömgren et al.[96] determined that small oil particles are highly toxic to mussels. The toxicity of the oil was primarily related to the size and concentration of the oil particles.[97] Data collected on the growth rate of mussels exposed to microencapsulated North Sea oil and to microencapsulated combinations of oil and dispersants reveal approximately the same decrease in growth rate (80 to 90%) in individuals within 170 h. The toxicity of the different oil-containing media (2 mg oil per liter) to mussel growth was as follows: microencapsulated pure oil>microencapsulated oil/dispersant mixtures>>oil mixed with dispersant>oil-in-water suspension. In conclusion, Strömgren[97] believes the use of dispersants has a greater toxic effect on mussels, but the subsequent recovery of the organisms appears to proceed more quickly than with untreated oil.

Strömgren,[97] La Roche et al.,[98] and Swedmark et al.[99] stress the role of dispersants in increasing the toxicity of oil. Oil treated with dispersants, according to these workers, produces a mixture more toxic than each component separately. The resultant toxicity of oil treated with dispersants derives partly from the increased flow of soluble hydrocarbons from the dispersed particles into the water and partly from the toxicity of the disperants.[97,100] The National Research Council (p. 146)[6] has recently refuted this claim, at least in terms of impacts on mollusks, remarking that "...laboratory research on mollusks shows that high concentrations of dispersed oil can be toxic, but that little defensible evidence (with measured concentrations) exists to suggest higher toxicities of oils and their components in the presence of dispersants."

Oysters, especially *Crassostrea virginica*, have also been the focus of research on petroleum hydrocarbon contamination for many years. Investigations on the effects of oil on *C. virginica* have assessed uptake of hydrocarbons and depuration,[101-103] growth and survival,[104] physiology,[105] and larval setting.[106] Petroleum hydrocarbons are readily assimilated by *C. virginica*, as well as other species of oysters and mussels.[107] The accumulation of these compounds and their retention in tissues subsequent to oil spills (e.g., the West Falmouth oil spill of 1969) have periodically resulted in the closure of shellfish beds for fear of public consumption and contamination.

Early life history stages of bivalves generally are more sensitive to oil pollution than adults. Renzoni,[108] for example, detailed the responses of gametes, developing eggs, and larvae of oysters (*C. angulata* and *C. gigas*) and mussels (*M. galloprovincialis*) to crude oils, oil derivatives, and oil-dispersant emulsions. Hydrocarbon concentrations of 1 to 1000 ppm were toxic to fertilization and hampered the swimming activity of the larvae. Only high hydrocarbon concentrations elicited toxic responses in the eggs and embryos. Smith and Hackney[106] suggested that spills of heavy and sticky oils on intertidal oyster shells would impact oyster spat settlement in intertidal habitats. Their experimental studies on the effects of oil-coated substrates on oyster spat (*C. virginica*) settlement revealed significantly lower spat densities on oil treatments vs. control and gas-treated clam shells in the high intertidal zone.

Bivalves exhibit variable tolerance to oil pollution. The hard clam (*Mercenaria mercenaria*) has been shown to withstand higher concentrations of oil than oysters (*C. virginica*).[109] Similarly, mussels (*M. edulis*) tolerate higher levels of oil contamination than cockles (*C. edule*) and scallops (*Pecten opercularis*).[99] While oil pollution does not necessarily lead to the abrupt demise of these species, it commonly contributes to sublethal effects, such as the necrosis, inflammation, and atrophy of gonadal cells observed in oysters following the *Amoco Cadiz* oil spill,[110] that can be detrimental to the long-term viability of the populations.

Animals other than *C. virginica* are adversely affected by thicker oils adhering to their shells and the substrate of intertidal habitats. For instance, when this type of oil adheres to the shells of the gastropod snails, *Littorina neritoides*, *L. saxatalis*, and *L. obtusata*, their volume and mass increase, causing them to be more easily dislodged by waves striking rocky shores.[55] The recovery of these populations on intertidal, rocky shores depends on the occurrences of successive spillages impacting the habitat as well as the toxicity of the crude oil components. Many individuals of the populations may be lost if the oil washed ashore contains highly toxic fractions with a high aromatic content.

Chan[111] attributed the extensive mortality of more than 4 million intertidal animals (mainly acorn barnacles) in San Francisco Bay to the spill of 3.18×10^6 l of Bunker C oil. Other oil spill events resulting in large kills of barnacles included the *Tampico Maru* accident[112] and the Santa Barbara spill[113] in 1957 and 1969, respectively. Due to the *Tampico Maru* spill, many *Balanus gladula* and *Chthamalus fissus* were killed, and *C. fissus* suffered heavy losses from the Santa Barbara spill as well.

The *Tampico Maru* oil spill also killed a significant number of crabs (*Pachygrapsus crassipes*) and lobsters (*Panulirus interruptus*).[112] Other species of lobsters may be even more sensitive to oil pollution than *P. interruptus*. When exposed to sublethal quantities of crude oil (i.e., 0.9 ml/l), adult *Homarus americanus* displayed depressed appetite and chemical excitability. Their ability to collect food also declined.[114] The larvae of *H. americanus* are even more sensitive to petroleum hydrocarbons and dispersing agents, especially immediately after molting.[115] Sublethal effects of crude oil emulsions on the lobster larvae occur at concentrations down to 1 ppm, and concentrations of 100 ppm are lethal.[116] Capuzzo et al.,[117] examining the physiological effects of physically and chemically dispersed Southern Louisiana crude oil and the ingestion of oil-laden food on larval lobsters (*H. americanus*), discerned reduced metabolism in the larvae exposed to the contaminants in the water and lower respiration in lobsters consuming oil-contaminated food. The accumulation of benzene, thiophene, toluene, and alkylbenzenes correlated with the suppression of respiratory activity in the larvae. Furthermore, recovery of normal respiratory activity did not take place immediately subsequent to the exposure of the larvae to uncontaminated seawater.[118] Larval decapod crabs (*Neopanope texana*) and shrimp (*Crangon crangon*) also have a high sensitivity to crude oil components.[119,120] In the case of *C. crangon*, the larvae are susceptible to hydrocarbon concentrations much lower than those that impact the adults.

Some long-term effects of oil contamination on crustaceans and other invertebrates have been reported in the literature. After the *Tampico Maru* disaster, sea urchins (*Strongylocentrotus franciscanus* and *S. purpuratus*), disappeared in the vicinity of the impacted site and did not recolonize the locale for 2 years. They did not become abundant until 4 years after the oil spill.[112] Krebs and Burns[121] monitored long-term reductions in population density, recruitment, female:male ratios of adult crabs, behavioral changes, and high mortality in overwintering populations of the fiddler crab *Uca pugnax* for 7 years following the spill of No. 2 fuel oil from the barge *Florida*. The crab populations recovered after the disappearance of naphthalenes and alkylated naphthalenes from contaminated bottom sediments.[118] Other examples of the long-term changes of benthos due to oil spills can be found in Elmgren et al.[122] and elsewhere.

Perhaps the most tolerant benthic macroinvertebrates to oil pollution are polychaetes.[6,38] Although polychaetes generally recover rapidly from oil spills and typically represent the initial species to colonize the benthic community after these spills, even they can be eliminated from impacted habitats. Thus, Levell[123] describes how Kuwait crude oil can adversely affect the lugworm *Arenicola marina*. Repeated oiling of bottom sediments harboring this polychaete may lead to its disappearance. While the oil does not always result in acutely lethal effects, behavioral aberrations arising from oil contamination (e.g., alterations in burrowing) may affect populations through their decreased ability to avoid predation. Olla et al.[124] surmise that such is the case in sand worms, *Nereis (Neanthes) virens* Sars, based on exposure of the worms in the laboratory to sublethal concentrations of Prudhoe Bay crude oil.

The effects of chronic oil pollution on benthic invertebrates (e.g., from refinery discharges) often differ from the effects of acute, short-term hydrocarbon exposure. Although the behavioral and physiological changes in these biota from short-term contaminant exposure are in many cases very temporary, with recovery to preimpact conditions taking place soon after reduction of contaminant levels, recovery from long-term pollutant exposure is usually more problematical. Sublethal effects of long-term exposure to oil contamination may be manifested at all levels of biological organization, and impaired feeding, growth, development, energetics, and recruitment of individuals can ultimately foster substantial modifications of community structure and dynamics.[118] Fletcher et al.[125] and McCain and Malins[126] indicate that benthic invertebrates and fishes continue to show sublethal responses throughout exposure to sediments laden with petroleum hydrocarbons. *Mytilus edulis*, for example, when subjected to long-term exposure of North Sea oil (30 μg/l), neither acclimated to the exposure conditions nor exhibited gradual recovery. In fact, the mussel experienced a progressive deterioration in physiological condition.[127]

Table 6 presents hydrocarbon levels observed in marine macroorganisms, including benthic invertebrates inhabiting estuarine enviromments. The organisms were sampled in areas impacted by both acute and chronic oil pollution. Morphological, cytological, and developmental abnormalities in marine organisms exposed to petroleum hydrocarbons are listed in Table 7.[118] Clearly, a wide range of

TABLE 6
Hydrocarbon Concentrations in Selected Marine Macroorganisms

Organisms	Area type[a]	Hydrocarbon type	Estimated hydrocarbon amount ($\mu g/g$)
Macroalgae			
Fucus	4	Bunker C[b]	40 dry
Enteromorpha	4	No. 2 fuel oil	429 wet
Sargassum	1	C_{14-30} range	1–5 wet
Higher plants			
Spartina	4	No. 2 fuel oil	15 wet
Mollusks			
Modiolus, mussel	4	No. 2 fuel oil	218 wet
Mytilus, mussel	4	No. 2 fuel oil[b]	36 dry
	4	Bunker C[b]	10 dry
	4	Bunker C, aromatics	74–100 wet
	3	n-C_{14-37}[b]	9 dry
Mya, clam	4	No.2 fuel oil	26 wet
Pecten, scallop	4	No. 2 fuel oil	7 wet
Littorina, snail	4	Bunker C, aromatics	46–220 wet
Mercenaria, clam	3	C_{16-32} range	160 dry
Crassostrea, oyster	2	Polycyclic aromatics	1 wet
Crustacea			
Hemigrapsus, crab	4	Bunker C[b]	8 dry
Mitella, barnacle	4	Bunker C[b]	8 dry
Lady crab	3	C_{14-30}	4 wet
Plankton	2	Benzopyrene	0.4 wet
Sargassum shrimp	1	C_{14-30}	3 wet
Lepas, barnacle	1	C_{14-30}	6 wet
Portunus, crab	1	C_{14-30}	34 wet
Planes, crab	1	C_{14-30}	11 wet
Fish			
Fundulus, minnow	4	No. 2 fuel oil	75 wet
Anguilla liver, ccl	4	No. 2 fuel oil	85 wet
Smelt	3	Benzopyrene	0–5 dry
Flatfish	2	C_{14-20}	4 wet
Flying fish	1	C_{14-20}	0.3 wet
Sargassum fish	1	C_{14-20}	1.6 wet
Pipefish	1	C_{14-20}	8.8 wet
Triggerfish	1	C_{14-20}	1.7 wet
Birds			
Herring gull, muscle	4	No.2 fuel oil	535 wet

TABLE 6 (CONTINUED)
Hydrocarbon Concentrations in Selected Marine Macroorganisms

Organisms	Area type[a]	Hydrocarbon type	Estimated hydrocarbon amount (μg/g)
Echinoderm			
Asterias, starfish	4	Bunker C, aromatics	20–147 wet
Luidia, starfish	2	$C_{14\text{-}30}$	3.5 wet

[a] 1, oceanic; 2, chronic pollution, coastal; 3, chronic pollution, harbor; 4, single spill.
[b] n-Alkanes only.

From *Petroleum in the Marine Environment,* National Academy of Sciences, Washington, D.C., 1975, p. 62. With permission.

abnormalities occurs in the organisms whether they are exposed to contaminants in natural or experimental systems.[128]

Fish

PAHs derived from oil spills in estuarine and marine environments accumulate in fish tissues, but subsequent metabolism and excretion tends to significantly lower concentrations over time.[129] Through mixed function oxygenase (MFO) activity, the hydrocarbons, as well as other lipophilic contaminants, are metabolized by these nektonic organisms; most of the MFO activity occurs in the liver, gills, and kidney.[130-132] Other tissues in fish have <10% of the activity seen in liver microsomes and consequently play much less of a role in metabolizing the compounds.[133] Hence, MFO activity in fish parallels that in mammals, being principally found in the liver. Uptake of PAHs by estuarine and marine fishes appears to be principally via the gills.[134] In assessing the impact of an oil spill on fish populations, therefore, it is necessary not only to determine the period of their exposure to the contaminants, the concentration of the contaminants, and the factors influencing bioaccumulation, but also the metabolic capacity of the species to process and eliminate the compounds.[118] Cytochrome P450 enzymes, for instance, appear to play a major role in fish metabolism of petoleum hydrocarbons.[135]

TABLE 7
Morphological, Cytological, and Developmental Abnormalities in Marine Organisms Exposed to Petroleum Hydrocarbons

Species	Hydro-carbon	Concentration and exposure time	Effect	References[118]
Invertebrates				
Homarus americanus	Crude oil	100–1000 µg l⁻¹, 15 d	Distended chromatophores	Forns, 1977
Mercenaria mercenaria	Phenol	100 µg l⁻¹, 24 h	Cytological damage to the gills and digestive gland	Fries and Tripp, 1977
Mytilus edulis	Aromatic	30 µl⁻¹, 34–182 d	Functional and structural changes in lysosomes	Lowe et al., 1981; Moore and Clark, 1982
Paracentrotus lividus	Benzo(a)-pyrene	4–5 µl⁻¹, 3 h	Abnormal cleavage in embryos	Ceas, 1974
Fish				
Clupea harengus	Benzene	0.9 µgl⁻¹, 24 h	Delayed embryonic development	Eldridge et al., 1977
Fundulus heteroclitus	Crude oil Naphthalene	1 µg g⁻¹ injected 200 µg l⁻¹, 15 d	Morphological abnormalities Ischemia Lateral line necrosis	Smith and Cameron, 1979 DiMichele and Taylor, 1978
Hypomesus pretiosus	Crude oil	54–113 µg l⁻¹, 3 h d⁻¹ for 15–21 h	Necrotic neurons in embryos	Hawkes and Stehr, 1982

Species	Oil	Conditions	Effects	Reference
Menidia beryllina	Crude oil	5–100 mg l⁻¹, 5–50% WSF[a] 21–30 d	Gill and olfactory hyperplasia; pancreatic atrophy and necrosis	Solangi and Overstreet, 1982
Menidia menidia	Crude oil	140 µg l⁻¹, 168 h	Hyperplasia	Gardner, 1975
Mugil cephalus	Crude oil	Experimental spill 4–5 µg g⁻¹), 56 d	Fin erosion	Minchew and Yarbrough, 1977
Parophrys vetulus	Crude oil	0.2–1.0% of the amount in sediments; 0.75–4 months	Hepatocellular lipid vacuolation; weight loss	McCain and Malins, 1982
Pleuronectes platessa	Crude oil	*Amoco Cadiz* (field sample)	Fin necrosis hyperplasia Decreased hepatocellular vacuolation	Haensly et al., 1982
			Delayed or suppressed ovarian development	Stott et al., 1983
Pseudopleuronectes americanus	Crude oil	Amounts found in sediments (1.02 kg); 4–5 months	Liver hypertrophy	Fletcher et al., 1982
Salmo gairdneri	Crude oil	200 µl l⁻¹, OWD[b]	Gill damage, chloride cell abnormalities	Engelhardt et al., 1981
Trinectes maculatus	Crude oil	5–100 mg l⁻¹ 5–50% WSF 38–60 d	Necrosis of olfactory tissue; pancreatic atrophy and necrosis	Solangi and Overstreet, 1982

[a] Water-soluble fraction.
[b] Oil-water dispersion.

From Capuzzo, J. M. and Kester, D. R., Eds., *Oceanic Processes in Marine Pollution*, Vol. 1, *Biological Processes and Wastes in the Ocean*, Robert E. Krieger Publishing, Malabar, FL, 1987, 3. With permission.

The greater capability of fish to metabolize and eliminate petroleum hydrocarbons than most other estuarine fauna does not render them free of the dangers of exposure to oil contamination. Many populations have developed morphological, cytological, and behavioral abnormalities after exposure to these compounds (Table 7). Longwell,[136] for instance, recorded increases in cytological deterioration, abnormal differentiation, and mitotic abnormalities in cod and pollack eggs due to the *Argo Merchant* oil spill. Linden,[137] Rabalais et al.,[138] and Boesch and Rabalais[139] registered a significant number of other developmental abnormalities in fishes exposed to oil pollution. In addition to histopathological disorders apparent in affected populations, biochemical alterations sometimes develop, as in populations of flatfish (*Pleuronectes platessa*) sampled in estuaries contaminated with oil from the *Amoco Cadiz* spill.[140]

Because of their mobility, fish tend to avoid oil-impacted habitats. Acute lethal effects in fish are almost always encountered only in the immediate vicinity of major oil spills where petroleum hydocarbon concentrations attain sufficiently high levels to impact them. Moreover, much of the mortality in fish populations reported in laboratory and field studies has been ascribed to the application of dispersants that produce oil-dispersant mixtures more toxic than the oil alone. Thus, many of the effects of oil on fish populations occur at sublethal concentrations of the hydrocarbons and are manifested as alterations in their feeding, migration, reproduction, swimming activity, schooling, and burrowing behavior.[69,118,141-143]

Light refined oils typically exert a greater impact on fish than crude and heavy fuel oils. Incidences of fishkills caused by spills of refined oils include the *Tampico Maru* spill in Southern California in 1957,[112] the *Florida* spill of No. 2 diesel fuel in West Falmouth, MA in 1969,[9] the release of high octane aviation gasoline, aviation jet fuel, aviation turbine fuel, diesel oil, and Bunker C oil from the tanker *R.C. Storer* on Wake Island,[144] and a large kill of Atlantic herring in Nova Scotia in 1969 from the release of intermediate oil containing high concentrations of aromatic hydrocarbons.[145]

Ichthyoplankton (eggs and larvae) of estuarine fishes are more sensitive to oil and dispersants than juveniles and adults, and experience greater mortality during spills. Many ichthyoplankters drift in surface waters for months. Their close proximity to polluting oil floating on the water surface places them at much greater risk to lethal and sublethal effects than later life history stages. Smith,[146] for example, observed 50 to 90% mortality of Pilchard (*Sardina pilcardus*) eggs in the region of the *Torrey Canyon* oil spill. Following the *Argo Merchant* spill, 46% of the pollock embryos and 25% of the cod embryos sampled at stations along Georges Bank and Nantucket shoals were dead. A substantial reduction in sand lance larvae (80%) also occurred in the affected area.[147]

Laboratory experiments assessing the toxicity of crude oil and dispersants on ichthyplankton generally corroborate field observations. Kühnhold[148] conducted laboratory studies on the effects of extracts of Iranian, Libyan, and Venezuelan crude oils on the eggs and larvae of cod (*Gadus morhua*) and the larvae of Baltic herring (*C. harengus*) and plaice (*Pleuronectes platessa*). He discerned delayed hatching and abnormal larvae from hatched eggs of impacted cod. Normal larvae of cod, herring, and plaice exposed to the oils exhibited aberrant behavior, necrosis, and death. Similarly, Wilson[149] delineated abnormalities in the embryos of Baltic herring (*C. harengus*), plaice (*P. platessa*), and sole (*Solea solea*) exposed to dispersants with aromatic solvents and kerosene. High mortalities of the larvae of these fishes, along with those of the lemon sole (*Microstomus knitt*) and haddock (*Melanogrammus aeglefinus*), were also noted. Linden[137,150] recognized an increase in both acute and lethal effects of Venezuelan crude on larval *C. harengus* when dispersants were applied, with the effects rising by three to four orders of magnitude if a highly toxic dispersant formulation was used. Mori et al.[151] found that the eggs of flounder and parrotfish had a lower tolerance to oil-dispersant mixtures than those of sea bass. According to Borseth et al.[152] the mortality of *P. platessa* eggs in their experiments approached 100% only when full strength water-soluble fractions of Statfjord A + B crude oil and Finasol OSR-5 mixtures were utilized. Despite these investigations, much more research needs to be performed on the vulnerability of early life stages of fish populations to oil.

Birds

Marine and coastal birds are more susceptible to harm from the physical properties of floating oil than most other organisms. The toxicity of oil constituents may be of secondary importance in the demise of many avifauna, although sublethal effects manifested in reproductive and physiological aberrations are not uncommon. Oil covering the plumage of seabirds accounts for the death of scores of individuals exposed to crude oil spills. As Clark[5] espouses, once the oil contaminates the plumage, the water-repellent properties of the feathers are lost. Hence, water accumulates between the feathers and skin of the birds, displacing air normally trapped there. Now devoid of the buoyancy and thermal insulation afforded by the air layer, the birds gradually become waterlogged and can sink and drown. The amelioration of the thermal insulation owing to the destruction of the fine structure of the feathers contributes to hypothermia of the individuals and, often, death.

Birds whose feathers are contaminated with oil usually attempt to preen them and, in the process, ingest it. The toxicity of the oil then can cause a wide range of physiological problems.[153,154] Some birds may suffer intestinal disorders, others kidney or liver failure, depending on the toxicity of the compounds swallowed. Oil-ingesting birds have greater reproductive problems in the form of depressed egg-laying and poor hatching success. Peakall et al.[155] demonstrate that even small amounts of oil (10 to 20 µl) create acute embryotoxicity in susceptible forms. Subsequent to applying Prudhoe Bay crude oil, Corexit 9527 dispersant, and oil-dispersant mixtures to mallard eggs, Albers[156] documented significant decreases in their hatchability relative to controls. Peakall et al.[157] verified a decline in the weight gain of young, recently hatched herring gulls after their treatment with oil and oil/dispersant. Table 8 shows the effects of oil and dispersant on various stages of the reproductive cycle of mallards, leach's petrels, and herring gulls.

Despite expressing apprehension over the sublethal effects of polluting oil on bird reproduction, some workers have directed their attention to the impact of oil on mortality of adult birds, which they consider to be a more pressing problem.[5] The greatest threat of oil may be to individual birds rather than to populations, as conveyed by Cormack.[11] He advances that no reductions in bird populations have ever been unequivocally attributed to oil pollution. However, certain populations are more sensitive than others, such as auks, divers, and sea ducks, with localized onshore breeding colonies being of particular concern. The National Academy of Sciences[158] has ascribed at least part of the blame of diminishing auk colonies to oil pollution.

The behavioral characteristics of seabirds largely determine their risk to oil pollution. Thus, birds that normally spend much of their time on the sea surface have a higher probability of encountering oil

TABLE 8
Studies of the Effect of Prudhoe Bay Crude Oil, Corexit 9527, and Combinations on Avian Reproduction

Stage of reproductive cycle	Species	Protocol	Combination studied	Finding	References
Hatchability of eggs	Mallard (*Anas platyrhynchos*)	Applied to egg surface with syringe	Oil; Corexit, 5:1 30:1 mixture	Toxicity ranking Corexit = 5:1 mixture Oil, 30:1 mixture	Albers, 1979
		Oil slick on water sprayed with Corexit	Oil, Corexit 10:1 combination	Corexit alone similar to control; oil and combination showed similar decrease in hatchability	Albers and Gay, 1982
Weight gain and survival of nestlings	Leach's Petrel (*Oceanodroma leucorhoa*)	Emulsion or oil painted on plumage or given internally	Oil, Corexit 10:1 combination	Combination applied externally to adults caused greater decrease of survival and weight gain of chicks than did oil alone	Butler et al., in press
	Mallard	Given in diet	Oil, Corexit 10:1 combination	Weight gain and survival were not affected	Eastin and Rattner, 1982

TABLE 8 (CONTINUED)
Studies of the Effect of Prudhoe Bay Crude Oil, Corexit 9527, and Combinations on Avian Reproduction

Stage of reproductive cycle	Species	Protocol	Combination studied	Finding	References
	Herring Gull (*Larus argentatus*)	Single internal dose	Oil, Corexit 10:1 combination	Corexit alone similar to control; oil and combination decreased weight gain to a similar extent	Peakall et al., 1982
		Single internal dose; birds food stressed	Oil, Corexit 10:1 combination	Oil and combination birds both lost weight faster than control	Peakall et al., 1985
		External, painting on feathers; not food stressed	Oil, Corexit 10:1 combination	Birds exposed to combination lost weight, oil and control birds maintained weight	

From Peakall, D. B., Wells, P. G., and Mackay, D., *Mar. Environ. Res.*, 22, 91, 1987. With permission.

and experiencing a problem. Auks (e.g., guillemots [murre], razor-bills, and puffins) and diving ducks (e.g., long-tailed ducks, scoters, velvet scoters, and eiders), which are gregarious populations, provide examples of species more likely to suffer heavy casualties from oil spills. Declining southern colonies of guillemots (*Urea aalege*), puffins (*Fratercula artica*), and razorbills (*Alca torda*) in the Atlantic, once coupled to polluting oil,[69,159] have now been attributed to climatic changes in the North Atlantic between 1850 and 1950. Other groups of birds regularly affected by oil, but in lesser numbers than those mentioned above, are grebes and divers (loons), shags and cormorants, mergansers, and gannets and pelicans.[5]

The total annual mortality of seabirds exceeds 100,000 individuals in regional areas. In the North Sea and North Atlantic, Tanis and Mörzer Bruyns[160] estimated that between 150,000 and 450,000 seabirds perish each year due to chronic oil pollution. Single oil spill events potentially kill substantial numbers of birds as well. For instance, a 750,000-l spill of No. 2 diesel oil into Puget Sound at Anacortes, WA caused the death of approximately 30,000 Brant, representing about 25% of the population in the Pacific Flyway of North America.[161] An oil spill in the Skaggerak in January 1981 resulted in the death of another 30,000 seabirds.[5]

Clark[5,159] and Peakall et al.[155] emphasize the need for highly effective dispersants to mitigate surface exposures and seabird casualties while maximizing the protection of populations from the damaging impacts of oil pollution. Since the major oiling of birds takes place at the sea surface, the rapid dispersing of oil into the water column can reduce the extent of oiling by a factor of 25 to 5000.[155] The risk of oil uptake underwater, therefore, remains very low, with most of the exposure and uptake occurring at the water-oil surface prior to the application of dispersant.

A highly effective dispersant will significantly lower oil exposure not only to surface-feeding birds, but also to diving birds. However, oil dispersion subsequent to the application of dispersant may increase the exposure of the birds to dispersed oil upon their submergence into the water column. Most often submergence is coupled to the feeding behavior of the animal which for the plunge divers (e.g., gannets, pelicans, and terns) and pursuit divers from the surface (e.g., auks, penguins, and sea ducks) places them at greater risk of exposure to the dispersed oil. Although surface feeders (e.g., gulls, petrels, and shear-waters) clearly benefit from the dispersal of oil into the water column, many of them (e.g., larids, phalaropes, and procellariiformes) enter the water column for short distances in search of food.[162] Nevertheless, these birds are still largely protected because of the appreciable diffusion resistance between the oil and their feathers that greatly reduces the extent of oiling.[155]

Mammals

Marine mammals are less likely to be impacted by oil pollution than seabirds, although Clark[5] writes of occasional reports of seal pups being oiled and possibly killed by crude or bunker oil. Once the fur of seals or other marine mammals become oiled, it loses the waterproofing and insulating properties that protect the animal from exposure, thereby often leading to the development of hypothermia and death. Seals devoid of thermal insulation cannot effectively pursue prey and are themselves prone to predation.

In its 1977 report, the Joint Group of Experts on the Scientific Aspects of Marine Pollution[69] outlined several oil spills that resulted in the demise of marine mammals. After the *Arrow* oil spill, 13 dead grey harbor seals were sighted in Chedabucto Bay. Larger numbers of sea lion pups purportedly died during the Santa Barbara oil spill when heavy oil slicks surrounded breeding colonies on San Miquel Island, although Brownell and LeBoeuf[163] state that this mortality could not be ascribed solely to the oil pollution. The literature does not support instances of heavy cetacean mortality due to oil spill events.

Based on laboratory and mesocosm toxicology experiments, as well as field observations, marine mammals may suffer a host of sublethal effects from oil spills, notably thermal and compensatory imbalances caused by oil coatings, changes in enzymatic activity in the skin, interferences with swimming, and eye irritation and lesions.[6,164] Effects of oiling on the fur of marine mammals is reviewed elsewhere.[6,38,164] The primary impacts from oiling of the fur are twofold: (1) direct toxicity of oil constituents following ingestion of the oil from the water or indirectly from grooming of the fur and (2) the loss of thermal insulation due to the negative effects of the oil on the water repellency of the fur.[6,38]

CASE STUDIES
THE *FLORIDA* OIL SPILL

The grounding of the barge *Florida* on the morning of September 16, 1969, on a rocky shoal off Fassett's Point, West Falmouth, MA, culminated in a spill of 650,000 to 700,000 l of Number 2 fuel oil into Buzzards Bay. As recounted by Sanders et al.,[9] strong SSW winds on that day churned the oil into an oil-water emulsion and drove the oil-tainted water into Wild Harbor River in North Falmouth. The oil contaminated a 6.5-km stretch of coastline and covered an area >400 ha. The highest concentrations of petroleum hydrocarbons occurred in the inter- and subtidal zones of Wild Harbor River, and they persisted there. Concentrations declined with increasing distance from the shore.

The loss of marine life was nearly immediate, with disturbance of the biota being most severe and persistent at the most heavily oiled sites and least severe at lightly oiled and more distant offshore stations. Small fish, benthic invertebrates, salt marsh fauna, and birds began to die in great numbers within 12 h after the spill; mortality peaked in Wild Harbor River, was less at nearshore subtital habitats, and least at more distant offshore locales. The greatest impact appeared to be on the benthic macrofauna which was nearly eradicated at heavily oiled sites within 2 d after the spill. In addition to the mass mortality of these groups immediately subsequent to the spill, chronic effects of the oil were manifested in long-term physiological and behavioral abnormalities of the organisms.

Analysis of sediments indicated that partly degraded oil persisted for more than 8 years in some areas, causing a depression in species diversity and abundance of biota. As a result, the faunal communities of Wild Harbor River exhibited instability in density, diversity, and species composition for years, with the populations showing only slight recovery 5 years after the spill at heavily oiled sites. Normal recovery of the benthos at moderately polluted stations did not take place for about 3 years. Benthic recovery in lightly oiled sediments was complete within 1 year. *Capitella*, an opportunistic polychaete which tends to colonize polluted habitats, numerically dominated the biologically denuded substrates of heavily oiled areas during the first 11 months after the spill, but then crashed. *Mediomastus ambiseta*, another capitellid polycheate, became a common element of moderately oiled sites nearly 1 year after the spill and persisted through the second year at these locales; however, it dropped in abundance at lightly oiled stations.

The salt marsh biotype was severely impacted by the oil spill as well. Heavily oiled salt marsh habitats appeared devoid of grasses and remained that way for long periods of time. Recovery of these areas was not complete by 1981, 12 years after the spill. Here, fiddler crabs (*Uca pugnax*) suffered heavy losses, and they had not completely recovered 7 years after the spill. The persistence of the oil in salt marsh sediments contributed to lethal impacts on the crabs soon after the spill and to sublethal effects over the long term. Adult crabs died when exposed to lightly weathered Number 2 fuel oil at concentrations of 1000 ppm and with more than 20% aromatics. Overwintering juveniles succumbed to the same oil in concentrations of 100 to 200 ppm. Impaired activity, loss of equilibrium, and death occurred in individuals at higher levels of oil residues. The life expectancy of the crabs gradually declined with increasing concentration of the residues.

Because of the devastation to the benthic environment, the shellfishery was adversely affected. In 1970 alone, mortality of softshelled clams amounted to 769 bushels, and that of native seed clams, 1135 bushels. Hydrocarbon contamination of the survivors prompted the closure of the industry.[69] Additionally, all seed and parent stocks of clams transplanted in Wild Harbor River in 1970 died from the oil. The shellfish harvest had not returned to the prespill levels as late as 1973. The slow recovery of the shellfishery was testimony to the pervasiveness of the chronic pollution that disrupted the benthic habitat.[165,166]

THE *ARROW* OIL SPILL

Running aground in Chedabucto Bay, Nova Scotia, on February 4, 1970, the tanker *Arrow* spilled approximately 10,000 tons of Bunker C oil into the cold waters (0 to 2°C) of the bay. Due to the low water temperature, the oil surfaced in ropelike pieces about 0.3 to 1 m long.[167] Strong wave action dispersed droplets of oil through the water column.[168] The oil spread rapidly and quickly coating 300 km of coastline with a tar-like covering of resistant oil to the mean high water mark. Even after 6 years, some of the oil still coated the shoreline, although wave action gradually reduced its coverage on the beaches. Significant concentrations of the oil were buried in sediments and could be detected there 10 years after the spill.

Both the flora and fauna of the bay were greatly impacted by the heavy refined oil, despite the fact that high seas swept much of the oil out to the Atlantic Ocean. Among the benthic macroflora, the rockweed *Fucus spiralis*, an intertidal algal species, disappeared from the rocky shores where oiling of the habitat precluded recolonization by the plant for at least 6 years.[80,169] The vascular salt marsh plant *Spartina alterniflora* likewise was eliminated from heavily oiled areas.[80] The species diversity of benthic plants decreased sharply after the spill and, similar to the diversity of benthic fauna, remained depressed at oiled locales 6 years later.

The loss of rockweed in intertidal zones caused a decline in the density of barnacles and other animals inhabiting impacted rocky shores. In intertidal habitats not altered by the elimination of rockweed, the faunal populations fared quite well.[169] Zooplankton possibly ingested as much as 20% of the Bunker C fuel oil droplets <0.1 mm in diameter in the water column, voiding the oil in fecal pellets that settled to the seafloor.[10] This process contributed to the contamination of bottom sediments and organisms.

The polluting oil was deemed to be directly responsible for the death of birds and mammals. An estimated 2000 birds died in Chedabucto Bay from the spill, but as many as 5000 more individuals drifted onto Sable Island, 320 km from the spill site. The absolute number of dead birds on Sable Island attributable to the *Arrow* spill has never been determined. Furthermore, many dead birds were buried in the oil and snow and may not have been accurately counted.[170] The oil also increased the mortality of grey seals and harbor seals,[69] while generating a variety of sublethal effects.

THE *EXXON VALDEZ* OIL SPILL

On March 24, 1989, the *Exxon Valdez* tanker, carrying 1.2 million barrels of crude oil struck rocks at Bligh Island in Prince William Sound along Alaska's south coast, rupturing its hull and spilling about 37.9 million l of oil — 240,000 barrels — into the pristine waters of the sound. A week later, the oil had spread over 233,100 ha of water, washed ashore, and coated hundreds of kilometers of shoreline. The estimated 35,000 t of oil entering Prince William

Sound could not be contained, and <4% of the oil was recovered 1 week after the spill. To this day, nearly two-thirds of the spilled oil has not been recovered or removed via evaporative processes.Within a few days of the spill, the oil converted to an emulsified mixture of oil and water (chocolate mousse) and could not be effectively treated with chemical dispersants. An unknown, but significant quantity of oil sank to the seafloor and was buried in the sediments. Experts have predicted that the oil may persist for more than 20 years in the sediments.[171] The low water temperatures in this subarctic environment have slowed microbial and photochemical degradation of the oil. Hence, the biota inhabiting bottom sediments will be exposed to hydrocarbon components for a protracted period of time. This long-term impact could be manifested in major changes in the trophic structure of the sound.

Shortly after the oil spill, birds, mammals, and fish were adversely affected. Numerous birds, sea otters, and fish perished from the oil. In addition, many hectares of habitat were destroyed due to the accumulation of oil in subtidal sediments and rocky intertidal areas. The well-being of the herring and salmon populations, which utilize Prince William Sound and comprise significant commercial fisheries, became a principal concern. Another major concern centered on the potential fouling of the salmon hatcheries in the sound and the danger that this fouling poses to the salmon fishery. Consequently, these finfishes will be monitored for years.

For months following the spill, cleanup efforts entailed the physical removal of tar-like goo and oil from stained rocks along the coastline. However, oiled bottom sediments, which could not be cleaned of the contaminated hydrocarbons, remained a focal point of contention regarding its role in the long-term recovery of the ecosystem. Historically, the environmental damage of previous major spills has been greater when larger quantities of the oil components accumulated in fine seafloor sediments below calm waters. Sequestered in these sediments, the oil can be a critical source of pollution to the system for years. The hope for recovery of the sound rests largely with degradation of this oil via photochemical and microbial processes and wave action. In the past, polluting oil on intertidal and subtidal benthic communities has accounted for the most persistent effects. Based on the volume of oil settling to the seafloor from the *Exxon Valdez,* the outlook for the recovery of Prince William Sound ecosystem does not appear to be very promising over the next decade or so. Not to be forgotten are the short-term impacts as well, especially the massive mortality of plankton, benthos, finfish, avifauna, and marine mammals incurred soon after the spill.

SUMMARY AND CONCLUSIONS

Acute and chronic oil pollution in estuarine and coastal marine environments represent a threat to aquatic communities and a potential danger to sensitive habitat areas. Of all the oil reaching the sea, about 45% is derived from river runoff, urban runoff, municipal wastes, and effluents from nonpetroleum industries. Activities related to oil transportation account for another 33% of the polluting oil, with only approximately 25% of this total ascribable to accidents and major spillages and 8% to normal operational losses. Natural oil seeps release roughly 20 to 23% of the oil found in the marine hydrosphere. Approximately 75% of the accidental oil spills in the U.S. occurs in coastal waters, primarily estuaries, enclosed bays, and wetlands, and most of the chronic oil pollution is associated with routine operations of oil refineries and oil installations as well as discharges of industrial and municipal wastes into these areas. Because these systems constitute some of the most productive areas on earth, the impact of oil on biota inhabiting them raises concern among estuarine and marine scientists.

Crude oil is comprised of a complex mixture of many thousands of gaseous, liquid, and solid organic compounds. Several classes of chemical compounds have been ascertained in crude oil, i.e., alkanes, cycloalkanes, alkenes, alkynes, aromatic hydrocarbons, and compounds containing most of the oxygen, nitrogen, and sulfur found in the oil. The hydrocarbon compounds comprise >75% of the weight of the oil. The toxicity of the oil tends to increase along the series from the alkanes, cycloalkanes, and alkenes to the aromatics. Among the toxic components of crude oil are benzene, toluene, xylene, and other low molecular weight aromatics, carboxylic acids, phenols, and sulfur compounds. During the first 24 to 48 h of an oil spill at sea, evaporative loss of volatile hydrocarbons removes the toxic lower molecular weight components, which can be detrimental to organisms.

Degradation of a spreading oil spill on the sea surface takes place via several abiotic and biotic processes. Important physical-chemical processes that alter the oil include evaporation, photochemical oxidation, and dissolution. Microbial degradation of petroleum hydrocarbons, principally by marine bacteria, represents the key biological method by which the oil is broken down. Through time, the polluting oil commonly converts to stable water-in-oil emulsions that promote the formation of viscous masses known as chocolate mousse, which are extremely persistent and can exacerbate the overall impact of the polluting oil. As a result of the aforementioned physical-chemical and microbial processes, the density of the oil increases, thereby enhancing its settlement to the seafloor. The sorption of hydrocarbons to particulate matter such as clay, silt, sand, shell fragments, and organic material also promotes sedimentation of the oil. Once the oil accumulates in seafloor sediments, the rate of microbial degradation of the oil decreases appreciably. The application of chemical dispersants to an oil slick soon after a spill creates dispersed oil droplets in the water column, thereby decreasing the potential for sedimentation or beaching of the oil.

Salt marshes, mangroves, seagrasses, mudflats, and estuarine waters are particularly sensitive to oil pollution. The oil easily degrades these critical habitat areas, and the toxicity of the oil can kill entire communities of organisms. The tendency of oil to be trapped in these habitats for long periods of time, being re-released at times of major episodic events (e.g., hurricanes and storm surges) which roil sediments, generates an inhospitable environment for organisms that may persist for years. Such long-term impacts are exemplified by the spill of oil from the barge *Florida* into Buzzards Bay in September 1969, which caused substantial loss of marine life for years in Wild Harbor River and contiguous waters; the grounding of the tanker *Arrow* in Chedabucto Bay, Nova Scotia, in February 1970, which greatly impacted biota of the bay for years; and the rupturing of the *Exxon Valdez* tanker on a reef in Prince William Sound, Alaska, and the release of large volumes of oil into coastal waters that have adversely affected biota and sensitive habitat areas of the sound to this day. These oil spill events clearly demonstrate the dangers of polluting oil to all major groups of estuarine organisms — phytoplankton, zooplankton, ichthyoplankton, benthos, fish, birds, and mammals.

REFERENCES

1. Odum, E. P., *Fundamentals of Ecology,* 3rd ed., W. B. Saunders, Philadelphia, 1974.
2. Hall, C. A. S., Howarth, R., Moore, B., III, and Vörösmarty, C. J., Environmental impacts of industrial energy systems in the coastal zone, *Annu. Rev. Energy,* 3, 395, 1978.
3. Carlberg, S. R., Oil pollution of the marine environment — with an emphasis on estuarine studies, in *Chemistry and Biogeochemistry of Estuaries,* Olausson, E. and Cato, I., Eds., John Wiley & Sons, Chichester, U.K., 1980, 367.
4. Abel, P. D., *Water Pollution Biology,* Ellis Horwood, Chichester, U.K., 1989.
5. Clark, R. B., *Marine Pollution 2nd ed.,* Clarendon Press, Oxford, 1989.
6. National Research Council, *Using Oil Spill Dispersants on the Sea,* National Academy Press, Washington, D.C., 1989.
7. Neff, J. M., *Polycyclic Aromatic Hydrocarbons in the Aquatic Environment: Sources, Fates, and Biological Effects,* Applied Science, London, 1979.
8. Press, F. and Siever, R., *Earth,* 4th ed., W. H. Freeman, New York, 1986.
9. Sanders, H. L., Grassle, J. F., Hampson, G. R., Morse, L. S., Garner-Price, S., and Jones, C. C., Anatomy of an oil spill: long-term effects from the grounding of the barge *Florida* off West Falmouth, Massachusetts, *J. Mar. Res.,* 38, 265, 1980.
10. Jordan, R. E. and Payne, J. R., *Fate and Weathering of Petroleum Spills in the Marine Environment,* Ann Arbor Science Publishers, Ann Arbor, MI, 1980.
11. Cormack, D., *Response to Oil and Chemical Marine Pollution,* Applied Science, London, 1983.
12. Murray, S. P., The effects of weather systems, currents, and coastal processes on major oil spills at sea, in *Pollutant Transfer and Transport in the Sea,* Vol. 2, Kullenberg, G., Ed., CRC Press, Boca Raton, FL, 1982, 169.
13. Elliott, A. J., Hurford, N., and Penn, C. J., Shear diffusion and the spreading of oil slicks, *Mar. Pollut. Bull.,* 17, 308, 1986.
14. Fay, J. A., Spread of oil slicks on a calm sea, in *Oil in the Sea,* Hoult, D. P., Ed., Plenum Press, New York, 1969, 53.
15. Stolzenback, K. L., Madsen, O., Adams, E., Pollack, A., and Cooper, C., A Review and Evaluation of Basic Techniques for Predicting the Behavior of Surface Oil Slicks, Tech. Rep. No. MIT SG 77-8, Massachusetts Institute of Technology, Cambridge, 1977.

16. Milgram, J. H., The role of physical studies before, during, and after oil spills, in *In the Wake of the Argo Merchant: A Symposium,* Center for Ocean Management Studies, University of Rhode Island, Kingston, 1978, 5.

17. Murray, S. P., Turbulent diffusion of oil in the ocean, *Limnol. Oceanogr.,* 17, 651, 1972.

18. Bishop, J. M., *Applied Oceanography,* John Wiley & Sons, New York, 1984.

19. Sivadier, H. O. and Mikolaj, P. G., Measurement of evaporation rates from oil slicks on the open sea, in *Proc. Conf. Prevention and Control of Oil Spills,* American Petroleum Institute, Washington, D.C., 1973, 475.

20. Harrison, W., Winnick, M. A., Kwang, P., and McKay, D., Crude oil spills, disappearance of aromatic and aliphatic components from small sea surface slicks, *Environ. Sci. Technol.,* 9, 231, 1975.

21. McAuliffe, C. D., Evaporation and solution of C_2 to C_{10} hydrocarbons from crude oils on the sea surface, in *Proc. Symp. Fate and Effects of Petroleum Hydrocarbons in Marine Ecosystems and Organisms,* Wolfe, D. A., Ed., Pergamon Press, New York, 1977, 363.

22. Johnson, J. C., McAuliffe, C. D., and Brown, R. A., Physical and chemical behavior of small crude oil slicks on the ocean, in *Chemical Dispersants for the Control of Oil Spills,* STP 659, McCarthy, L. T., Jr., Lindblom, G. P., and Walter, H. F., Eds., American Society for Testing and Materials, Philadelphia, 1978, 141.

23. McAuliffe, C. D., Organism Exposure to Volatile Hydrocarbons from Untreated and Chemically Dispersed Crude Oils in Field and Laboratory, 9th Arctic Marine Oil Program Technical Seminar, Environment Canada, Environmental Protection Service, Edmonton, Alberta, 1986, 497.

24. McAuliffe, C. D., Organism exposure to volatile/soluble hydrocarbons from crude oil spills — a field and laboratory comparison, in *Proc. 1987 Oil Spill Conf.,* American Petroleum Institute, Washington, D.C., 1987, 275.

25. Goldberg, E. D., *The Health of the Oceans,* UNESCO Press, Paris, 1976.

26. Parker, C. A., Freegarde, M., and Hatchard, C. G., The effects of some chemical and biological factors on the degradation of crude oil at sea, in *Water Pollution by Oil,* Hepple, P., Ed., Elsevier, London, 1971, 237.

27. Hansen, H. P., Photodegradation of hydrocarbon surface films, *Rapp. P.-v. Reun. Cons. Int. Explor. Mer.,* 171, 101, 1977.

28. Klein, A. E. and Pilpel, N., The effects of artificial sunlight upon floating oils, *Water Res.,* 8, 79, 1974.

29. McAuliffe, C. D., Smalley, A. E., Groover, R. D., Welsh, W. M., Pickle, W. S., and Jones, G. E., Chevron Main Pass Block 41 oil spill: chemical and biological investigation, in *Proc. 1975 Conf. on Prevention and Control of Oil Pollution,* American Petroleum Institute, Washington, D.C., 1975, 555.

30. Bridie, A. L., Wander, T. H., Zegveld, W., and van der Heijde, Formation, prevention, and breaking of sea water in crude oil emulsions (chocolate mousses), *Mar. Pollut. Bull.,* 11, 343, 1980.

31. Payne, J. R. and Phillips, C. R., Petroleum Spills in the Marine Environ-
    ment, in *The Chemistry and Formation of Water-in-Oil Emulsions and Tar
    Balls,* Lewis Publishers, Chelsea, MI, 1985.

32. Conomos, T. J., Movement of spilled oil as predicted by estuarine nontidal
    drift, *Limnol. Oceanogr.,* 20, 159, 1975.

33. Galt, J. A., Investigation of physical processes, in The *Amoco Cadiz* Oil
    Spill — A Preliminary Scientific Report, Hess, W. N., Ed., National Oce-
    anic and Atmospheric Administration/Environmental Protection Agency,
    Washington, D.C., 1978, 7.

34. Bassin, J. J. and Ichiye, T., Flocculation behavior of suspended sediments
    and oil emulsions, *J. Sed. Petrol.,* 47, 671, 1977.

35. Zobell, C. E., Microbial degradation of oil: present status, problems, and
    perspectives, in *Microbial Degradation of Oil Pollutants, Publ. No. LSU-
    SG-73-01,* Ahearn, D. G. and Meyers, S. P., Eds., Louisiana State Univer-
    sity Center for Wetland Resources, Baton Rouge, 1973.

36. Kaneko, T., Roubal, G., and Atlas, R. M., Bacterial populations in the
    Beaufort Sea, *Arctic,* 31, 97, 1977.

37. Horowitz, A. and Atlas, R. M., Continuous open flow-through system as a
    model for degradation in the Arctic Ocean, *Appl. Environ. Microbiol.,* 33,
    647, 1977.

38. National Research Council, *Oil in the Sea: Inputs, Fates, and Effects,*
    National Academy Press, Washington, D.C., 1985.

39. Rontani, J. S., Bertrand, J. S., Blanc, F., and Giusti, G., Accumulation of
    some monoaromatic compounds during the degradation of crude oil by
    marine bacteria, *Mar. Chem.,* 18, 1, 1986.

40. Westlake, D. W. S., Microorganisms and the degradation of oil under
    northern marine conditions, in Oil and Dispersants in Canadian Seas —
    Research Appraisal and Recommendations, Rep. EPS-3-EC-82-2, Sprague,
    J. B., Vandermeulen, J. H., and Wells, P. G., Eds., Environment Canada,
    Ottawa, 1982, 47.

41. Mitchell, R., Togel, S., and Chet, I., Bacterial chemoreception: an important
    ecological phenomenon inhibited by hydrocarbons, *Water Res.,* 6, 1137,
    1972.

42. Kennish, M. J., *Ecology of Estuaries,* Vol. 2, CRC Press, Boca Raton, FL,
    1990.

43. Gilfillan, E. S., Page, D. S., Hanson, S. A., Foster, J. C., Hotham, J., Vallas,
    D., Pendergast, E., Hebert, S., Pratt, S. D., and Gerber, R., Tidal area
    dispersant experiment, Searsport, Marine: an overview, *Proc. 1985 Oil Spill
    Conf.,* American Petroleum Institute, Washington, D.C., 1985, 553.

44. Gunkel, W. and Gassman, G., Oil, oil dispersants and related substances in
    the marine environment, *Helgol. Meeresunters.,* 33, 164, 1980.

45. Daling, P. S. and Brandvik, P. J., A Study of the Formation and Stability of
    Water-in-Oil Emulsions, Proc. 11th AMOP Semin., Environment Canada,
    Ottawa, 1988, 153.

46. Green, D. R., Buckley, J., and Humphrey, B., Fate of Chemically Dispersed Oil in the Sea; A Report on Two Field Experiments, Rep. No. 4-EC-82-5, Environment Canada, Ottawa, 1982.

47. Oudot, J., Rates of microbial degradation of petroleum components as determined by computerized capillary gas chromatography and computerized mass spectrometry, *Mar. Environ. Res.,* 13, 277, 1984.

48. Massie, L. C., Ward, A. P., and Davies, J. M., The effects of oil exploration and production in the northern North Sea. II. Microbial biodegradation of hydrocarbons in water and sediments, 1978-1981, *Mar. Environ. Res.,* 15, 235, 1985.

49. Oudot, J. and Dutrieux, E., Hydrocarbon weathering and biodegradation in a tropical estuarine ecosystem, *Mar. Environ. Res.,* 27, 195, 1989.

50. Lee, R. F. and Benson, A. A., Fates of petroleum in the sea: biological aspects, in *Proc. Workshop on Inputs, Fates, and Effects of Petroleum in the Marine Environment,* Vol. 2, National Academy of Sciences, Washington, D.C., 1973, 541.

51. Anderson, J. W., Neff, J. M., Cox, B. A., Tatem, H. E., and Hightower, G. M., Effects of oil on estuarine animals: toxicity, uptake and depuration, and respiration, in *Pollution and Physiology of Marine Organisms,* Vernberg, F. J. and Vernberg, W. B., Eds., Academic Press, New York, 1974, 285.

52. Blumer, M., Sanders, H. L., Grassle, J. F., and Hampson, G. R., A small oil spill, *Environment,* 13, 2, 1971.

53. Vandermeulen, J. H. and Gordon, D. C., Jr., Reentry of 5-year old stranded Bunker C fuel oil from a low energy beach into the water, sediments, and biota of Chedabucto Bay, Nova Scotia, *J. Fish Res. Bd. Can.,* 33, 2002, 1976.

54. Stebbings, R. E., Recovery of a salt marsh in Brittany 16 months after heavy pollution by oil, *Environ. Pollut.,* 1, 163, 1970.

55. Cowell, E. B., Oil pollution of the sea, in *Marine Pollution,* Johnson, R., Ed., Academic Press, London, 1976, 353.

56. Baker, J. M., Cruthers, J. H., Little, D. I., Oldham, J. H., and Wilson, C. M., Comparison of the fate and ecological effects of dispersed and non-dispersed oil in a variety of marine habitats, in *Oil Spill Chemical Dispersants: Research, Experience, and Recommendations,* STP 840, Allen, T. E., Ed., American Society for Testing and Materials, Philadelphia, 1984, 239.

57. Delaune, R. A., Smith, C. J., Patrick, W. H., Fleeger, J. W., and Tolley, M. D., Effect of oil on salt marsh biota: methods for restoration, *Environ. Pollut. Ser. A.,* 36, 207, 1984.

58. Lane, P. A., Vandermeulen, J. H., Crowell, M. J., and Patriquin, D. G., Impact of experimentally dispersed crude oil on vegetation in a northwestern Atlantic salt marsh — preliminary observations, *Proc. 1987 Oil Spill Conf.,* American Petroleum Institute, Washington, D.C., 1987, 509.

59. George, J. D., Mortality at Southend, *Mar. Pollut. Bull.,* 1, 187, 1970.

60. Dekker, R. and van Moorsel, G. W. N. M., Effects of Different Oil Doses, Dispersants, and Dispersed Oil on Macrofauna in Model Tidal Flat Ecosystems, paper presented at TNO Conf. Oil Pollution, Dordrecht, 1987 (abstract).

61. American Petroleum Institute, *Effects of a Dispersed and Undispersed Crude Oil on Mangroves, Seagrasses, and Corals,* Publ. 4460, American Petroleum Institute, Washington, D.C., 1987.

62. Teas, H. J., Duerr, E. O., and Wilcox, J. R., Effects of South Louisiana crude oil and dispersants on *Rhizophora* mangroves, *Mar. Pollut. Bull.,* 18, 122, 1987.

63. American Society for Testing and Materials, Ecological considerations for the use of chemical spill response — coral reefs, in *Annual Book of Standards,* Vol. 11.04, Standard F931-85, Seagrasses, American Society for Testing and Materials, Philadelphia, 1987.

64. Baca, B. J. and Getter, C. D., The toxicity of oil and chemically dispersed oil to the seagrass *Thalassia testudinum,* in *Oil Spill Chemical Dispersants: Research, Experience, and Recommendations,* STP 840, Allen, T. E., Ed., American Society for Testing and Materials, Philadelphia, 1984, 314.

65. Thorhaug, A. and Marcus, J. H., Effects of dispersant and oil on subtropical and tropical seagrasses, in *Proc. 1985 Oil Spill Conf.,* American Petroleum Institute, Washington, D.C., 1985, 497.

66. Thorhaug, A. and Marcus, J. H., Preliminary mortality effects of seven dispersants on subtropical/tropical seagrasses, in *Proc. 1987 Oil Spill Conf.,* American Petroleum Institute, Washington, D.C., 1987, 223.

67. Thorhaug, A. and Marcus, J. H., Oil spill clean-up: the effect of three dispersants on three subtropical/tropical seagrasses, *Mar. Pollut. Bull.,* 18, 124, 1987.

68. Thorhaug, A., Marcus, J., and Booker, F., Oil and dispersed oil on subtropical and tropical seagrasses in laboratory studies, *Mar. Pollut. Bull.,* 17, 357, 1986.

69. IMCO/FAO/UNESCO/WMO/WHO/IAEA/UN Joint Group of Experts on the Scientific Aspects of Marine Pollution (GESAMP), Impact of Oil on the Marine Environment, Reports and Studies No. 6, Food and Agriculture Organization, Rome, 1977.

70. Evans, D. R. and Rice, S. D., Effects of oil on marine ecosystems: a review for administrators and policy makers, *Fish. Bull., U.S.,* 72, 625, 1974.

71. Nelson-Smith, A., *Oil Pollution and Marine Ecology,* Paul Elek Scientific Books, London, 1972.

72. Tokuda, H., Fundamental studies on the influence of oil pollution upon marine organisms. IV. The toxicity of mixtures of oil products and oil-spill emulsifiers to phytoplankton, *Nippon Suisan Gakkaishi,* 45, 1289, 1979.

73. Fabregas, J., Herrero, C., and Veiga, M., Effect of oil and dispersant on growth and chlorophyll *a* content of the marine microalga *Tetraselmis suecica, Appl. Environ. Microbiol.,* 47, 445, 1984.

74. Hsaio, S. I. C., Kittle, D. W., and Foy, M. G., Effects of crude oils and the oil dispersant Corexit on primary production of Arctic marine phytoplankton and seaweed, *Environ. Pollut.,* 15, 209, 1978.

75. Chan, K.-Y. and Chiu, S. Y., The effects of diesel oil and oil dispersants on growth, photosynthesis, and respiration of the *Chlorella salina, Arch. Environ. Contam. Toxicol.,* 14, 325, 1985.

76. Falk-Peterson, I. B., Lönning, S., and Jakobsen, R., Effects of oil and dispersants on plankton organisms, *Astarte,* 12, 45, 1983.

77. Spooner, M. F. and Corkett, C. J., Effects of Kuwait oils on feeding rates of copepods, *Mar. Pollut. Bull.,* 10, 197, 1979.

78. Foy, M. G., Acute Lethal Toxicity of Prudhoe Bay Crude Oil and Corexit 9527 to Arctic Marine Fish and Invertebrates, Technology Devel. Rep. EPS-4-EC-82-3, Environment Canada, Ottawa, 1982.

79. Foster, M., Neushul, M., and Zingmark, R., The Santa Barbara oil spill. II. Initial effects on intertidal and kelp bed organisms, *Environ. Pollut.,* 2, 115, 1971.

80. Thomas, M. L. H., Effects of Bunker C oil on intertidal and lagoonal biota in Chedabucto Bay, Nova Scotia, *J. Fish. Res. Bd. Can.,* 30, 83, 1973.

81. Ganning, B. and Billing, U., Effects on community metabolism of oil and chemically dispersed oil on Baltic Bladder Wrack, *Fucus vesiculosus,* in *Ecological Aspects of Toxicity Testing of Oils and Dispersants,* Beynon, L. R. and Cowell, E. B., Eds., Applied Science, Essex, U.K., 1974, 53,

82. Thelin, I., Effects in culture of two crude oils and one oil dispersant on zygotes and germlings of *Fucus serratus, Linnaeus, Fucales,* and *Phaeophyceae, Bot. Mar.,* 24, 515, 1981.

83. Lowe, D. M. and Pipe, R. K., Mortality and quantitative aspects of storage cell utilization in mussels, *Mytilus edulis,* following exposure to diesel oil hydrocarbons, *Mar. Environ. Res.,* 22, 243, 1987.

84. Widdows, J., Physiological responses to pollution, *Mar. Pollut. Bull.,* 16, 129, 1985.

85. Clifton, H. E., Kvenvolden, K. A., and Rapp, J. B., Spilled oil and infaunal activity — modification of burrowing behavior and redistribution of oil, *Mar. Environ. Res.,* 11, 111, 1984.

86. Mayo, D. W., Page, D. S., Cooley, J., Sorenson, E., Bradley, F., Gilfillan, E. S., and Hanson, S. A., Weathering characteristics of petroleum hydrocarbons deposited in fine clay marine sediments, Searspot, Maine, *J. Fish. Res. Bd. Can.,* 35, 552, 1978.

87. Wade, T. L. and Quinn, J. G., Incorporation, distribution, and fate of saturated petroleum hydrocarbons in sediments from a controlled marine ecosystem, *Mar. Environ. Res.,* 3, 15, 1980.

88. Delaune, R. D., Hambrick, G. A., III, and Patrick, W. H., Jr., Degradation of hydrocarbons in oxidized and reduced sediments, *Mar. Pollut. Bull.,* 11, 103, 1980.

89. Hambrick, G. A., III, Delaune, R. D., and Patrick, W. H., Jr., Effect of estuarine sediment pH and oxidation-reduction potential on microbial hydrocarbon degradation, *Appl. Environ. Microbiol.,* 40, 365, 1980.

90. Aller, R. C., The effects of animal-sediment interactions on geochemical processes near the sediment-water interface, in *Estuarine Interactions,* Wiley, M. L., Ed., Academic Press, New York, 1978, 157.

91. Rhoads, D. C. and Boyer, L. F., The effects of marine benthos on physical properties of sediments: a successional perspective, in *Animal-Sediment Relations: The Biogenic Alteration of Sediments,* McCall, P. L. and Tevesz, M. J. S., Eds., Plenum Press, New York, 1982, 3.

92. Aller, R. C., The effects of macrobenthos on chemical properties of marine sediment and overlying water, in *Animal-Sediment Relations: The Biogenic Alteration of Sediments,* McCall, P. L. and Tevesz, M. J. S., Eds., Plenum Press, New York, 1982, 53.

93. Mulligan, H. F., Mathieson, A. C., Jones, G. E., Borror, A. C., Loder, T. C., Sawyer, P. J., and Harris, L. G., Impact of an Oil Refinery on the New Hampshire Marine Environment, in The Impacts of an Oil Refinery Located in Southeastern New Hampshire: Preliminary Study, Unpub. Tech Rep., University of New Hampshire, Durham, 1974, chap. 10.

94. Neff, J. M., Bothner, M. H., Maciolek, N. J., and Grassle, J. F., Impacts of exploratory drilling for oil and gas on the benthic environment of Georges Bank, *Mar. Environ. Res.,* 27, 77, 1989.

95. National Oceanic and Atmospheric Administration, A Summary of Data on Tissue Contamination from the First Three Years (1986-1988) of the Mussel Watch Project, NOAA Tech. Mem. NOS OMA 49, Rockville, MD, 1989.

96. Strömgren, T., Nielsen, M. V., and Ueland, K., The short term effect of microencapsulated hydrocarbons on shell growth of *Mytilus edulis, Mar. Biol.,* 91, 33, 1986.

97. Strömgren, T., Effect of oil and dispersants on the growth of mussels, *Mar. Environ. Res.,* 21, 239, 1987.

98. La Roche, G., Eisler, R., and Tarzwell, C. M., Bioassay procedures for oil and oil dispersant toxicity evaluation, *Water Pollut. Control Fed.,* 42, 1982, 1970.

99. Swedmark, M., Granmo, Å., and Kollberg, S., Effects of oil dispersants and oil emulsions on marine animals, *Water Res.,* 7, 1649, 1973.

100. Nes, H. and Nordland, S., Effectiveness and Toxicity Experiments with Oil Dispersants, PFO-Prosjekt No. 1405, Oslo, Norway.

101. Blumer, M., Souza, G., and Sass, J., Hydrocarbon pollution of edible shellfish by an oil spill, *Mar. Biol.,* 5, 195, 1970.

102. Stegeman, J. J. and Teal, J. M., Accumulation, release, and retention of petroleum hydrocarbons by the oyster *Crassostrea virginica, Mar. Biol.,* 22, 37, 1973.

103. Neff, J. M., Cox, B. A., Dixit, D., and Anderson, J. W., Accumulation and release of petroleum-derived aromatic hydrocarbons by four species of marine animals, *Mar. Biol.,* 38, 279, 1976.

104. Mackin, J. G. and Sparks, A. K., A Study of the Effect on Oysters of Crude Oil Loss from a Wild Well, Vol. 7, Institute of Marine Science, University of Texas, College Station, 1961, 230.

105. Heitz, J. R., Lewis, L., Chambers, J., and Yarbrough, J. D., The acute effects of Empire Mix crude oil on enzymes in oysters, shrimp, and mullet, in *Pollution and Physiology of Marine Organisms,* Vernberg, F. J. and Vernberg, W. B., Eds., Academic Press, New York, 1974, 311.

106. Smith, C. M. and Hackney, C. T., The effects of hydrocarbons on the setting of the American oyster, *Crassostrea virginica,* in intertidal habitats in southeastern North Carolina, *Estuaries,* 12, 42, 1989.

107. Clark, R. C., Jr. and Finley, J. S., Paraffin hydrocarbon patterns in petroleum-polluted mussels, *Mar. Pollut. Bull.,* 4, 172, 1973.

108. Renzoni, A., Influence of crude oil, derivatives, and dispersants on larvae, *Mar. Pollut. Bull.,* 4, 9, 1973.

109. Hawks, A. L., A review of the nature and extent of damage caused by oil pollution at sea, *Trans. North Am. Wildl. Nat. Res. Conf.,* 26, 343, 1961.

110. Berthou, F., Balouët, G., Bodennec, G., and Marchand, M., The occurrence of hydrocarbons and histopathological abnormalities in oysters for seven years following the wreck of the *Amoco Cadiz* in Brittany (France), *Mar. Environ. Res.,* 23, 103, 1987.

111. Chan, G. L., A study of the effects of the San Francisco oil spill on marine organisms, in *Proc. Joint Conf. on Prevention and Control of Oil Spills,* American Petroleum Institute, Washington, D.C., 1973, 741.

112. North, W. J., Neushul, M., and Clendenning, K. A., Successive biological changes observed in a marine cove exposed to a large spillage of mineral oil, *Symp. Pollut. Mar. Microorg. Prod. Petrol.,* 1964, 333.

113. Straughan, D. and Abbott, B. C., The *Santa Barbara* oil spill: ecological changes and natural oil leaks, in *Water Pollution by Oil,* Hepple, P., Ed., Institute of Petroleum, London, 1971, 257.

114. Atema, J. and Stein, L. S., Sublethal Effects of Crude Oil on Lobster (*Homarus americanus*) Behavior, Tech. Rep., Woods Hole Oceanographic Institution, Woods Hole, MA, 1972.

115. Engel, R. H. and Neat, M. J., Toxicity of oil dispersing agents determined in a circulating aquarium system, in *Proc. Joint Conf. on Prevention and Control of Oil Spills,* American Petroleum Institute, Washington, D.C., 1971, 297.

116. Wells, P. G., Influence of Venezuela crude oil on lobster larvae, *Mar. Pollut. Bull.,* 3, 105, 1972.

117. Capuzzo, J. M., Lancaster, B. A., and Sasaki, G., The effects of petroleum hydrocarbons on lipid metabolism and energetics of larval development and metamorphosis in the American lobster (*Homarus americanus*), *Mar. Environ. Res.,* 14, 201, 1984.

118. Capuzzo, J. M. and Kester, D. R., Biological effects of waste disposal: experimental results and predictive assessments, in *Oceanic Processes in Marine Pollution,* Vol. 1, *Biological Processes and Wastes in the Ocean,* Capuzzo, J. M. and Kester, D. R., Eds., Robert E. Krieger Publishing, Malabar, FL, 1987, 3.

119. Katz, L. M., The effects of the water soluble fraction of crude oil on larvae of the decapod crustacean *Neopanope texana* (Sayi), *Environ. Pollut.*, 5, 199, 1973.

120. Portmann, J. E., Results of acute toxicity tests with marine organisms using a standard method, in *Marine Pollution and Sea Life,* Ruivo, M., Ed., Fishing News (Books) Ltd., West Byfleet, Surrey, 1972, 212.

121. Krebs, C. T. and Burns, K. A., Long-term effects of an oil spill on populations of the salt-marsh crab *Uca pugnax, Science,* 197, 484, 1977.

122. Elmgren, R., Hanson, S., Larsson, U., Sundelin, B., and Boehm, P. D., The *Tsesis* oil spill: acute and long-term impact on the benthos, *Mar. Biol.,* 73, 51, 1983.

123. Levell, D., The effects of oil and dispersant on the lugworm *Arenicola marina*, in *Marine Ecology and Oil Pollution,* Baker, J. M., Ed., Applied Science, London, 1975.

124. Olla, B. L., Bejda, A. J., Studholme, A. L., and Pearson, W. H., Sublethal effects of oiled sediment on the sand worm, *Nereis (Neanthes) virens:* induced changes in burrowing and emergence, *Mar. Environ. Res.,* 13, 121, 1984.

125. Fletcher, G. L., King, M. J., Kiceniuk, J. W., and Addison, R. F., Liver hypertrophy in winter flounder following exposure to experimentally oiled sediments, *Comp. Biochem. Physiol.,* 73C, 457, 1982.

126. McCain, B. B. and Malins, D. C., Effects of petroleum hydrocarbons on selected demersal fish and crustaceans, in *Ecological Stress and the New York Bight: Science and Management,* Mayer, G. F., Ed., Estuarine Research Federation, Columbia, SC, 1982, 315.

127. Widdows, J., Bakke, T., Bayne, B. J., Donkin, P., Livingstone, D. R., Lowe, D. M., Moore, M. N., Evans, S. V., and Moore, S. L., Responses of *Mytilus edulis* on exposure to the water accommodated fraction of North Sea oil, *Mar. Biol.,* 67, 15, 1982.

128. Malins, D. C., Alterations in the cellular and subcellular structure of marine teleosts and invertebrates exposed to petroleum in the laboratory and field: a critical review, *Can. J. Fish. Aquat. Sci.,* 39, 877, 1982.

129. Malins, D. C. and Hodgins, H. O., Petroleum and marine fishes: a review of uptake, disposition, and effects, *Environ. Sci. Technol.,* 15, 1272, 1981.

130. Lindstrom-Seppa, P., Koivusaari, U., and Hanninen, O., Metabolism of xenobiotics by vendace (*Coregonus albula*), *Comp. Biochem. Physiol.,* 68C, 121, 1981.

131. Stegeman, J. J., Polynuclear aromatic hydrocarbons and their metabolism in the marine environment, in *Polycyclic Hydrocarbons and Cancer,* Vol. 3, Belboin, H. V. and Ts'O, P. O. P., Eds., Academic Press, New York, 1981, 1.

132. Eisler, R., Polycyclic Aromatic Hydrocarbon Hazards to Fish, Wildlife, and Invertebrates: A Synoptic Review, Biol. Rep. 85(1.11), U. S. Fish and Wildlife Service, Washington, D.C., 1987.

133. Neff, J. M., Polycyclic aromatic hydrocarbons, in *Fundamentals of Biological Toxicology: Methods and Applications,* Rand, G. M. and Petrocelli, S. R., Eds., Hemisphere Publishing, New York, 1985, 416.

134. Lee, R. F., Sauerheber, R., and Dobbs, G. H., Uptake, metabolism, and discharge of polycyclic aromatic hydrocarbons by marine fish, *Mar. Biol.,* 17, 201, 1972.

135. Stegeman, J. J., Detecting the biological effects of deep-sea waste disposal, *Oceanus,* 33, 54, 1990.

136. Longwell, A. C., A genetic look at fish eggs and oil, *Oceanus,* 20, 45, 1977.

137. Linden, O., The influence of crude oil and mixtures of crude oil/dispersants on the ontogenetic development of the Baltic herring, *Clupea harengus membras* L., *Ambio,* 5, 136, 1976.

138. Rabalais, S. C., Arnold, C. R., and Wohlschlag, N. S., The effects of IXTOCI oil on the eggs and larvae of red drum (*Sciaenops ocellata*), *Tex. J. Sci.,* 33, 33, 1981.

139. Boesch, D. F. and Rabalais, N. N., Eds., *Long-Term Environmental Effects of Offshore Oil and Gas Development,* Elsevier Applied Sciences, New York, 1987.

140. Neff, J. M., The use of biochemical measurements to detect pollutant-mediated damage to fish, in *Aquatic Toxicology and Hazard Assessment, 7th Symp.,* Spec. Tech. Publ. No. 854, Cardwell, R. D., Purdy, R., and Bahner, Eds., American Society for Testing and Materials, Philadelphia, 1985, 155.

141. Gardner, G. R., Chemically induced lesions in estuarine or marine teleosts, in *The Pathology of Fishes,* Ribelin, W. E. and Migaki, G., Eds., University of Wisconsin Press, Madison, 1975, 657.

142. Berge, J. A., Johannessen, K. I., and Reiersen, L.-O., Effects of water soluble fraction of North Sea crude oil on the swimming activity of the sand goby *Pomatoschistus minutus* (Pallas), *J. Exp. Mar. Biol. Ecol.,* 68, 159, 1983.

143. Pearson, W. H., Woodruff, D. L., and Sugarman, P. C., The burrowing behavior of sand lance, *Ammodytes hexapterus*: effects of oil-contaminated sediment, *Mar. Environ. Res.,* 11, 17, 1984.

144. Gooding, R. M., Oil Pollution on Wake Island from the Tanker, *R. C. Stoner,* NOAA/NMFS Spec. Sci. Rep. (Fish.), 1971, 636.

145. Zitko, V. and Tibbo, S. N., Fish kill caused by an intermediate oil from coke ovens, *Bull. Environ. Contam. Toxicol.,* 6, 24, 1971.

146. Smith, J. E., *Torrey Canyon, Pollution, and Marine Life,* Cambridge University Press, Cambridge, 1970.

147. Grose, P. L. and Mattson, J. S., The *Argo Merchant* Oil Spill: A Preliminary Scientific Report, U. S. Department of Commerce and National Oceanic and Atmospheric Administration, Boulder, CO, 1977.

148. Kühnhold, W. W., An Examination of the Toxicity of Extracts of Crude Oil and Crude Oil Emulsions to Eggs and Larvae of Cod and Herring, Ph.D. thesis, University of Kiel, Kiel, Germany, 1972.

149. Wilson, K. W., Toxicity of oil-spill dispersants to embryos and larvae of some marine fish, in *Marine Pollution and Sea Life,* Ruivo, M., Ed., Fishing News (Books) Ltd., West Byfleet, Surrey, 1972.

150. Linden, O., Acute effects of oil and oil/dispersant mixture on larvae of Baltic herring, *Ambio,* 4, 130, 1975.

151. Mori, K., Kobayashi, T., and Fujishima, T., Effects of the toxicity of mineral oil and solvent emulsifier upon the eggs of marine fish, *Bull. Fac. Fish. Mie Univ.,* 10, 15, 1983.

152. Borseth, J., Aunaas, T., Ekker, M., and Zachariassen, K. E., A comparison of *in vivo* and *in situ* exposure of marine fish eggs to Statfjord A + B crude oil topped at 15°C and this oil premixed with Finasol OSR-5, Proc. Int. Semin. Chemical and Natural Dispersion of Oil on the Sea, Heimdal, Norway, 1986 (abstract).

153. Leighton, F. A., The pathophysiology of petroleum oil toxicity to birds: a review, in *The Effects of Oil on Birds: A Multidiscipline Symposium,* Rosie, D. and Barnes, S. N., Eds., Tri-State Bird Rescue and Research, Washington, D.C., 1983.

154. Holmes, W. N., Petroleum pollutants in the marine environment and their possible effects on seabirds, *Rev. Environ. Toxicol.,* 1, 251, 1984.

155. Peakall, D. B., Wells, P. G., and Mackay, D., A hazard assessment of chemically dispersed oil spills and seabirds, *Mar. Environ. Res.,* 22, 91, 1987.

156. Albers, P. H., Effects of Corexit 9527 on the hatchability of mallard eggs, *Bull. Environ. Contamin. Toxicol.,* 23, 661, 1979.

157. Peakall, D. B., Hallett, D. J., Bend, J. R., Foureman, G. L., and Miller, D. S., Toxicity of Prudhoe Bay crude oil and its aromatic fractions to nestling herring gulls, *Environ. Res.,* 27, 206, 1982.

158. *Petroleum in the Marine Environment,* National Academy of Sciences, Washington, D.C., 1975.

159. Clark, R. B., Impact of chronic and acute oil pollution on sea birds, in *Background Papers for a Workshop on Inputs, Fates, and Effects of Petroleum in the Marine Environment,* National Academy of Sciences, Washington, D.C., 1973, 619.

160. Tanis, J. J. C. and Mörzer Bruyns, M. F., The impact of oil pollution on seabirds in Europe, Proc. Int. Conf. Oil Pollution in the Sea, Rome, 1968, 67.

161. Shiang-Chia, F., Diesel spill at Anacortes, *Mar. Pollut. Bull.,* 2, 105, 1971.

162. Ashmole, N. P., Seabird ecology and the marine environment, in *Avian Biology,* Vol. 1, Farner, D. S. and King, J. R., Eds., Academic Press, New York, 1971, 224.

163. Brownell, R. J., Jr. and LeBoeuf, B. J., California sea lion mortality: natural or artifact?, in *Biological and Oceanographical Survey of the Santa Barbara Oil Spill, 1969-1970,* Vol. 1, Straughan, D., Ed., Allan Hancock Foundation, University of Southern California, Los Angeles, 1971, 287.

164. Engelhardt, F. R., Effects of petroleum on marine mammals, in *Petroleum Effects in the Arctic Environment*, Englehardt, F. R., Ed., Elsevier Applied Science, New York, 1985, 217.

165. Souza, G., Report of the Shellfish Warden, Annual Report of the Finances of the Town of Falmouth for the Year Ending December 31, 1970, 161.

166. Souza, G., Report of the Shellfish Officer, Annual Report of the Finances of the Town of Falmouth for the Year Ending December 31, 1973, 174.

167. Barber, F. G., Report of the Task Force: Operation Oil (Cleanup of *Arrow* Oil Spill in Chedabucto Bay), in *Proc. 1975 Conf. Prevention and Control of Oil Polllution*, Vol. 3, American Petroleum Institute, Washington, D.C., 1970, 35.

168. Forrester, W. D., Distribution of suspended oil particles following the grounding of the tanker *Arrow*, *J. Mar. Res.*, 29, 151, 1971.

169. Roberts, L., The legacy of past spills, *Science*, 244, 23, 1989.

170. Brown, R. G., Seabirds and oil pollution, in *Oil and the Canadian Environment — Proceedings of the Conference, 16 May 1973*, Mackay, D. and Harrison, W., Eds., University of Toronto, Ontario, 1973.

171. Roberts, L., Long, slow recovery predicted for Alaska, *Science*, 244, 22, 1989.

3 Polynuclear Aromatic Hydrocarbons

INTRODUCTION

Among the most carcinogenic, mutagenic, and toxic compounds found in estuaries are polynuclear aromatic hydrocarbons (PAHs), also known as polycyclic aromatic hydrocarbons and polycyclic organic matter. These ubiquitous compounds occur in bottom sediments, overlying waters, and biota of estuaries, with the primary repositories being systems located near urban industrialized areas. Consisting of hydrogen and carbon arranged in the form of two or more fused benzene rings in linear, angular, or cluster arrangements with substituted groups possibly attached to one or more rings,[1,2] PAHs encompass a wide range of chemicals. Thousands of them occur in the environment, all differing in the number and position of aromatic rings. Sources of PAHs in estuaries include sewage and industrial effluents, petroleum spills, creosote oil, combustion of fossil fuels, and forest and brush fires.[3-5] Direct discharges, urban and agricultural runoff, and groundwater flow deliver substantial quantities of PAHs to aquatic environments.[6-8] Jackim and Lake[9] stress that the major source of PAHs is the pyrolysis of organic matter. An estimated 2.3×10^5 metric tons of PAHs enter aquatic environments each year, principally from petroleum spills and atmospheric deposition (Table 1).[1]

Two general types of PAHs have been characterized: (1) fused-ring compounds (e.g., alkyl naphthalenes, phenanthrene, and benzo(a)pyrene); and (2) linked-ring compounds (e.g., biphenyls).

TABLE 1
Major Sources of PAHs in Atmospheric and Aquatic Environments

Ecosystem and sources	Annual input (metric tons)
Atmosphere	
Total PAHs	
Forest and prairie fires	19,513
Agricultural burning	13,009
Refuse burning	4,769
Enclosed incineration	3,902
Heating and power	2,168
Benzo(a)pyrene	
Heating and power	
Worldwide	2,604
U.S. only	475
Industrial processes	
(mostly coke production)	
Worldwide	1,045
U.S. only	198
Refuse and open burning	
Worldwide	1,350
U.S. only	588
Motor vehicles	
Worldwide	45
U.S. only	22
Aquatic environments	
Total PAHs	
Petroleum spillage	170,000
Atmospheric deposition	50,000
Wastewaters	4,400
Surface land runoff	2,940
Biosynthesis	2,700
Total benzo(a)pyrene	700

From Eisler, R., Polycyclic Aromatic Hydrocarbon Hazards to Fish, Wildlife, and Invertebrates: A Synoptic Review, Biol. Rep. 85 (1.11), U.S. Fish and Wildlife Service, Washington, D.C., 1987.

The fused-ring varieties usually are more abundant.[10] Figure 1 shows the basic structures of PAHs typically found in the marine environment. Illustrated in the figure are the parent compounds and the alkyl derivatives that have been recovered from estuaries. These PAHs display many of the chemical and physical properties identified in one- and two-ring aromatic compounds; i.e., they

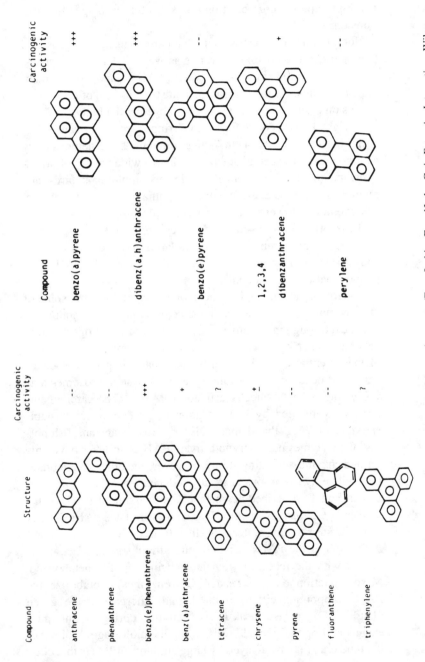

FIGURE 1. Examples of PAHs typically found in the marine environment. (From Jackim, E. and Lake, C., in *Estuarine Interactions*, Wiley, M. L., Ed., Academic Press, New York, 1978, 415. With permission.)

absorb ultraviolet (UV) light, are oxidized by chlorine, ozone, and UV light, and undergo electrophilic substitution in specific positions on the ring.[9,11]

Not all PAH compounds are potent carcinogens, mutagens, and teratogens, and their toxicity varies as well. Eisler (p. iii)[1] states: "Unsubstituted lower molecular weight PAH compounds, containing 2 or 3 rings, exhibit significant acute toxicity and other adverse effects on organisms, but are noncarcinogenic; the higher molecular weight PAHs, containing 4 to 7 rings, are significantly less toxic, but many of the 4- to 7-ring compounds are demonstrably carcinogenic, mutagenic, or teratogenic to a wide variety of organisms, including fish and other aquatic life, amphibians, birds, and mammals." Anthracenes, fluorenes, naphthalenes, and phenanthrenes exemplify lower molecular weight unsubstituted PAH compounds that are acutely toxic to some organisms. Benzo(c)phenanthrene, 3-methylcholanthrene, benzo(a)pyrene, and dibenzo(a,i)pyrene typify high molecular weight PAH compounds that are carcinogenic.

Aquatic organisms rapidly metabolize most PAHs, but some of these compounds become carcinogenic or mutagenic or both when activated through metabolism.[12-14] Metabolized carcinogenic products have taken on increasing environmental importance because of the potential ecological and human health impacts associated with their occurrence in natural systems. Tissue microsomes play a key role in metabolically activating PAHs.[15] The carcinogenic process is initiated by hydroxylation or production of unstable epoxides of PAHs that damage DNA.[16] Crustaceans and fish possess the enzymes necessary to activate PAHs; mollusks have only limited enzyme-activating ability, however, and other invertebrates metabolize hydrocarbons at greatly reduced levels.[9]

Van Engel[14] described the metabolic processing of PAHs by the blue crab, *Callinectes sapidus*. This crab assimilates PAHs directly from the surrounding water as well as from the food it consumes. A fraction of the PAHs passes through the intestinal tract and is voided with the feces, the remainder being rapidly metabolized. Mixed-function oxygenases (MFO) (enzymes) metabolize the PAH compounds either in the hepatopancreas or the green gland.[17,18] Oxygenases peak in concentration during the intermolt stage and drop to minimal levels during the molt stage. Molting of the blue crab may be delayed by large amounts of PAHs that elicit elevated levels of oxygenase activity for their detoxification. This activity inhibits production of ecdysone by the crab, thereby restricting molting and limb regeneration.[19]

PAHs do not undergo significant biomagnification in food chains due to their rapid degradation and depuration in most animals and low intestinal absorption in higher animals.[9] Nevertheless, questions still abound regarding the potential carcinogenic and mutagenic risks to man through consumption of estuarine and marine foods containing PAH contaminants. Some of this uncertainty involves carcinogenicity itself — threshold levels, synergistic or co-carcinogenic effects, and cellular recovery processes — in the face of the PAH compounds.[9] Other unknown factors concern the bioavailability of specific compounds and the partitioning of the compounds between aqueous and nonaqueous phases in many estuaries, all of which strongly affect the uptake, bioaccumulation, and subsequent depuration of the estuarine organisms.[20]

DISTRIBUTION

PAHs In Sediments

PAHs tend to concentrate in seafloor sediments owing to their relative insolubility in water and strong adsorption to particulate matter.[8] Although most PAHs accumulate in bottom sediments of estuaries, they also disperse into the water column, concentrate in aquatic biota, or undergo chemical oxidation and biodegradation.[1,21] Only about 33% of the PAHs occur in dissolved form, with most associated with particulate matter.[22] Dissolved PAHs in the water column appear to degrade rapidly by photooxidation,[23] a process facilitated by higher temperatures, dissolved oxygen levels, and incidences of solar radiation.[1,21,24,25] Once in bottom sediments, PAH compounds are subject to biotransformation and biodegradation by benthic organisms.[23]

Natural sources of PAHs in sediments, such as direct biosynthesis by microbes and plants and diagenesis of sediment organic matter,[26-29] are small relative to anthropogenic inputs. Oil seeps and forest fires accounted for most of the PAHs entering aquatic systems prior to the Industrial Revolution.[30] With the advent of industrialization and, in particular, the combustion of large amounts of fossil fuels (and later through large oil spills), a diversity of PAH compounds in substantial quantities was released to the environment. Abrupt increases in PAH levels in bottom sediments mark the beginning of the Industrial Revolution.[31-33]

PAHs exist in estuarine sediments worldwide, being derived from oil spills,[3] deposition of airborne particulates,[34] municipal and industrial wastewater discharges,[35] runoff from streets and highways,[36] riverine transport,[37] and commercial and recreational boating activities.[38] Estuaries located in proximity to urban and industrial centers are major repositories of PAH compounds.[38-40] The partitioning behavior of PAHs onto sediments and suspended particulates must be taken into consideration when assessing the extent of exposure of estuarine biota to the pollutants. In connection with this type of assessment are sorption-desorption equilibria, parameters of importance in evaluating the fate, transport, and toxicological impact of the contaminants.[29]

The bottom sediments of some U.S. estuaries contain significant concentrations of PAHs. The Elizabeth River, a subestuary of Chesapeake Bay, is heavily contaminated with these organic compounds.[41] Sediment concentrations of PAHs as high as 390 μg/g dry weight have been recorded at one heavily contaminated site; highest levels of PAHs in the system correspond to sediments near a former creosote wood preservation facility.[42] Weeks and Warriner[43] chronicled a gradient of concentrations of PAHs in bottom sediments of the Elizabeth River, with values gradually rising upriver. In the area of highest contamination, benzo(a)pyrene levels in the top 2 cm of sediment in a core sample approached 100 μg/g and increased to 200 μg/g at a depth of 30 cm, where the total PAH concentration equalled 13,000 μg/g.

Hogchokers, *Trinectes maculatus,* and several species of sciaenid fishes sampled in the river had integumental lesions and cataracts, which are manifestations of the toxic effects of the PAHs.[44] When exposed to sediments collected from the river, one sciaenid species, *Leiostomus xanthurus* (spot), developed lesions and cataracts in the laboratory; others died.[43,45] Spot, a catadromous species, occupies Chesapeake Bay and its subestuaries between April and December.[46] This benthic feeder can be contaminated with xenobiotics both in the water column and sediments of the river. The hogchoker likewise is a bottom-feeding fish. Both spot and hogchokers collected from the Elizabeth River, in addition to exhibiting fin loss and dermal lesions, displayed hyperplasia of gill tissue and necrosis of the pancreatic tissue of the liver.[43] Fishes collected at heavily contaminated sites had the highest incidences of abnormalities (e.g., dermal lesions, cataracts, and fin erosion).[5]

Bender et al.[47] summarized the abnormalities observed in fishes at 11 sites along the Elizabeth River in relation to sediment PAH contamination levels. At the most contaminated sites, 30% of the toadfishes (*Opsanus tau*) and 11% of the hogchokers (*T. maculatus*) suffered fin erosion. Cataract incidence in weakfish (*Cynoscion regalis*), Atlantic croaker (*Micropogonias undulatus*), and spot (*L. xanthurus*) amounted to 21, 18, and 10%, respectively. The most common lesions, fin and skin erosion, and cataracts occurred with greatest frequency in individuals from heavily contaminated regions. In addition, selected finfish species sampled at the more contaminated locales exhibited lower biomasses and total number of individuals compared to other stations.

Bender et al.[47] also explored the use of shellfish to monitor PAH contamination in Chesapeake Bay. They demonstrated the feasibility of utilizing three bivalves — oysters (*Crassostrea virginica*), hard clams (*Mercenaria mercenaria*), and brackish water clams (*Rangia cuneata*) — in contaminated estuarine sediments. Analysis of residue levels in these mollusks indicated that they are good monitors of pollution inputs along salinity gradients from 25 to 0.5‰. All three species appear to be essential to monitoring PAHs over a broad salinity regime. For instance, *C. virginica* and *M. mercenaria* live in polyhaline and mesohaline waters of the estuary, whereas *R. cuneata* prefers upper-mesohaline and oligohaline regions. Hence, *R. cuneata*, together with either *C. virginica* or *M. mercenaria*, must be monitored to assess PAH contamination in sediments along an extensive salinity gradient of the estuary.

Huggett et al.[48] discussed the general distribution of PAHs in Chesapeake Bay sediments. A few distinct PAHs occur in higher concentrations than other organic compounds in the sediments. A general trend of declining concentrations of these compounds from north to south in the main stem of the bay is apparent (Table 2). This distribution results from the greater human population densities surrounding the northern perimeter of the estuary and the anthropogenic impacts associated with these population numbers. Huggett et al.[48] discerned relatively constant levels of PAHs at their sampling stations over the period 1979 to 1986.

PAH contamination of bottom sediments has been noted in other

TABLE 2
Concentrations of Selected PAH Compounds in Sediments from Chesapeake Bay[a]

Station and Year

Compound	LE5.5[b]					WE4.2					LE3.6				
	79S[c]	79F[d]	84	85[e]	86	79S	79F	84	85	86	79S	79F	84	85	86
Phenanthrene	11	47	22	100	25	5	8	26	32	28	10	24	28	29	27
Fluoranthene	29	52	51	410	54	26	16	54	58	60	16	59	63	56	51
Pyrene	34	46	40	380	52	21	18	49	67	57	12	58	64	55	48
Benzo(a)fluorene	13	25	16	130	23	7	4	13	13	21	3	13	24	15	18
Benzo(a)anthracene	18	30	21	140	19	12	9	28	17	23	5	30	29	16	20
Chrysene/triphenylene	37	47	35	170	31	18	16	39	34	37	7	39	44	29	34
Benzo(e)pyrene	2	1	23	93	16	2	11	25	17	24	5	2	27	17	23
Benzo(a)pyrene	22	18	23	130	19	18	12	26	19	33	4	35	33	19	31
Pyrene	26	8	42	36	9	21	22	44	34	38	11	39	46	21	42
Benzo(g,h,i)perylene	15	6	18	46	9	15	8	31	23	20	3	17	28	12	26

	LE2.3				CB5.1				CB4.3C				CB3.3C[f]			
	79S	79F	84	86	79S	79F	84	86	79S	79F	84	86	79S	79F	84	86
Phenanthrene	19	42	54	64	17	49	47	50	44	68	26	11	280	220	300	240
Fluoranthene	34	70	89	88	35	81	85	74	60	82	42	14	370	220	370	300
Pyrene	29	57	72	87	33	73	79	66	43	70	38	4	360	220	370	290
Benzo(a)fluorene	10	16	24	34	10	27	24	23	11	38	13	2	120	98	150	77
Benzo(a)anthracene	10	23	28	25	10	32	26	20	8	25	16	>1	100	92	120	94
Chrysene/triphenylene	14	35	51	48	15	46	51	39	12	41	96	13	150	140	210	150
Benzo(e)pyrene	9	13	31	27	8	3	35	24	5	2	19	3	3	89	150	99
Benzo(a)pyrene	8	13	39	37	7	33	36	27	1	31	20	3	64	100	150	110
Perylene	9	14	43	46	5	37	59	51	9	65	260	150	110	220	220	140
Benzo(g,h,i)perylene	7	5	27	21	5	18	35	19	4	21	21	10	38	56	96	79

a Concentrations in μg/kg.
b Southernmost sampling site.
c S = Summer.
d F = Fall.
e Northern Chesapeake Bay not sampled in 1985.
f Northernmost sampling site.

From Huggett, R. J., de Fur, P. O., and Bieri, R. H., *Mar. Pollut. Bull.*, 19, 454, 1988. With permission.

estuarine systems as well, such as Penobscot Bay in the Gulf of Maine,[49] Boston Harbor,[50] Oregon Bay,[13,20] Puget Sound,[35,51,52] Tamar estuary,[53] and the Rhone River estuary.[54] Johnson et al.[49] encountered 13 of 16 U.S. Environmental Protection Agency (EPA) priority PAHs in the sediments of Penobscot Bay. The concentration of total PAHs in the sediments ranged from 286 to 8794 ppb dry wt; although the composition of PAHs was uniform throughout the bay, a distinct gradient in the contaminant concentration existed seaward from the head, with highest PAH levels at the entrance to Belfast Harbor and the mouth of the Penobscot River to the north and lowest PAH levels at the seaward margin of the bay to the south. While PAH concentrations in sediments at the head of the estuary commonly exceeded 5900 ppb dry wt, those at the seaward margin were <1000 ppb dry wt. The authors hypothesized a pyrolytic origin for much of the PAH contamination in the bottom sediments. Freshwater runoff probably introduced a large fraction of the PAH compounds into the bay. These compounds were originally particulate combustion products transported atmospherically and deposited in watersheds draining into the system. In spite of its remote location from highly industrialized regions, Penobscot Bay has sediments with PAH levels comparable to those of these regions. Table 3 compares PAH measurements in surficial sediments from the northeastern U.S., Europe, and elsewhere.

Boston Harbor, a water body used extensively for shipping, sport fishing, and recreation, is one of the most heavily polluted harbors on the East Coast of the U.S. Municipal and industrial sewage has grossly tainted at least one third of the harbor, and much of the seafloor contains deposits of decaying organic matter and oil residues.[50] The concentration of total PAHs in bottom sediments of the harbor amounts to 120 mg/kg dry wt, which decreases to 160 μg/kg dry wt at a distance of 64 km seaward.[20] Similar to the progressive diminution of PAH concentrations in seafloor sediments when moving away from Boston Harbor, Windsor and Hites[55] documented a decline in PAH values in seafloor sediments with increasing distance from the Gulf of Maine. They ascertained PAH concentrations ranging from 200 to 870 μg/kg in the Gulf of Maine and only 18 to 160 μg/kg in deep ocean sediments. PAH concentrations invariably decrease in sediments removed from urbanized areas.[20] LaFlamme and Hites,[27]

TABLE 3
Comparison of Total PAH Concentrations in Marine and Freshwater (FW) Surficial Sediments

Location	Total PAH (ppb, wet weight)	No. of stations	Depth (m)	References[49]
North America				
Penobscot Bay	286–8,974[a]	49	9.2–126.3	Johnson et al., 1985
Casco Bay	215–14,425	32	2–43	Larsen et al., 1983
Gulf of Maine	543[a]	1	—	Laflamme and Hites, 1978
Murray Basin	540	1	282	Windsor and Hites, 1979
Jordan Basin	500	1	265	Windsor and Hites, 1979
Wilkinson Basin	540–870	1	215	Windsor and Hites, 1979
Franklin Basin	200	1	225	Windsor and Hites, 1979
North Atlantic				
Continental rise	160	1	4,150	Windsor and Hites, 1979
Continental slope	120	1	1,830	Windsor and Hites, 1979
Abyssal plain	18–97	2	5,250, 5,465	Windsor and Hites, 1979
Abyssal plain	55[a]	1	—	Laflamme and Hites, 1978
Charles River, MA	87,000[a]	1	—	Laflamme and Hites, 1978
	120,000	1	—	Windsor and Hites, 1979
Massachusetts Bay	160–3,400	3	90, 130, 155	Windsor and Hites, 1979
Boston Harbor, MA	8,500	1	6	Windsor and Hites, 1979

TABLE 3 (CONTINUED)
Comparison of Total PAH Concentrations in Marine and Freshwater (FW) Surficial Sediments

Location	Total PAH (ppb, wet weight)	No. of stations	Depth (m)	References[49]
Buzzards Bay, MA	800	1	17	Hites et al., 1977
	4,000–5,000	3	—	Youngblood and Blumer, 1975
	803[a]	1	17	Laflamme and Hites, 1978
Falmouth Marsh, MA	800	1	Intertidal	Youngblood and Blumer, 1975
New Bedford Harbor, MA	63,000	1	—	Youngblood and Blumer, 1975
Pettaquamscutt River, RI	10,000	1	—	Hites et al., 1980b
New York Bight	5,830[a]	1	28	Laflamme and Hites, 1978
Pennsylvania Creek (FW)	100	4	0.3	Herbes, 1981
Lake Erie (FW)	530–3,750	7	—	Eadie et al., 1982
Adirondack Lakes (FW)	4,070–12,807	2	—	Heit et al., 1981
Alaska	5–113[a]	2	Intertidal	Laflamme and Hites, 1978
Mono Lake, California (FW)	157–399[a]	2	5–10	Laflamme and Hites, 1978
Europe				
Tamar Estuary (FW)	4,900	8	—	Readman et al., 1982
Southampton Water Estuary	91,000–1,791,000	19	—	Knap et al., 1982

			Intertidal	
Severn Estuary drainage system	1,600–25,700[a]	9		John et al., 1979
Mediterranean	198–372	2	6	Mille et al., 1981
Côte Bleue	1,232–232,000	3	3–10	Mille et al., 1982
Les Embiez	13,000–15,000	2	3–10	Mille et al., 1982
Monaco	5,200–12,100	2	3–10	Mille et al., 1982
Baltic Sea	258[a]	1	164	Poutanen et al., 1981
South Baltic Sea	50–2,550[a]	7	—	Law and Andrulewicz, 1983
Gulf of Finland	437[a]	1	60	Poutanen et al., 1981
Western Norway	284–99,452[a]	6	—	Bjorseth et al., 1979
Neckar, Rhine, and Danube Rivers (FW)	600–44,560[a]	73	—	Hagenmaier and Kaut, 1981
Other				
Walvis Bay, Africa	68[a]	1	—	Laflamme and Hites, 1978
Cariaco Trench	1756[a]	1	—	Laflamme and Hites, 1978
Amazon River system (FW)	ND–544	4		Laflamme and Hites, 1978
South Georgia Island	100	1	18	Platt and Mackie, 1979

[a] ppb, dry weight.

From Johnson, A. C., Larsen, P. F., Gadbois, D. F., and Humason, A. W., *Mar. Environ. Res.*, 15, 1, 1985. With permission.

focusing on the global distribution of PAHs, conveyed that PAH values in most regions rise near urban centers due largely to the input of combustion products.

Barrick and Prahl[52] investigated the sources and distributions of PAHs in Puget Sound by taking 96 sediment samples from 24 sites and analyzing each PAH fraction by means of gas chromatography (GC). Results of analyses on 24 ^{210}Pb-dated cores from greater Puget Sound revealed significant anthropogenic sources of the contaminants, with highest PAH levels in samples collected within a few kilometers of riverine systems draining coal-bearing strata, urban areas in central Puget Sound, and industrial facilities in northern Puget Sound. Air and water transport do not seem to widely disperse PAH compounds through the sound, and a substantial fraction of the compounds are removed to sediments in close proximity to major sources. This is not to deemphasize the role of runoff or atmospheric fallout which represent major routes of entry of combustion-derived PAHs into the sound. Conspicuous features of sediment cores extracted from the seafloor near urban centers are subsurface maxima in combustion-derived PAH levels; [210]P-dating of the cores indicates a mid-1950s period of deposition of these subsurface maxima.

The concentration of total combustion-derived PAHs — defined by Barrick and Prahl[52] as the sum of the concentration of nine individual compounds, notably fluoranthene, pyrene, benzo(a)anthracene, chrysene, benzofluoroanthenes, benzo(a)pyrene, benzo(e)pyrene, indeno(c,d)pyrene, and benzo(ghi)perylene — in surface sediments (0 to 2 cm) ranged from 8 to 73 µg/g OC (16 to 2400 ng/g) dry sediment. Values of total combustion PAHs vary over one order of magnitude in midchannel surface sediments of the sound, with highest figures registered in proximity to urban centers (Figure 2). The median concentration of total combustion-derived PAHs in surface sediments amounts to approximately 30 µg/g OC. Since 1870, total combustion-derived PAH concentrations have ranged from 10 to 50 µg/g at the stations surveyed. In general, stations distant from urban centers have highest combustion-derived PAH concentrations in surface sediments, and the values gradually decline with increasing depth in the sample cores, reflecting lower anthropogenic loading in the past. This pattern differs from that of cores taken from nearby urban centers, as described above. Prior to major

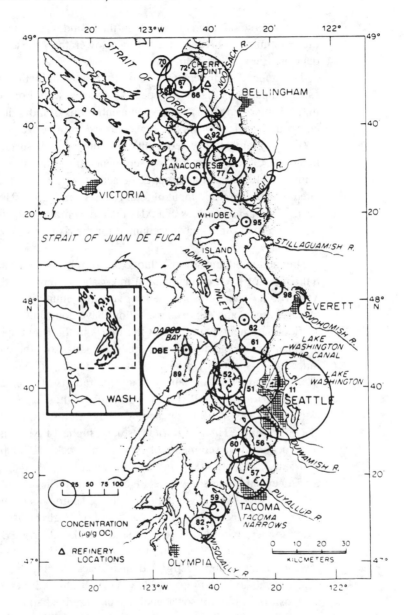

FIGURE 2. Total combustion-derived PAH at sampling stations (numbers) in Puget Sound. The radius of the circle about each station is proportional to the total concentration (µg/g OC) of the PAHs. (From Barrick, R. C. and Prahl, F. G., *Est. Coastal Shelf Sci.*, 25, 175, 1987. With permission.)

anthropogenic additions of PAHs in the sound, concentrations of the contaminants averaged 2.5 times lower than those recently deposited in surface sediments.

The general sediment profile of PAH contamination in U.S. estuaries parallels that of European systems as well. For example, in the Tamar (England) estuary, concentrations of individual PAH compounds reflect an exponential decrease from between 100 and 1000 ng (grams of dry sediment) in contemporary surface sediments to <30 ng (grams of dry sediment) in sediments deposited prior to 1940. Greater anthropogenic activity since 1940 has translated into an increased input of total PAHs from 0.23 to 21 mg/m^2/ year.[53] Pyrogenic sources of PAHs related to motor vehicle emissions and road runoff have accounted for the rising PAH levels in the estuary since 1940.

Dredged-spoil and sewage sludge disposal have been a substantial source of PAH contamination in coastal systems. In the New York Bight region, oil spills, sewer overflows, municipal wastewater discharges, and atmospheric fallout have contributed PAHs to seafloor sediments, but most of the PAHs in the Bight apex have originated from the dumping of dredged spoils from the outer and inner harbors which contain pollutant products of man's activities.[12] O'Connor et al.[12] tabulated concentrations of selected PAH compounds and total PAHs in sediments of the Hudson-Raritan estuary and New York Bight, and in sewage sludge from the region (Table 4). In the Hudson-Raritan estuary, highest PAH concentrations occurred in sediments of Newtown Creek (182,000 ng/g), Gowanus Canal (16,400 ng/g), and the lower Bay (9900 ng/g). The Christiaensen Basin showed relatively high PAH levels (6000 ng/ g) in the New York Bight region, with the sediments of the outer Bight having the lowest values (22 ng/g). High measurements likewise were recorded in sewage sludge (20,400 ng/g). Boehm[56] uncovered much higher concentrations of PAH compounds in sewage sludge (total 47 µg/g) than in background or nonwaste deposits (50 to 500 ng/g) in New York Bight sediments. PAH composition differs in sewage sludge and dredged-spoil deposits; the sewage sludge is rich in petroleum PAHs, and the dredged material, pyrogenic PAHs.[57] Hence, the sewage sludge incorporated greater quantities of naphthalenes, fluorenes, and dibenzothiophenes ascribable to petroleum sources. In contrast, the dredged material contained greater amounts of fluoranthenes, benzan-

TABLE 4
Selected PAH and Total PAH Concentrations in Sediments from the Hudson-Raritan Estuary and the New York Bight and in Sewage Sludge[a]

Material and location	Naph-thalene	Phenan-threne	Anthra-cene	Benzo(a)-anthracene	Total PAH
Sediment					
Hudson-Raritan estuary					
15 km north of the Battery	60	120	60	330	2,000
Pierhead Channel	200	300	200	500	3,200
Gowanus Canal	100	1,000	500	3,000	16,400
Newtown Creek	120,000	14,600	9,600	5,600	182,000
Lower Bay	100	600	300	2,000	9,900
New York Bight Region					
Christiaensen Basin	800	500	300	1,000	6,000
Sewage sludge dumpsite	80	70	40	200	1,100
Outer Bight	0.6	3	N.D.[b]	3	22
Sewage sludge	2,200	4,400	1,100	1,000	20,400

[a] Concentrations in ng/g, dry wt.
[b] Not detected.

From O'Connor, J. M., Klotz, J. B., and Kneip, T. J., in *Ecological Stress and the New York Bight: Science and Management,* Mayer, G. F., Ed., Estuarine Research Federation, Columbia, SC, 1982, 631. With permission.

thracenes, chrysenes, and benzopyrenes originating primarily from high-temperature combustion of fossil fuels.

PAH-bearing particulates enter the overlying water via the roiling and resuspension of bottom sediments during storms and other disturbances of the seafloor. Particle reworking and sediment irrigation by bioturbating macrofauna may profoundly affect sediment biogeochemistry and remobilization of pollutants in these shallow coastal environments.[58] Once placed in suspension, these particulates can be transported significant distances by tidal currents. For example, during flood tide, resuspended particulates rich in high molecular weight PAHs have been detected in transit from the inner bight apex to the mouth of the Hudson-Raritan estuary. Resuspended particulates high in PAHs from dredged-material deposits have also been found in mid- and bottom water in the New York Bight apex.[57] Information is deficient, however, on the benthos-water column and estuary-shelf transport of the resuspended PAH contaminants.

PAHs IN BIOTA
Accumulation, Depuration, and Toxicity

Recent monitoring programs have confirmed the widespread occurrence of PAHs in estuarine and coastal marine organisms. For some of these organisms (e.g., bivalve mollusks) which cannot properly metabolize and excrete PAH compounds because of inefficient or missing MFO systems,[59,60] high PAH levels tend to accumulate in their tissues. Thus, the highest total PAH concentrations recorded by O'Connor et al.[12] in biota of the New York Bight region were those of the blue mussel, *Mytilus edulis* (Table 5). Mussels collected at Sandy Hook (New Jersey) had total PAH levels equalling 250 ng/g compared to levels of 120 ng/g in mussels sampled at Shark River (New Jersey). Total PAH concentrations in lobsters taken from Raritan Bay were also elevated in comparison to other biota, ranging from 25 to 77 ng/g. Finfishes exhibited the lowest PAH concentrations due to their ability to metabolize and clear the compounds quickly. Atlantic mackerel (*Scomber scombrus*) from the New York Bight apex had total PAH concentrations of 10 ng/g, which exceeded the values obtained on winter flounder (*Pseudopleuronectes americanus*) from the Christianesen Basin (8 ng/g) and Raritan Bay (5 ng/g). Similar figures were observed in striped bass (*Morone saxatilis*) from Montauk Point (New York) (19 ng/g) and the Hudson River (8 ng/g). Despite the high anthropogenic impact in the New York Bight region, PAH contamination in many populations remains relatively low.[12]

Eisler[1] compiled statistics on PAH concentrations in marine fin- and shellfish (Table 6). As expected, highest PAH concentrations generally corresponded to shellfish collected in areas adjacent to industrialized bayfronts. Mussels with the greatest PAH levels also suffered from cellular proliferative disorders that resembled neo-

TABLE 5
Selected PAH and Total PAH Concentrations in Finfish and Shellfish from the Hudson River and New York Bight[a]

Species (location)	Naph-thalene	Phen-anthrene	Anthra-cene	Bi-phenyl	Total PAH
Atlantic mackerel *(Scomber scombrus)* (New York Bight apex)	N.D.	10	N.D.	N.D.	10
Winter flounder *(Pseudopleuronectes americanus)* (Christiaensen Basin)	2	N.D.	N.D.	6	8
Winter flounder *(P. americanus)* (Raritan Bay)	2	1	N.D.	N.D.	5
Striped bass *(Morone saxatilis)* (Montauk Point)	7	N.D.	N.D.	N.D.	19
Striped bass *(M. saxatilis)* (Hudson River)	4	N.D.	N.D.	4	8
Lobster *(Homarus americanus)* (New York Bight)	7	N.D.	N.D.	N.D.	7
Lobster *(H. americanus)* (Raritan Bay)	5	5	N.D.	N.D.	25
Lobster *(H. americanus)* (Raritan Bay)	7	N.D.	N.D.	N.D.	77
Blue mussel *(Mytilus edulis)* (Sandy Hook)	6	6	N.D.	4	250
Blue mussel *(M. edulis)* (Shark River)	20	10	1	40	120

[a] Concentrations in ng/g, dry wt.

From O'Connor, J. M., Klotz, J. B., and Kneip, T. J., in *Ecological Stress and the New York Bight: Science and Management,* Mayer, G. F., Ed., Estuarine Research Federation, Columbia, SC, 1982, 631. With permission.

plastic conditions in vertebrates.[61] Shellfish sampled from more remote areas had the lowest abundances of PAHs.

Seasonal differences in the quantity of PAH contamination were evident among the shellfish. The American oyster (*Crassostrea virginica*), for instance, accumulated larger quantities of PAH compounds in tissues during the cooler months when they stored lipids and glycogen in preparation for spawning.[62] The softshell clam (*Mya arenaria*), in turn, accumulated larger amounts of PAHs in the spring and summer than in the fall and winter.[63] Once contaminated, clams experienced substantial depuration of unsubstituted 3- and 4-ring PAHs subsequent to their removal to clean seawater for 24 h. PAHs that tended to resist depuration over the 24-h period included the higher molecular weight 5-, 6-, and 7-ring compounds.[61]

American lobsters (*Homarus americanus*) sampled at the site of a major oil spill in Nova Scotia in 1979 contained higher levels of PAHs than lobsters from coastal control sites.[60] Elevated tissue levels of PAHs were also documented in lobsters gathered in vicinity of a coking facility in Nova Scotia in 1980, where concentrations of chrysene, fluorene, phenanthrene, pyrene, benz(a)anthracene, benzo(a)pyrene, benzo(e)pyrene, and benzo(b)fluoranthene in bottom sediments exceeded 100 mg/kg dry wt.[64] Sirota and Uthe[60] discovered higher PAH residues in larger lobsters (mean weight = 3.6 kg) collected offshore than in smaller lobsters (mean weight = 0.6 kg) obtained inshore. Larger or older lobsters may have a greater capability of metabolizing and excreting PAHs than smaller or younger lobsters.

Finfish rapidly metabolize PAHs, as stated above, resulting in generally lower accumulations of these compounds in their tissues than those in shellfish.[59] They do not appear to accumulate higher molecular weight PAHs,[65] which are typically more carcinogenic than the lower molecular weight varieties. Their skin probably affords some degree of protection.

The bioconcentration and toxicity of PAHs are quite variable among estuarine organisms. Whereas most of these organisms accumulate PAHs from estuarine waters or sediments, uptake of the contaminants is species specific.[1] Those groups that cannot metabolize PAHs, such as algae and mollusks, tend to accumulate greater quantities of the compounds. In addition to this species specific effect, bioconcentration factors generally rise with increasing

TABLE 6
PAH Concentrations in Selected Shellfish and Finfish from Estuarine and Coastal Marine Waters[a]

Taxonomic group, compound, and other variables	Concentration	References[1]
Shellfish		
Rock crab, *Cancer irroratus*		
Edible portions, 1980		
New York Bight		Humason and Gadbois, 1982
Total PAHs	1600 FW	
BaP	1 FW	
Long Island Sound		
Total PAHs	1290 FW	
BaP	ND	
American oyster		
Crassostrea virginica, soft parts,		
South Carolina, 1983,		
residential resorts		Marcus and Stokes, 1985
Total PAHs		
Spring months		
Palmetto Bay	520 FW	
Outdoor Resorts	247 FW	
Fripp Island	55 FW	

TABLE 6 (CONTINUED)
PAH Concentrations in Selected Shellfish and Finfish from Estuarine and Coastal Marine Waters[a]

Taxonomic group, compound, and other variables	Concentration	References[1]
Summer months		
Palmetto Bay	269 FW	
Outdoor Resorts	134 FW	
Fripp Island	21 FW	
American lobster, *Homarus americanus*		
Edible portions, 1980		
New York Bight		
Total PAHs	367 FW	Humason and Gadbois, 1982
BaP	15 FW	
Long Island Sound		
Total PAHs	328 FW	
BaP	15 FW	
Softshell clam, *Mya arenaria*		
Coos Bay, Oregon, 1978-1979		
Soft parts		
Contaminated site		Mix, 1982
Total PAHs	555 FW	
Phenanthrene = PHEN	155 FW	
FL	111 FW	
Pyrene = PYR	62 FW	

BaP	55 FW	
Benz(a)anthracene = BaA	42 FW	
Chrysene = CHRY	27 FW	
Benzo(b)fluoranthene = BbFL	12 FW	
Others	<10 FW	
Uncontaminated site		
Total PAHs	76 FW	
PHEN	12 FW	
FL	10 FW	
Others	<10 FW	Mix and Schaffer, 1983
Bay mussel, *Mytilus edulis*		
Oregon, 1979-1980		
Soft parts, total PAHs		
Near industrialized area	106–986 FW	
Remote site	27–274 FW	
Sea Scallop, *Placopecten magellanicus*		
Baltimore Canyon, East Coast U.S.		
Muscle		
BaA	1 FW	
BaP	<1 FW	
PYR	4 FW	Brown and Pancirov, 1979
New York Bight, 1980		
Edible portions		
Total PAHs	127 FW	
BaP	3 FW	Humason and Gadbois, 1982
Clam, *Tridacna maxima*		
Australia, 1980-1982, Great Barrier Reef		
Soft parts, total PAHs		
Pristine areas	<0.07 FW	
Powerboat areas	Up to 5 FW	Smith et al., 1984

TABLE 6 (CONTINUED)
PAH Concentrations in Selected Shellfish and Finfish from Estuarine and Coastal Marine Waters[a]

Taxonomic group, compound, and other variables	Concentration	References[1]
Mussel, *Mytilus* sp.		
Greenland		
Shell	60 FW	Harrison et al., 1975
Soft parts	18 FW	
Italy		
Shell	11 FW	
Soft parts	130–540 FW	
Bivalve mollusks, 5 spp.		
Edible portions	6 (max. 36) FW	Stegeman, 1981
Decapod crustaceans, 4 spp.		
Edible portions	2 (max. 8) FW	
Softshell clam, *Mya arenaria*, soft parts		
Coos Bay, Oregon		
1976-1978		
Near industrialized areas	6–20 FW	Mix and Schaffer, 1983
Remote areas	1–2 FW	
1978-1979		
Near industrialized areas	9 FW	
Remote areas	4 FW	
Finfish		
Fish, muscle		
Baltimore Canyon,		
East Coast, U.S., 5 spp.		
BaA	Max. 0.3 FW	Brown and Pancirov, 1979
BaP	Max. <5 FW	
PYR	Max. <5 FW	

Smoked		
FL	3 FW	
PYR	2 FW	
Nonsmoked		
FL	Max. 1.8 FW	
PYR	Max. 1.4 FW	EPA 1980
Winter flounder, *Pseudopleuronectes americanus*		
Edible portions, 1980		
New York Bight		
Total PAHs	315 FW	
BaP	21 FW	
Long Island Sound		
Total PAHs	103 FW	
BaP	ND	Humason and Gadbois, 1982
Windowpane, *Scophthalmus aquosus*		
Edible portions, 1980		
New York Bight		
Total PAHs	536 FW	
BaP	4 FW	
Long Island Sound		
Total PAHs	86 FW	
BaP	ND[b]	
Red hake, *Urophycus chuss*		
Edible portions, 1980		
New York Bight		
Total PAHs	412 FW	
BaP	22 FW	

TABLE 6 (CONTINUED)
PAH Concentrations in Selected Shellfish and Finfish from Estuarine and Coastal Marine Waters[a]

Taxonomic group, compound, and other variables	Concentration	References[1]
Long Island Sound		
Total PAHs	124 FW	
BaP	5 FW	
BaP		
Fish		
Marine, edible portions	Max. 3 FW	
9 spp.	15 FW	Stegeman, 1981
Greenland	65 FW[c]	Harrison et al., 1975
Italy	5–8 DW[d]	
Steak, charcoal broiled		
Ribs, barbecued	11 DW	Barnett, 1976

[a] Concentrations in μg/kg.
[b] Not detected.
[c] Fresh weight.
[d] Dry weight.

Modified from Eisler, R., Polycyclic Aromatic Hazards to Fish, Wildlife, and Invertebrates: A Synoptic Review, Biol. Rep. 85(1.11), U. S. Fish and Wildlife Service, Washington, D.C., 1987.

molecular weight of the PAH, with higher concentrations of organic matter, and with greater lipid content of the organism; moreover, various endogenous and exogenous factors influence bioconcentration factors.[9,22,66,67] The uptake of naphthalene, phenanthrene, fluorene, and their methylated derivatives by oysters (*Crassostrea virginica*) and clams (*Rangia cuneata*) in the laboratory, for example, provided evidence of increased accumulation with increasing methylation and PAH molecular weight.[68] *Crassostrea virginica* exposed to benzo(a)pyrene over a 14-d period had a bioconcentration factor of 242.[23] Neff[3] and the U.S. EPA[23] reported bioconcentration factors for benzo(a)pyrene in *R. cuneata* ranging from 9 to 236 over a 24-h exposure period. The bioconcentration factors for chrysene, naphthalene, and phenanthrene in *R. cuneata* over a 24-h exposure period equalled 8, 6, and 32, respectively.[3]

Most marine bivalve mollusks depurate or biotransform 50% of their accumulated PAHs during a period of 2 to 16 d.[1] Some significant differences exist between certain species; thus, Jackim and Lake[9] disclosed that little or no depuration takes place in the hard clam (*Mercenaria mercenaria*), while the oyster (*C. virginica*) releases as much as 90% of its accumulated PAHs in 14 d, with the remainder being eliminated slowly over a protracted time interval. The oyster may retain traces of PAH compounds indefinitely. Neff[67] recounted the following percent losses of PAHs in *C. virginica* 7 d after exposure to the contaminants: benzo(a)pyrene (0%), benz(a)anthracene (32%), fluoranthene (66%), anthracene (79%), dimethylnaphthalene (90%), naphthalene (97%), and methylnaphthalene (98%). The blue mussel (*Mytilus edulis*) accumulated high concentrations of PAHs (approaching 1000 times control levels) from heavily contaminated sediments during a 28-d exposure period, but it rapidly depurated them during a 5-week postexposure period.[69]

Shellfish may sequester PAHs to a point at which they cannot be safely consumed by man. Miller et al.,[70] examining PAH uptake by the pink shrimp (*Penaeus duorarum*), specified that the organism concentrated chrysene to levels of 9 µg/kg fresh weight in the abdomen and 48 µg/kg fresh weight in the cephalothorax when exposed to 1.0 µg chrysene per liter for 28 d. Even though they were transferred to unpolluted seawater for 28 more d, the shrimp still contained PAH levels deemed to be potentially unsafe for human consumption.

Table 7 lists toxicities of selected PAHs to aquatic organisms, some of which attain high concentrations in estuaries. Crustaceans are the most susceptible group to PAH toxicity. Finfishes, which metabolize PAH compounds more effectively than most other biotic groups, exhibit the lowest toxicity values. The PAH levels required to create acutely toxic conditions for estuarine organisms appear to be several orders of magnitude higher than the values encountered in natural waters. Whereas PAH concentrations in heavily tainted bottom sediments may approach levels acutely toxic to some estuarine organisms, their limited bioavailability probably makes them less toxic than PAHs in solution.[1,3]

NOAA Mussel Watch Project

Since 1986, the National Oceanic and Atmospheric Administration (NOAA) has utilized mussels and oysters as sentinel organisms to monitor chemical contaminants (i.e., polyaromatic hydrocarbons, polychlorinated biphenyls, chlorinated pesticides, organotin compounds, and 12 trace elements) in estuarine and coastal marine waters. The concentrations of more than 40 organic contaminants have been measured. In addition, normalizing parameters (i.e., lipid content, TOC, etc.) have been surveyed in bivalve mollusks and surface sediments. At a number of sampling sites, detailed histopathological observations have been performed on bivalve samples to document pathological abnormalities such as neoplasia.[71,72] The principal goals of the Mussel Watch Project, a component of the NOAA National Status and Trends Program, are to monitor spatial distributions and temporal trends of the contaminant concentrations in these coastal environments and to determine biological responses to the contamination.[71]

Mussel and oyster samples were collected at a total of 177 estuarine and coastal marine sites in the U.S. during the first 3 years of the Mussel Watch Project, with samples taken at 132 of these sites in all years. Sampling stations in estuaries and embayments were approximately 20 km apart, compared to about 100 km along open coastlines. Appendix A of the NOAA technical report on the first 3 years of the Mussel Watch Project supplies the listings of all

TABLE 7
Toxicities of Selected PAHs to Aquatic Organisms

PAH compound, organism, and other variables	Concentration in medium[a]	Effect[b]	References[1]
Benzo(a)pyrene			
Sandworm, *Neanthes arenceodentata*	>1,000	LC-50 (96 h)	Neff, 1979
Chrysene			
Sandworm	>1,000	LC-50 (96 h)	
Dibenz(a,h)anthracene			
Sandworm	>1,000	LC-50 (96 h)	Neff, 1979
Fluoranthene			
Sandworm	500	LC-50 (96 h)	
Fluorene			
Grass shrimp, *Palaemonetes pugio*	320	LC-50 (96 h)	
Amphipod, *Gammarus pseudoliminaeus*	600	LC-50 (96 h)	
Sandworm	1,000	LC-50 (96 h)	Neff, 1979
Sheepshead minnow, *Cyprinodon variegatus*	1,680	LC-50 (96 h)	
Naphthalene			
Copepod, *Eurytemora affinis*	50	LC-30 (10 d)	Neff, 1979
Pink salmon, *Oncorhynchus gorbuscha*, fry	920	LC-50 (24 h)	
Dungeness crab, *Cancer magister*	2,000	LC-50 (96 h)	Neff, 1985
Grass shrimp	2,400	LC-50 (96 h)	Neff, 1979
Sheepshead minnow	2,400	LC-50 (24 h)	
Brown shrimp, *Penaeus aztecus*	2,500	LC-50 (24 h)	
Amphipod, *Elasmopus pectenicrus*	2,680	LC-50 (96 h)	
Coho salmon, *Oncorhyncus kisutch*, fry	3,200	LC-50 (96 h)	Neff, 1985
Sandworm	3,800	LC-50 (96 h)	Neff, 1979
Mosquitofish, *Gambusia affinis*	150,000	LC-50 (96 h)	
1-Methylnaphthalene			
Dungeness crab, *Cancer magister*	1,900	LC-50 (96 h)	
Sheepshead minnow	3,400	LC-50 (24 h)	
2-Methylnaphthalene			
Grass shrimp	1,100	LC-50 (96 h)	Neff, 1985

TABLE 7 (CONTINUED)
Toxicities of Selected PAHs to Aquatic Organisms

PAH compound, organism, and other variables	Concentration in medium[a]	Effect[b]	References[1]
Dungeness crab	1,300	LC-50 (96 h)	
Sheepshead minnow	2,000	LC-50 (24 h)	Neff, 1979
Trimethylnaphthalenes			
Copepod, *Eurytemora affinis*	320	LC-50 (24 h)	
Sandworm	2,000	LC-50 (96 h)	
Phenanthrene			
Grass shrimp	370	LC-50 (24 h)	
Sandworm	600	LC-50 (96 h)	EPA, 1980
1-Methylphenanthrene			
Sandworm	300	LC-50 (96 h)	

[a] Concentrations in μg/l.
[b] m = months, d = days, h = hours.

Modified from Eisler, R., Polycyclic Aromatic Hazards to Fish, Wildlife, and Invertebrates: A Synoptic Review, Biol. Rep. 85(1.11), U.S. Fish and Wildlife Service, Washington, D.C., 1987.

of the sampling sites as well as their sampling dates and location maps.[71]

In the NOAA report,[71] PAH compound concentrations have been aggregated into two classes: low molecular weight PAHs (LMWpah) and high molecular weight PAHs (HMWpah). Compounds comprising the LMWpah class include acenaphthene, anthracene, biphenyl, dimethylnaphthalene, fluorene, naphthalene, 1- and 2-methylnaphthalene, 1-methylphenanthrene, and phenanthrene. Those constituting the HMWpah class are benz(a)anthracene, benzo(a)pyrene, benzo(e)pyrene, chrysene, dibenzanthracene, fluoranthene, perylene, and pyrene. Figure 3 illustrates the mean percentages that each compound contributes to the aggregate total. As depicted in Figure 3, phenanthrene is responsible for the largest fraction of PAHs in the aggregate class LMWpah, and fluoranthene, the largest fraction of PAHs in the aggregate class HMWpah. The LMWpah aggregate class consists of 2-ring (i.e., acenaphthene, biphenyl, dimethylnapthalene, 1- and

2-methylnaphthalene, and naphthalene) and 3-ring (i.e., anthracene, fluorene, 1-methylphenanthrene, and phenanthrene) polyaromatic hydrocarbons. The HMWpah aggregate class embodies 4-ring (i.e., benz(a)anthracene, fluoranthene, and pyrene) and 5-ring (i.e., chrysene, benzo(a)pyrene, benzo(e)pyrene, perylene, and dibenz(a,h)anthracene) polyaromatic hydrocarbons.

Temporal trends in PAH contaminant concentrations in mollusks were followed at several sites during the 3-year survey period. Sites where LMWpah concentrations in mollusks decreased between 1986 and 1988 included the Hudson-Raritan estuary (New York), New York Bight (New Jersey), Tampa Bay (Florida), San Simeon Point (California), and Point St. George (Oregon). Those locales where LMWpah measurements increased over this interval were Buzzards Bay (Massachusetts), Chesapeake Bay, Mississippi Sound (Mississippi), Matagorda Bay (Texas), and Corpus Christi (Texas). HMWpah values in mollusk tissues declined in the Hudson-Raritan estuary, Tampa Bay, St. Andrew Bay (Florida), and Sinclair Inlet (Washington). Increasing HMWpah levels were obtained in samples collected in Long Island Sound (Connecticut), Chesapeake Bay, and Choctawhat Bay (Florida). The HMWpah compounds originate principally from fossil fuel combustion; the LMWpah compounds generally derive from relatively fresh, unburned petroleum. Efforts to lower PAH contamination in U.S. estuarine and coastal marine waters in future years cannot rely simply on the banning of their production and use, as in the case of certain chlorinated compounds (e.g., DDT and PCBs). Both point- and nonpoint control measures must continue to be implemented in order to minimize their input to these aquatic environments.

PAHs IN WATER

As is evident from the aforementioned discussion, PAHs are not evenly distributed among sediments, biota, and waters of estuaries and the coastal ocean. Unequivocally, these compounds accumulate near point sources, and their concentrations drop essentially logarithmically with increasing distance from the sources. Hence, upon entering the aquatic environment, most PAHs attain peak levels in riverine, estuarine, and coastal marine waters.

Because PAHs strongly adsorb to suspended inorganic and organic particles which settle to the bottom, as iterated above, PAII

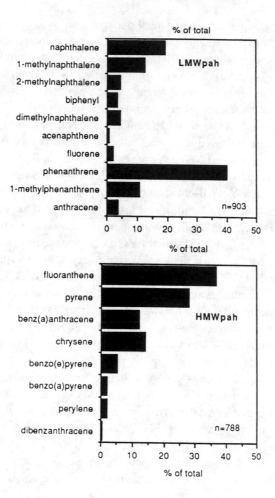

FIGURE 3. Mean percentage contributions of individual PAH compounds to low molecular weight (LMWpah) and high molecular weight (HMWpah) contaminant classes. (From National Oceanic and Atmospheric Administration, A Summary of Data on Tissue Contamination from the First Three Years (1986-1988) of the Mussel Watch Project, NOAA Tech. Mem. NOS OMA 49, National Oceanic and Atmospheric Administration, Rockville, MD, 1989.)

concentrations in seafloor sediments usually exceed those in the water column by a factor of 1000 or more.[4] PAHs are hydrophobic and have low aqueous solubility, accounting for their rapid adsorption onto solid surfaces. Rates of adsorption vary depending on the composition of the particle. Clay particles, for example, adsorb PAH contaminants less readily than organic particles.[73,74] Subsequent to their deposition in seafloor sediments, a small amount of the PAHs may reenter the water column via leaching or biological activity.

Rainwater or seawater unaffected by industrialization or other anthropogenic activity usually contains <0.1 µg/l total PAHs. Somewhat higher values (1 to 5 µg/l) are observed in rivers located in heavily urbanized areas. Waters receiving liquid sewage or sewage sludge understandably exhibit higher PAH levels. Although liquid domestic sewage contains <1 µg/l total PAHs,[4] industrial sewage and sewage sludge can have much more, ranging from 5 to 15 µg/l and 1 to 30 mg/kg, respectively.[75-77] Added to this are PAH compounds derived from oil spillage and natural oil seepage; more than 6×10^6 metric tons of oil comprised of up to several percent PAHs enter the marine environment each year by these pathways. Not to be forgotten are nonpoint anthropogenic sources of PAH compounds, notably runoff from roadways containing compounds leached from asphalt, carbon black on road surfaces, and used crankcase oil. Surface runoff, together with the deposition of airborne PAHs, is responsible for a substantial fraction of high molecular weight PAHs in estuarine and marine environments. Oil spillage supplies the largest fraction of total PAHs to these environments.[4] Approximately 2.3×10^5 metric tons of total PAHs enter aquatic environments annually, primarily from the sources mentioned above.[3]

Because of its carcinogenicity in mammals,[78-80] benzo(a)pyrene has been the target of much research on environmental PAH contamination of estuarine and marine waters. About 70 metric tons of benzo(a)pyrene enter the aquatic environment each year, leading to its broad geographic distribution.[3] Combustion effluents generate a large quantity of airborne benzo(a)pyrene that ultimately settle to water surfaces via fallout or rainout.[80]

PAH TRANSFORMATION

Several processes act on PAHs to transform and degrade them, thereby affecting their environmental concentrations. According to Neff,[4] the most important of these processes are photooxidation, chemical oxidation, and biological transformation by bacteria, fungi, and animals. The following information draws heavily from Neff's work.

PHOTODEGRADATION

Molecular oxygen, sunlight, or both are required by most natural methods of oxidative transformation of PAH compounds. When incorporated into anoxic bottom sediments, therefore, PAH contaminants become highly stable and persistent. Photoinduced transformation of PAHs in seawater may take place by direct photolysis reactions,[78] as well as by photooxygenated reactions involving singlet oxygen, ozone, OH radicals, and other oxidizing agents. These chemical reactions transform PAHs in solution or the PAHs are sorbed to waterborne particles.[81]

CHEMICAL OXIDATION

Chlorination and ozonation are two methods that destroy aqueous PAHs. Sodium hypochlorite has proven to be effective in oxidizing most PAHs.[82,83] These compounds may be removed from water more efficiently by ozonation than by chlorination. Ozone reacts with aqueous PAHs to yield aromatic aldehydes, carboxylic acids, and guinones.[4,81,84] Chlorine- and ozone-mediated PAH degradation can be successfully applied in water treatment facilities, although the high volatilization and reaction rates of ozone with organics create difficult conditions for maintaining sufficient ozone concentrations for effecting total removal of the PAH compounds.[4,83,84]

BIOTIC TRANSFORMATION

Some bacteria and fungi utilize PAHs as a carbon source, oxidizing them to carbon dioxide and water. Bacterial oxidation of PAH compounds produces dihydrodiols,[85] and through a series of additional reactions, yields catechols and ultimately carbon dioxide and water. Because of the serious threat that oil spills pose to estuarine and marine habitats, numerous investigations have been conducted on bacterial degradation of petroleum hydrocarbons, including PAH compounds. In general, the higher the PAH molecular weight, the lower the rate of PAH degradation.[86]

Fungi metabolize PAHs in a manner similar to that observed in mammals. They employ an enzyme system known as the cytochrome P-450-dependent MFO system to metabolize PAHs. MFO activity has also been delineated in invertebrates, but it appears to be restricted principally to populations of annelids and arthropods.[3,87] Members of the phyla Mollusca and Echinodermata possess either no or low levels of MFO activity. Among fish populations, MFO activity is significant, with most of the activity occurring in the liver and somewhat lower activity in the gills and kidney.[88] Various endogenous and exogenous factors affect PAH concentrations, enzyme activity, and MFO components in aquatic fauna. Among endogenous factors influencing PAH metabolism are age, sex, nutritional status, and period of the molt cycle (in arthropods). Exogenous factors of importance include temperature, season, current, prior history of exposure to inducers or inhibitors of different components of the microsomal PAH-metabolizing system, and possibly salinity. Aquatic fauna exhibit both inter- and intraspecific differences in their response to these factors.[4]

PAHs IN ESTUARINE SYSTEMS

CASE STUDIES
Chesapeake Bay

Huggett et al.[48] reported on PAH samples collected in Chesapeake Bay between 1979 and 1986. As part of an assessment of animal-sediment bioaccumulation, Foster and Wright[88] measured the concentrations of unsubstituted PAH compounds (UPAH) in sediments, Baltic clams (*Macoma balthica*), and clam worms (*Nereis succinea*) in the bay. Most of the PAHs sampled in sediments of the estuary appeared to originate from anthropogenic sources, primarily combustion or high temperature pyrolysis of carbonaceous fuels and not from oil spills or releases. Naturally occurring PAHs, constituting <20% of the total PAH concentrations in bottom sediments, consisted principally of perylene produced from unknown precursors by geochemical processes, aromatized diterpenoids (notably retene) generated by early diagenesis of pine resins, and aromatized triterpenoids (e.g., trimethyltetrahydrochrysenes, tetramethyl-octahydrochrysenes, and picene derivatives).[48]

The concentrations of the most abundant PAHs, as discussed previously, generally diminish downestuary along a north-to-south gradient away from the densest human populations (Table 2). Between 1979 and 1986, the relative composition of the PAHs remained constant at the sampling stations. The concentrations of the PAHs during this time interval also remained relatively constant.[89]

UPAHs attain high levels in bottom sediments of the estuary, most conspicuously in areas near heavy industrial and urban activity.[88] The use of Baltic clams (*M. balthica*) and clam worms (*N. succinea*) has proven to be valuable in developing indices of exposure and bioavailability of UPAH in Chesapeake Bay water quality programs.[90] They are excellent sentinel organisms because of their nutritional requirement of sediment and their broad distribution in the estuary. Furthermore, recreationally and commercially harvested fish and shellfish utilize them as a source of food.

Sixteen UPAH compounds were monitored in the clams and worms at seven locations in the Maryland waters of Chesapeake Bay from the Baltimore Harbor area to the mouth of the Potomac River. These consisted of, in order of increasing molecular weight, naphthalene, acenaphthylene, acenaphthene, fluorene, phenanthrene/anthracene (unresolved chromatographic retention), fluoranthene, pyrene, benz(a)anthracene, chrysene, benzo(b)- and benz(a)anthracene, chrysene, benzo(b)- and benzo(k)fluoranthenes, benzo(a)pyrene, indeno(123-cd)pyrene, benzo(ghi)perylene, and dibenz(ah)anthracene. Analysis of biotic samples revealed UPAH at all bay stations, ranging in total concentrations (t-UPAH) from 0.551 μg/g fraction organic carbon (ppm-oc) to 178 ppm-oc. Lowest UPAH loadings occurred in sandy sediments low in organic carbon; highest loadings existed near Hawkins Point, at the entrance to Baltimore Harbor. At Hawkins Point, all 16 UPAHs were detected.

The total UPAH concentrations in both the clams and worms appeared to be principally a function of the pollutant concentrations in bottom sediments. The most abundant contaminants in the animals were naphthalene, phenanthrene/anthracene, fluoranthene, and pyrene. UPAH profiles of the clams paralleled those of the worms. With the exception of one sampling site (i.e., Hawkins Point), clams and worms collected throughout the estuary had no apparent differences in bioaccumulation of t-UPAH when body concentrations were normalized to total extractable lipids. *Nereis* sampled at Hawkins Point contained less than half of the concentration of t-UPAH observed in *Macoma*, possibly reflecting the bivalve's greater enzymatic activities to metabolize xenobiotics in polluted habitats.[87] Hence, *Nereis* at Hawkins Point may have been more successful at biotransforming UPAH to oxygenated products.

Animal-sediment bioaccumulation factors were estimated to determine the bioavailability of UPAH to *Macoma* and *Nereis*. These factors indicate the magnitude of UPAH exposure to the benthos while they inhabited the contaminated sediments. A significant determinant of animal-sediment bioaccumulation factors for these two macrofauna was the amount of organic carbon in the sediments. For example, in sediments classified as mud or sandy-mud with >0.5% organic carbon content, bioaccumulation factors for t-UPAH consistently ranged from 0.495 to 2.20 for *Macoma* and 0.407 to 1.77 for *Nereis*. In sandy sediments having <0.5% organic carbon, however, bioaccumulation factors for t-UPAH ranged from 8.06 to 106 for *Macoma* and from 4.13 to 98.1 for *Nereis*. These animal-sediment bioaccumulation factors (BAF) were calculated using the following relationship

$$BAF = \frac{\text{Animal UPAH ppm / fraction lipid}}{\text{Sediment UPAH ppm / fraction organic carbon}} \quad (1)$$

As is apparent from the values given above, the bioaccumulation factors of both *Macoma* and *Nereis* were much higher for coarse-grained sediments than for fine-grained sediments. Nonetheless, both serve as an index of exposure and bioavailability of UPAH in the various habitats of Chesapeake Bay.

Narragansett Bay

Lake et al.[39] studied PAH assemblages in the sediments of Narragansett Bay, Rhode Island, focusing on the origin, concentration, and distribution of the compounds by utilizing GC and GC-mass spectrometric (MS) techniques of analysis. Sediment samples were collected along a 20-km transect, oriented roughly north-south, commencing at a sewage outfall site near Providence, RI. Analysis of the samples revealed rapidly diminishing concentrations of PAHs with increasing distance from Providence. The city was deemed to be the likely source for much of the PAHs along the sampling transect; however, at least three possible origins exist for the PAHs in the estuary.

The predominant source of PAHs in the transect sediments is fossil fuel combustion material.[39] Urban runoff, along with sewage discharge and atmospheric deposition, accounts for the input of the combustion products into the estuary. A second source of PAH compounds, petroleum oils, enters the bay via sewage systems. PAHs derived from the petroleum peak in sediments of the upper bay transect and decrease along the lower, less polluted (southern) end of the transect. Sediments sampled in proximity to tarred piers and docks contain PAHs released from the coal tar used to coat the pilings and docks. This third source of PAHs is much less quantitatively significant than the combination of inputs of PAHs from petroleum and combustion processes.

ENVIRONMENTAL CYCLE OF PAH COMPOUNDS

The transport pattern of PAHs in the estuarine environment entails the input of PAH compounds from natural background and anthropogenic sources, as well as their bioaccumulation and biodegradation by biota, adsorption by inorganic and organic matter, and storage in bottom sediments. A fraction of these compounds eventually reaches man via food chain pathways. Background PAHs refer to the direct biosynthesis of the compounds by microbes (bacteria and fungi) and plants.[21] These PAH assemblages are compositionally simple and mostly immobile,[4] remaining in the organisms producing them in riverine, estuarine, and oceanic ecosystems.[21] By far, the largest quantity of PAHs found in estuaries derives from anthropogenic sources, primarily via high-temperature pyrolysis of organic materials.[4] A potentially large input of PAHs to estuaries results from atmospheric deposition onto the water surfaces, even though photooxidation decomposes some of the PAH compounds while still in the atmosphere. Another component is released to estuaries in solid or liquid by-products of the pyrolytic process. Domestic pyrosynthesis of PAHs that contribute to the estuarine pool of PAH compounds include the use of internal combustion engines, heating of homes with wood or fossil fuels, waste incineration, and cigarette smoking. Industrial pyrosynthesis of PAHs which may affect estuarine systems are petroleum refining, synfuel production from coal, coal coking, and the production of carbon blacks, creosote, coal tar, and related materials from fossil fuels.[4]

Estuaries receive additional PAH loading from domestic and industrial waste effluents either directly or from riverine influx. Oil waste comprises a portion of these effluents. Of greater magnitude in some areas is the input of oil from oil spills or its entry from natural oil seeps.

Nonpoint source runoff also can transport high PAH loads to estuaries, particularly in regions having heavy atmospheric fallout on nearby terrestrial habitats. However, it is difficult to accurately estimate this component. While it appears likely that leaching removes PAHs from soils, microbial degradation decomposes some of these compounds, and complex surface and groundwater flow often obscure the influx of the PAH assemblages into estuarine systems.

Once in the estuary, PAH compounds sorb easily to particulate matter, settle through the water column, and concentrate in bottom sediments. Organisms take up PAHs from the sediments as well as from the water and food they consume. The ability or inability of the organisms to metabolize the contaminants influences the transfer of PAH compounds from depot lipid stores, biological membranes, and macromolecules in the organisms to the aqueous environmental phase. For biota exhibiting active release of PAHs, the compounds undergo metabolic transformation to polar water-soluble metabolites that are excreted more readily. For biota showing passive release of PAH from tissues, an equilibrium distribution occurs between the aqueous phase and the aforementioned lipophilic compartments in contact with it.[4] These different abilities of the biota to metabolize PAHs have an important bearing on the biological transport, bioaccumulation, disposition, and toxicity of the contaminants in estuaries.[91]

Benthic organisms that lack the enzyme system (MFO activity) needed to metabolize xenobiotics, such as PAH compounds, may accumulate significant concentrations of them. Shellfish, which are an important source of food for man (e.g., *Mya arenaria*, *Mytilus edulis*, and *Ostrea edulis*), fall into this category. People consuming shellfish tainted with PAH may retain a certain residual in their body organs, which increases carcinogenic risk, albeit slightly. Much of the PAH ingested is degraded by metabolic processes or eliminated in urine and feces. The excreted PAH reenters the environment through the municipal sewer, thereby completing the cycle.[21]

Neff (p. 422)[4] summarizes the cycling of PAH contaminants in these systems:

> "...The cycle of PAH in the aquatic environment appears to be relatively simple. PAH entering water from various sources quickly become adsorbed on organic and inorganic particulate matter and large amounts are deposited in bottom sediments. Leaching or biological activity in the sediments may return a small fraction of these PAH to the water column. PAH are readily accumulated by aquatic biota reaching levels higher than those in the ambient medium. Relative concentrations of PAH in aquatic ecosystems are generally highest in the sediments, intermediate in aquatic biota, and lowest in the water column. Routes of removal of PAH from the aquatic environment include volatilization from the water surface (mainly low molecular weight PAH), photooxidation, chemical oxidation, microbial metabolism, and metabolism by higher metazoans."

SUMMARY AND CONCLUSIONS

PAHs, the nearly ubiquitous trace contaminants in estuarine and marine environments, consist of carbon and hydrogen arranged in the form of two or more fused aromatic (benzene) rings.[1,4] Thousands of PAH compounds exist, many of which are known to be carcinogenic, mutagenic, or teratogenic to mammals. The substantial quantities of PAHs in the environment and the potential threat that they pose to man have spurred numerous studies. A natural background of PAH occurs in the environment ascribable to biosynthesis, primarily by microbes and plants. However, the largest fraction is attributable directly or indirectly to human activities. The principal sources of PAHs in estuaries include incomplete combustion of fossil fuels, industrial and sewage effluents, accidental spillage or natural seepage of oil, grass and forest fires, riverborne influx, nonpoint source runoff of materials from terrestrial habitats, and *in situ* diagenesis of organic matter in sediments. The fate of PAHs in estuaries are the following: (1) dispersal into the water column; (2) concentration in biota; (3) incorporation in bottom sediments; and (4) chemical oxidation and biodegradaton. Photooxidation, chemical oxidation, and biological transformation by microbes and aquatic animals are the principal degradative processes for PAHs in estuarine and marine environments.[3]

An estimated 2.3×10^5 metric tons of PAHs enter the aquatic environment from all sources.[3] The most potent and ubiquitous of the PAH carcinogens is 3,4-benzpyrene (BaP), although it constitutes only about 1 to 20% of the group.[21] It continues to be the most intensely studied PAH. The aquatic environment receives approximately 70 metric tons of BaP each year from all sources.[3] The high molecular weight PAHs, containing four to seven rings, tend to be carcinogenic, mutagenic, and teratogenic to a wide diversity of organisms; however, they are significantly less toxic than the unsubstituted lower molecular weight PAH compounds, containing two or three rings, which are noncarcinogenic.

PAHs have a strong affinity for particulate matter; only about one third of all PAHs occur dissolved in the water column and photooxidation probably degrades them rather rapidly.[22] As PAHs settle to the seafloor, they undergo bioaccumulation, biotransformation, and biodegradation in benthic organisms.[23] In anoxic sediments, however, PAHs degrade very slowly and may persist indefinitely.[3,21]

Certain aquatic biota — those lacking an MFO (enzyme) activity — accumulate considerable amounts of PAHs. Notable in this respect are bivalve mollusks, such as *Mya arenaria*, *Mytilus edulis*, and *Ostrea edulis*. MFO activity among invertebrates appears to be restricted mainly to some members of the Arthropoda and Annelida.[3,4] Finfish, in general, seem to be able to metabolize PAH compounds quite effectively in contrast to shellfish. As in mammals, most MFO activity in fish takes place in the liver.

Because bivalve mollusks concentrate PAH compounds in their tissues due to deficient MFO activity, a few of them have proven to be effective as sentinel organisms in monitoring the chemical contaminants. For example, since 1986, NOAA has employed mussels and oysters as sentinel organisms to survey spatial distributions and temporal trends of PAH concentrations in estuarine and coastal marine waters nationwide. Between 1986 and 1988, mussels and oysters were sampled at 177 estuarine and coastal marine sites in the U.S. by NOAA to assess PAH as well as other chemical contaminants.

In addition to NOAA nationwide surveys of PAHs using sentinel organisms, detailed independent investigations of the xenobiotics have been conducted on several U.S. estuaries. Studies of PAHs in Chesapeake Bay show that their highest concentrations exist in bottom sediments in proximity to heavy industrial and urban areas. They originate principally from anthropogenic sources, most importantly from the combustion of high temperature pyrolysis of carbonaceous fuels. A conspicuous gradient in PAH concentrations exists along the longitudinal axis of the bay, with highest values found in the northern reaches in areas of greatest human population densities. From there, they gradually decline downestuary. Two sentinel organisms, Baltic clams (*Macoma balthica*) and clam worms (*Nereis succinea*), have been successfully utilized in developing indices of exposure and bioavailability of unsubstituted PAHs in Chesapeake Bay water quality programs.

Results of research on the origin, concentration, and distribution of PAHs in Narragansett Bay somewhat parallel those of Chesapeake Bay. Highest levels of PAHs in bottom sediments are apparent in the northernmost segment of the estuary at a sewage outfall site for the City of Providence, RI. Progressing downestuary, along a north-to-south transect, the quantity of PAHs in bottom sediments gradually diminishes. As in the case of Chesapeake Bay, the main source of PAHs in bottom sediments is fossil fuel combustion. However, two other sources of PAHs contribute to the pool of PAH compounds. Included here are petroleum oils that enter the bay in sewage effluent and coal tar on pilings and docks that release PAHs. Of these two sources, coal tar accounts for smaller quantities of the xenobiotics in the estuary.

REFERENCES

1. Eisler, R., Polycyclic Aromatic Hydrocarbon Hazards to Fish, Wildlife, and Invertebrates: A Synoptic Review, Biol. Rep. 85(1.11), U.S. Fish and Wildlife Service, Washington, D.C., 1987.
2. Sims, R. C. and Overcash, R., Fate of polynuclear aromatic compounds (PNAs) in soil-plant systems, *Residue Rev.,* 88, 1, 1983.
3. Neff, J. M., *Polycyclic Aromatic Hydrocarbons in the Aquatic Environment: Sources, Fates, and Biological Effects,* Applied Science, London, 1979.
4. Neff, J. M., Polycyclic aromatic hydrocarbons, in *Fundamentals of Aquatic Toxicology,* Rand, G. M. and Petrocelli, S. R., Eds., Hemisphere, New York, 1985, 416.
5. Klauda, R. J. and Bender, M. E., Contaminant effects on Chesapeake Bay finfishes, in *Contaminant Problems and Management of Living Chesapeake Bay Resources,* Majumdar, S. K., Hall, L. W., Jr., and Austin, H. M., Eds., Pennsylvania Academy of Science, Easton, 1987, 321.
6. Van Vleet, E. S. and Quinn, J. G., Contribution of chronic petroleum inputs to Narragansett Bay and Rhode Island Sound sediments, *J. Fish. Res. Bd. Can.,* 35, 536, 1978.
7. Bedding, N. D., McIntyre, A. E., Perry, R., and Lester, J. N., Organic contaminants in the aquatic environment. I. Sources and occurrence, *Sci. Tot. Environ.,* 32, 411, 1982.
8. Kerr, R. P. and Capone, D. G., The effect of salinity on the microbial mineralization of two polycyclic aromatic hydrocarbons in estuarine sediments, *Mar. Environ. Res.,* 26, 181, 1988.
9. Jackim, E. and Lake, C., Polynuclear aromatic hydrocarbons in estuarine and nearshore environments, in *Estuarine Interactions,* Wiley, M. L., Ed., Academic Press, New York, 1978, 415.
10. Carlberg, S. R., Oil pollution of the marine environment — with an emphasis on estuarine studies, in *Chemistry and Biogeochemistry of Estuaries,* Olausson, E. and Cato, I., John Wiley & Sons, Chichester, U.K., 1980, 367.
11. Clar, E., *Polycyclic Hydrocarbons,* Vol. 2, Academic Press, New York, 1964.
12. O'Connor, J. M., Klotz, J. B., and Kneip, T. J., Sources, sinks, and distribution of organic contaminants in the New York Bight ecosystem, in *Ecological Stress and the New York Bight: Science and Management,* Mayer, G. F., Ed., Estuarine Research Federation, Columbia, SC, 1982, 631.
13. Mix, M. C., Polycyclic aromatic hydrocarbons in the aquatic environment: occurrence and biological monitoring, in *Reviews in Environmental Toxicology,* Vol. 1, Hodgson, E., Ed., Elsevier, Amsterdam, 1984, 51.
14. Van Engel, W. A., Factors affecting the distribution and abundance of the blue crab in Chesapeake Bay, in *Contaminant Problems and Management of Living Chesapeake Bay Resources,* Majumdar, S. K., Hall, L. W., Jr., and Austin, H. M., Eds., Pennsylvania Academy of Science, Easton, 1987, 177.

15. Miller, J. A. and Miller, E. C., Chemical carcinogenesis: mechanisms and approaches to its control, *J. Natl. Cancer Inst.*, 47, 5, 1971.

16. Ames, B. N., Sims, P., and Grover, P. L., Epoxides of carcinogenic polycyclic hydrocarbons are frameshift mutagens, *Science*, 176, 47, 1972.

17. Lee, R. F., Ryan, C., and Neuhauser, M. L., Fate of petroleum hydrocarbons taken up from food and water by the blue crab, *Callinectes sapidus, Mar. Biol.*, 37, 363, 1976.

18. Singer, S. C. and Lee, R. F., Mixed function oxygenase activity in blue crab, *Callinectes sapidus:* tissue distribution and correlation with changes during molting and development, *Biol. Bull.*, 153, 377, 1977.

19. Hale, R. C., Mixed-function-oxygenase enzyme systems: purpose and possible deleterious interactions with organic pollutants in the blue crab, *J. Shellfish Res.*, 3, 92, 1983.

20. McLeese, D. W., Ray, S., and Burridge, L. E., Accumulation of polynuclear aromatic hydrocarbons by the clam *Mya arenaria*, in *Wastes in the Ocean*, Vol. 6, *Nearshore Waste Disposal*, Ketchum, B. H., Capuzzo, J. M., Burt, W. V., Duedall, I. W., Park, P. K., and Kester, D. R., John Wiley & Sons, New York, 1985, 81.

21. Suess, M. J., The environmental load and cycle of polycyclic aromatic hydrocarbons, *Sci. Tot. Environ.*, 6, 239, 1976.

22. Lee, S. D. and Grant, L., Eds., *Health and Ecological Assessment of Polynuclear Aromatic Hydrocarbons*, Pathotex Publishing, Park Forest South, IL, 1981.

23. U.S. Environmental Protection Agency, Ambient Water Quality Criteria for Polynuclear Aromatic Hydrocarbons, Tech. Rep. No. 440/5-80-069, U.S. Environmental Protection Agency, Washington, D.C., 1980.

24. McGinnes, P. R. and Snoeyink, V. L., Determination of the Fate of Polynuclear Aromatic Hydrocarbons in Natural Water Systems, Water Research Center Tech. Rep. UILU-WRC-74-0080, Res. Rept. 80, University of Illinois at Urbana-Champaign, 1974.

25. Bauer, J. E. and Capone, D. G., Degradation and mineralization of the polycyclic aromatic hydrocarbons anthracene and naphthalene in intertidal marine sediments, *Appl. Environ. Microbiol.*, 50, 81, 1985.

26. Gerarde, H. W. and Gerarde, D. F., The Ubiquitous Hydrocarbons, U.S. Assoc. Food Drug Office, 25 6, 1, 1962.

27. LaFlamme, R. E. and Hites, R. A., The global distribution of polycyclic aromatic hydrocarbons in recent sediments, *Geochim. Cosmochim. Acta*, 42, 289, 1978.

28. Wakeham, S. G., Shaffner, C., and Giger, W., Polycyclic aromatic hydrocarbons in recent lake sediments. II. Compounds derived from biogenic precursors during early diagenesis, *Geochim. Cosmochim. Acta*, 44, 415, 1980.

29. Wijayaratne, R. D. and Means, J. C., Sorption of polycyclic aromatic hydrocarbons by natural estuarine colloids, *Mar. Environ. Res.*, 11, 77, 1984.

30. Blumer, M. and Youngblood, W. W., Polycyclic aromatic hydrocarbons in soils and recent sediments, *Science,* 188, 53, 1975.

31. Grimmer, G. and Boehnke, H., Profile analysis of polycyclic aromatic hydrocarbons and metal content in sediment layers of a lake, *Cancer Lett.,* 1, 75, 1975.

32. Hites, R. A., LaFlamme, R. E., and Farrington, J. W., Sedimentary polycyclic aromatic hydrocarbons: the historical record, *Science,* 198, 829, 1977.

33. Hinga, K. R., Pilson, M. E. Q., and Almquist, G., The degradation of 7,12-dimethylbenz(a)anthracene in an enclosed marine ecosystem, *Mar. Environ. Res.,* 18, 79, 1986.

34. Prahl, F. G., Crecelius, E., and Carpenter, R., Polycyclic aromatic hydrocarbons in Washington coastal sediments: an evaluation of atmospheric and riverine routes of introduction, *Environ. Sci. Technol.,* 18, 687, 1984.

35. Barrick, R. C., Flux of aliphatic and polycyclic aromatic hydrocarbons to central Puget Sound from Seattle (Washington) primary sewage effluent, *Environ. Sci. Technol.,* 16, 682, 1982.

36. Hoffman, E. J., Mills, G. L., Latimer, J. S., and Quinn, J. G., Urban runoff as a source of polycyclic aromatic hydrocarbons to coastal waters, *Environ. Sci. Technol.,* 18, 580, 1984.

37. Herrmann, R. and Hubner, D., Behaviour of polycyclic aromatic hydrocarbons in the Exe estuary, Devon, *Neth. J. Sea Res.,* 15, 362, 1982.

38. Guerin, W. F. and Jones, G. E., Estuarine ecology of phenanthrene-degrading bacteria, *Est. Coastal Shelf Sci.,* 29, 115, 1989.

39. Lake, J. L., Norwood, C., Dimock, C., and Bowen, R., Origins of polycyclic aromatic hydrocarbons in estuarine sediments, *Geochim. Cosmochim. Acta,* 43, 1847, 1979.

40. Olsen, C. R., Cutshall, N. H., and Larsen, I. L., Pollutant-particle associations and dynamics in coastal marine environments: a review, *Mar. Chem.,* 11, 501, 1982.

41. Bieri, R. H., Hein, C., Huggett, R. J., Shou, P., Slone, H., Smith, C., and Su, C.-W., Toxic Organic Compounds in Surface Sediments from the Elizabeth and Patapsco Rivers and Estuaries, Tech. Rep., Virginia Institute of Marine Science, Gloucester Point, 1982.

42. Roberts, M. H., Jr., Hargis, W. J., Jr., Strobel, C. J., and De Lisle, P. F., Acute toxicity of PAH contaminated sediments to the estuarine fish, *Leiostomus xanthurus, Bull. Environ. Contam. Toxicol.,* 42, 142, 1989.

43. Weeks, B. A. and Warinner, E., Effects of toxic chemicals on macrophage phagocytosis in two estuarine fishes, *Mar. Environ. Res.,* 14, 327, 1984.

44. Roberts, M. H., Jr., Sved, D. W., and Felton, S. P., Temporal changes in AHH and SOD activities in feral spot from the Elizabeth River, a polluted sub-estuary, *Mar. Environ. Res.,* 23, 89, 1987.

45. Hargis, W. J., Jr., Roberts, M. H., Jr., and Zwerner, D. E., Effects of contaminated sediments and sediment-exposed effluent water on an estuarine fish: acute toxicity, *Mar. Environ Res.,* 14, 337, 1984.

46. Hildebrand, S. F. and Schroeder, W. C., *Fishes of Chesapeake Bay*, T. F. H. Publications, Neptune, NJ, 1972.

47. Bender, M. E., Hargis, W. J., Jr., Huggett, R. J., and Roberts, M. H., Jr., Effects of polynuclear aromatic hydrocarbons on fishes and shellfish: an overview of research in Virginia, *Mar. Environ. Res.*, 24, 237, 1988.

48. Huggett, R. J., de Fur, P. O., and Bieri, R. H., Organic compounds in Chesapeake Bay sediments, *Mar. Pollut. Bull.*, 19, 454, 1988.

49. Johnson, A. C., Larsen, P. F., Gadbois, D. F., and Humason, A. W., The distribution of polycyclic aromatic hydrocarbons in the surficial sediments of Penobscot Bay (Maine, USA) in relation to possible sources and to other sites worldwide, *Mar. Environ. Res.*, 15, 1, 1985.

50. Shiaris, M. P. and Jambard-Sweet, D., Polycyclic aromatic hydrocarbons in surficial sediments of Boston Harbor, Massachusetts, USA, *Mar. Pollut. Bull.*, 17, 469, 1986.

51. Malins, D.C., Krahn, M. M., Meyers, M. S., Rhodes, L. D., Brown, D. W., Krone, C. A., McCain, B. M., and Chan, S.-L., Toxic chemicals in sediments and biota from a creosote-polluted harbor: relationships with hepatic neoplasms and other hepatic lesions in English sole (*Parophrys vetulus*), *Carcinogenesis*, 6, 1463, 1985.

52. Barrick, R. C. and Prahl, F. G., Hydrocarbon geochemistry of the Puget Sound region. III. Polycyclic aromatic hydrocarbons in sediments, *Est. Coastal Shelf Sci.*, 25, 175, 1987.

53. Readman, J. W., Mantoura, R. F. C., and Rhead, M. M., A record of polycyclic aromatic hydrocarbon (PAH) pollution obtained from accreting sediments of the Tamar estuary, U. K.: evidence for non-equilibrium behaviour of PAH, *Sci. Tot. Environ.*, 66, 73, 1987.

54. Milano, J. C., Fache, B., and Vernat, J. L., Polyaromatic hydrocarbons in the Mediterranean: the Rhone River as a vector of pollution, *J. Rech. Oceanogr.*, 11, 50, 1986.

55. Windsor, J. G. and Hites, R. A., Polycyclic aromatic hydrocarbons in Gulf of Maine sediments and Nova Scotia soils, *Geochim. Cosmochim. Acta*, 43, 27, 1979.

56. Boehm, P. D., New York Bight Benthic Sampling Survey: Coprostanol, Polynuclear Aromatic Hydrocarbons and Polychlorinated Biphenyls in Sediments (1980 1981), Tech. Rep., National Marine Fisheries Service, U.S. National Oceanic and Atmospheric Administration, Sandy Hook, NJ, 1981.

57. Boehm, P. D., Drew, S., Dorsey, T., Yarko, J., Mosesman, N., Jefferies, A., Pilson, D., and Fiest, D., Organic pollutants in New York Bight suspended particulates, in *Wastes in the Ocean*, Vol. 6, *Nearshore Waste Disposal*, Ketchum, B. H., Capuzzo, J. M., Burt, W. V., Duedall, I. W., Park, P. K., and Kester, D. R., Eds., John Wiley & Sons, New York, 1985, 251.

58. Kersten, M., Geobiological effects on the mobility of contaminants in marine sediments, in *Pollution of the North Sea: An Assessment,* Salomons, W., Bayne, B. L., Duursma, E. K., and Foerstner, U., Eds., Springer-Verlag, Berlin, 1988, 36.

59. Lawrence, J. F. and Weber, D. F., Determination of polycyclic aromatic hydrocarbons in some Canadian commercial fish, shellfish, and meat products by liquid chromatography with confirmation by capillary gas chromatography-mass spectrometry, *J. Agric. Food Chem.,* 32, 789, 1984.

60. Sirota, G. R. and Uthe, J. F., Polynuclear aromatic hydrocarbon contamination in marine shellfish, in *Chemical Analysis and Biological Fate: Polynuclear Aromatic Hydrocarbons,* Cook, M. and Dennis, A. J., Eds., Battelle Press, Columbus, OH, 1981, 329.

61. Mix, M. C., Polynuclear Aromatic Hydrocarbons and Cellular Proliferative-Disorders in Bivalve Molluscs from Oregon Estuaries, Rep. 600/4-82-026, U.S. Environmental Protection Agency, Washington, D.C., 1982.

62. Marcus, J. M. and Stokes, T. P., Polynuclear aromatic hydrocarbons in oyster tissue around three coastal marinas, *Bull. Environ. Contam. Toxicol.,* 35, 835, 1985.

63. Mix, M. C. and Schaffer, R. L., Concentrations of unsubstituted polycyclic aromatic hydrocarbons in softshell clams from Coos Bay, Oregon, USA, *Mar. Pollut. Bull.,* 14, 94, 1983.

64. Sirota, G. R., Uthe, J. F., Screedharan, A., Matheson, R., Musial, C. J., and Hamilton, K., Polynuclear aromatic hydrocarbons in lobster (*Homarus americanus*) and sediments in the vicinity of a cooking facility, in *Polynuclear Aromatic Hydrocarbons: Formation, Metabolism, and Measurement,* Cooke, M. and Dennis, A. J., Eds., Battelle Press, Columbus, OH, 1983, 1123.

65. West, W. R., Smith, P. A., Stoker, P. W., Booth, G. M., Smith-Oliver, T., Butterworth, B. E., and Lee, M. L., Analysis and genotoxity of a PAC-polluted river sediment, in *Polynuclear Aromatic Hydrocarbons: Mechanisms, Methods, and Metabolism,* Cooke, M. and Dennis, A. J., Eds., Battelle Press, Columbus, OH, 1984, 1395.

66. Southworth, G. R., Beauchamp, J. J., and Schneider, P. K., Bioaccumulation potential of polycyclic aromatic hydrocarbons in *Daphnia pulex, Water Res.,* 12, 973, 1978.

67. Neff, J. M., Accumulation and release of polycyclic aromatic hydrocarbons from water, food, and sediment by marine animals, in Symposium: Carcinogenic Polynuclear Aromatic Hydrocarbons in the Marine Environment, Rep. 600/9-82-013, Richards, N. L. and Jackson, B. L., Eds., U.S. Environmental Protection Agency, Washington, D.C., 1982, 385.

68. Neff, J. M., Cox, B. A., Dixit, D., and Anderson, J. W., Accumulation and release of petroleum derived aromatic hydrocarbons by four species of marine animals, *Mar. Biol.,* 38, 279, 1976.

69. Lake, J., Hoffman, G. L., and Schimmel, SC, Bioaccumulation of Contaminants from Black Rock Harbor Dredged Material by Mussels and Polychaetes, Tech. Rep. D-85-2, U.S. Environmental Protection Agency, 1985.

70. Miller, D. L., Corliss, J. P., Farragut, R. N., and Thompson, H. C., Jr., Some aspects of the uptake and elimination of the polynuclear aromatic hydrocarbon chrysene by mangrove snapper, *Lutjanus griseus,* and pink shrimp, *Penaeus duorarum,* in Symposium: Carcinogenic Polynuclear Aromatic Hydrocarbons in the Marine Environment, Rep. 600/9-82-013, Richards, N. L. and Jackson, B. L., Eds., U.S. Environmental Protection Agency Washington, D.C., 1982, 321.

71. National Oceanic and Atmospheric Administration, A Summary of Data on Tissue Contamination from the First Three Years (1986-1988) of the Mussel Watch Project, NOAA Tech. Mem. NOS OMA49, National Oceanic and Atmospheric Administration, Rockville, MD, 1989.

72. Freitas, S. T., Contaminant profiles in bivalve molluscs from U.S. coastal waters, 10th Biennial International Estuarine Research Conference Abstracts, Estuarine Research Federation, Baltimore, 1989, 26.

73. Meyers, P. A. and Quinn, J. G., Association of hydrocarbons and mineral particles in saline solutions, *Nature,* 244, 23, 1973.

74. Herbes, S. E., Partitioning of polycyclic aromatic hydrocarbons between dissolved and particulate phases in natural waters, *Water Res.,* 11, 493, 1977.

75. Borneff, J. and Kunte, H., Carcinogenic substances in water and soil. XVII. Concerning the origin and estimation of the polycyclic aromatic hydrocarbons in water, *Arch. Hyg. (Berlin),* 149, 226, 1965.

76. Grimmer, G., Bohnke, H., and Borwitzky, H., Profile analysis of polycyclic aromatic hydrocarbons in sewage sludge by gas chromatography, *Fresenius Z. Anal. Chem.,* 289, 91, 1978.

77. Nicholls, T. P., Perry, R., and Lester, J. N., The influence of heat treatment on the metallic and polycyclic aromatic hydrocarbon content of sewage sludge, *Sci. Tot. Environ.,* 12, 137, 1979.

78. Zepp, R. G. and Schlotzhauer, P. F., Photoreactivity of selected aromatic hydrocarbons in water, in *Polynuclear Aromatic Hydrocarbons, Third International Symposium on Chemistry and Biology — Carcinogenesis and Mutagenesis,* Jones, P. W. and Leber, P., Eds., Ann Arbor Science, Ann Arbor, MI, 1979, 141.

79. Heidelberger, C., Studies on the mechanisms of carcinogenesis by polycyclic aromatic hydrocarbons and their derivatives, in *Polynuclear Aromatic Hydrocarbons: Chemistry, Metabolism and Carcinogenesis,* Freudenthal, R. I. and Jones, P. W., Eds., Raven Press, New York, 1976, 1.

80. Giam, C. S., Ray, L. E., Anderson, R. S., Fries, C. R., Lee, R., Neff, J. M., Stegeman, J. J., Thomas, P., and Tripp, M. R., Pollutant responses in marine animals: the program, in *Pollutant Studies in Marine Animals,* Giam, C. S. and Ray, L. E., Eds., CRC Press, Boca Raton, FL, 1987, 1.

81. National Academy of Sciences, *Particulate Polycyclic Organic Matter,* National Academy Press, Washington, D.C., 1972.

82. Harrison, R. M., Perry, R., and Wellings, R. A., Effect of water chlorination upon levels of some polynuclear aromatic hydrocarbons in water, *Environ. Sci. Technol.,* 10, 1151, 1976.

83. Harrison, R. M., Perry, R., and Wellings, R. A., Chemical kinetics of chlorination of some polynuclear aromatic hydrocarbons under conditions of water treatment processes, *Environ. Sci. Technol.,* 10, 1156, 1976.

84. Meineke, I. and Klamberg, H., On the degradation of polycyclic aromatic hydrocarbons. I. Reaction products of ozonolysis of polycyclic aromatic hydrocarbons in aqueous systems, *Fresenius Z. Anal. Chem.,* 293, 201, 1978.

85. Gibson, D. T., Biodegradation of aromatic petroleum hydrocarbons, in *Fate and Effects of Petroleum Hydrocarbons in Marine Ecosystems and Organisms,* Wolfe, D. A., Ed., Pergamon Press, New York, 1977, 36.

86. Herbes, S. E. and Schwall, L. R., Microbial transformation of polycyclicaromatic hydrocarbons in pristine and petroleum-contaminated sediments, *Appl. Environ. Microbiol.,* 35, 306, 1978.

87. Lee, R. F., Mixed function oxygenases (MFO) in marine invertebrates, *Mar. Biol. Lett.,* 2, 87, 1981.

88. Foster, G. D. and Wright, D. A., Unsubstituted polynuclear aromatic hydrocarbons in sediments, clams, and clam worms from Chesapeake Bay, *Mar. Pollut. Bull.,* 19, 459, 1988.

89. de Fur, P., Organic Compounds from Chesapeake Bay and Its Virginia Tributaries, Virginia Institute of Marine Science Technical Report to the Virginia State Water Control Board, Gloucester Point, 1986.

90. Wright, D. A., Foster, G. D., and Whitlow, S. I., Chesapeake Bay Water Quality Monitoring Component, Ref. No. [UMCEES]CBL, 86-135, Report to the Maryland Department of Health, Office of Environmental Programs, University of Maryland, College Park, 1986.

91. Varanasi, U., Ed., *Metabolism of Polycyclic Aromatic Hydrocarbons in the Aquatic Environment,* CRC Press, Boca Raton, FL, 1989.

4 Chlorinated Hydrocarbons

INTRODUCTION

Estuaries are sites of accumulation of stable chlorinated compounds used by man as biocides (insecticides, herbicides, and fungicides) and as substances for a variety of industrial applications. These chlorinated hydrocarbons or organochlorine compounds resist breakdown in aquatic habitats and may persist for many years in the estuarine environment. Having been detected in water, suspended sediment, bottom sediment, and biota of estuaries, chlorinated hydrocarbons are a cause of grave concern because of their potential toxicity to estuarine organisms and man. Moreover, these fat-soluble compounds tend to become biomagnified in the food chain; i.e., organisms occupying higher trophic levels contain greater concentrations of a particular organochlorine compound than those located on lower trophic levels. Heavy body burdens of DDT (1,1,1-trichloro-2,2-bis (p-chlorophenyl)ethane) in eagles, ospreys, and other species of birds, for example, testify to this effect.

Prior to the introduction of DDT and related organochlorine biocides (synthesized from petrochemicals and chlorine), early insecticide formulations consisted of naturally derived substances, such as nicotine, pyrethrum, and rotenone. In addition to DDT, an extensive list of chlorinated compounds was developed for use as commercial pesticides, including but not limited to aldrin, chlordane, endrin, heptochlor, dieldrin, perthane, and toxaphene. A second group of pesticides, organophosphates, while exhibiting greater overall toxicity than chlorinated hydrocarbon pesticides, degrades much more rapidly, usually being broken down into harmless compounds within 1 week.[1] Little is known regarding the impact of phosphorus-containing biocides (among them dipterex, fenthion, gluthion, malathion, naled, parathion, and ronnel) on estuarine and marine organisms. Similar to DDT and, other organochlorines, however, these compounds act as nerve poisons and, when active, can be extremely effective in controlling population levels of insects. Wetlands fringing estuarine waterbodies often represent ideal breeding sites for mosquitos and other insects.[2] In the past, some of these areas have received considerable quantities of insecticides. Furthermore, the runoff of pesticides and herbicides from farmland regions upstream of estuaries can pose serious environmental hazards to constituent biota, especially the stable, long-lived organochlorine compounds.

Aside from organochlorine biocides, polychlorinated biphenyls (PCBs) are a major class of synthetic chlorinated organic compounds having a wide range of industrial applications due to their unique set of properties. For example, their low dielectric constant and high heat capacity once made them extremely valuable to the electric utility industry for use in electrical capacitors and transformers.[3] Furthermore, because of the chemical stability of these compounds, they were widely utilized in the past as additives to adhesives, hydraulic oils, and paints.[4] The widespread application of PCBs and their resultant toxicological impacts on aquatic and terrestrial organisms, however, soon ushered in a series of strong regulatory controls in the U.S. Despite these controls, investigations revealed global PCB contamination of aquatic systems. Perhaps most alarming was the bioaccumulation potential of these compounds in aquatic food chains and the resultant dangers to man.[5]

PCBs, although initially synthesized about 1880, did not undergo commercial production until 1929. Between 1929 and 1970, approximately 1×10^6 metric tons of PCBs were produced, but after 1970, global production declined substantially, primarily due to growing environmental apprehension in the U.S., which ultimately led to the termination of industrial production of PCBs in 1977. Worldwide production of PCBs in 1978 had dropped to about 10% of that in 1970.[6]

A similar pattern of production is evident for DDT. First described in 1874, DDT did not become an important commercial product until after 1939 when its insecticidal properties were proven by Paul Muller.[4] Production of DDT on a large scale commenced in 1944, and global production between 1944 and 1980 equalled about 3×10^6 metric tons. North American and western European countries dominated the production of DDT between the mid-1940s and mid-1960s. Subsequent to 1965, however, production decreased drastically there, with most of the production shifting to other countries, especially those in subtropical and tropical latitudes. Historical diminution in production of both PCBs and DDT is closely linked to the acquisition of data on the environmental fate of these organochlorine compounds.

Among the best known chlorinated hydrocarbon contaminants in the marine biosphere are DDT and PCBs. These persistent, stable, and toxic materials have been found in the tissues of many estuarine organisms.[7] Their accumulation in bottom sediments is the principal cause of the continued contamination of biota inhabiting these coastal ecosystems.[8] Because of their potential carcinogenicity and mutagenicity to man, DDT and PCBs have been monitored intensely during the past 2 decades.[9] This chapter examines the sources, distribution, and biotic impacts of DDT (and other organochlorine pesticides) and PCBs in estuaries.

ORGANOCHLORINE PESTICIDES

DDT

This synthetic pesticide, applied successfully during World War II, was put into global use in 1945 to ameliorate the spread of diseases transmitted by insects and to mitigate insect-related problems in aquaculture and forestry.[4,10] Production and use of DDT increased during the 1950s and 1960s, and by 1971, 2×10^{12} g of the pesticide had been manufactured, mostly in the U.S.[11] By the 1960s, the insidious effects of DDT and its metabolites, DDD and DDE, in estuarine and marine flood webs began to surface. As a direct result of the environmental dangers posed by DDT, its production was severely curtailed; by 1972, its use in the northern hemisphere virtually ceased.[11] The impacts of this pesticide on estuarine and marine organisms was manifested most conspicuously among upper trophic level organisms, particularly marine birds. Although DDT was not applied directly to estuarine or coastal marine waters, land runoff, anthropogenic disposal, precipitation, and airborne particulates accounted for the bulk of its translocation to these environments.

The brown pelican (*Pelicanus occidentalis*) on islands off southern California and the osprey (*Pandion haliaetus*) of Chesapeake Bay experienced diminishing population sizes and increased mortality in the late 1960s and late 1970s, respectively. DDT released from the White's Point sewer outfall (at a rate of about 100 metric tons/ year) raised DDT concentrations in waters nearby Los Angeles to 370 ppm.[10] This anthropogenic input caused the failure of breeding of brown pelicans, accounting for the virtual disappearance of the species from this area. At Auacapa Island, the abundance of brown pelicans dropped significantly between 1969 and 1972 in response to the thinning of eggshells.[6] In addition, negative effects of sewage disposal on benthic invertebrates and demersal fishes in the Los Angeles-San Diego region were ascribed to the accumulation of DDT, PCBs, and trace metals in bottom sediments.[11]

Even with the demise of aerial spraying of DDT by 1972, the persistence of this organochlorine pesticide and its metabolites in the environment continued to affect osprey populations years later. Hence, between 1975 and 1982, 11 dead or moribund ospreys in Chesapeake Bay contained variable concentrations of organochlorine residues, with some of the birds harboring potentially harmful DDE levels.[12] Measurements on 33 osprey eggs during 1977 and 1978 indicated that they were 10 to 24% thinner than during the pre-DDT era and comparable to those in the early 1970s. Osprey eggs taken from the eastern shore of the bay during 1977 and 1978 had mean DDE values ranging from 1.9 to 2.8 ppm, compared to DDE concentrations of 1.4 to 3.5 ppm in osprey eggs removed from the western shore in 1977.[13] A reduction in the thickness of bald eagle eggs of Chesapeake Bay in the 1970s paralleled that of the osprey.[14]

DDT residues in estuaries peaked in 1968 and diminished markedly after 1970.[6] The resiliency of estuarine and marine environments to the impacts of DDT input did not occur immediately subsequent to its banned use in the U.S. in 1972. Randers[15] contended that DDT concentrations in fish would continue to increase for 2 to 50 years, even if DDT usage decreased or remained unchanged from 1972 levels.[16] The compound resists physical, chemical, and biological degradation, and consequently, its complete elimination from estuaries may take decades. The primary

pathways of eliminating DDT and other organochlorine compounds from estuaries are microbial or chemical decomposition,
photodegradation, metabolism, and vaporization.[4] Its most common
breakdown derivative, DDE (1,1-dichloro-2,2-bis(*p*-chloro-
phenyl)ethylene) can be equally toxic to many organisms. DDE is
also fat soluble and generaly more stable than DDT; because it
remains the most common breakdown product in seafloor sediments, this organochlorine compound may be even of greater
concern to estuarine communities than DDT itself. DDD (1,1-
dichloro-2,2-bis (*p*-chlorophenyl)ethane) appears to be only moderately toxic to a limited group of organisms.[4,17] Figure 1 (a–c) shows
the chemical structure of DDT, DDE, and DDD.

Phytoplankton populations take up DDT residues when available
in the water column, and they serve as a major source of the
organochlorine contaminant for filter-feeding invertebrates and
other higher trophic level organisms. At the time of significant influx
of DDT into marine waters, Goldberg et al.[18] reported average DDT
concentrations in phytoplankton amounting to 0.01 µg/g. DDE and
DDD purportedly inhibit the growth of marine phytoplankton.[19,20]

Because of the biomagnification of organochlorines in estuarine
and marine food chains and the potential harm of these compounds
to humans consuming contaminated food from the sea, a much
greater effort has been expended on monitoring chlorinated pesticide levels in fin- and shellfish. Butler and Schutzmann,[21] for
example, recorded residue levels of pesticides in estuarine fish from
144 estuaries in the U.S. They concluded that the biotic residues of
DDT, dieldrin, endrin, and toxaphene were much higher in samples
taken during the period of 1965 to 1970 than during the interval of
1972 to 1976. Butler et al.[22] also documented a decline in the
concentration and frequency of biotic residues of chlorinated pesticides in estuarine mollusks over the period of 1972 to 1977. DDT
was detected in samples taken from only 2 of the 87 estuaries
investigated in 1977, and 10 other pesticides that were monitored
could not be detected at all in the samples. In contrast, most samples
collected in 1965 contained detectable levels of DDT and a greater
frequency of occurrences of other pesticides.

In a synthesis of trends of chlorinated pesticide (including DDT)
and PCB contamination in U.S. coastal fish and shellfish, Mearns et
al.[9] registered more than 35,000 biotic samples from at least 300

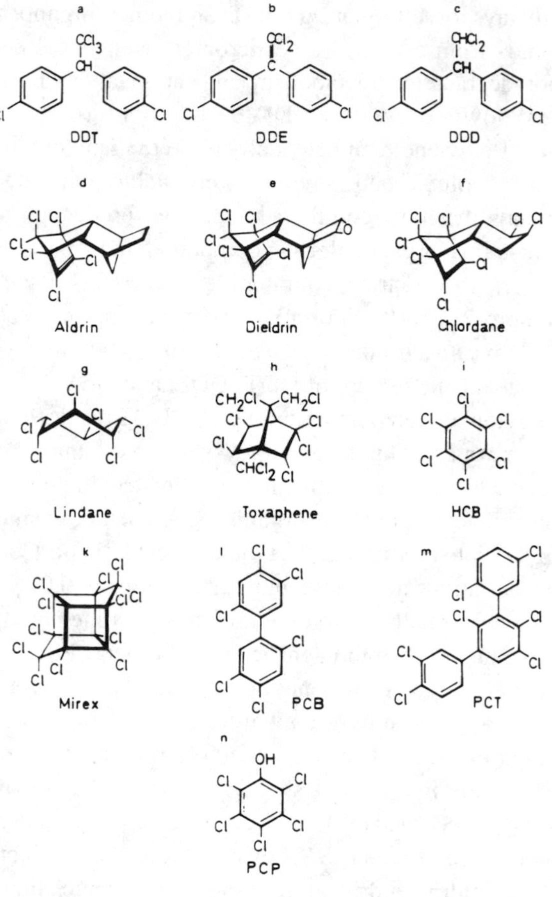

FIGURE 1. Chemical structure of common organochlorine compounds encountered in estuaries. (From Reutergardh, L., in *Chemistry and Biogeochemistry of Estuaries,* Olausson, E. and Cato, I., Eds., John Wiley & Sons, Chichester, U.K., 1980, 349. With permission.)

separate surveys conducted between 1940 and 1985. Based on this database, regional and nationwide patterns of fish and shellfish contamination for DDT, dieldrin, and to a lesser extent, other pesticides, as well as PCBs were mapped for three heavily sampled periods, specifically 1965 to 1972, 1976 to 1977, and 1984. The review covered seven classes of historically important chlorinated pesticides: (1) DDT and structurally related compounds (e.g.,

methoxychlor); (2) cyclodiene pesticides (e.g., chlordanes, endosulfan, and dieldrin); (3) hexachlorocyclohexane (primarily lindane);
(4) hexachlorocyclopentadiene pesticides (principally kepone and
mirex); (5) chlorinated camphenes (i.e., toxaphene); (6) carboxylic
acid herbicides (e.g., 2,4-D and dacthal); and (7) chlorinated benzenes and phenols (e.g., pentachlorophenol) (Table 1). More than
200 individual bays, estuaries, and other distinct embayments

---

TABLE 1
List of Target Chemical Residues Surveyed by Mearns et al.[a]

Residue	Use and occurrence
Polychlorinated biphenyls (PCBs)	Dielectric fluid in capacitors; transformer fluid; lubricants; hydraulic fluids; plasticizers; cutting oil extenders; carbonless paper. Banned in 1976. Total is either sum of chlorination mixtures (Arochlors®) or number.
Total PCB	
Arochlor® 1016	
Arochlor® 1242	
Arochlor® 1248	
Arochlor® 1254	
Arochlor® 1260	
PCBs by chlorination number (2-10)	
DDT and structurally related chemicals	
DDE (o-p and p-p')	Insecticides; DDT metabolites
DDD (TDE; o-p and p-p')	Insecticides; DDT metabolites
DDT (o-p and p-p')	Insecticides; parents of DDD/DDE
Total DDT	Sum of parent and metabolites
Kelthane (Dicofol)	Acaracide; parent of DDE
Methoxychlor	Insecticide
Cyclodiene pesticides	
Technical chlordane	Insecticide; mix of constituents
Chlordane (trans- and cis-)	Insecticides; major constituents of technical chlordane
Nonachlor (trans- and cis-)	Insecticides; minor constituents of technical chlordane
Oxychlordane	Chlordane metabolite
Heptachlor	Insecticide; minor constituent of technical chlordane
Heptachlor epoxide	Metabolite of heptachlor
Endosulfan (I and II)	Insecticides; 7:3 mixture of stereoisomers
Endosulfan sulfonate	Metabolite of endosulfans
Aldrin	Insecticide
Dieldrin	Insecticide; Aldrin metabolite
Endrin	Insecticide
Hexachlorocyclohexane insecticides	
a-BHC	Constituent of BHC insecticide mix
Y-BHC (Lindane)	Insecticide; BHC constituent

TABLE 1 (CONTINUED)
List of Target Chemical Residues Surveyed by Mearns et al.[a]

Residue	Use and occurrence
Hexachlorocyclopentadiene pesticides	
Kepone	Acaricide, larvicide, fungicide, ant bait
Mirex	Insecticide (fire ant control)
Chlorinated camphenes	
Toxaphene	Insecticide (cotton)
Carboxylic acid derivatives	
2,4-D (2,4-DEP)	Weed herbicide (in cereals)
2,4,5-T	Wood plant herbicide
DCPA (dacthal)	Pre-emergence weed herbicide
Chlorinated benzenes and phenols	
HCB (hexachlorobenzene)	Fungicide
PCP (pentachlorophenol)	Wood preservative

[a] Modified from Schmitt, C. J., Zajiik, J. L., and Ribick, M. A., *Arch. Environ. Contam. Toxicol.,* 14, 225, 1985.

From Mearns, A. J., Matta, M. B., Simecek-Beatty, D., Buckman, M. F., Shigenaka, G., and Wert, W. A., PCB and Chlorinated Pesticide Contamination in U.S. Fish and Shellfish: A Historical Assessment Report, NOAA Tech. Memo. NOS OMA 39, Seattle, 1988.

yielded data for the survey on finfish (i.e. hagfish, lampreys, sharks, rays, and all bony fishes) and shellfish (i.e., clams, oysters, mussels, snails, octopus, and squid) populations.

Results of the survey by Mearns clearly show that DDT concentrations have decreased dramatically in U.S. coastal fish and shellfish since the early 1970s. On a national basis, the decline may approach 100-fold. Other organochlorine compounds — PCBs, dieldrin, and toxaphene — by comparison exhibited decreasing contamination over the same period for selected species and sites. DDT was a most widespread and prevalent organochlorine contaminant in estuarine invertebrates and fishes in the late 1960s and early 1970s. During this time period, DDT concentrations averaged about 0.024 ppm (wet weight)[23] in bivalves and 0.7 to 1.1 ppm in whole fish.[24] Areas having the highest levels of DDT contamination were Southern California, central California, Mobile Bay, and San Francisco Bay. One short-lived, "hot spot" region existed in Florida. A broad, low-level gradient of DDT contamination also occurred along the East Coast of the U.S. radiating north and south of Delaware Bay and possibly secondarily from Long Island, New York.

DDT in Bivalves

During the National Pesticide Monitoring Program survey conducted from 1965 to 1972, 63% of the 8095 samples of clams, mussels, and oysters collected from 180 sites contained DDT metabolites. Hot spot regions occurred in California and Florida, with five sites in California and one site in Florida having time-averaged site mean values >0.5 ppm wet weight.[23] A resampling program in 1977 revealed a reduction of total DDT concentrations below 0.01 ppm wet weight at nearly all sites except one location in the Point Muger Lagoon in Southern California (0.12 ppm wet weight) and at six locations in upper Delaware Bay (0.033 ppm wet weight).[22] These data reflect a marked nationwide drop in DDT contamination of bivalves during the early 1970s.

Results of the EPA Mussel Watch Program of 1976 to 1978 support the findings of the 1977 National Pesticide Monitoring Program, although these samples were analyzed for DDE and not total DDT. The 1986 NOAA Mussel Watch Program obtained a median concentration of 0.003 ppm wet weight for total DDT at 145 sites.[25] This value represents about a 10-fold reduction from the 180-site median of 0.024 ppm wet weight recorded during the 1965 to 1972 estuarine mollusk survey.[9]

DDT in Fish

Trends in DDT concentrations in estuarine fish paralleled those in estuarine shellfish. Highest measurements of total DDT, exceeding 0.5 ppm wet weight, were observed in samples from sites in the Northeast (i.e., Delaware Bay and the Chesapeake-Delaware Canal) and in the Southwest (i.e., Southern California) during the period of 1972 to 1976. The median concentration of total DDT in whole estuarine fish equaled 0.014 ppm and ranged from 0.01 to 2.1 ppm wet weight at this time. In fish muscle, the median value of total DDT between 1972 and 1975 amounted to 0.11 ppm for collections made at 31 areas. A lower median concentration of 0.013 ppm wet weight was measured in fish muscle from specimens taken from 19 areas in

1976 and 1977.[9] It is apparent, therefore, that the total DDT levels dropped by a factor of 10 in estuarine fish muscle between the early and mid-1970s.

Higher amounts of DDT exist in fish liver than in muscle; for example, the EPA-NOAA Cooperative Estuarine Monitoring Program (CEMP) surveys of 1976 and 1977 revealed that total DDT concentrations in fish liver surpassed those in muscle tissue by a factor of 13 (i.e., 0.147 ppm wet weight in liver compared to 0.011 ppm wet weight in muscle).[26] The median concentration of total DDT in fish liver during the CEMP surveys of 1976 and 1977 and the NOAA Southern California Coastal Water Resource Project[27] equaled 0.22 ppm wet weight. This value is nearly four times greater than the median of 0.06 ppm wet weight obtained in fish liver during the 1984 National Status and Trends Program.[25] Hence, in less than a decade, the DDT concentrations in fish liver decreased significantly on a nationwide basis.[9]

CYCLODIENE PESTICIDES

Aldrin (Figure 1d), dieldrin (Figure 1e), chlordane (Figure 1f), lindane (Figure 1g), and endrin belong to a group of related pesticides termed cyclodienes. They have been used most extensively in agriculture, although chlordane was adapted after 1940 as a pesticide to control termites, ants, and other insects in domestic areas as well as to control agricultural pests.[4] While cyclodiene pesticides proved to be effective agents in the control of insects, they retained the dubious distinction of being very toxic to aquatic organisms, particularly fish. Of all the chlorinated hydrocarbon pesticides, endrin is the most toxic. In addition to these substances, lindane (Figure 1k), toxaphene (Figure 1h), hexachlorobenzene (Figure 1i), and mirex (Figure 1k) have served as insecticides or fungicides, and have secondarily impacted estuarine organisms.

Aldrin

Various U.S. surveys of estuarine shellfish and finfish have detected low concentrations of aldrin. For instance, the National Pesticide Monitoring Program surveys of several thousand samples from 1965 to 1972 and 1977 found no shellfish with aldrin concentrations above the detection limit of 0.01 ppm wet weight.[22,26] Estuarine fish samples taken from 144 sites also showed no detectable amounts of aldrin.[21] At sampling sites of the National Status and

Trends Benthic Surveillance Project along the Gulf of Mexico and Pacific Coast during 1984, the highest concentration of aldrin recorded in livers of estuarine fish occurred in spot from Sapelo Sound, Georgia (0.009 ppm wet weight).[9] The low concentrations of aldrin in the tissues of estuarine biota are not surprising because this pesticide has not been used by the agricultural industry since 1974. Moreover, it is also quickly metabolized to dieldrin in most animals, with dieldrin being metabolized further to water-soluble components that are excreted.[4]

Dieldrin

This pesticide is more toxic to estuarine organisms than aldrin, and it has been detected more frequently in estuarine samples than other cyclodiene contaminants. Based on data from the National Pesticide Monitoring Program, only DDT had greater frequency of detection in shellfish samples collected during the period from 1965 to 1972. However, the Federal Food and Drug Administration action limit of 0.3 ppm wet weight was not exceeded by any of the samples, even though some approached the limit. A National Pesticide Monitoring Program resurvey of shellfish from 87 sites in 1977 did not catalog any samples with dieldrin concentrations above the detection limit of 0.01 ppm wet weight.[22] While no shellfish samples exhibited dieldrin levels in excess of the 0.01 ppm wet weight detection limit in 1977, juvenile estuarine fish from 25 of 144 sampling sites of the National Pesticide Monitoring Program had dieldrin concentrations exceeding this limit between 1972 and 1978. The 1984 National Status and Trends Benthic Surveillance Project survey of fish livers found concentrations of dieldrin above 0.001 ppm wet weight at 58% of the sampling sites investigated. Contamination was generally greater in inland fish than estuarine and marine forms.[9]

Endrin

A less common contaminant than dieldrin on a nationwide basis, endrin has been monitored in more than 10,000 shellfish and finfish samples. The National Pesticide Monitoring Program survey of 1965 to 1972 uncovered endrin in bivalves from sites in 2 of 15 states.[23] Some samples from San Antonio Bay and Neuces Bay in Texas and parts of Monterey Bay and San Francisco Bay in California had endrin in concentrations >0.01 ppm wet weight.[9] However, endrin levels declined in the early 1970s in estuarine mollusks, and a National Pesticide Monitoring Program resurvey in 1977 did not detect this pesticide in any bivalve.[22] Juvenile estuarine fish surveys conducted during the 1972 to 1976 National Pesticide Monitoring Program discovered endrin in fish at 2 of 144 sampling sites, both in Texas. Endrin contamination has been detected mainly in biota from the Gulf of Mexico and parts of California and secondarily from the mid- and South Atlantic regions.[9]

Lindane

The National Pollutant Monitoring Program reported lindane (Figure 1g) concentrations below the 0.01 ppm wet weight detection limit in more than 8000 shellfish samples and more than 1500 finfish samples. The cyclodiene pesticide was recorded in 44% of the 64 mussel watch samples collected in California during the surveys of 1980 to 1981 through 1985 to 1986. Based on the findings of the National Status and Trends Benthic Surveillance Project survey, lindane does not appear to pose a nationwide contamination threat.[9]

Chlordane

Chlordane, used as an agricultural and domestic pest control and insecticide, is mainly composed of polychloromethanoindenes.[28] National Pesticide Monitoring Program estuarine bivalve monitoring activities from 1965 to 1972 and in 1977 did not detect chlordane (Figure 1f) above the detection limit of 0.01 ppm wet weight in more than 8000 samples.[22,26] During the National Pesticide Monitoring Program surveys from 1972 to 1976, however, chlordane was discerned in 39 samples of whole juvenile fish from five states, the highest concentration being recorded in samples from the island of Kauai in Hawaii. Analyses of fish livers from specimens collected by NOAA at 48 National Status and Trends sites in 1984 disclosed that two target compounds of chlordane, α-chlordane and trans-nonachlor, occurred in an average concentration of 0.038 ppm wet weight.[25] In general, trans-nonachlor was present in greater quantities than α-chlordane. In nongovernmental studies, Hayes and Phillips,[29] as part of the California Mussel Watch, documented a high frequency of chlordane compounds in shellfish. Kawano et al.,[30] evaluating chlordane in samples of marine organisms collected in 1981 and 1982 from the Gulf of Alaska and Bering Sea, found high levels of this contaminant, second only to those of DDT and PCBs.

Results of these two studies indicate that chlordane compounds should continue to be monitored in estuarine and coastal marine organisms.

OTHER PESTICIDES

Kepone

Kepone, also known as chlordecone, was discovered to be a major contaminant of the James River estuary in 1975.[31] Entering this tidal river principally from point sources (e.g., chemical plant discharges, runoff from contaminated landfills, and sewage effluents), kepone was responsible for restrictions on the commercial harvest of some finfishes in the James River for more than a decade. In December 1975, all forms of fishing in the James River were banned, and since that time the ban has been modified several times. The most severe restrictions were imposed on striped bass whose harvest was prohibited year-round. By the mid-1980s, concentrations of kepone in crabs and fish generally ranged from 0.2 to 0.8 ppm wet weight and those in oysters usually below 0.1 ppm wet weight.[9] These contaminant levels are significantly lower than the levels in specimens collected during the mid-1970s. At that time, for example, some fish had kepone concentrations above 7.0 ppm wet weight, and concentrations >1.0 ppm wet weight were commonly observed in many fish populations. Kepone continues to be sampled in bottom sediments from Hopewell to the river mouth; however, it only remains an important contaminant in the lower James River.

Mirex

Closely related to kepone, mirex (Figure 1k) is a pesticide used frequently in the control of ants. It has been utilized most intensely in the southeastern U.S., and serious concern over its application in this region has led to speculation on possible impacts on estuarine organisms of the southeastern U.S. Mearns et al.[9] recounted that of the more than 10,000 shellfish and finfish samples analyzed for mirex, no shellfish or adult fish contained mirex in concentrations above the detection limit of 0.01 ppm wet weight. The NOAA Benthic Surveillance Project of 1984 chronicled the concentrations of mirex in fish livers at 44 sites ranging from < 0.001 ppm wet weight to 0.003 ppm wet weight.[25]

The U.S. Environmental Protection Agency (EPA) banned the use of mirex in 1978 due, in part, to its potential harm to nontarget biota.

Prior to this ban, about 2.5×10^5 kg of mirex were sold in the southeastern U.S., mainly to control fire ants (*Solenopsis* spp.). Fish and wildlife in nine southeastern states and the Great Lakes, particularly Lake Ontario, experienced severe impacts, such as adverse effects on reproduction, early growth, and development; high biomagnification and persistence; delayed mortality of aquatic and terrestrial fauna; and population alterations.[32]

Toxaphene

Historically, toxaphene (Figure 11) has been an important contaminant in streams and lakes, but recent surveys suggest that it may also be a significant estuarine contaminant.[9] Consisting of a mixture of chlorinated camphenes, toxaphene has been widely used as an insecticide. Highest readings of this organochlorine pesticide have been obtained on biota of southern Georgia and southern Laguna Madre, Texas, with secondary occurrences reported on organisms of Oso Bay, Texas, East Bay, California, and the San Francisco Bay-Delta area. The National Pesticide Monitoring Program mollusk surveys from 1965 to 1972 ascertained toxaphene in concentrations <0.25 ppm wet weight in samples at all but 8 of 89 sites investigated, but locations in Georgia, Texas, and California exhibited high levels (e.g., concentrations of 54 ppm wet weight in oysters from Terry Creek near Brunswick, GA and 11 ppm in bay mussels from Hedionda Lagoon, San Diego County, California).[9]

Mearns et al.[9] reviewed three case histories of toxaphene contamination in estuaries, specifically several systems in Brunswick, GA, Laguna Madre, Texas, and the delta island of San Francisco Bay, California. In the late 1960s and early 1970s, discharges of toxaphene from a pesticide plant heavily contaminated marine organisms in at least seven estuaries surrounding Brunswick, GA. Earliest reports of toxaphene contamination in this region were filed in 1967. Large numbers of crustaceans (4 species), mollusks (3 species), and fish (more than 20 species) monitored for toxaphene had levels >1 ppm wet weight and some >10 ppm wet weight.[33,34] Although these levels decreased by 1974 due to control of toxaphene discharges at the source, contamination remained in the parts per million wet weight range in many organisms.

The Arroyo Colorado and southern Laguna Madre in Texas were sites of high toxaphene contamination of biota in the 1970s. Some of the most toxaphene-contaminated fish in the U.S. inhabit this region. Intense pesticide use in four counties bordering the Arroyo Colorado, comprising one third of the total pesticide use in Texas during 1979, accounted for much of the toxaphene discharged into the system. Hence, the concentrations of toxaphene in the muscle tissue of spotted sea trout (*Cynoscion nebulosus*) reached 5 ppm wet weight and that in whole menhaden (*Brevoortia patronus*), 8 ppm wet weight in 1974. In addition to toxaphene, DDT, DDE, DDD, dieldrin, chlordane, endrin, dacthal, and a variety of nonorganochlorine pesticides (i.e., DEF, ethion, ethyl parathion, methyl parathion, and trithion) have been identified in biota of the Arroyo Colorado-Laguna Madre system, reflecting the high degree of pesticide usage in the region.

The delta inland of San Francisco Bay is another area showing high toxaphene contamination in biota. Between 1965 and 1984, the San Francisco Bay-Delta complex was the subject of at least 15 surveys for pesticides or PCBs.[9] This system also provided some of the earliest fish samples (1964) for toxaphene analysis. Toxaphene contamination in the gonads of striped bass (*Morone saxatilis*) ranged from 1.1 to 1.9 ppm wet weight. The levels in mussels from Richmond in Chesapeake Bay approached 0.2 ppm wet weight.[35]

The concentration of toxaphene deemed to be safe for protecting marine life is 0.07 µg/l.[36] Highly hazardous to nontarget estuarine and marine organisms, toxaphene has been responsible for numerous fishkills subsequent to application. It persists in soil and water, having documented half-lives of 9 to 11 years. While it poses a threat to many aquatic organisms, toxaphene does not represent a significant hazard to warm-blooded animals, including waterfowl and other wildlife frequenting estuarine habitats.

Toxaphene served as a favored pesticide during the 1950s and 1960s. However, because of its persistence in aquatic systems, acute toxicity to biota, and bioaccumulation and biomagnification effects, toxaphene was discontinued in fish eradication programs in the 1960s and 1970s.[36] Nevertheless, production of toxaphene, mainly for use against insect pests of cotton, continued to increase in the early 1970s. By 1980, production dropped substantially, and in November 1982, the EPA cancelled the registration of toxaphene for most uses, signaling the end of large-scale application of the pesticide. Despite its restrictive utilization in the 1980s, toxaphene remained a significant contaminant at some estuarine sites during the decade.[9] Its lengthy persistence in aquatic systems is largely responsible for these observed impacts.

CHLORINATED BENZENES AND PHENOLS
Hexachlorobenzene

Hexachlorobenzene (HCB) (Figure 1i) was not a target chemical in national pesticide monitoring programs prior to the NOAA National Status and Trends Benthic Surveillance Project commencing in 1984. Measurable concentrations of this fungicide have been documented in biota from the New York Bight,[37] Upper Chesapeake Bay,[38] Galveston Bay,[39,40] Santa Monica Bay,[41] Commencement Bay,[9] and Elliott Bay.[9] Liver samples of fish taken during the National Status and Trends Benthic Surveillance Project had a median concentration of 0.0013 ppm wet weight HCB and a range of 0.001 to 0.037 ppm wet weight HCB. Only 4.5% of the fish livers exhibited concentrations above 0.01 ppm wet weight HCB. Samples from Puget Sound, specifically from Commencement Bay (0.037 ppm wet weight), Nisqually Reach (0.01 ppm wet weight), and Elliott Bay (0.007 ppm wet weight) contained the highest HCB levels. Comparable levels ranging from 0.024 to 0.060 ppm wet weight HCB were recorded in flatfish livers collected near two Southern California outfall sites in Santa Monica Bay.[41] Based on these data, hexachlorobenzene does not appear to be an important contaminant in estuarine fish in the U.S.

Pentachlorophenol

A major metabolite of HCB is pentachlorophenol (PCP) (Figure 1n), perhaps best known as a wood and textile preservative (fungicide), but also used as a slimicide in the paper pulp industry and as an insecticide for controlling termites and other woodboring insects. Despite being highly toxic to aquatic organisms, PCP had not been a target chemical in national surveys of estuarine and marine contaminants. When monitored in biota during site-specific studies, PCP concentrations have generally ranged from about 0.008 to 0.02 ppm wet weight. For instance, PCP levels in blue crabs (*Callinectes sapidus*) and brown shrimp (*Penaeus aztecus*) from San Luis Pass, Texas, amounted to 0.002 and 0.017 ppm wet weight, respectively.[41] Oysters from Galveston Bay had PCP levels ranging from 0.003 to

0.008 ppm wet weight. At the site of a wood-treatment operation in Eagle Harbor, Puget Sound, clams screened for PCP contained concentrations ranging from 0.003 to 0.008 ppm wet weight.[42] Investigations of marine fish from the Penobscot River in Maine, the Raritan River in New Jersey, Cape Fear River in North Carolina, and the Willamette River in Oregon revealed levels of PCA (pentachloroanisole), a metabolite of PCP, in the range of 0.03 to 0.07 ppm wet weight.[9]

POLYCHLORINATED BIPHENYLS

INTRODUCTION

PCBs (Figure 11) are a group of synthetic halogenated aromatic hydrocarbons that have been linked to a number of environmental and public health concerns. Consisting of a mixture of chlorinated biphenyls that contain a varying number of substituted chlorine atoms on aromatic rings,[43] PCBs have been scrutinized extensively since the 1970s when their presence in the aquatic environment was initially perceived as a serious contamination problem.[44] First produced industrially in 1929, PCBs rapidly gained widespread use because of their unique physical-chemical properties, including chemical and thermal stability, miscibility with organic compounds, high dielectric constant, nonflammability, and low cost. When commercial production of PCBs ceased in 1977, an estimated 0.54 billion kg of these compounds had been produced in the U.S. for use in the manufacture of adhesives, caulking compounds, additives to hydraulic oils, capacitor fluid, fluorescent light ballasts, paints, varnishes, fire retardants, plastics, carbonless copying paper, newsprint, and other materials.[45,46] The electric utility industry, recognizing the remarkable insulating capacity and nonflammable characteristics of PCBs, employed them as insulating fluids in electrical transformers and capacitors. Aside from these applications, PCBs have been utilized as fillers for casting waxes, as extenders in pesticides, and as control agents in road construction. Their most common use in the U.S., however, has been in the manufacture of electrical capacitors and transformers.

PHYSICAL-CHEMICAL PROPERTIES

Sawhney,[3] Hutzinger et al.[45] Brinkman and DeKok,[47] and Cairns et al.[48] give excellent descriptions of the physical-chemical properties of PCBs. Biphenyl is the basic structural unit of these synthetic chlorinated organic compounds.[3] By the controlled chlorine substitution of the biphenyl molecule, a wide variety of PCBs can be developed, including a total of 209 possible chlorobiphenyls (congeners), the structures of which are given by Ballschmiter and Zell.[49]

Individual chlorobiphenyls are prepared by the phenylation or arylation of aromatic compounds. Industrial preparations have entailed the chlorination of biphenyl in the presence of a catalyst (e.g., iron filings or iron chloride).[3] The complex mixture of chlorobiphenyls constituting the purified PCB product contains 18 to 79% chlorine.[50]

Ten PCB congener groups exist, owing to ten possible degrees of chlorination of the biphenyl molecule, namely mono-, di-, tri-, tetra-, penta-, hexa-, hepta-, octa-, nona-, and decachlorobiphenyl (Figure 2).[50,51] Within these groups, 209 possible isomers can be delineated (Table 2). Mathematical and computer-assisted techniques aid in the identification and quantification of the isomers.[50,52,53] One of the most viable methods of chemical analysis for PCBs is gas-liquid chromatography (GLC) aided by an electron capture detector.[54]

In the U.S., PCBs have been manufactured under the trade name Arochlor®; by increasing the chlorine content, the consistency of Arochlors change from colorless mobile oils (Arochlor® 1016, 1221, 1232, 1242, and 1248) to viscous liquids (Arochlor® 1254), sticky resins (Arochlor® 1260 and 1262) and white powders (Arochlor® 1268 and 1270). PCBs have also been marketed under several other trade names, notably Chloretol, Dyknol, Inerteem, Noflamol, and Pyranol. In foreign countries, PCBs have been marketed as Delor (Czechoslovakia), Phenoclor and Pyralene (France), Apirolio and Fenclor (Italy), Kanechlor and Santotherm (Japan), Soval (Russia), and Chophen (West Germany).[50] Although commercial production of PCBs in the U.S. ceased in 1977, older transformers and capacitors in particular still contain significant amounts of PCBs as dielectric fluids.[50,51]

Sawhney[3] reviews the physical-chemical properties of PCBs, and the following discussion draws heavily from his work. PCBs,

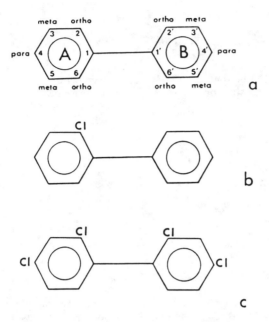

FIGURE 2. a = Structure of biphenyl (modified from Safe[51]); b = 2-monochlorobiphenyl; c = 2,2',4,4'-tetra-chlorobiphenyl. (From Eisler, R., Polychlorinated Biphenyl Hazards to Fish, Wildlife, and Invertebrates: A Synoptic Review, Biol. Rep. 85 (1.7), U.S. Fish and Wildlife Service, Washington, D.C., 1986.)

extremely stable organic compounds, are very persistent in aquatic environments. Bacteria degrade PCBs, with the rate of degradation contingent upon the degree of chlorination and the position of the chlorine atom on the biphenyl molecule.[50] Higher chlorinated biphenyl compounds are transformed at a slower rate by bacteria than lower chlorinated biphenyls.[55]

In contrast to most individual chlorobiphenyls that exist as solids at room temperature, commercial preparations generally occur as resins or viscous fluids with densities greater than that of water (Table 3). Their transport and distribution in aquatic habitats depends in part on sorption reactions, solubility, and vaporization. Haque and Schmedding[56] specified that as the chlorine content of chlorobiphenyls increases not only do the compounds sorb more tightly onto sorbent surfaces, but they also sorb in greater quantities to the surfaces. Sediment sorption of PCBs has been related to the

TABLE 2
Number of Possible Isomers of PCBs

Chlorine substitution	Number of possible isomers
Mono-	3
Di-	12
Tri-	24
Tetra-	42
Penta-	46
Hexa-	42
Hepta-	24
Octa-	12
Nona-	3
Deca-	1
Total	209

From Cairns, T., Doose, G. M., Froberg, J. E., Jacobson, R. A., and Sigmund, E. G., in *PCBs and the Environment,* Waid, J. S., Ed., CRC Press, Boca Raton, FL, 1986, 1. With permission.

TABLE 3
Physical Characteristics of Four Common Arochlors®

Arochlor®	Dielectric constant (at 1000 c)		Distillation range(°C)	Density
	25°C	100°C		
1242	5.8	4.9	325–366	1.38
1248	5.6	4.6	340–375	1.44
1254	5.0	4.3	365–390	1.53
1260	4.3	3.7	385–420	1.62

total organic carbon associated with the grains.[57] As the organic carbon content of the sediments rises so does the amount of PCBs sorbed to the particles.[58] Highest concentrations of PCBs typically occur in sediments having a large fraction of clays, organic matter, or microparticulates.[43,55,59] In the marine realm, the North Atlantic Ocean represents the principal sink for PCBs, accounting for 50 to 80% of these compounds in the environment.[43] Lacustrine sediments serve as a major repository of PCBs on continents.[43,50]

PCBs have low aqueous solubilities which are valuable indicators of their environmental fate; however, they are freely soluble in biological lipids.[43,50] Absolute solubility values range from 2.7 ppb for Arochlor® 1260 to 3500 ppb for Arochlor® 1221.[58] Table 4 lists aqueous solubilities of individual PCB isomers and commercially available mixtures of PCBs. As the degree of chlorination increases, the solubility of PCB isomers decreases in water.[3] For individual chlorobiphenyls, solubilities range from 0.007 ppm for octachloro-biphenyl to approximately 6 ppm for monochlorobiphenyl.[45] As explained by Sawhney,[3] environmental conditions strongly influence PCB solubilities. For example, the concentration of organic matter, as noted above, influences the sorption of PCBs on sediment surfaces which ultimately decreases their solution concentration.

With increasing chlorination in PCBs, vapor pressures decline. Vaporation rates differ 200-fold from Arochlor® 1221 to 1260 (Table 5). The vaporization rate drops markedly if PCBs are sorbed on sediment surfaces. Some factors purportedly affecting the vaporization of sorbed PCBs are the surface area of the grain, organic matter content, type of clay, and pH of the medium.[3]

ENVIRONMENTAL EFFECTS

In 1976, the Toxic Substances Act was passed enabling the EPA to ban the production of PCBs in the U.S. and to regulate their disposal. This banning occurred because PCBs were linked to acute and chronic health effects in man, such as: (1) cancer; (2) liver damage; (3) skin lesions; and (4) reproductive disorders. Their toxic effects were documented.[55] In addition, these synthetic compounds were found to be deleterious to aquatic life, especially organisms at higher trophic levels (e.g., fish) that serve as a source of food for humans.

TABLE 4
Solubility, K_{oc} and K_{ow} of Several PCBs

Compound	Solubility (ppb)	log S	Ref.[a]	K_{oc}	log K_{oc}	Ref.[a]	K_{ow}	log K_{ow}	Ref.[a]
Biphenyl	7,500	3.88	43	2,512	(3.40)		7,540	3.88	18
Monochlorobiphenyls									
2-	5,900	3.77	27	2,951	(3.47)		14,790	(4.17)	
3-	3,500	3.54	27	4,168	(3.62)		21,877	(4.34)	
4-	1,190	3.08	26, 27	7,943	(3.90)		79,400	4.90	50
Dichlorophenyls									
2,4-	1,400	3.15	27	7,244	(3.86)		41,686	(4.62)	
2,2'-	1,500	3.18	27	6,918	(3.84)		39,810	(4.60)	
2,4'-	1,260	3.10	23, 27	8,000	3.90	8	45,708	(4.66)	
4,4'-	80	1.90	27	42,658	(4.63)		346,736	(5.54)	
Trichlorobiphenyls									
2,4,4'-	85	1.93	27	40,738	(4.61)		323,593	(5.51)	
2',3,4-	78	1.89	27	43,652	(4.64)		346,736	(5.54)	
Tetrachlorobiphenyls									
2,2'5,5'-	36	1.56	23, 27	47,000	4.67	8	602,559	(5.78)	
2,2',3,3'-	34	1.53	27	72,443	(4.86)		645,654	(5.81)	

Compound										
2,2',3,5'-	170	2.23		27	26,915	(4.43)		194,984	(5.29)	
2,2',4,4'-	66	1.82	24, 25, 27	27	47,863	(4.68)		158,400	5.20	49
2,3',4,4'-	58	1.76		27	52,480	(4.72)		436,555	(5.64)	
2,3',4,5'-	41	1.61		27	64,565	(4.81)		562,341	(5.75)	
3,3',4,4'-	180	2.26		27	25,633	(4.41)		186,208	(5.27)	
Pentachlorobiphenyls										
2,2',3,4,5'-	22	1.34		27	95,324	(4.98)		870,693	(5.94)	
2,2',4,5,5'-	31	1.49		27	76,948	(4.89)		691,830	(5.84)	
Hexachlorobiphenyl										
2,4,5,2',4',5'-	0.95	-0.02		23	1,200,000	6.08	43	5,248,000	6.72	15
Arochlors®										
1221	3,500	3.54		19	4,123	(3.62)		12,300	4.09	25
1232	1,450	3.16		12	7,092	(3.85)		41,686	(4.62)	
Used capacitor fluid	698	2.84		19	10,725	4.03		70,794	(4.85)	
1016	332	2.52	10, 12, 14		17,684	(4.25)		202,000	5.31	14, 18
1242	288	2.46	12, 13, 16, 17, 20		12,400	4.09	44	196,500	5.29	14, 18
1248	54	1.73		27	54,626	(4.74)		562,000	5.75	12
1254	42	1.62	15, 20, 21, 22		63,914	(4.82)		1,288,000	6.11	15
1260	2.7	0.43		12	349,462	(5.54)		4,073,800	(6.61)	

[a] See Reference 53.

From Chou, S. F. J. and Griffin, R. A., in *PCBs and the Environment*, Vol. 1, Waid, J. S., Ed., CRC Press, Boca Raton, FL, 1986, 101. With permission.

TABLE 5
Vaporization Rates of Six Arochlors® Measured at 100°C[2]

Arochlor® (12.28 cm² surface)	Vaporization rate (g/cm²/h)
1221	0.00174
1232	0.000874
1242	0.000338
1248	0.000152
1254	0.000053
1260	0.000009

From Sawhney, B. L., in *PCBs and the Environment,* Vol. 1, Waid, J. S., Ed., CRC Press, Boca Raton, FL, 1986, 47. With permission.

Several characteristics make PCBs a threat to marine food webs. They are highly stable in marine waters, have a high specific gravity, and have a high affinity for solids. With a low solubility in water and high solubility in fats and oils, they tend to partition out of the aquatic ecosystem into organismal tissue, becoming concentrated in lipid-rich areas. PCBs bioaccumulate in the food chain principally because of their great stability and persistence and because they are poorly metabolized by biological systems.[60]

Despite the ban on production of PCBs in the U.S. in 1977, PCB contamination continues to be a universal problem.[61] The compounds occur in most organisms, as well as in the air, water, soils, and sediments worldwide. The National Academy of Sciences[55] estimates that 82×10^6 kg of PCBs exist in the natural environment, in disposal sites, in deployed transformers, and in other containers.

Aquatic environmental surveys have focused on the uptake of PCBs by finfish and shellfish suitable for human consumption. Fish exposed to PCBs experience a higher incidence of fin erosion, epidermal lesions, blood anemia, and altered immune response. Studies have indicated a lower disease resistance in fish exposed to PCBs.[62] The U.S. Food and Drug Administration (FDA) effectively controls public exposure to orally ingested PCBs, originally setting their upper limit in the edible portions of fish and shellfish at 5.0 ppm, and later lowering the tolerance level of PCBs to 2.0 ppm in August 1984, as a result of new toxicity data and declining incidences.

Because of their high affinity for particulate matter, notably organic matter and sediments, PCBs readily adsorb to particulate surfaces and subsequently settle to the river- or seabed. Consequently, the highest concentrations of these compounds are expected in bottom sediments which serve as a transient reservoir of the compounds.[63] Far lower concentrations occur in the water column,[64] especially surface waters with low particulate loads. Bottom sediments may contain substantial concentrations of PCBs despite nearly undetectable levels being present in the water column. The compounds persist in the estuarine and marine environments; the half-life of these hydrocarbons in environmental samples ranges from 8 to 15 years.[65]

The highest values of PCBs have been obtained near point sources, most typically in contaminated sediments in proximity to point sources.[43,66] PCBs tend to be more widespread in fish than in sediments because of the great mobility of the nektonic organisms and bioaccumulation effects. The FDA considers fish to be the single most important dietary contributor of PCBs in humans.

PCB Concentrations in Water

Table 6 amasses PCB ranges recorded for rivers, estuaries, coastal waters, and the open ocean. As is evident from this table of data, PCB levels vary considerably, with concentrations in estuaries and coastal waters exceeding those in the open ocean. The highest levels, as expected, exist in rivers exposed to industrial sources of the contaminants.[67]

In the water column, PCBs partition between particulate and soluble phases. Under normal suspended loads in coastal waters (i.e., 0.5 to 10 g/m^3), Pavlou and Dexter[68] concluded that <20% of the PCBs would be sorbed on inorganic particulate matter. However, in extreme cases of excessively high concentrations of either suspended matter or PCBs, the quantities of PCBs in solution may be less than those in the particulate fraction. The Rhine-Meuse estuary and its influent systems exemplify the condition whereby high turbidity results in larger amounts of PCBs sorbed to particulates than in solution in the water column.[69]

Uncertainties exist concerning the partitioning of PCBs among various particulate fractions in natural waters. Affinities of PCBs for specific particulates — colloids, humates, sedimentary grains —

TABLE 6
Ranges in PCB Concentrations (ng/l) Reported for Open Oceans, Coastal Waters, and Estuaries or Rivers

Area	Location	Range in PCBs (ng/l)	References[67]
Open oceans	North Atlantic	<1 –150	Harvey et al, 1974
		0.4 – 41	Harvey et al.,1973
	Sargasso Sea	0.9 – 3.6	Bidleman and Olney, 1974
	North-South Atlantic	0.3 – 8.0	Harvey and Steinhauer, 1976
	Mediterranean Sea	0.2 – 8.6	Elder and Villeneuve, 1977
Coastal waters	Southern California	2.3 – 36	Scura and McClure, 1975
	Northwest Mediterranean	1.5 – 38	Elder et al., 1976
	Atlantic coast, U.S.	10 –700	Sayler et al., 1978
	Baltic coasts	0.3 –139	Brugmann and Luckas, 1978
		0.1 – 28	Ehrhardt, 1981
	Dutch coast	0.7 – 8	Duinker and Hillebrand, 1979
Estuaries/rivers	Wisconsin rivers, U.S.	<10 –380	Dennis, 1976
	Rhine-Meuse system, Holland	10 –200	Duinker and Hillebrand, 1979
	Tiber estuary, Italy	9 –1000[a]	Puccetti and Leoni, 1980
	Brisbane River, Australia	ND – 50	Shaw and Connell, 1980
	Hudson River, U.S.	<100 – 2.8×10^6	Nisbet, 1976

Note: ND, not detectable (no limits quoted).

[a] As decachlorinated biphenyl equivalents.

From Phillips, D. J. H., and *PCBs in the Environment,* Vol. 2, Waid, J. S., Ed., CRC Press, Boca Raton, FL, 1986, 127. With permission.

have not been clearly elucidated, and much needs to be learned regarding the transfer of PCBs from one phase to another.[70] A number of studies have established the high affinities of PCBs for phytoplankton and zooplankton.[70,71-75] The hydrophobic, lipophilic nature of these compounds is responsible for their concentration and persistence in the plankton. Phytoplankton constitutes the basis of grazing food webs in estuaries and oceans and, consequently, a root cause of or significant contributor to the amplification of PCBs up the food chain.

PCB levels tend to be higher in surface films than subsurface waters.[68,76,77] The surface microlayer is critical to the transport of PCBs from sediment and subsurface waters to the atmosphere.[77] Surface-active materials accumulate, particularly in areas where currents converge, to form slicks or films. They are especially common in coastal waters.[78] Concentrating at the sea surface, surface-active agents can alter the properties of the air-seawater interface, affecting the exchange of substances between the atmosphere and oceans or estuaries.[78] Many other contaminants (e.g., DDT, DDE, lindane, etc.), in addition to PCBs, concentrate in surface films. By means of volatilization or codistillation, PCBs together with some of these other contaminants, can pass across the air-seawater interface to the atmosphere.[77]

Progressive contamination by PCBs in the northern hemisphere is evident from the higher concentrations found in North Atlantic waters relative to the South Atlantic. Indeed, PCB values in open ocean surface waters in the midlatitudes of the northern hemisphere are much greater than those recorded in the southern hemisphere. Gradual contamination of North Atlantic waters by PCBs attributable to their extensive use by the U.S. and Europe over the years probably accounts for the observed ocean geographical distribution of these organochlorine compounds.[78]

PCB Concentrations in Sediments

Suspended and bottom sediments act as a sink for PCBs and other hydrophobic organic pollutants. The strong affinity of PCBs for sediments presumably results from equilibrium sorption or partitioning, which is a function of the aqueous solubility of PCB isomers and the attractiveness of the sedimentary matrix to PCBs.[79] The concentration of organic matter, as discussed above, is fundamentally important in the PCB-sediment association. The sorption capacity of marine and estuarine sediment for PCBs rises with increases in the amount of organic matter.

The use of bottom sediments to quantify PCBs in aquatic environments offers several advantages over that of overlying water samples. First, sediment samples provide a time-averaged indication of PCB contamination at a site. Second, difficulties of sample handling diminish when bottom samples are utilized. In spite of these advantages, several factors can obfuscate precise interpretation of PCBs in seafloor sediments. For instance, the samples taken may not be representative of the general seafloor due to practical deficiencies of sampling with remote coring or graph devices that can disturb the bottom and bias a sample. The investigator must also consider other factors possibly affecting PCB concentrations in an area, including sedimentation rates, grain size, content of organic matter, and the depth of the sediment sample.[67]

PCB concentrations in bottom sediments vary widely from very low levels in relatively uncontaminated marine areas[80,81] to elevated readings in highly polluted estuarine and coastal marine systems (Table 7).[69,82-84] Seafloor sediments largely untainted by anthropogenic inputs of PCB compounds usually contain concentrations of less than 25 ng/g dry weight. It is not uncommon for heavily contaminated systems to have PCB levels >1000 ng/g dry weight in sediments (e.g., Hudson River, New York Bight, and Escambia Bay). In those systems where PCB contamination occurs concomitantly with high rates of sedimentation (e.g., the Christiaensen Basin of the New York Bight and sections of the Husdon River), the organochlorine compounds can be buried rapidly and, in effect, removed from the general PCB cycle in a rather brief interval of time. Bioturbation by benthic organisms tend to counter this burial process, but ultimately, significant quantities of the compounds will accumulate at depth. Thus, West and Hatcher[84] discerned substantial amounts of PCBs at depths >35 cm in the Christiaensen Basin, and Bopp et al.[85] detected PCBs at depths >40 cm in the Hudson River.

In New Jersey, a state whose waters have been subjected to considerable quantities of PCBs in the past, sediments of Raritan Bay and the New York Bight are important sources of PCB contamination of estuarine biota. PCBs in bottom sediments of these regions originate principally from the Hudson River plume, dredged-material dumping, and sewage sludge.[86] The mean value of PCBs in sediments of Raritan Bay, as cited by Stainken and Rollwagen,[87] equals 100 ng/g dry sediment. PCB residues in the bay sediments range from 3.4 to 2035 ng/g dry sediment. According to McCormick et al.,[88] mixtures of Arochlor® 1016 and 1242 appear to be present in all sediments.

Sewage sludge and dredged-material dumping are responsible for most of the PCB contamination of New York Bight sediments.[84,89,90] Bopp et al.[85] estimated that 70 and 30% of the total PCB loading in the New York Bight during the late 1970s were ascribable to dredged

TABLE 7

Ranges in Concentration of PCBs (ng/g dry weight) Reported for Sediments in Areas Ranging from Relatively Uncontaminated to Highly Contaminated

Area	PCB conc.	References[67]
Mediterranean Sea	0.8–9	Elder et al., 1976
Gulf of Mexico	0.2–35	Nisbet, 1976
Chesapeake Bay	4–400	Sayler et al., 1978
Lake Superior	5–390	Eisenreich et al., 1979
Tiber estuary	28–770	Puccetti and Leoni, 1980
Rhine-Meuse estuary	50–1,000	Duinker and Hillebrand, 1977
New York Bight	0.5–2,200	West and Hatcher, 1980
Palos Verdes Peninsula	30–7,900	Pavlou et al., 1977
Hudson River	tr–6,700	Nisbet, 1976
Escambia Bay	190–61,000	Nisbet, 1976

Note: tr, trace.

From Phillips, D. J. H., and *PCBs in the Environment,* Vol. 2, Waid, J. S., Ed., CRC Press, Boca Raton, FL, 1986, 127. With permission.

material and sewage-sludge dumping, respectively. An estimated 25% of the PCBs in dredged material is derived from upstream wastewater discharges; these PCBs settle to the seafloor subsequent to adsorption onto harbor sediments. A much smaller proportion of the contaminants, probably <10% of the dredged material enters the New York Bight from other riverine and estuarine PCB sources.[91]

PCBs in Aquatic Organisms

When PCBs are removed from the general PCB cycle through deep burial in seafloor sediments, assessment of present-day levels of contamination is better served via the use of bioindicators than sediments. Certain species of organisms accumulate pollutants in their tissues in amounts proportionate to their ambient environment; these species proffer a time-averaged indication of pollutant bioavailability in the environment and, consequently, are potentially valuable monitoring tools.[67] Phillips[67] outlines the attributes of an ideal indicator (or sentinel) organism:

1. It accumulates the pollutant studied in proportion to the average levels present in the ambient waters.

2. It should be sedentary in order to be representative of the area in which it is collected.
3. It should be abundant throughout the study area and should tolerate brackish water.
4. It should tolerate the presence of high levels of the pollutant studied without being toxicologically affected.
5. It should be of reasonable size, giving adequate tissue for analysis.
6. It should be easy to sample and should preferably be hardy enough to survive in the laboratory.
7. It should be relatively long-lived, to permit time-integration of pollutant levels over several months or years.

While indicator organisms have great practical value in quantifying pollutants in aquatic environments, a variety of factors, such as the organism's age, sex, condition, and mobility, must be considered since they may affect the correlation between the concentrations of a pollutant in the organism's tissues and those in the environment. The lipid content of an organism is especially important in monitoring PCBs in aquatic systems.[67] Hence, both field surveys and laboratory analyses must be carefully planned and executed to minimize biasing that will preclude accurate profiles of PCB abundance at specific sites.

Of the myriad of potential estuarine and marine bioindicator organisms, probably the most suitable for environmental monitoring of lipophilic compounds are benthic invertebrates that utilize glycogen as their principal energy depot.[92] Particularly since the inception of the U.S. Mussel Watch Program in 1976, the blue mussel (*Mytilus edulis*) has become a key bioindicator in pollution monitoring studies, including those involving assessment of chlorinated hydrocarbons, petroleum hydrocarbons, radionuclides, and trace metals.[93-95] The value of mussels in pollution monitoring programs is the direct result of their broad geographical distribution, population dynamics, physiology, and ability to sequester a diverse number of contaminants. Preston,[96] borrowing from the work of Goldberg,[97,98] proposes seven reasons why *M. edulis* is among the foremost bioindicators for pollution monitoring purposes:

1. The species is wide ranging in the northern hemisphere and should be readily introducible to those areas where it does not occur.
2. Mussels are a bay and sheltered coastal species and will, therefore, record environmental levels well in such localities which are frequently those most exposed to contamination.
3. *Mytilus edulis* has been extensively studied both experimentally and ecologically with respect to its ability to record pollutant levels.
4. It is a filter-feeder, and its visceral mass is compact and easily extractable from the shell. Being a filter-feeder means that it tends to concentrate contaminants to readily measurable levels.
5. Contaminant uptake is generally reversible, and storage of organic compounds occurs with little breakdown. Mussels, therefore, give an indication of the mean, prevailing contamination.
6. Mussels accumulate a wide variety of contaminants, including many metals and radionuclides, saturated and unsaturated hydrocarbons, and chlorinated hydrocarbons.
7. Chemical analysis of mussels is reasonably straightforward and economical.

An example of the usefulness of mussels for research contaminants in estuarine and marine waters is provided by the surveys of deKock[99] in the Netherlands. Employing *M. edulis*, surveys were conducted of Cu bioavailability in and around the eastern Scheldt estuary, Hg bioavailability in the water column of the Ems-Dollard estuary, and Zn bioavailability along the North Sea coast. In addition, two PCBs were monitored in aquatic habitats to evaluate the effectiveness of legal restrictions to PCB use.

Bivalves other than mussels likewise have been the subject of monitoring projects involving PCBs. For instance, Vreeland[100] studied the uptake of PCB compounds by the American oyster, *Crassostrea virginica*, discovering that uptake was proportional to PCB concentration and varied with degree of isomer chlorination. The hard clam, *Mercenaria mercenaria*, experiences preferential uptake of lower chlorinated biphenyls.[101] Langston[102,103] ascertained selective accumulation of PCB components in the cockle, *Cerastoderma* (=*Cardium*) *edule*, and the baltic clam, *Macoma balthica*, which parallels the pattern of accumulation in *C. virginica*.[104]

Factors Affecting PCB
Concentrations in
Aquatic Biota

Aside from the lipid levels of an organism, which is the most important factor controlling PCB levels in biota, the condition of the organism, its size, and season of sampling all may potentially influence PCB concentrations in a species. Phillips[67] cited four basic mechanisms that can cause PCB levels to vary with organism size: (1) the greater surface area to volume ratio of small individuals enables them to accumulate higher PCB concentrations than larger individuals; (2) the lipid content of a species may be size dependent; (3) slow PCB excretion in a species during ontogeny can lead to size- or age-dependent PCB variations; and (4) changes in diet, migration, and the attainment of sexual maturity during ontogeny, with an increase in individual size, potentially influence the amount of PCBs in organismal tissue. Among estuarine fauna exhibiting size related PCB variations are the horseshoe crab (*Limulus polyphemus*), sand shrimp (*Crangon septemspinosa*), and polychaete (*Nereis virens*), with smaller individuals taking up greater quantities of PCBs.[105,106] In contrast, the hard clam, *Mercenaria mercenaria*, shows no difference in PCB concentrations with size or age,[101] as do many other bivalves.[67]

Estuarine fishes, more so than bivalve mollusks, have PCB levels dependent upon organismal size. In many of these fishes, lipid content, as well as PCB concentrations, is length dependent. Phillips[107] hypothesizes that inconsistent increases in lipid levels with age or size in bivalve mollusks accounts for the observed differences in the two groups. Positive correlations between fish length and PCB concentrations have been documented in the sheepshead minnow (*Cyprinodon variegatus*),[108] Atlantic herring (*Clupea harengus*),[109] and striped mullet (*Mugil cephalus*),[110] as well as a host of other species.[67]

Uptake of PCBs by
Estuarine Organisms:
National Surveys

Several national programs have been undertaken during the past 25 years to survey PCBs in bivalves and finfishes. The National Pesticide Monitoring Program analyzed PCBs in oysters from Chesapeake Bay in 1971 and 1972, as well as from Escambia Bay (Florida) in 1970, 1971, and 1972.[23] The highest recorded PCB concentration in an oyster from the Chesapeake Bay system was 2.8 ppm wet weight for a specimen from the Elizabeth River. Later surveys conducted by the states of Maryland and Virginia from 1976 to 1986 indicated significant reductions of PCB contamination in the bivalves of the bay amounting to as much as a 10-fold decrease in concentrations over that recorded by the National Pesticide Monitoring Program in 1971 and 1972. PCB levels from 1976 to 1986 were low and variable in the main body of the upper estuary. In Escambia Bay, the mean concentration of PCBs in oysters from 1970 to 1972 equalled 0.165 ppm wet weight.[9] Wilson and Forester,[111] surveying PCB levels in the oysters of East Bay and Escambia Bay from 1969 to 1976, showed that the mean annual Arochlor® 1254 concentration in the bivalves declined from approximately 12 ppm wet weight in 1969 to 2 ppm wet weight in 1976. A resurvey by the National Estuarine Mollusk Program of the National Pesticide Monitoring Program in 1977 revealed a substantial drop in PCB contamination such that the detection limit of 0.05 ppm wet weight was no longer appropriate.[9,21]

Dramatic declines of PCB contamination in bivalves during the past 2 decades have been uncovered in other impacted systems as well. Mussels monitored for total PCBs at the base of the White Point outfalls on the Palos Verdes Peninsula showed a 20-fold drop in the annual mean total PCB contamination from approximately 0.4 ppm wet weight in 1971 to <0.02 ppm wet weight in 1982. This marked reduction reflected improved source control of sewage-borne PCBs. In Narragansett Bay, Rhode Island, Arochlor® 1254 concentrations in mussels fluctuated in the range of 0.4 to 0.5 ppm dry weight between 1976 and 1979,[112] but the 1986 NOAA Mussel Watch

Program divulged total PCB levels of about 0.2 ppm dry weight. Measurements of Arochlor® 1242 in Boston Harbor mussels revealed peak readings in 1978 and lower readings from 1976 to 1977 and in 1981 to 1982. Mussels monitored in Beaufort, NC, had higher PCB concentrations in 1981 (0.15 ppm dry weight) than in 1976 to 1977 (0.05 ppm dry weight). Seasonal variations of PCB concentrations in bivalves are superimposed over long-term data trends and may complicate the assessment of temporal changes in PCB contamination of the biota.

Based on results of the National Pesticide Monitoring Program from 1970 to 1972 and in 1977,[23] the EPA Mussel Watch Program from the mid-1970s,[112] and the 1986 NOAA Mussel Watch Program,[113] the most contaminated sites of PCBs in regard to bivalves occur along the northeast coast of the U.S. and in the harbors of Southern California. The EPA Mussel Watch Survey of 86 nationwide sites in 1976 documented a median value for the PCB mixture, Arochlor® 1242, of 0.009 ppm wet weight and a range of 0.0008 to 2.09 ppm wet weight. Similar values were listed in the 1986 NOAA Mussel Watch Survey of 144 sites, with a median reading of total PCBs amounting to 0.015 ppm wet weight and a range of 0.0009 to 0.68 ppm wet weight. Because of analytical, site, and species differences between these two national surveys, however, comparisons of the two databases cannot be seriously attempted unless some correction or adjustment of the data is made.[9]

PCB contamination of finfish generally is greater than that of shellfish. Total PCB concentrations as high as 6.59 ppm wet weight have been noted in whole estuarine or coastal fish, but the mean value in coastal fishes is slightly <0.10 ppm wet weight.[9] Nevertheless, PCB contamination in estuarine fish is widespread, and its toxicity to this group has been a major cause of concern.[114] Questions still abound, however, regarding PCB toxicity of estuarine and marine ecosystems.[115]

Similar to other forms of estuarine and marine life, fish displaying the highest concentrations of PCBs inhabit urban embayments of the northeastern and northwestern regions of the U.S., as well as Southern California.[9] Results of a national survey of PCBs in whole fish conducted during the Juvenile Estuarine Fish Monitoring Program of the National Pesticide Monitoring Program from 1972 to 1976 detected PCBs in 331 of 1524 composite samples taken at 144 stations.[21] The median value of PCBs was < 0.10 ppm wet weight. Whole fish samples collected at 58% of the stations had PCB levels less than the detection limit of 0.05 ppm wet weight. High values were chronicled in the Chesapeake-Delaware canal (mean PCB concentration = 1.9 ppm wet weight). By comparison, the 1976 to 1977 National Pesticide Monitoring Survey of whole inland U.S. fish sampled at 99 locations cataloged a mean PCB concentration of 0.88 ppm wet weight which is nearly 10 times greater than the mean level recorded in juvenile estuarine fish.[116]

During the past 15 years, more PCB determinations have been made of fish muscle and liver than of whole fish. Approximately 90% of the muscle tissue samples analyzed for PCBs during the NOAA/EPA Cooperative Estuarine Monitoring Program had PCB levels below the detection limit of 0.4 ppm wet weight.[26] Gadbois and Maney,[117] reporting on fillet samples of a mix of pelagic and predatory fishes from 15 coastal and estuarine areas during 1979 and 1980, demonstrated highest mean concentrations of PCBs in samples of the New York Bight apex (1.1 ppm wet weight) and East Bay, Florida (0.42 ppm wet weight) and lowest mean concentrations at Catalina Island, offshore of Los Angeles (<0.04 ppm wet weight) and Candeleur Sound, east of New Orleans (0.05 ppm wet weight). Mearns et al.[9] compared PCB levels in muscle or flesh of several species of flatfish sampled in 1980. They calculated that total PCB levels in coastal flatfish from Atlantic and Pacific coastal regions were greater than (>0.1 ppm wet weight) those of the Gulf of Mexico (<0.1 ppm wet weight). Flatfish taken from New Bedford Harbor, Massachusetts, contained the highest concentrations of PCBs in their muscle tissue. Samples gathered in 1970, 1976, and 1980 had mean PCB levels of 7.98, 3.15, and 2.12 ppm wet weight, respectively. These values far exceeded the mean level of 0.18 ppm wet weight in flatfish muscle from the New York Bight. However, it is important to note the declining trend in PCB contamination of New Bedford Harbor fish from nearly 8.0 ppm wet weight to almost 2.0 ppm wet weight from 1970 to 1976. Nevertheless, the concentrations of PCBs in flatfish muscle of New Bedford Harbor continued to exceed the U.S. FDA action limit of 2.0 ppm wet weight for PCBs in edible tissue for all years. On the Pacific Coast, flatfish muscle samples from Elliott Bay, Washington, and the Palos Verdes Peninsula, Southern California, exhibited the highest PCB concentrations among the coastal ecosystems sampled. The mean concentrations of Elliott Bay samples did not exceed 1.0 ppm wet weight in any year and those of the Palos Verdes Peninsula surpassed 2.0 ppm wet weight during only one year, peaking at 2.58 ppm wet weight in 1971.

Fish livers generally have greater PCB concentrations than muscle or whole fish. For example, the PCB concentration of a liver excised from a starry flounder from the Duwamish River near Seattle, WA, in 1977 equalled 160 ppm wet weight,[18] far above any PCB value for fish muscle or whole fish during a national survey program.[118] The median PCB concentration of fish livers in samples collected during the 1984 National Status and Trends Benthic Surveillance Project at 42 Pacific and northeastern U.S. coastal sites amounted to 0.58 ppm wet weight.[9] This value is nearly identical to the median from all site means of Butler[26] and Sherwood[118] (0.6 ppm

TABLE 8
Total PCB Concentrations in Fish Livers from Nine Estuarine and Coastal Marine Systems[a]

| | tPCB, ppm, ww | |
Area	1976–77	1984
Western Long Island Sound	0.62	0.81
Lower Chesapeake Bay, Virginia	0.62	0.28
Duwamish River/Elliot Bay, Washington	26.7	4.23
Nisqually Reach, Washington	0.31	0.49
Columbia River, Oregon	0.24	0.20
Coos Bay, Oregon	<0.20	<0.15
Southern San Francisco Bay, California	0.22	1.23–2.30
Palos Verdes/San Pedro Canyon, California	18.63	2.27
Dana Point, California	0.07	0.38
Median	0.31	0.49
Range: minimum	<0.20	0.15
maximum	26.7	4.23

[a] Values in ppm wet weight.

From Mearns, A. J., Matta, M. B., Simecek-Beatty, D., Buchman, M. F., Shigenaka, G., and Wert, W. A., PCB and Chlorinated Pesticide Contamination in U.S. Fish and Shellfish: A Historical Assessment Report, NOAA Tech. Memo. NOS OMA 39, Seattle, 1988.

wet weight) for 1976 and 1977 surveys. Systems with the highest mean PCB values in fish livers, based on results of the 1984 National Status and Trends Benthic Surveillance Project, included Elliott Bay, Washington (4.23 ppm wet weight), Boston Harbor, Massachusetts (2.62 ppm wet weight), and Commencement Bay, Washington (2.30 ppm wet weight). As expressed by Mearns et al.,[9] data collected since the mid-1970s do not reflect any substantial decrease in PCB contamination of estuarine and coastal marine fish livers, and in fact the levels may actually have increased. Table 8 compares mean PCB concentrations in fish livers from nine estuarine and coastal marine systems.

Based on the findings of at least 300 separate environmental surveys in which a minimum of 35,000 samples of invertebrates and fishes were analyzed for chlorinated pesticides and PCBs, Mearns et al.[9] presented the following conclusions on PCB concentrations in fish and shellfish. In general, the highest PCB contamination of these biotic groups occur in urban embayments of the Atlantic and Pacific Coasts. The biota of the Gulf Coast, while possibly being undersampled, appear to have less PCB contamination than other coastal regions of the U.S. An exception to this is the Pensacola-Escambia Bay area which exhibits elevated PCB readings in shellfish tissues. Sites along the Atlantic Coast that harbor flatfish and shellfish with high PCB concentrations include Boston Harbor, New Bedford Harbor, the Hudson River, Long Island Sound, and the New York Bight. Along the Pacific Coast, flatfish and shellfish with the highest levels of PCBs have been recovered from Elliott Bay, Santa Monica Bay, Palos Verdes Peninsula, San Pedro Harbor, and San Diego Harbor. Data collected from 1940 to the mid-1980s indicate that, by the early 1980s, PCB levels had declined in some estuarine and coastal flatfish and shellfish.

Effects of PCBs on Biotic Groups

Phytoplankton: Because phytoplankton populations form the basis of many food chains in estuarine and marine ecosystems, their sensitivity to PCB toxicity has been scrutinized extensively both in the laboratory and in natural environments. It has become increasingly evident over the years that phytoplankton sensitivity to PCBs varies greatly both within and between species.[119] Indeed, the same phytoplankton species inhabiting different environments may even display different tolerances to PCB exposure. For example, Fisher et al.[120] conveyed that diatoms isolated from the open ocean had a lower tolerance to PCB exposure than isolates of the same species from estuarine environments. Despite the variable sensitivities of phytoplankton to PCB toxicity, the generally deleterious effects of PCBs on estuarine and marine populations have been well documented.

Several environmental factors, notably temperature, cell density, osmotic pressure, interspecific competition, presence of other toxic substances, and geographic location, affect phytoplankton sensitivity to PCBs in estuaries.[121-128] While a given phytoplankton species may preferentially absorb, retain, and tolerate even high levels of PCBs, most marine diatoms investigated to date are sensitive to concentrations as low as 0.1 ptm (parts per thousand million).[121] Certain algal populations exhibit extreme sensitivity to PCBs. The growth of *Scenedesmus obliguus*, for example, is inhibited at 0.25 ppm PCBs,[129] that of *S. quadricauda* at 5 ppb.[130] More sensitive still, *S. obtusiusculus* is inhibited at 300 ppb PCBs.[131] Some algal species differ from diatoms in that they accumulate PCBs to a greater amount than that present in the environment; these plants are resistant species to PCB toxicity. PCB inhibition in phytoplankton may be manifested in a disruption of chloroplast membranes, reduction in chlorophyll index, termination of cell division, and alteration of photosynthetic processes.[128] Clearly, these changes can have a marked impact on phytoplankton communities in estuaries, manifested in the alteration of species composition, abundance, and distribution.

Zooplankton: A significant fraction of the PCBs accumulating in copepods of the genus *Acartia* derives from their consumption of phytoplankton.[132] Selective accumulation of PCB components by zooplankton has been invoked by Clayton et al.,[133] who examined the zooplankton of Puget Sound. They determined that at most sites zooplankton selectively absorbed and accumulated PCB components of higher chlorination. Table 9, which lists zooplankton concentration factors (PCBs in liquids/PCBs in water) for chlorobiphenyls of different chlorine content in specific regions of Puget Sound, demonstrates this effect. As stated previously, smaller zooplankton tend to accumulate greater quantities of PCBs than larger zooplankton. Hence, selective accumulation of PCB components at

TABLE 9
Partitioning Values or Concentration Factors (PCBs in lipids/PCBs in water) for Zooplankton in Different Regions of Puget Sound

Region	Concentration factor × 10⁶			Dominant fauna
	N = 4ᵃ	N = 5	N = 6	
Elliot Bay	1.06 (±0.81)	1.42 (±1.06)	2.17 (±1.64)	Euphausiids
Main Basin	1.07 (±0.61)	1.90 (±1.25)	3.18 (±2.32)	Copepods
Whidbey Basin (Port Gardner)	0.80 (±0.38)	1.09 (±0.51)	1.47 (±0.75)	Copepods/ euphausiids
Hood Canal	0.98 (±0.56)	0.74 (±0.52)	0.43 (±0.32)	Ctenophores
Sinclair Inlet	1.12 (±0.68)	3.61 (±2.02)	6.90 (±3.45)	Ctenophores
Admiralty Inlet and Straits of Juan de Fuca	0.34 (±0.21)	0.28 (±0.18)	0.29 (±0.18)	Copepods

ᵃ N = number of chlorines/biphenyl.
ᵇ Values in parentheses are standard deviations.

From Phillips, D. J. H., in *PCBs and the Environment*, Vol. 2, Waid, J. S., Ed., CRC Press, Boca Raton, FL, 1986, 127. With permission.

higher trophic levels may hinge on selectivity at the primary consumer level.[67]

Benthic Invertebrates: Biota inhabiting bottom sediments of estuaries are typically exposed to much higher levels of PCBs than planktonic and pelagic organisms. However, it has not been unequivocally established for most benthic fauna whether the uptake of PCB contaminants occurs through pore water or direct contact with the sediment.[134] Of all benthic organisms, mussels have been most useful in identifying "hot spots" of PCB contamination and in monitoring PCB levels.[135] Table 10 presents PCB ranges in mussels, *Mytilus galloprovincialis*, from the Mediterranean Sea. The data suggest higher levels of PCBs in the northwestern basin. Comparable values were obtained on the East and West Coasts of the U.S. during the 1976 Mussel Watch Program[136] and in the North Sea during surveys from 1972 to 1976.[137-139] PCB concentrations in mussels, as well as several other invertebrate groups from the highly impacted waters of Puget Sound, Washington, are given in Table 11.

TABLE 10
PCB Concentrations (μg/kg wet) in Mussels[a] from the
Mediterranean Sea

Year	Region	PCB range	References[135]
1970	Ebro Delta	400 –1800	Baluja et al., 1973
1980	Ebro Delta	34 – 200[b]	Risebrough et al., 1983
1972	French Coast	110 –1920	DeLappe et al., 1972
1973–74	French Coast	40 –2700[b]	Marchand et al., 1976
1974	French Coast	200 –1100	Risebrough et al., 1976
~1978	French Coast	128[b]	Monod and Arnoux, 1979
1977–78	Ligurian Sea (Italy)	93 – 233[c]	Contardi et al., 1979
1976–78	Tuscan Iss. (Italy)	20 – 59	Bolognari et al., 1979
1976–79	North Adriatic (Italy)	41 – 100[c]	Fossato and Craboledda, 1981
1974–76	North Adriatic (Yugoslavia)	12 – 71	Nazansky et al., 1979
1977–78	Central Adriatic (Yugoslavia)	10 – 179	Dujmov et al., 1979
1977–78	South Adriatic (Yugoslavia)	2.2 – 148[b]	Vilicic et al., 1979
1976–77	Ionian Sea (Italy)	61– 100	Amico et al., 1979
1975–79	North Aegean (Greece)	261– 575	Kilikidis et al., 1981
1983	Algeria	7.3– 630	Fowler and Villeneuve, unpublished

[a] Refers to *Mytilus galloprovincialis* which is often considered to be the same species as *M. edulis*. In the case of Algeria some samples were *Perna perna*.
[b] Values calculated using 0.2 dry/wet weight ratio.
[c] Range of mean values.

From Fowler, S. W., in *PCBs and the Environment,* Vol. 3, Waid, J. S., Ed., CRC Press, Boca Raton, FL, 1986, 209. With permission.

These PCB measurements parallel those of invertebrates of the Mediterranean Sea. As is apparent in other coastal waters worldwide, a trend of decreasing PCBs in mussels of the Mediterranean Sea through time is consistent with diminishing production and use of PCBs in countries bordering the Sea.[135]

In laboratory experiments using Arochlor® 1254, Lowe et al.[140] witnessed significant reduction in the growth of young oysters (*Crassostrea virginica*) exposed to 5 μg PCB/l of water. The uptake and concentration of PCBs by benthic invertebrates, such as *C. virginica*, are dependent upon PCB levels in sediments and the overlying water. For example, *Arenicola marina* and *Nereis diver-*

TABLE 11
Range of PCB Concentrations (µg/kg) in Invertebrate Tissue from Puget Sound

Organism	Duwamish River (Elliott Bay)	Hylebos Waterway (Commencement Bay)	Reference areas
Polychaete	250	66–260	45
Mussel	92–210	72	10–30
Clam	50–180	54–120	<2–30
Shrimp	480	800	26–54
Crab			
Hepatopancreas	9600	3600	130
Muscle	76	47–58	U[a] (10)

[a] U indicates undetected at detection limit shown.

From Ginn, T. C. and Barrick, R. C., in *Oceanic Processes in Marine Pollution,* Wolfe, D. A. and O'Connor, T. P., Eds., Robert E. Krieger Publishing, Malabar, FL, 1988, 157. With permission.

sicolor exposed to sediments containing 1 µg PCB (Arochlor® 1254) per gram for 5 d contained 0.24 ± 0.08 and 0.36 ± 0.12 µg PCB per gram worm (wet weight), respectively. The uptake of PCBs was much more rapid from water than from sediments.[141] However, the worms accumulated most of their PCBs from the sediments.[141] More research must be conducted on the partitioning of PCBs between bottom sediments and overlying water in estuaries and whether the PCBs are taken up directly by the benthos from the sediment through contact with the body, indirectly via the pore water, or through ingestion of sediment and subsequent absorption in the gut.[134]

Fish: Some juvenile and adult estuarine and marine fish accumulate high concentrations (ppm) of PCBs.[142-144] Having a high affinity for lipids, PCBs have also been found to accumulate to high concentrations in fish eggs and embryos.[134,145,146] Spot, *Leiostomus xanthurus*, exposed to 1 µg Arochlor® 1254/l of water for 56 d, accumulated a maximum PCB concentration of 3.7×10^4 times that in the water.[147] Hansen et al.[148] observed significant mortality of pinfish, *Lagodon rhomboides* exposed to 32 µg PCB 1016/l of water for 42 d. Coho salmon, *Oncorhynchus kisutch*, experienced toxicosis when exposed to PCB 1254 at levels up to 14,500 μg/kg body weight per day after 260 d.[149] PCBs are generally considered to be toxic to fish and a cause of various abnormalities in them.[147,150,151]

Fish accumulate highly variable levels of PCBs, related to species-specific differences in percent lipids, diet, feeding habits, migration, PCB excretion, and other factors.[152] Sex-based variation in PCB accumulation is documented for certain species, such as the fathead minnow, *Pimephales promelas*; the females of this species have greater lipid contents than the males, and consequently they accumulate about two times the amount of PCBs.[153] Within a given species, PCB levels often vary due to age, weight, and size differences.[67] For example, Narbonne[154] revealed that 3.5-year-old grey mullet, *Chelon labrosus*, fed a dry fish diet contaminated with 50 μg Phenoclor DP6 per gram of food always accumulated higher concentrations than 2.5-year-old individuals fed the identical diet. Zitko[109] and Califano et al.[155] discovered a coupling between the weight of some species and the amount of PCBs in their tissues. The species they studied accumulated PCBs on a weight ratio basis. Correlations between PCB concentration and fish length have been established for the Atlantic herring, *Clupea harengus*, and striped mullet, *Mugil cephalus*.[110]

Fowler,[135] investigating PCB concentrations in fishes of the coastal waters of the Mediterranean Sea, found that PCB levels varied by more than four orders of magnitude (Table 12). As emphasized by Fowler,[135] not all of the variation is attributable to environmental levels. However, fishes inhabiting areas in proximity to industrialized sectors tended to have higher concentrations of PCBs. Such is the case for species collected from waters along the northern coastline of Italy and from river mouths in Spain.

Birds: Because of the potential threat of PCBs to birds and waterfowl frequenting estuarine habitats, various studies have dealt with PCB accumulation in these organisms[156,157] Stout[157] summarized surveys undertaken during the National Pesticide Monitoring Program on mallard ducks, *Anas platyrhynchos*, and the black duck, *A. rubripes*. She also reviewed results of monitoring programs on the brown pelican, *Pelecanus occidentalis*. Peakall[156] provided an overview of PCB accumulation and effects on birds, both under laboratory and field conditions, utilizing examples from the Baltic Sea and Great Lakes.

Table 12
PCB Concentrations (ng/kg wet) in Muscle Tissue of Fish from the Mediterranean

Species	Location	Date	PCB[a]	References[67]
Mullus barbatus	Ligurian Sea (Italy)	1977	188–1,486	Contardi et al., 1979
	North and Central Adriatic (Italy)	1976–79	69–211	Fossato and Craboledda, 1981
	Central Adriatic (Yugoslavia)	1975–79	<1–497	Dujmov et al., 1979
	North Adriatic (Yugoslavia)	1973	3	Revelante and Gilmartin, 1975
	Sicily coast	1976–77	17–373	Amico et al., 1979
	Augusta Bay, Sicily	1980	300	Castelli et al., 1983
	North Aegean	1975–79	703	Kilikidis et al., 1981
	Saronikos Gulf	1975–82	8–138[b]	Voutsinou-Taliadouri and Satsmadjis, 1982
		1975–76	4–1,100	Satsmadjis and Gabrielides, 1979
	Eastern Turkey	~1980	2	Bastruck et al., 1980
	Israel	1975–79	60	Ravid et al., 1985
M. surmuletus	Israel	1975–79	69	Ravid et al., 1985
Mugil auratus	Eastern Turkey	~1980	10	Basturk et al., 1980

Table 12 (Continued)
PCB Concentrations (ng/kg wet) in Muscle Tissue of Fish from the Mediterranean

Species	Location	Date	PCB[a]	References[67]
M. cephalus	North and Central Adriatic (Italy)	1972	870	Viviani et al., 1974
Sardina pilchardus	North and Central Adriatic (Italy)	1970	37–1,060	Viviani et al., 1973
		1972	620	Viviani et al., 1974
	North Adriatic (Yugoslavia)	1973	2–19	Revelante and Gilmartin, 1975
	France	1975	51–309	Alzieu, 1976
	Spain	1970	540–6,900	Baluja et al., 1973
	Augusta Bay, Sicily	1980	2,300–6,100	Castelli et al., 1983
Engraulis encrasicholus	Ligurian Sea (Italy)	1977–78	88–232	Contardi et al., 1979
	Sicily coast	1976–77	9–176	Amico et al., 1979
	North and Central Adriatic (Italy)	1970	510–960	Viviani et al., 1973
		1972	370	Viviani et al., 1974
		1976–79	119–162	Fossato and Craboledda, 1981
	North Adriatic (Yugoslavia)	1977	11–23	Revelante and Gilmartin, 1975
Thunnus thynnus	France	1975	95–407	Alzieu, 1976
		1977	6–89	Fowler and Elder, unpublished
	Sicily coast	1976–77	9–44	Amico et al., 1979
	North and Central Adriatic (Italy)	1976–79	344	Fossato and Craboledda, 1981
	North Aegean	1975–79	2,613	Kilikidis et al., 1981
Euthynnus alletteratus	Ligurian Sea (Italy)	1977–78	191–1,020	Contardi et al., 1979
Sarda sarda	Ligurian Sea (Italy)	1977–78	1,133–14,020	Contardi et al., 1979

Species	Location	Year	Value	Reference
Xiphias gladius	North Aegean	1975–79	364	Kilikidis et al., 1981
Boops boops	Israel	1975–79	74	Ravid et al., 1985
B. salpa	Libya	1982	2.5	Fowler and Villeneuve, unpublished
Scorpaena scrofa	Libya	1982	3.9	Fowler and Villeneuve, unpublished
Sprattus sprattus	North and Central Adriatic (Italy)	1970	620–920	Viviani et al., 1973
Gobius paganellus		1972	100	Viviani et al., 1974
Pleuronectes flesus		1972	250	Viviani et al., 1974
Squalus acanthias		1972	720	Viviani et al., 1974
Esox lucius		1972	350	Viviani et al., 1974
Anguilla anguilla		1972	720	Viviani et al., 1974
	Augusta Bay, Sicily	1980	2,500	Castelli et al., 1983
Saurida undosquamis	Israel	1975-79	236	Ravid et al., 1985
Merluccius merluccius		1975-79	16	Ravid et al., 1985
Trachurus mediterraneus		1975-79	63	Ravid et al., 1985
Upeneus moluccensis		1975-79	151	Ravid et al., 1985
Pagellus acarne		1975-79	151	Ravid et al., 1985
P. erythrinus		1975-79	188	Ravid et al., 1985
Maena maena		1975-79	91	Ravid et al., 1985
Dentex macrophthalmus		1975-79	195	Ravid et al., 1985

[a] In some cases the ranges given are ranges of mean values. Single values represent either means of several determinations or a concentration based on a single measurement of a pooled sample containing tissues from several individuals.

[b] Computed from original data using 0.3 dry/wet weight ratio.

From Fowler, S. W., in *PCBs and the Environment*, Vol. 3, Waid, J. S., Ed., CRC Press, Boca Raton, Fla., 1986, 209. With permission.

TABLE 13
Accumulation of PCBs in Birds from Field Experiments

Avian species	Residue level in prey	Residue level (ppm, wet weight)					References[56]
		Whole body	Muscle	Liver	Brain	Eggs	
Great-crested grebe (Podiceps cristatus)	0.17 (0.16–0.18)	—	—	11	—	—	Koeman et al., 1972
	0.12 (0.04–0.35)	32	—	—	—	—	Mowrer et al., 1982
Brown pelican (Pelecanus occidentalis)	0.8–1.3	—	—	—	—	13	Thompson et al., 1977
	0.19 (0.17–0.23)	—	—	—	—	2.2	
White pelican	0.05–0.11	2.3	3.1	4.5	—	1.7	Greichus et al., 1973
Double-crested cormorant (Phalacrocorax auritus)		3.6	2.3	2.0	—	5.7	
White-crested cormorant (Phalacrocorax carbo)	1.6	6.2	—	—	4.4	2.9	Greichus et al., 1977
White-tailed sea eagle (Haliaetus albicilla)	0.03–0.18	—	150–240	130	29–70	—	Bagge, 1975
Common guillemot (Uria aalge)	0.27 (0.01–1.0)	—	—	—	—	7.9–21	Jensen et al., 1969
	0.01–2.0	3.4	—	0.4	—	—	Holdgate, 1971
Herring gull (Larus argentatus)	2.2	—	—	—	—	124–157	Norstrom et al., 1978

From Peakall, D. B., in PCBs and the Environment, Vol. 2, Waid, J. S., Ed., CRC Press, Boca Raton, Fla., 1986, 31.

PCBs appear to cause problems of reproduction and behavioral abnormalities in some birds.[158] In regard to reproduction, PCBs have a more direct effect on egg hatchability than egg production, leading to lower percentages of hatching success.[156] Unlike DDT, PCBs probably do not reduce eggshell thickness;[156] however, they may impact reproductive success by generating behavioral abnormalities. Such is the case in ring doves which, when fed 10 ppm Arochlor® 1254, experienced a marked drop of reproductive success.[159] In contrast, mallard ducks fed 25 ppm Arochlor® 1254 exhibited no behavioral abnormalities and no decline in breeding success.[160]

Consistent with the findings on other biota, PCBs in birds primarily accumulate in organ tissues having a high lipid content. Table 13 shows PCB values recorded in different organ tissues of several species of birds. Highest readings have been found in the liver and muscle of the white-tailed sea eagle, *Haliaetus albicilla*, and the herring gull, *Larus argentatus*. The herring gull is an effective monitor of environmental pollution.[161] Because it has a broad distribution throughout the northern hemisphere, principally consumes fish, and bioaccumulates PCBs, the herring gull serves as a good monitor of PCBs.[157] PCB levels in herring gulls from various geographical areas are compiled in Table 14. With the exception of the Lake Ontario population of gulls that suffered poor reproduction, no adverse effects ascribable to PCB contamination could be ascertained in the birds from other locations.

The National Pesticide Monitoring Program tabulated PCB contamination in black ducks and mallards in the U.S. for several years between 1969 and 1979 by determining concentrations in duck wings (Table 15). Between 1969 and 1972, PCB contamination in ducks was essentially constant in the Atlantic Flyway, decreased by 48 to 50% in the Central and Pacific Flyways, and increased in the Mississippi River Flyway. From 1972 to 1976, however, PCB values decreased in the Atlantic Flyway from 1.36 ± 0.15 to 0.52 ± 0.08 ppm.[162] PCB readings in mallards decreased in all flyways over the years 1976 to 1979.[157]

Declining populations of the brown pelican promulgated extensive surveys of this popular waterbird by the U.S. Fish and Wildlife Service in the 1960s, 1970s, and 1980s. During the period from 1969 to 1977, PCB concentrations in brown pelican eggs in South Caro-

TABLE 14
PCB Levels in Gulls and Their Eggs

Species	Location	Tissue	Year of collection	Residual level (ppm wet weight)	References[56]
Herring	Norway (west coast)	Egg	1972	3.1–12.6	Fimreite et al., 1977
	Baltic (Gdansk Bay)	Muscle	1975–76	23–150	Falandysz, 1980
	Finland	Muscle	1972–74	0.68–38	Särkkä et al., 1978
	East Scotland	Muscle	1971–75	0.2–1.2	Bourne and Bogan, 1972
	Camague, France	Egg	1972	16–160	Mendola et al., 1977
	New Brunswick, Canada	Egg	1969–72	3.1–8.2	Gilbertson and Reynolds, 1974
	Lake Ontario	Egg	1974–75	74–261	Gilman et al., 1977
	Maine	Egg	1977	(0–32.0) 7.76 (30)	Vermeer and Reynolds, 1970
	Virginia	Egg	1977	(0.13–16.70) 9.06 (28)	Szaro et al., 1979
	East Frisian Is., Germany	Egg	1975	26.5	Szaro et al., 1979
			1971	5.5	Szaro et al., 1979
	Denmark (North Sea)	Egg	1971	2.1 (1.3–2.6)	Becker et al., 1980
	Baltic		1972	92 (21 - 199)	
	Alberta	Egg	1969	0.87 (11)	Jorgensen and Kraul, 1974

From Peakall, D. B., in *PCBs and the Environment,* Vol. 2, Waid, J. S., Ed., CRC Press, Boca Raton, FL, 1986, 31. With permission.

lina varied greatly and no obvious trend could be established.[163] Lower mean values were obtained from 1978 to 1980 for the same region.[157]

Pelican eggs along the East Coast of Florida increased significantly in PCBs from a mean of 2.7 ppm in 1969–1970 to 6.1 ppm in 1974.[164] Along the West Coast of Florida, the rise was less marked from 0.70 to 1.18 ppm during the period of 1969–1970 to 1974, respectively. Similarly, the mean PCB concentrations in brown

TABLE 15
PCB Levels in Duck Wings (1969-79)[a]

Species	Flyway	Year	Pools	Incidence	PCBs mean (ppm) Detect.[b] /all	SE
Black	Atlantic	1969	42		(1.4)	0.16
		1972	44		(1.4)	0.15
		1976	32	100[a]	0.52	0.08
		1979	24	100[a]	0.63	0.09
Mallard	Atlantic	1969	19		(1.3)	0.46
		1972	21		(1.2)	0.23
		1976	20	100[a]	0.52	0.18
		1979	29	100[a]	0.45	0.07
	Mississippi	1969	51		(0.44)	0.061
		1972	61		(0.66)	0.30
		1976	69	61[a]	0.23/ 0.14	0.03
		1979	64	98[b]	0.11/ 0.11	0.02
	Central	1969	49		(0.20)	0.039
		1972	56		(0.10)[c]	0.013
		1976	56	13[a]	0.15/ 0.02	0.01
		1979	54	90[b]	0.06/ 0.05	0.01
	Pacific	1969	51		(0.20)	0.014
		1972	55		(0.11)[c]	0.009
		1976	50	14[a]	0.16/ 0.02	0.04
		1979	44	93[b]	0.07/ 0.07	0.02

Note: ppm: mg/g wet weight; each sample consisted of 25 wings.

[a] Data from Cain, 1981; White and Heath, 1976.
[b] Arithmetic means; for 1976 and 1979 "Detect." Values include only samples with detectable PCB levels; "all" values are means using 0.00 for undetected values. Parentheses indicate means for 1969 and 1972 not comparable to later years because of differences in analysis. a,b: Incidences differ ($p <0.05$) within a flyway when letters (a,b) differ.
[c] 1972 mean differed ($p<0.05$) from 1969 mean in same flyway; means of detectable levels in 1976 and 1979 not significantly different ($p. >0.05$).

From Stout, V. F., in *PCBs in the Environment,* Vol. 1, Waid, J. S., Ed., CRC Press, Boca Raton, FL, 1986, 163. With permission.

pelican eggs in Louisiana only increased from 2.6 to 3.9 ppm between 1971 and 1976.[165]

Other birds inhabiting estuarine habitats have also been monitored for PCB uptake. For example, PCB levels in Clapper Rails (*Rallus longirostris*) in the salt marshes of New Jersey increased 160% between 1967 and 1973.[166] The black-crowned night heron (*Nycticorax nycticorax*) incurred reproductive problems, including congenital malformations, reduced hatching success, and decreased survival of hatchlings at the San Francisco Bay National Wildlife Refuge due to PCB contamination. In addition, a possible impact of PCBs on embryonic growth of the herons was suggested.[167]

Peakall (p. 43)[156] stated, "Based on laboratory experiments, it is clear that PCBs are not sufficiently toxic in either acute or chronic feeding studies to pose a hazard to birds." Furthermore, laboratory research on birds reveal marked interspecific variations in reproductive problems and behavioral abnormalities caused by PCBs. In respect to PCB metabolites, little information exists on the toxic impact on birds. Additional laboratory and field investigations must be conducted, therefore, on the exposure birds and waterfowl to PCBs and PCB metabolites in the estuarine environment.

Mammals: Mammals, like birds, obtain most PCBs through the food chain via the consumption of contaminated prey.[5] Dolphins and seals, for instance, can accumulate high PCB levels in their tissues, principally deriving them from their diet.[67] PCBs appear to adversely affect ovulation and development in mammals, leading to declining populations in some regions.[158] In the Baltic Sea, for example, population abundances of the otter (*Lutra lutra*), ringed seal (*Phusa hispida*), common seal (*Phoca vitulina*), gray seal (*Halichoerus grypus*), and common porpoise (*Phocoena phocoena*) have dropped dramatically during the last 30 years, and PCBs are strongly suspected as an agent in the decline.[168] While the usual pregnancy rate of female ringed seals is 80%, only 28% of the females in the Baltic Sea have become pregnant each year since about 1970.[158] According to Clark,[158] occlusions block one or both uterine horns in approximately 40% of the females, rendering them infertile. PCBs in the blubber lipids of pregnant females amount to 56 ppm compared to 77 ppm in nonpregnant females. These values are not substantially different than PCBs in the blubber lipids of pregnant grey seals and common seals (73 ppm) and nonpregnant females (110 ppm).[158]

Effects other than environmental pollution have contributed to diminishing populations of seals in the Baltic Sea since 1900, namely hunting, disturbances, and stress. However, after the 1950s hunting was very rare, yet the seal populations continued to rapidly decline. Disturbances and stress imposed on seal populations may have been partly and locally responsible for the drop in the abundances of these animals. Nevertheless, they are unlikely causes of the universal decimation of the seals throughout the ecosystem. More likely, effects of PCB and DDT on the reproductive capacity of Baltic seals reduced their numbers along the coast.

Biotic Conclusions: The monitoring of PCB concentrations in aquatic organisms over the years indicates that PCBs have declined in the most heavily polluted areas while remaining stable or even increasing in regions of previously low or zero levels. For example, PCB readings fell substantially in the Hudson River during the 1980s subsequent to years of massive dumping of Arochlor® 1016. This decrease is manifested in PCB concentrations in striped bass (*Morone saxatilis*) which dropped from a mean of 18.1 ppm in 1978 to 4.8 ppm in 1981. Elsewhere in habitats characterized by low-level pollution of PCBs, as connoted by the National Pesticide Monitoring Program, PCB levels in biota are variable from year to year and distinct trends cannot be established.[157]

Although a significant fraction of PCBs released to the aquatic environment ultimately accumulate in bottom sediments, biological and physical disturbances of the benthic regime roil the sediment surface, fostering their remobilization. They then can be recycled among the various trophic levels of organisms comprising the communities. Consequently, they dissipate slowly from the organisms, following an exponential curve, with a rapid initial decline in concentration followed by a protracted, gradual drop.[169] Burial in bottom sediments of the deep sea ultimately removes them from circulation and contamination of the biosphere.[170]

POLYCHLORINATED TERPHENYLS

Little is known regarding the impact of polychlorinated terphenyls (PCTs) on estuarine biota. These organochlorine compounds (Figure 1m), with similar usages as PCBs, have been identified in organisms at high trophic levels.[171] For instance, grey seals accumulate PCTs, but more research needs to be conducted on their overall effects on these animals.

CASE STUDIES

Chesapeake Bay

Despite the heavy concentration of industrial activity in Cheaspeake Bay at Baltimore and Norfolk, a number of organochlorine contaminants are now regionally dispersed in the estuary. PCBs and certain herbicides can be included in this group. PCB contamination in Chesapeake Bay has existed for at least 30 years.[172] Nonetheless, most data on organochlorine compounds in the system prior to the late 1970s involved pesticides (e.g., DDT, DDE, dieldrin, etc.). A significant amount of research on organochlorine compounds in Chesapeake Bay biota in the late 1960s and early 1970s dealt with insecticide contamination of upper trophic level organisms, including estuarine birds. Through this research, it became increasingly clear that certain avifauna (for example, the osprey and bald eagle) were susceptible to the damaging effects, in particular, of DDE which caused eggshell thinning and a decrease in population abundance. Surveys of organochlorine poisoning in adult birds and adverse reproductive effects in waterfowl, such as black ducks and mallards, uncovered no major impacts. Hence, in 1978, the mean DDE concentration in black duck eggs along the bay equalled 0.1 ppm.[173] By comparison, the mean concentrations of DDE in osprey eggs amounted to 1.9 to 2.8 ppm on the eastern shore of the bay in 1977 and 1978, respectively, and 1.4 to 3.5 ppm on the western shore in 1977. These values are significantly less than the mean DDE reading of about 10 ppm recorded on bald eagle eggs in the Chesapeake region in 1978 and 1979.[13,174]

Apart from birds, finfishes (notably the striped bass) were monitored for DDT, DDD, DDE, chlordane and other contaminants. Levels of chlordane, DDT, DDD, and DDE in juvenile striped bass from the Nanticoke and Potomac Rivers amounted to <0.06 µg/g wet weight. In the Potomac River, total organochlorine residues in striped bass ranged from 0.21 to 0.40 µg/g wet weight, which exceeded those in the Nanticoke River (0.06 to 0.09 µg/g wet weight).[175] Mehrle et al.[175] considered these values to be insignificant residue levels, and much less of a problem than PCBs, the most prevalent organic contaminant residue in this species in Chesapeake Bay and elsewhere.[7,31,176]

Although PCBs are widely distributed in Chesapeake Bay, as demonstrated by analyses of bottom sediments (Figure 3),[177] high

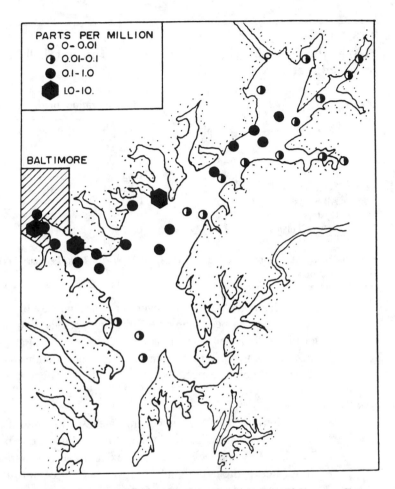

FIGURE 3. PCB concentrations in the surface sediments of the upper Chesapeake Bay. (From Munson, T.O., in Upper Bay Survey, Vol. 2, Final Report to the Maryland Department of Natural Resources, Westinghouse Electric Corporation, Ocean Sciences Division, Annapolis, MD, 1976. With permission.)

readings occur in industrialized regions, such as Baltimore Harbor,[172] where Tsai et al.[178] ascertained concentrations in surface sediments of >2 mg/kg. Surveys of commercially important fin- and shellfish in Maryland and Virginia waters reveal PCBs in tissues about an order of magnitude less than the FDA action level of 2 μg/ g wet weight (Table 16).[179,180] Monitoring programs performed by both the Maryland Department of Health and the Virginia Institute of Marine Science confirm these findings. In highly populated areas

TABLE 16
**Polychlorinated Biphenyl (Arochlor® 1254) in Fish and Shellfish from
Maryland Waters**

Species	No. samples	Concentration (mg/l)	
		Mean	Range
Rockfish	44	0.23	0–0.58
Seatrout	12	0.05	0.02–0.13
White perch	13	0.21	0–0.42
All Finfish	80	0.20	0–0.58
Oysters	115	0.02	0–0.07
Softshell clams	13	0.02	0–0.06

From Eisenberg, M., Mallman, R., and Tubiash, H. S., *Mar. Fish. Rev.,* February 21-25, 1980.
With permission.

around Hampton Roads, however, a general increase of PCB concentrations has been perceived in some species.[172] In areas of localized, "hot spot" contamination of PCBs (e.g., the Patapsco and Elizabeth Rivers),[181] more data on PCB concentrations should be collected on biota, especially commercially important species.[182]

SAN FRANCISCO BAY

Phillips and Spies[183] examined the database available on chlorinated hydrocarbons (organochlorine pesticides and PCBs) in the San Francisco Bay and Delta. Less data exist on the concentrations of chlorinated hydrocarbons in sediments than in organisms of the estuary. The most conspicuous feature of the sediments is the high site-to-site variability of chlorinated hydrocarbon concentrations. In regard to biota, the bulk of the contamination data has been obtained on benthic invertebrates, nearly exclusively bivalves, and teleosts. Much less information has been gathered on upper trophic level populations, with some values reported on birds and seals.

Organochlorines in Sediments

The sediments in some areas of the estuarine system are highly contaminated with chlorinated hydrocarbons. Law and Goerlitz[184] surveyed DDT and metabolites, as well as PCBs and chlordane, in surface sediments of 26 streams discharging into the bay. A number

TABLE 17
Chlorinated Hydrocarbon Residues of Bottom Sediments of Impacted Streams Draining into San Francisco Bay[a]

| | | | | Contaminant | | |
Stream	Chlordane	DDD	DDE	o,p'-DDT	p,p'-DDT	PCB
Belmont Creek	660	41	17	89	200	52
San Francisquito Creek[b]	670	160	43	20	150	430
Los Gatos Creek[c]	280	33	25	11	32	170
Union Creek	200	45	16	2.6	2.8	140
Miller Creek	310	16	11	13	8.4	35
San Rafael Creek	800	120	61	38	51	350
Corte Madera Creek[d]	140	12	11	8.7	48	81
Arroyo Corte Madero del Presidio	140	34	7.6	11	16	24

[a]   ng/g dry wt.
[b,c] Downstream site.
[d]   Upstream site.

of these streams had very high levels of contaminants in their sediments, especially those draining into the South Bay. Bottom sediments of several streams flowing into the western perimeter of San Pablo and Central Bays, and Union Creek, emptying into Suisun Bay, also contained large quantities of chlorinated hydrocarbons. Table 17 records measurements of chlorinated hydrocarbon residues in bottom sediments of some impacted systems.

In a more recent monitoring program, Chapman et al.[185] analyzed bay sediments for DDT and metabolites, PCBs, and chlordane-related compounds. Three areas were sampled, including Islais Creek (on the eastern side of the San Francisco peninsula), a site off Oakland, and San Pablo Bay. Islais Creek displayed the highest levels of organochlorines in sediments and San Pablo Bay the lowest, with the area off Oakland exhibiting intermediate values. The elevated readings in Islais Creek are explained, in part, by wet weather sewage overflows and urban runoff that enter at the inner end.[183]

Additional data on organochlorines in bay sediments were collected during the National Status and Trends Program of NOAA in the mid-1980s.[25] The concentrations of DDT and metabolites in sediment samples taken in 1984 at four locations in the bay (i.e., San Pablo Bay, Southampton Shoal, Oakland, and Hunter's Point) by NOAA were <6 ng/g dry weight. At San Pablo Bay and Southampton Shoal, the sum of the parent compound and its metabolites fell below 1 ng/g dry weight. Somewhat higher readings (3 to 6 ng/g dry weight) occurred in sediments at Oakland and Hunter's Point. PCB levels in sediments were somewhat higher: (1) San Pablo Bay, 9 ng/g dry weight; (2) Southampton Shoal, 12 ng/g dry weight; (3) Oakland, 61 ng/g dry weight; and (4) Hunter's Point, 40 ng/g dry weight. By comparison, the other locations monitored on the West Coast of the U.S. for PCBs in sediments had similar levels. Much higher concentrations were registered in industrialized areas; for example, sediment samples in Elliot Bay nearby Seattle had much greater quantities of PCBs, equalling 330 ng/g dry weight.[183]

In an independent investigation, the EPA determined the amount of PCBs, DDT, and metabolites in sediments at ten sites in South Bay.[183] A high degree of spatial heterogeneity of contaminants was apparent among the samples, with PCB levels ranging from below detection limits to 34 ng/g dry weight. The sum of DDT and its metabolites, generally <10 ng/g dry weight, approached 90 ng/g dry weight at one location in proximity to Coyote Point.

Probably the highest concentrations of PCBs in sediments have been observed in the vicinity of the Port of Stockton in the Delta (i.e., Mormon Channel, Mormon Slough, and the Port of Stockton turning basin.)[183] Here, PCB levels quantified as sums of Arochlors® 1242, 1254, and 1260, ranged from 7100 to 17,800 ng/g dry weight. Phillips and Spies[183] mentioned that sediments of open areas of the estuary tended to be more contaminated by PCBs than DDT and its metabolites which often were present in concentrations <10 ng/g dry weight. Greater organochlorine contamination existed along the margins of the estuary. In areas of restricted water circulation abutting the shoreline, a considerable quantity of organochlorine compounds can accumulate. Of all organochlorine compounds, PCBs may be the most ubiquitous in bottom sediments of the estuary and, at elevated levels, possibly contribute to detrimental effects on biota.

Organochlorines in Biota

Much of the chlorinated hydrocarbon contamination of the biota in San Francisco Bay originates from the heavy agricultural use of pesticides in the Central Valley. Here, more than 20,000 metric tons of pesticides — about 500 different types — are utilized for agricultural purposes annually, which represents 10% of the total pesticides applied each year to crops nationwide.[182] Despite bans on the use of the most toxic and environmental persistent pesticides, their residues continue to enter the estuary via the leaching of soils, posing a hazard to aquatic organisms. The California State Mussel Watch surveys in recent years have uncovered significant amounts of chlordane, DDE, dieldrin, and toxaphene in some areas of the bay; these surveys have also delineated the baywide contamination of PCBs. Other pesticides found in the California State Mussel Watch samples, albeit in smaller concentrations than the aforementioned organochlorines, include unmetabolized DDT, HCB, HCH isomers, and chlorbenside.[35]

Because of the vigilant use of mussels as bioindicators of chlorinated hydrocarbon pollution in California estuaries, as well as elsewhere, much of the data available on chlorinated hydrocarbons in the biota of San Francisco Bay relates to these bivalves. Studies by Hayes and Phillips,[29] Stephenson et al.,[35] Farrington et al.,[169] Goldberg et al.,[136] and Hayes and Phillips[187] show persistent, measurable levels of DDT in California in 1970. Little evidence exists for any significant reduction in total DDT concentrations in the bay since 1970, suggesting that DDT and its metabolites are being released slowly from Central Valley soils, enabling the maintenance of moderate levels of contamination.[183] Girvin et al.,[188] monitoring total DDT levels in four bivalve species shortly after restrictions were placed on DDT usage, reported values ranging from 7 to 34 ng/g wet weight. Transplanted mussels, *Mytilus californianus*, and native bay mussels, *M. edulis*, monitored by the State Mussel Watch Program between 1979 and 1986 commonly had mean total DDT concentrations >100 ng/g dry weight. At some sampling sites, the concentrations of these contaminants in the mussels exceeded 1000 ng/g dry weight and at one locale, 22,000 ng/g dry weight.[29,35,186] The most heavily contaminated sites were in the Sante Fe Channel in

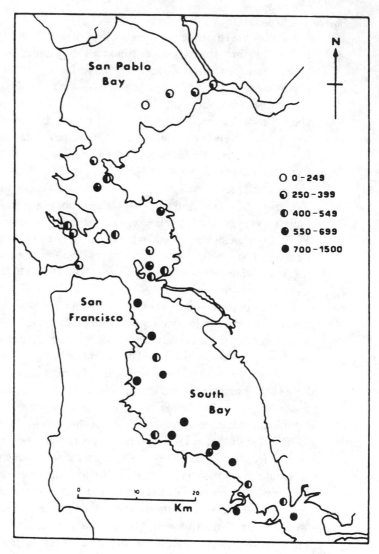

FIGURE 4.   PCB concentrations (ng/g dry wt) in whole soft parts of native bay mussels (*Mytilis edulis*) in San Francisco Bay. (From Risebrough, R. W., Chapman, J. W., Okazaki, R. K., and Schmidt, T. T., Toxicants in San Francisco Bay and Estuary, Report of the Association of Bay Area Governments, Berkeley, CA, 1978.)

Richmond Harbor owing to a pesticide formulation and packaging plant that once operated on the banks of the Lauritzen Canal that discharged into the Sante Fe Channel.

Analysis of transplanted and native mussels in the bay during the State Mussel Watch Program indicated baywide PCB contamination, with PCB levels varying from approximately 150 to more than 1000 µg/g dry weight in the organisms.[182] Girvin et al.[188] documented concentrations of 29 to 152 ng/g wet weight in four species of bivalves in the Central and South Bays. Risebrough et al.[189] in a comprehensive survey of native bay mussels, disclosed ubiquitous PCB contamination in the estuary; lowest PCB concentrations were discerned in mussels from San Pablo Bay (mainly 0 to 400 ng/g dry weight) and highest concentrations in mussels from South Bay (typically 500 to 1500 ng/g dry weight) (Figure 4). Data provided by Hayes and Phillips,[29] Stephenson et al.,[35] and Hayes et al.[186] on transplanted or native mussels in San Francisco Bay between 1979 and 1986 revealed mean PCB levels ranging from about 50 to 1800 ng/g dry weight; most of the stations sampled had mussels with mean concentrations between 100 and 1000 ng/g dry weight. As in the case of pesticide contamination, no substantial long-term reduction in PCB contamination can be ascertained from the data.

Finfishes of Central Valley waters often retain elevated concentrations of organochlorines in their tissues.[190] Analysis of fish muscle samples from these waters reveal that at times the levels of total chlordane and DDT, dieldrin, endosulfan, hexachlorobenzene, lindane (γ-HCH), and toxaphene have surpassed the guidelines for predator protection as promulgated by the National Academy of Sciences and the guidelines for public health as set forth by the FDA.[183] The National Academy of Sciences guideline for PCBs in fish (500 ng/g wet weight) and that of the FDA (2000 ng/g wet weight) have also been occasionally exceeded. In the late 1960s, most fish sampled in the estuary contained total DDT residues amounting to 100 to 200 ng/g wet weight.[191] Results of the investigations on eight species of fish by Earnest and Benville[191] confirmed a positive correlation between total DDT and the lipid content of the fishes.

Two species under considerable scrutiny over the years for chlorinated hydrocarbon contamination are the striped bass (*Morone saxatilis*) and the starry flounder (*Platichthys stellatus*). Diminishing stocks of striped bass have led some workers to infer a causal relationship between organochlorine contamination and increased mortality of the fish. Total DDT concentrations in the axial muscle of striped bass collected during the early 1980s varied from about 0.5

to 1.0 μg/g wet weight. Somewhat higher values occurred in the ovaries and livers of the samples.[192-194] Comparing total DDT, PCBs, and toxaphene in striped bass from San Francisco Bay and those from Coos River, Oregon, Crosby et al.[192] noted that all of the contaminants were higher in San Francisco Bay samples. Whipple et al.[194] drew similar conclusions for aldrin, chlordane, dacthal, dieldrin, and hexachlorobenzene.

PCBs have been implicated in reproductive problems of starry flounder in the bay. Spies et al.[195] linked PCB concentrations in eggs of this species with its poor embryological success in certain areas of the estuary. Other studies underscore the variable levels of PCBs in starry flounder throughout the system. For instance, PCB levels in livers of this fish from San Pablo Bay, Southampton Shoal, and Hunter's Point equalled 1191, 3734, and 6990 ng/g dry weight, respectively.[113] In general, starry flounder from San Francisco Bay were more contaminated than those from waters of Oregon and Washington.[183]

Ohlendorf and Miller[196] investigated organochlorines in the waterfowl of San Francisco Bay; they focused on organochlorine exposure of northern pintails (*Anas acuta*), northern shovelers (*Anas clypeata*), lesser scaups (*Aythya affinis*), and canvasbacks (*Aythya valisineria*). During their wintering in the bay catchment, these four species seemed to accumulate significant amounts of DDE, PCBs, and hexachlorobenzene, but the uptake of the organochlorines did not appear to reduce the waterfowl populations. Analysis of DDE in the wings of northern pintails taken from the Sacramento-San Joaquin River Delta showed intermediate concentrations when compared to that in northern pintails from four other California locales.[13]

DDE measurements on eggs of the Forster's terns, snowy egrets, black-crowned night herons, and Caspian terns collected in southern San Francisco Bay during 1982 amounted to 1.92, 2.04, 2.84, and 6.93 ppm, respectively. The thickness of night heron eggs was correlated negatively with DDE concentrations. About 11% of the night heron eggs analyzed had DDE levels in excess of 8 ppm, which could have adversely affected reproduction of the species.[13,197,198] Furthermore, PCB concentrations in eggs of night herons in South Bay at the Bair Island National Wildlife Refuge, ranging up to 52 ppm, correlated inversely with embryonic weights, which were significantly lower than those of a control population of birds at the Patuxent Wildlife Research Center in Maryland.[1] Caspian tern eggs likewise contained higher PCB levels than those from other systems; for example, PCB concentrations were significantly higher in Caspian tern eggs from San Francisco Bay (4.85 ppm) than in eggs of the same species from Monterey Bay (1.83 ppm) and San Diego Bay (1.70 ppm). PCB contamination in shorebirds, night herons, as well as other biota, of the estuary may have far-reaching implications for the overall health of the system. As Phillips and Spies[183] remark, "Although San Francisco Bay and Delta are not as contaminated by PCBs as certain other (generally smaller) coastal areas of the USA, PCBs are widespread throughout the local estuary and show no signs of temporal reductions over the last decade. Evidence for their detrimental impacts is accumulating, and although most of the data cannot be considered conclusive at present, it appears probable that PCBs are exerting measurable damage to biological resources in the estuary."

Little information is available on organochlorines in mammals of the bay. Risebrough et al.[189] provided results of organochlorine analyses of dead harbor seals (*Phoca vitulina*) from San Francisco Bay. PCB levels in several seals were high; in one of them, PCB concentrations equalled 500 μg/g lipid weight in blubber, 12,000 μg/g in liver, and 31,000 μg/g in muscle. Large amounts of DDE were also found in the tissues of this seal.[183]

Puget Sound

A major repository of various wastes (e.g., municipal sewage, coal combustion waste, industrial wastewater, storm water runoff, and particles from metal smelters) for more than a century, Puget Sound has been a site of accumulation of organochlorine compounds since the 1930s.[199] Among organochlorines recovered from bottom sediments of the sound are DDT, PCBs, and CBD (chlorinated butadienes). Chlorinated hydrocarbons in bottom sediments of the sound reached peak levels in the 1950s and 1960s during the time of high industrial production and release to the environment. Bans or restrictions on the production and discharge of these chemicals have reduced their concentrations in surface sediments over the last two decades (Figure 5).[199]

According to Ginn and Barrick,[144] the most frequently detected

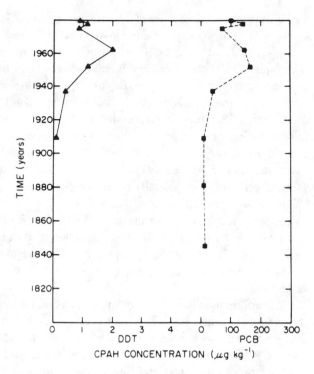

FIGURE 5.   Profile of mean concentrations of DDT (solid tri-
angle) and PCBs (solid square) in three sediment cores from central
Puget Sound as a function of age. Values given in µg/kg dry wt.
(From Ginn, T. C. and Barrick, R. C., in *Oceanic Processes in
Marine Pollution,* Vol. 5, *Urban Wastes in Coastal Marine Envi-
ronments,* Wolfe, D. A. and O'Connor, T. P., Eds., Robert E.
Kreiger Publishing, Malabar, FL, 1988, 157. With permission.)

organochlorine compounds in invertebrates, fish, birds, and mammals of Puget Sound are PCBs and DDT and its metabolites. Biota
in nearby industrialized, urban areas of the sound have higher
concentrations of these contaminants than those in proximity to
nonurban regions. Data comparing PCB concentrations in invertebrate groups of Elliott Bay, Commencement Bay, and control sites
corroborate this pattern (Table 11). Similarly, muscle tissues of fish
collected in both Elliott and Commencement Bays retain elevated
levels of PCBs in excess of 190 µg/kg wet weight mean concentrations in English sole (*Parophrys vetulus*) (Table 18). PCBs appear to
be elevated in bottom sediments of Commencement Bay to approxi-

TABLE 18
**Comparison of Average Concentrations of Organic Compounds in
Muscle Tissue from English Sole[a]**

Contaminant	Elliott Bay	Commencement Bay[b]		
		Hylebos Waterway	City Waterway	Discovery Bay
PCB	900	570	190	<13
DDT	7	20	17	<5
Hexachlorobenzene	U (20)[c]	110	U (1)	U (1)
Napthalene	U (1.6)	U (2)	510[d]	U (2)
Phenanthrene	U (5)[e]	U (5)	U (5)	U (5)
Benzo(a)pyrene	U (13)	U (40)	U (40)	U (40)
Phthalate esters	360[f]	420	<560	<300

[a] Values in μg/kg wet wt.
[b] See literature cited in Table 15.4 of Reference 144.
[c] U indicates undetected at detection limit shown.
[d] Value reported for a single English sole sample.
[e] Phenanthrene was detected in salmon tissue only (470 μg/kg; one sample).
[f] Higher concentrations of phthalates were found in single cod (7200 μg/kg) and salmon (51,000 μg/kg) samples from Elliott Bay.

From Ginn, T. C. and Barrick, R. C., in *Oceanic Processes in Marine Pollution,* Vol. 5, *Urban Wastes in Coastal Marine Environments,* Wolfe, D. A., and O'Connor, T. P., Eds., Robert E. Krieger Publishing, Malabar, FL, 1988, 157. With permission.

mately the same extent as that in the muscle tissue of fish. Blubber of harbor seals from southern Puget Sound have PCB values as high as 750 mg/kg wet weight.[200] The mean PCB concentration of seals in southern Puget Sound is ten times greater than that of seals in northern Puget Sound and is among the highest values reported on seals worldwide.[144] PCB contamination, while present in all biota examined in Puget Sound, is pronounced in marine mammals, suggesting biomagnification in the food chain.[144] Of approximately 100 marine mammals from Puget Sound analyzed for PCBs and DDT, by Calambokidis et al.,[200] harbor porpoises, killer whales, as well as harbor seals, display high levels of PCBs. These animals bioconcentrate PCBs and DDT from the fish they consume. Calambokidis et al.[200] conclude that because of declining DDT levels in Puget Sound, the principal pollutant threat to marine mammals in the sound is PCB contamination. In addition, no other chemical contaminants measured in the tissues of marine mammals, including heavy metals, polynuclear aromatic hydrocarbons (PAHs), and other chlorinated organic compounds, approach concentrations great enough to pose a serious health threat.[201] Arndt[202] documented high pup mortalities of harbor seals in south Puget Sound and speculated that PCBs may have been responsible. The seal populations having the highest pup mortalities were those bioaccumulating PCBs and the DDT metabolite DDE.[203] These factors may account for the slower increase in abundance of harbor seal populations in Puget Sound compared to outer-coast populations.[200]

Food chain effects of PCB accumulation are also likely to be manifested in certain species of marine birds. For example, the adipose (fat) tissues of great blue herons collected from Elliott and Commencement Bays have PCB levels ranging from 14 to 80 mg/kg wet weight. By comparison, the PCBs in liver tissue of these birds range from 0.19 to 3.2 mg/kg wet weight. These values are greater than those of birds sampled in control areas.[144]

The high concentrations of PCBs in tissues of marine birds and mammals, such as blue herons and harbor seals, provide classic examples of biomagnification effects up the food chain. The high mortality of harbor seal pups, together with observed impacts of organochlorines on the hatching success of marine birds,[204] is evidence of significant adverse effects on organisms occupying the highest trophic levels of the system.[201] It is necessary to continue to monitor chlorinated hydrocarbons in biota of Puget Sound and to make comparison studies with other systems, perhaps even remote from U.S. coastal areas,[205] to properly assess long-term chemical pollution effects.

SUMMARY AND CONCLUSIONS

Chlorinated hydrocarbons are perceived as a chemical hazard to estuarine organisms as well as to man. The accumulation of these compounds in sediments, water, and biota of estuaries derives from anthropogenic activities. Some of the organochlorines originate from herbicides, fungicides, and insecticides (especially in agricultural regions) that enter estuarine habitats via surface runoff and groundwater drainage. Other substances, such as PCBs commonly gain entry into these shallow coastal environments directly from manufacturing sites or via disposal of waste materials. In contrast to those pollutants that degrade rapidly and consequently pose a more limited hazard to estuarine biota, chlorinated hydrocarbons are very persistent and tend to accumulate in the fatty tissues of organisms. Moreover, they undergo biomagnification such that the upper trophic level consumers exhibit the highest concentrations of the contaminants. Hence, adverse effects of organochlorines often appear more pronounced among fish and mammal populations inhabiting estuaries. The growing societal awareness of the pollution problems associated with chlorinated hydrocarbons has spurred numerous monitoring programs in estuarine and marine waters. For example, during the last 2 decades, the National Status and Trends Program of the NOAA and the nationwide Mussel Watch Program have collected numerous samples to assess organochlorine contamination in aquatic biota. It has been possible to delineate regional and nationwide patterns of fin- and shellfish contamination of organochlorine compounds through these programs.

This chapter addresses the effects of organochlorine pesticides in estuaries, including DDT and its metabolites, the cyclodiene pesticide group — aldrin, dieldrin, endrin, lindane, and chlordane — as well as kepone, mirex, and toxaphene. The chlorinated benzenes and phenols, hexachlorobenzene and pentachlorophenol, and polychlorinated biphenyls are also examined. Much of the work on chlorinated hydrocarbons in estuaries has focused on DDT, its metabolites, and PCBs, which are nearly ubiquitous in these systems. Due to bioaccumulation of DDT up the food chain and the potential threat to humans consuming DDT-tainted food products from the sea, monitoring efforts on these compounds have concentrated on surveys of fin- and shellfish.

PCBs, a group of synthetic halogenated aromatic hydrocarbons, pose a direct toxic danger to estuarine and marine organisms. Moreover, adventitious concentrations of PCBs can produce sublethal reproductive effects that may ultimately reduce population sizes. This latter effect is evident among Baltic seal populations. PCBs, being lipophilic, tend to concentrate in the fat tissues of aquatic organisms. Because they are poorly metabolized, the PCBs pass up the food chain to organisms that serve as a source of food for man. These organochlorines have been coupled to both acute and chronic health effects in humans (e.g., cancer, liver disease, skin lesions, and reproductive disorders). As in the case of DDT and its metabolites, PCBs have been monitored carefully in fin- and shell-fish on a nationwide basis. Despite a ban in commercial production of PCBs since 1977, there is no clear evidence of a large-scale nationwide decrease of these contaminants in aquatic environments during the last 15 years except in proximity to known industrial sources and other "hot spot" areas.

REFERENCES

1. Stickney, R. R., *Estuarine Ecology of the Southeastern United States and Gulf of Mexico,* Texas A & M University Press, College Station, 1984.
2. Dale, P. E. R. and Hulsman, K., A critical review of salt marsh management methods for mosquito control, *Rev. Aquat Sci.,* 3, 281, 1990.
3. Sawhney, B. L., Chemistry and properties of PCBs in relation to environmental effects, in *PCBs and the Environment,* Vol. 1, Waid, J. S., Ed., CRC Press, Boca Raton, FL, 1986, 47.
4. Reutergardh, L., Chlorinated hydrocarbons in estuaries, in *Chemistry and Biogeochemistry of Estuaries,* Olausson, E. and Cato, I., Eds., John Wiley & Sons, Chichester, U.K., 1980, 349.
5. Shaw, G. R. and Connell, D. W., Factors controlling bioaccumulation in food chains, in *PCBs and the Environment,* Vol. 1, Waid, J. S., Ed., CRC Press, Boca Raton, FL, 1986, 135.
6. Jernelov, A., The history of chlorinated hydrocarbon pollution in the marine environment, in *Oceanography: The Past,* Sears, M. and Merriman, D., Eds., Springer-Verlag, New York, 1980, 414.
7. O'Connor, J. M., Klotz, J. B., and Kneip, T. J., Sources, sinks, and distribution of organic contaminants in the New York Bight ecosystem, in *Ecological Stress and the New York Bight: Science and Management,* Mayer, G. F., Ed., Estuarine Research Federation, Columbia, SC, 1982, 631.
8. Young, D. R., Gossett, R. W., and Heesen, T. C., Persistence of chlorinated hydrocarbon contamination in a California marine ecosystem, in *Oceanic Processes in Marine Pollution,* Vol. 5, *Urban Wastes in Coastal Marine Environments,* Wolfe, D. A. and O'Connor, T. P., Eds., Robert E. Krieger Publishing, Malabar, FL, 1988, 33.

9. Mearns, A. J., Matta, M. B., Simecek-Beatty, D., Buchman, M. F., Shigenaka, G., and Wert, W. A., PCB and Chlorinated Pesticide Contamination in U.S. Fish and Shellfish: A Historical Assessment Report, NOAA Tech. Memo. NOS OMA 39, National Oceanic and Atmospheric Administration, Seattle, WA, 1988.

10. Nybakken, J. W., *Marine Biology: An Ecological Approach,* Harper & Row, New York, 1982.

11. Thurman, H. V. and Webber, H. H., *Marine Biology,* Charles E. Merrill Publishing, Columbus, OH, 1984.

12. Wiemeyer, S. N., Schmeling, S. K., and Anderson, A., Environmental pollutant and necropsy data for ospreys from the eastern United States, 1975-1982, *J. Wildl. Dis.,* 23, 279, 1987.

13. Ohlendorf, H. M. and Fleming, W. J., Birds and environmental contaminants in San Francisco and Chesapeake Bays, *Mar. Pollut. Bull.,* 19, 487, 1988.

14. Wiemeyer, S. N., Lamont, T. G., Bunck, C. M., Sindclar, C. R., Gramlich, F. J., Fraser, J. D., and Byrd, M. A., Organochlorine pesticide, polychlorobiphenyl, and mercury residues in bald eagle eggs — 1969-79 — and their relationship to shell thinning and reproduction, *Arch. Environ. Contam. Toxicol.,* 13, 529, 1984.

15. Randers, J., System simulation to test environmental policy: DDT, *Int. J. Environ. Study,* 4, 51, 1972.

16. Hansen, P. -D., Von Westernhagan, H., and Rosenthal, H., Chlorinated hydrocarbons and hatching success in Baltic herring spring spawners, *Mar. Environ. Res.,* 15, 59, 1985.

17. Johnston, R., Mechanisms and problems of marine pollution in relation to commercial fisheries, in *Marine Pollution,* Johnston, R., Ed., Academic Press, London, 1976, 3.

18. Goldberg, E. D., Butler, P., Meier, P., Menzer, D., Risebrough, R. W., and Stickel, L. F., Chlorinated Hydrocarbons in the Marine Environment, Tech. Rep., Panel on Monitoring Persistent Pesticides in the Marine Environment, National Academy of Science, Washington D.C., 1971.

19. Menzel, D.W., Andersson, J., and Randke, A., Marine phytoplankton vary in their response to chlorinated hydrocarbons, *Science,* 167, 1724, 1970.

20. Bowes, G. W., Uptake and metabolism of 2,2bis(*p*-chlorophenyl)-1,1,1-trichloroethane (DDT) by marine phytoplankton and its effect on growth and chloroplast electron transport, *Plant Physiol., Lancaster,* 49, 172, 1972.

21. Butler, P. A. and Schutzmann, R. L., Residues of pesticides and PCBs in estuarine fish, 1972-1976 — national pesticide monitoring program, *Pestic. Monit. J.,* 12, 51, 1978.

22. Butler, P. A., Kennedy, C. D., and Schutzmann, R. L., Pesticide residues in estuarine mollusks, 1977 versus 1972 — national pesticide monitoring program, *Pestic. Monit. J.,* 12, 99, 1978.

23. Butler, P. A., Organochlorine residues in estuarine mollusks, 1965-72. National Pesticide Monitoring Program, *Pestic. Monit. J.,* 6, 238, 1973.

24. Schmitt, C. J., Ribick, M. A., Ludke, J. L., and May, T. W., National Pesticide Monitoring Program: Organochlorine Residues in Freshwater Fish, 1976-79, Resource Publ. 152, U.S. Department of the Interior, Fish and Wildlife Service, Washington, D.C., 1983.

25. National Oceanic and Atmospheric Administration, National Status and Trends Program for Marine Environmental Quality: Progress Report and Preliminary Assessments of Findings of the Benthic Surveillance Project — 1984, Tech. Rep., OAD/NOS/NOAA, Department of Commerce, Rockville, MD, 1987.

26. Butler, P. A., EPA-NOAA Cooperative Estuarine Monitoring Program, Final Rep., U.S. Environmental Protection Agency, Gulf Breeze, FL, 1978.

27. Sherwood, M. J., Fin erosion, liver condition, and trace contaminant exposure in fishes from three coastal regions, in *Ecological Stress and the New York Bight: Science and Management,* Mayer, G. F., Ed., Estuarine Research Federation, Columbia, SC, 1982, 359.

28. Sovocool, G. W., Lewis, R. G., Harless, R. L., Wilson, N. K., and Zehr, R. D., Analysis of technical chlordane by gas chromatography/mass spectrometry, *Anal. Chem.,* 49, 734, 1977.

29. Hayes, S. P. and Phillips, P. T., California State Mussel Watch Marine Water Quality Monitoring Program 1984 to 1985, Water Quality Monitoring Rep., No. 86-3WQ, California Water Research Control Board, Sacramento, 1986.

30. Kawano, M. S., Matsushida, T., Inoue, H., Tanaka, H., and Tatsukawa, R., Biological accumulation of chlordane compounds in marine organisms from the northern North Pacific and Bering Sea, *Mar. Pollut. Bull.,* 17, 512, 1986.

31. Klauda, R. J. and Bender, M. E., Contaminant effects on Chesapeake Bay finfishes, in *Contaminant Problems and Management of Living Chesapeake Bay Resources,* Majumdar, S. K., Hall, L. W., Jr., and Austin, H. M., Eds., Pennsylvania Academy of Science, Easton, 1987, 321.

32. Eisler, R., Mirex Hazards to Fish, Wildlife, and Invertebrates: A Synoptic Review, Biol. Rep. 85, U.S. Fish and Wildlife Service, Washington, D.C., 1985.

33. Reimold, R. J. and Durant, C. J., Survey of Toxaphene Levels in Georgia Estuaries, Tech. Rep. Ser. 72-2, Georgia Marine Science Center, Skidaway Island, 1972.

34. Reimold, R. J. and Durant, C. J., Toxaphene content of estuarine fauna and flora before, during, and after dredging toxaphene-contaminanted sediments, *Pestic. Monit. J.,* 8, 44, 1974.

35. Stephenson, M., Smith, D., Ichikawa, G., Goetzl, J., and Martin, M., State Mussel Watch Program Preliminary Data Report 1985-86, Tech. Rep., State Water Resources Control Board, California Department of Fish and Game, Monterey, 1986.

36. Eisler, R. and Jacknow, J., Toxaphene Hazards to Fish, Wildlife, and Invertebrates: A Synoptic Review, Biol. Rep. 85(1.4), U.S. Fish and Wildlife Service, Washington, D.C., 1985.

37. MacLeod, W. D., Jr., Ramos, L. S., Friedman, A. J., Burrows, D. G., Prohaska, P. G., Fisher, D. L., and Brown, D. W., Analysis of Residual Hydrocarbons and Related Compounds in Selected Sources, Sinks, and Biota of the New York Bight, NOAA Tech. Memo. NOS OMPA-6, Rockville, MD, 1981.

38. Gossett, R. W., Brown, D. A., and Young, D. R., Predicting the bioaccumulation of organic compounds in marine organisms using octanol/water partition coefficients, *Mar. Pollut. Bull.*, 14, 387, 1983.

39. Murray, H. E., Neff, G. S., Hrung, Y., and Giam, C. S., Determination of benzo(a)pyrene, hexachlorobenzene, and pentachlorophenol in oysters from Galveston Bay, Texas, *Bull. Environ. Contam. Toxicol.*, 25, 663, 1980.

40. Murray, H. E., Ray, L. F., and Giam, C. S., Analysis of marine sediment, water, and biota for selected organic pollutants, *Chemosphere*, 10, 1327, 1981.

41. Young, D. R. and Gossett, R., Chlorinated benzenes in sediments and organisms, in Biennial Report 1979-80, Coastal Water Research Project, Bascom, W., Ed., Southern California Coastal Water Research Project, Long Beach, 1980, 181.

42. Yale, B., Joy, J., and Johnson, A., Chemical contaminants in clams and crabs from Eagle Harbor, Washington State, with emphasis on polynuclear aromatic hydrocarbons, Water Quality Investigation Section Rep., Washington Department of Ecology, Olympia, 1984.

43. U.S. Environmental Protection Agency, Ambient Water Quality Criteria for Polychlorinated Biphenyls, EPA Rep. 440/5-80-068, EPA, Washington, D.C., 1980.

44. Brown, M. P., Werner, M. B., Sloan, R. J., and Simpson, K. W., Polychlorinated biphenyls in the Hudson River, *Environ. Sci. Technol.*, 19, 656, 1985.

45. Huntzinger, O., Safe, S., and Zitko, V., *The Chemistry of PCBs*, CRC Press, Cleveland, 1974.

46. Whelan, E. M., *Toxic Terror*, Jameson Books, Ottawa, IL, 1985.

47. Brinkman, U. A. T. and DeKok, A., Production properties and usage, in *Halogenated Biphenyls, Terphenyls, Napthalenes, Dibenzodioxins, and Related Products*, Elsevier/North Holland, Amsterdam, 1980, 1.

48. Cairns, T., Doose, G. M., Froberg, J. E., Jacobson, R. A., and Sigmund, E. G., Analytical chemistry of PCBs, in *PCBs and the Environment*, Vol. 1, Waid, J. S., Ed., CRC Press, Boca Raton, FL, 1986, 1.

49. Ballschmiter, K. and Zell, M., Analysis of polychlorinated biphenyls (PCB) by glass capillary gas chromatography, *Fresenius Z. Anal. Chem.*, 302, 20, 1980.

50. Eisler, R., Polychlorinated Biphenyl Hazards to Fish, Wildlife, and Inverte-
    brates: A Synoptic Review, Biol. Rep. 85 (1.7), U.S. Fish and Wildlife
    Service, Washington, D.C., 1986.

51. Safe, S., Polychlorinated biphenyls (PCBs) and polybrominated biphenyls
    (PBBs): biochemistry, toxicology, and mechanism of action, *Crit. Rev.
    Toxicol.,* 13, 319, 1984.

52. Dunn, W. J., III, Stalling, D. L., Schwartz, T. R., Hogan, J. W., and Petty, J.
    D., Pattern recognition for classification and determination of polychlori-
    nated biphenyls in environmental samples, *Anal. Chem.,* 56, 1308, 1984.

53. Schwartz, T. R., Campbell, R. D., Stalling, D. L., Little, R. L., Petty, J. D.,
    Hogan, J. W., and Kaiser, E. M., Laboratory data base for isomer-specific
    determination of polychlorinated biphenyls, *Anal. Chem.,* 56, 1303, 1984.

54. Holden, A.V., The reliability of PCB analysis, in *PCBs in the Environment,*
    Vol. 1, Waid, J. S., Ed., CRC Press, Boca Raton, FL, 1986, 85.

55. National Academy of Sciences, Polychlorinated Biphenyls, Department of
    Community Assessment of PCBs in the Environment, Environmental Studies
    Board, Committee of Natural Resources, National Research Council/Na-
    tional Academy of Science, Washington, D.C., 1979.

56. Haque, R. and Schmedding, D., Studies on the adsorption of selected
    polychlorinated biphenyl isomers on several surfaces, *J. Environ. Sci.
    Health,* B11, 129, 1979.

57. Weber, W. J., Voice, T. C., Pirbazari, M., Hunt, G. E., and Ulanoff, D. M.,
    Sorption of hydrophobic compounds by sediments, soils, and suspended
    solids. II. Sorbent evaluation studies, *Water Res.,* 17, 1443, 1983.

58. Chou, S. F. J. and Griffin, R. A., Solubility and soil mobility of polychlori-
    nated biphenyls, in *PCBs and the Environment,* Vol. 1, Waid, J. S., Ed., CRC
    Press, Boca Raton, FL, 1986, 101.

59. Duinker, J. C. Hillebrand, M. T. M., and Boon, J. P., Organochlorines in
    benthic invertebrates and sediments from the Dutch Wadden Sea; identifica-
    tion of individual PCB components, *Neth. J. Sea Res.,* 17, 19, 1983.

60. Cantlon, J. E., The PCB problem: an overview, in *PCBs: Human and
    Environmental Hazards,* D'Itri, F. M. and Kamin, M. A., Eds., Butterworths,
    Boston, 1983, 5.

61. Zabik, M. J., The photochemistry of PCBs, in *PCBs: Human and Environ-
    mental Hazards,* D'Itri, F. M. and Kamin, M. A., Eds., Butterworths, Boston,
    1983, 141.

62. Stolen, J., The Effects of the PCB, Arochor 1254, and ethanol on the humoral
    immune response of a marine teleost to a sludge bacterial isolate of *E. coli,*
    in *Marine Pollution and Physiology: Recent Advances,* Vernberg, F. J.,
    Thurberg, F. P., Calabrese, A., and Vernberg, W. B., Eds., University of South
    Carolina Press, Columbia, 1985, 419.

63. Jensen, S., Report of a new chemical hazard, *New Sci.,* 32, 612, 1966.

64. Dennis, D. S., PCBs in the surface waters and bottom sediments of the major
    drainage basins of the U.S., in Conf. Proc. National Conference on PCBs,
    EPA-560/6-004, U.S. Environmental Protection Agency, Chicago, 1975.

65. Swain, W. R., An overview of the scientific basis for concern with polychlorinated biphenyls in the Great Lakes, in *PCBs: Human and Environmental Hazards,* D'ltri, F. M. and Kamin, M. A., Eds., Butterworths, Boston, 1983, 11.

66. Nisbet, I. C. and Sarofim, A. F., Rates and routes of transport of PCBs in the environment, *Environ. Health Perspect.,* 1, 21, 1972.

67. Phillips, D. J. H., Use of organisms to quantify PCBs in marine and estuarine environments, in *PCBs and the Environment,* Waid, J. S., Ed., CRC Press, Boca Raton, FL, 1986, 127.

68. Pavlou, S. P., and Dexter, R. N., Distribution of polychlorinated biphenyls (PCB) in estuarine ecosystems: testing the concept of equilibrium partitioning in the marine environment, *Environ. Sci. Technol.,* 13, 65, 1979.

69. Duinker, J. and Hillebrand, M. T. J., Behaviour of PCB, pentachlorobenzene, hexachlorobenzene, α-HCH, γ-HCH, β-HCH, dieldrin, endrin and *p,p'*DDD in Rhine-Meuse Estuary and the adjacent coastal area, *Neth. J. Sea Res.,* 13, 256, 1979.

70. Harding, L. W., Jr. and Phillips, J. H., Jr., Polychlorinated biphenyls: transfer from microparticulates to marine phytoplankton and the effects on photosynthesis, *Science,* 202, 1189, 1978.

71. Ware, D.M. and Addison, R.F., PCB residues in plankton from the Gulf of St. Lawrence, *Nature,* 246, 519, 1973.

72. Hiraizumi, Y., Takahashi, M., and Nishimura, H., Adsorption of polychlorinated biphenyl onto sea bed sediment, marine plankton, and other adsorbing agents, *Environ. Sci. Technol.,* 13, 580, 1979.

73. Fowler, S. W. and Elder, D. L., Chlorinated hydrocarbons in pelagic organisms from the open Mediterranean Sea, *Mar. Environ. Res.,* 4, 87, 1980.

74. Wyman, K. D. and O'Connors, H. B., Jr., Implications of short-term PCB uptake by small estuarine copepods (Genus *Acartia*) from PCB-contaminated water, inorganic sediments, and phytoplankton, *Est. Coastal Mar. Sci.,* 11, 121, 1980.

75. Hamdy, M. K. and Gooch, J. A., Uptake, retention, biodegradation, and depuration of PCBs by organisms, in *PCBs and the Environment,* Waid, J. S., Ed., CRC Press, Boca Raton, FL, 1986, 63.

76. Liss, P. S., Chemistry of the sea surface microlayer, in *Chemical Oceanography,* Vol. 2, Riley, J. P. and Skirrow, G., Eds., Academic Press, New York, 1975, 193.

77. Södergren, A. and Larsson, P., Transport of PCBs in aquatic laboratory model ecosystems from sediment to the atmosphere via the surface microlayer, *Ambio,* 11, 41, 1982.

78. Gross, M. G., *Oceanography: A View of the Earth,* 3rd ed., Prentice-Hall, Englewood Cliffs, N J, 1982.

79. Beller, H. R. and Simoneit, B. R. T., Polychlorinated biphenyls and hydrocarbons: distributions among bound and unbound lipid fractions of estuarine sediments, in *Organic Marine Geochemistry,* Sohn, M. L., Ed., Chemical Symposium Ser. No. 305, 1986, 198.

80. Elder, D. L., Villeneuve, J. P., Parsi, P., and Harvey, G. R., Polychlorinated biphenyls in seawater, sediments, and over-ocean air of the Mediterranean, in *Activities of the International Laboratory of Marine Radioactivity, Monaco, 1975*, Rep. Tech. Doc. IAEA-187, International Atomic Energy Agency, Vienna, 1976, 136.

81. Nisbet, I. C. T., Criteria Document for PCBs, Rep. No. 440/9-76-021, U.S. Environmental Protection Agency, Washington, D.C., 1976.

82. Pavlou, S. P., Dexter, R. N., Hom, W., and Krogslund, K. A., Polychlorinated Biphenyls in Puget Sound: Baseline Data and Methodology, Spec. Rep. No. 74, Ref. No. M77-36, University of Washington, Seattle, 1977.

83. Puccetti, G. and Leoni, V., PCB and HCB in the sediments and waters of the Tiber Estuary, *Mar. Pollut. Bull.*, 11, 22, 1980.

84. West, R. H. and Hatcher, P. G., Polychlorinated biphenyls in sewage sludge and sediments of the New York Bight, *Mar. Pollut. Bull.*, 11, 126, 1980.

85. Bopp, R. F., Simpson, H. J., Olsen, C. R., and Kostyk, N., Polychlorinated biphenyls in sediments of the tidal Hudson River, New York, *Environ. Sci. Technol.*, 15, 210, 1981.

86. Swanson, R. L., Champ, M. A., O'Connor, T., Park, P. K., O'Connor, J., Mayer, G. F., Stanford, H. M., and Erdheim, E., Sewage-sludge dumping in the New York Bight apex: a comparison with other proposed ocean dump-sites, in *Wastes in the Ocean*, Vol. 6, *Nearshore Waste Disposal*, Ketchum, B. H., Capuzzo, J. M., Burt, W. V., Duedall, I. W., Park, P. K., and Kester, D. R., Eds., John Wiley & Sons, New York, 1985, 461.

87. Stainken, D. M. and Rollwagen, J., PCB residues in bivalves and sediments of Raritan Bay, *Bull. Environ. Contam. Toxicol.*, 23, 690, 1979.

88. McCormick, J. M., Multer, H. G., and Stainken, D. M., A review of Raritan Bay research, *Bull. N. J. Acad. Sci.*, 29, 47, 1984.

89. O'Connor, J. M., Klotz, J. B., and Kneip, T. J., Sources, sinks, and distribution of organic contaminants in the New York Bight ecosystem, in *Ecological Stress and the New York Bight: Science and Management*, Mayer, G. F., Ed., Estuarine Research Federation, Columbia, SC, 1982, 631.

90. Heaton, M. G. and Dayal, R., Metals in interstitial waters of the New York Bight dredged-material deposit, in *Wastes in the Ocean*, Vol. 6, *Nearshore Waste Disposal*, Ketchum, G. H., Capuzzo, J. M., Burt, W. V., Duedall, I., Park, P. K., and Kester, D. R., Eds., John Wiley & Sons, New York, 1985, 235.

91. Schubel, J. R., Pritchard, D. W., Abood, K. A., Casper, T. C., Gross, M. G., Hunkins, K., Malone, T. C., and Terry, D. W., Influences of the Hudson-Raritan estuary on the New York Bight, Working Paper 82-5, Marine Science Research Center, State University of New York, Stony Brook, 1982.

92. Boon, J. P. and Duinker, J. C., Monitoring of cyclic organochlorines in the marine environment, in *Monitoring in the Marine Environment*, Part 1, Kramer, C. J. M. and Hekstra, G. P., Eds., Environ. Monit. Assess., 7(Spec. iss.), 189, 1986.

93. *The International Mussel Watch*, National Academy of Sciences, Washington, D.C., 1980.

94. Goldberg, E. D., Koide, M., Hodge, V., Flegal, A. R., and Martin, J., U.S. Mussel Watch: 1977-1978 results on trace metals and radionuclides, *Est. Coastal Shelf Sci.,* 16, 69, 1983.

95. Goldberg, E. D., The mussel watch concept, *Environ. Monit. Assess.,* 7, 91, 1986.

96. Preston, M. R., Marine pollution, in *Chemical Oceanography,* Vol. 9, Riley, J. P., Ed., Academic Press, London, 1989, 53.

97. Goldberg, E. D., *The Health of the Oceans,* UNESCO, Paris, 1976.

98. Goldberg, E. D., *Siren,* No. 23, 1984.

99. deKock, W. C., Monitoring bioavailable marine contaminants with mussels (*Mytilus edulis* L.) in the Netherlands, in Monitoring in the Marine Environment, Part 2, Kramer, C. J. M. and Hekstra, G. P., Eds., *Environ. Monit. Assess.,* 7(Spec. iss.), 209, 1986.

100. Vreeland, V., Uptake of chlorobiphenyls by oysters, *Environ. Pollut.,* 6, 135, 1974.

101. Deubert, K. H., Rule, P., and Corte-Real, I., PCB residues in *Mercenaria mercenaria* from New Bedford Harbor, 1978, *Bull. Environ. Contam. Toxicol.,* 27, 683, 1981.

102. Langston, W. J., Accumulation of polychlorinated biphenyls in the cockle *Cerastoderma edule* and the tellin *Macoma balthica, Mar. Biol.,* 45, 265, 1978.

103. Langston, W. J., Persistence of polychlorinated biphenyls in marine bivalves, *Mar. Biol.,* 46, 35, 1978.

104. Nimmo, D. R., Hansen, D. J., Couch, J. A., Cooley, N. R., Parrish, P. R., and Lowe, J. I., Toxicity of Aroclor 1254 and its physiological activity in several estuarine organisms, *Arch. Environ. Contam. Toxicol.,* 3, 22, 1975.

105. Neff, J. M. and Giam, C. S., Effects of Aroclor 1016 and Halowax 1099 on juvenile horseshoe crabs *Limulus polyphemus,* in *Physiological Responses of Marine Biota to Pollutants,* Vernberg, F. J., Calabrese, A., Thurberg, P., and Vernberg, W. B., Eds., Academic Press, New York, 1977, 21.

106. McLeese, D. W., Meltcalfe, C. D., and Pezzack, D. S., Uptake of PCBs from sediment by *Hereis virens* and *Crangon septemspinosa, Arch. Environ. Contam. Toxicol.,* 9, 507, 1980

107. Phillips, D. J. H., *Quantitative Aquatic Biological Indicators: Their Use to Monitor Trace Metal and Organochlorine Pollution,* Applied Science, London, 1980.

108. Hansen, D. J., Schimmel, SC, and Forester, J., Effects of Aroclor 1016 on embryos, fry, juveniles, and adults of sheepshead minnows (*Cyprinodon variegatus*), *Trans. Am. Fish. Soc.,* 104, 584, 1975.

109. Zitko, V., Polychlorinated biphenyls and organochlorine pesticides in some freshwater and marine fishes, *Bull. Environ. Contam. Toxicol.,* 6, 464, 1971.

110. Young, D. R. and Mearns, A. J., Pollutant Flow Through Food Webs, Annual Report, Southern California Coastal Water Research Project, Los Angeles, 1978, 185.

111. Wilson, A. J. and Forester, J., Persistence of Aroclor 1254 in a contaminated estuary, *Bull. Environ. Contam. Toxicol.*, 19, 637, 1978.
112. Farrington, J. W., Risebrough, R. W., Parker, P. L., Davis, A. C., deLapps, B., Winters, J. K., Boatwright, D., and Frew, N. M., Hydrocarbons, Polychlorinated Biphenyls, and DDE in Mussels and Oysters from the U.S. Coast, 1976-78. The Mussel Watch, Tech. Rep., WHOI-82-42, Woods Hole Oceanographic Institution, Woods Hole, MA, 1982.
113. National Oceanic and Atmospheric Administration, National Status and Trends Program for Marine Environmental Quality: Progress Report — A Summary of Selected Data on Chemical Contaminants in Tissues Collected During 1984, 1985, 1986, Tech. Rep., OAD/OMA/NOS/NOAA, Department of Commerce, Rockville, MD, 1987.
114. Murty, A. S., *Toxicity of Pesticides to Fish*, Vol. 2, CRC Press, Boca Raton, FL, 1986.
115. Editorial, A need for reevaluation of PCB toxicity, *Mar. Pollut. Bull.*, 20, 247, 1989.
116. Schmitt, D. J., Zajiik, J. L., and Ribick, M. A., National Pesticide Monitoring Program: residues of organochlorine chemicals in freshwater fish, 1980-81, *Arch. Environ. Contam. Toxicol.*, 14, 225, 1985.
117. Gadbois, D. F. and Maney, R. D., Survey of polychlorinated biphenyls in selected finfish species from United States coastal waters, *Fish. Bull., U.S.*, 81, 389, 1983.
118. Sherwood, M. J., Fin erosion, liver condition, and trace contaminant exposure in fishes from three coastal regions, in *Ecological Stress and the New York Bight: Science and Management*, Mayer, G. F., Ed., Estuarine Research Federation, Columbia, SC, 1982, 359.
119. Cosper, E. M., Snyder, B. J., Arnold, L. M., Zaikowski, L. A., and Wurster, C. F., Induced resistance to polychlorinated biphenyls confers cross-resistance and altered environmental fitness in a marine diatom, *Mar. Environ. Res.*, 23, 207, 1987.
120. Fisher, N. S., Graham, L. B., and Wurster, C. G., Geographic differences in phytoplankton sensitivity to PCBs, *Nature*, 241, 548, 1973.
121. Fisher, N. S. and Wurster, C. F., Individual and combined effects of temperature and PCBs on the growth of three species of phytoplankton, *Environ. Pollut.*, 5, 205, 1973.
122. Fisher, N. S., Graham, L. B., Carpenter, E. J., and Wurster, C. F., Geographic differences in phytoplankton sensitivity to PCBs, *Nature*, 24, 548, 1973.
123. Mosser, J. L., Fisher, N. S., and Wurster, C. F., Polychlorinated biphenyls and DDT alter species composition in mixed cultures of algae, *Science*, 176, 533, 1972.
124. Fisher, N. S. and Wurster, C. F., Impact of pollutants on plankton communities, *Environ. Conserv.*, 1, 189, 1974.
125. Cole, D. R. and Plapp, F. W., Jr., Inhibition of growth and photosynthesis in *Chlorella pyrenoidosa* by a polychlorinated biphenyl and several insecticides, *Environ. Entomol.*, 3, 217, 1974.

126. Fisher, N. S., On the differential sensitivity of estuarine and open-ocean diatoms to exotic chemical stress, *Am. Nat.*, 111, 871, 1977.

127. Mosser, J.L., Teng, T.C., Walther, W.G., and Wurster, C.F., Interactions of PCBs, DDT, and DDE in a marine diatom, *Bull. Environ. Contam. Toxicol.*, 12, 665, 1974.

128. Mahanty, H. K., Polychlorinated biphenyls: accumulation and effects upon plants, in *PCBs and the Environment*, Vol. 2, Waid, J. S., Ed., CRC Press, Boca Raton, FL, 1986, 1.

129. Mahanty, H. K., Kanamycin-resistant *Scenedesmus obliquus* showing simultaneous resistance to other antibiotics and polychlorinated biphenyls, *Nature*, 277, 562, 1979.

130. Glooschenko, V. and Glooschenko, W., Effect of PCB compounds on growth of Great Lakes phytoplankton, *Can. J. Bot.*, 53, 653, 1974.

131. Larsson, C. M. and Tillberg, J. G., Effects of the commercial polyclorinated biphenyl mixture Aroclor 1242 on growth, viability, phosphate uptake, respiration, and oxygen evolution in *Scenedesmus*, *Physiol. Plant.*, 33, 256, 1975.

132. Wyman, K. D. and O'Connors, H. B., Jr., Implications of short-term PCB uptake by small estuarine copepods (Genus *Acartia*) from PCB-contaminated water, inorganic sediments, and phytoplankton, *Est. Coastal Mar. Sci.*, 11, 121, 1980.

133. Clayton, J. R., Jr., Pavlou, S. P., and Breitner, N. F., Polychlorinated biphenyls in coastal marine zooplankton: bioaccumulation by equilibrium partitioning, *Environ. Sci. Technol.*, 11, 676, 1977.

134. Harding, G. C. and Addison, R. F., Accumulation and effects of PCBs in marine invertebrates and vertebrates, in *PCBs and the Environment*, Vol. 2, Waid, J. S., Ed., CRC Press, Boca Raton, FL, 1986, 9.

135. Fowler, S. W., PCBs and the environment: the Mediterranean marine ecosystem, in *PCBs and the Environment*, Vol. 3, Waid, J. S., Ed., CRC Press, Boca Raton, FL, 1986, 209.

136. Goldberg, E. D., Bowen, V. T., Farrington, J. W., Harvey, G., Martin, J. H., Parker, P. L., Risebrough, R. W., Robertson, W., Schneider, E., and Gamble, E., The mussel watch, *Environ. Conserv.*, 5, 101, 1978.

137. International Council for the Exploration of the Sea, Report of Working Group for the International Study of the Pollution of the North Sea and Its Effects on Living Resources and Their Exploitation, Coop. Res. Rep. No. 39, International Council for the Exploration of the Sea, Copenhagen, 1974.

138. International Council for the Exploration of the Sea, The ICES Co-ordinated Monitoring Programme in the North Sea, 1974, Coop. Res. Rep. No. 58, International Council for the Exploration of the Sea, Copenhagen, 1977.

139. International Council for the Exploration of the Sea, The ICES Co-ordinated Monitoring Programmes 1975 and 1976, Coop. Res. Rep. No. 72, International Council for the Exploration of the Sea, Copenhagen, 1977.

140. Lowe, J. I., Parrish, P. R., Patrick, J. M., Jr., and Forrester, J., Effects of the polychlorinated biphenyl Aroclor 1254 on the American oyster, *Crassostrea virginica*, *Mar. Biol.*, 17, 209, 1972.

141. Fowler, S. W., Polikarpov, G. G., Elder, D. L., Parsi, P., and Villeneuve, J. P., Polychlorinated biphenyls: accumulation from contaminanted sediments and water by the polychaete *Nereis diversicolor*, *Mar. Biol.*, 48, 303, 1978.

142. Klauda, R. J., Peck, T. H., and Rice, G. K., Accumulation of polychlorinated biphenyls in Atlantic tomcod (*Microgadus tomcod*) collected from the Hudson River estuary, New York, *Bull. Environ. Contam. Toxicol.*, 27, 829, 1981.

143. Stein, J. E., Horn, T., and Varanasi, U., Simultaneous exposure of English sole (*Parophrys vetulus*) to sediment-associated xenobiotics. I. Uptake and disposition of ^{14}C-polychlorinated biphenyls and ^{3}H-benzo[a]pyrene, *Mar. Environ. Res.*, 13, 97, 1984.

144. Ginn, T. C. and Barrick, R. C., Bioaccumulation of toxic substances in Puget Sound organisms, in *Oceanic Processes in Marine Pollution,* Wolfe, D. A. and O'Connor, T. P., Eds., Robert E. Krieger Publishing, Malabar, FL, 1988, 157.

145. Van den Broek, W. L. F., Seasonal levels of chlorinated hydrocarbons and heavy metals in fish and brown shrimps from the Medway estuary, Kent, *Environ. Pollut.*, 19, 21, 1979.

146. Jessop, B. M. and Vithayassai, C., Creel surveys and biological studies of the striped-bass fisheries of the Shubenacadie, Gaspereau, and Annapolis Rivers, 1976, Fish. Mar. Serv. Manuscript Rep., No. 1532, Department of Fisheries and Oceans, Ottawa, Canada, 1979.

147. Hansen, D. J., Parrish, P. R., Lowe, J. I., Wilson, A. J., Jr., and Wilson, P. D., Chronic toxicity, uptake, and retention of Aroclor 1254 in two estuarine fishes, *Bull. Environ. Contam. Toxicol.*, 6, 113, 1971.

148. Hansen, D. J., Parrish, P. R., and Forrester, J., Aroclor 1016: toxicity to and uptake by estuarine animals, *Environ. Res.*, 7, 363, 1974.

149. Mayer, F. L., Mehrle, P. M., and Sanders, H. O., Residue dynamics and biological effects of polychlorinated biphenyls in aquatic organisms, *Arch. Environ. Contam. Toxicol.*, 5, 501, 1977.

150. Mauck, W. L., Mehrle, P. M., and Mayer, F. L., Effects of the polychlorinated biphenyl Aroclor 1254 on growth, survival, and bone development in brook trout (*Salvelinus fontinalis*), *J. Fish. Res. Bd. Can.*, 35, 1084, 1978.

151. Sangalang, G. B., Freeman, H. C., and Crowell, R., Testicular abnormalities in cod (*Gadus morhua*) fed Aroclor 1254, *Arch. Environ. Contam. Toxicol.*, 10, 617, 1981.

152. Steen, W. C., Paris, D. F., and Baughman, G. L., Partitioning of selected polychlorinated biphenyls to natural sediments, *Water Res.*, 12, 655, 1978.

153. DeFoe, D. L., Veith, G. D., and Carlson, R. W., Effects of Aroclor 1248 and 1260 on the fathead minnow (*Pimephales promelas*), *J. Fish. Res. Bd. Can.*, 35, 997, 1978.

154. Narbonne, J. F., Polychlorinated biphenyl accumulation in grey mullets (*Chelon labrosus*): effect of age, *Bull. Environ. Contam. Toxicol.*, 22, 65, 1979.

155. Califano, R. J., O'Connor, J. M., and Peters, L. S., Uptake, retention, and elimination of PCB (Aroclor 1254) by larval striped bass (*Morone saxatilis*), *Bull. Environ. Contam. Toxicol.*, 24, 467, 1980.

156. Peakall, D. B., Accumulation and effects on birds, in *PCBs and the Environment,* Vol. 2, Waid, J.S., Ed., CRC Press, Boca Raton, FL, 1986, 31.

157. Stout, V. F., What is happening to PCBs?: elements of effective environmental monitoring as illustrated by an analysis of PCB trends in terrestrial and aquatic organisms, in *PCBs and the Environment,* Vol. 1, Waid, J. S., Ed., CRC Press, Boca Raton, FL, 1986, 163.

158. Clark, R. B., *Marine Pollution, 2nd ed.*, Clarendon Press, Oxford, 1989.

159. Peakall, D. B. and Peakall, M. L., Effect of a polychlorinated biphenyl on the reproduction of artificially and naturally incubated dove eggs, *J. Appl. Ecol.*, 10, 863, 1973.

160. Custer, T. W. and Heinz, G. H., Reproductive success and next attentiveness of mallard ducks fed Aroclor 1254, *Environ. Pollut. (Ser. A)*, 21, 313, 1980.

161. Gilman, A. P., Peakall, D. B., Hallett, D. J., Fox, G. A., and Nordstrom, R. J., Herring gulls (*Larus argentatus*) as monitors of contaminants in the Great Lakes, in *Animals as Monitors of Environmental Pollutants,* National Academy of Sciences, Washington, D.C., 1979, 280.

162. White, D. H., Nationwide residues of organochlorine compounds in wings of adult mallards and black ducks, 1976-77, *Pestic. Monit. J.*, 13, 12, 1979.

163. Blus, L. J., Further intrepetation of the relation of organochlorine residues in brown pelican eggs to reproductive success, *Environ. Pollut.*, 28(A), 15, 1982.

164. Blus, L. J., Lamont, T. G., and Neely, B. S., Jr., Effects of organochlorine residues on eggshell thickness, reproduction, and population status of brown pelicans (*Pelecanus occidentalis*) in South Carolina and Florida, 1969-76, *Pestic. Monit. J.*, 12, 172, 1979.

165. Blus, L., Cromartie, E., McNease, L., and Joanen, T., Brown pelican: population status, reproductive success, and organochlorine residues in Louisiana, 1971-1976, *Bull. Environ. Contam. Toxicol.*, 22, 128, 1979.

166. Klaas, E. E. and Belisle, A. A. Organochlorine pesticide and polychlorinated biphenyl residues in selected fauna from a New Jersey salt marsh — 1967 vs. 1973, *Pestic. Monit. J.*, 10, 149, 1977.

167. Hoffman, D. J., Rattner, B. A., Bunck, C. M., Krynitsky, A., Ohlendorf, H. M., and Lowe, R. W., Association between PCBs and lower embryonic weight in black-crowned night herons in San Francisco Bay, *J. Toxicol. Environ. Health,* 19, 383, 1986.

168. Olsson, M., PCBs in the Baltic environment, in *PCBs and the Environment,* Vol. 3, Waid, J. S., Ed., CRC Press, Boca Raton, FL, 1986, 181.

169. Farrington, J. W., Goldberg, E. D., Risebrough, R. W., Martin, J. H., and Bowen, V. T., U.S. "Mussel Watch" 1976-1978; an overview of the trace metal DDE, PCB, hydrocarbon, and artificial radionuclide data, *Environ. Sci. Technol.*, 17, 490, 1983.

170. Elder, D. L. and Fowler, S. W., Polychlorinated biphenyls: penetration into the deep ocean by zooplankton fecal pellet transport, *Science,* 197, 459, 1977.

171. Renberg, L., Sundström, G., and Reutergardh, L., Polychlorinated terphenyls (PCT) in Swedish white-tailed eagles and in grey seals, a preliminary study, *Chemosphere,* 6, 477, 1978.

172. Helz, G. R. and Huggett, R. J., Contaminants in Chesapeake Bay: the regional perspective, in *Contaminant Problems and Management of Living Chesapeake Bay Resources,* Majiuemdar, S. K., Hall, L. W., Jr., and Austin, H. M., Eds., Pennsylvania Acadmy of Science, Easton, 1987, 270.

173. Halestine, S. D., Mulhern, B. M., and Stafford, C., Organochlorine and heavy metal residues in black duck eggs from the Atlantic Flyway, 1978, *Pestic. Monit. J.,* 14, 53, 1980.

174. Olendorf, H. M., The Chesapeake Bay's birds and organochlorine pollutants, *Trans. North Am. Wildl. Nat. Res. Conf.,* 46, 259, 1981.

175. Mehrle, P. M., Haines, T. A., Hamilton, S., Ludke, J. L., Mayer, F. L., and Ribick, M. A., Relationship between body contaminants and bone development of East-Coast Striped bass, *Trans. Am. Fish. Soc.,* 111, 231, 1982.

176. Peters, L. S. and O'Connor, J. M., Factors affecting short-term PCB and DDT accumulation by zooplankton and fish from the Hudson estuary, in *Ecological Stress and the New York Bight: Science and Management,* Mayer, G. F., Ed., Estuarine Research Federation, Columbia, SC, 1982, 451.

177. Munson, T. O., Biochemistry, in Upper Bay Survey, Vol. 2, Final Report to the Maryland Department of Natural Resources, Westinghouse Electric Corporation, Ocean Sciences Division, Annapolis, MD, 1976.

178. Tsai, C. F., Welch, J., Kewi-yang, C., Shaeffer, J., and Cronin, L. E., Bioassay of Baltimore Harbor sediments, *Estuaries,* 2, 141, 1979.

179. Eisenberg, M., Mallman, R., and Tubiash, H. S., Polychlorinated biphenyls in fish and shellfish in the Chesapeake Bay, *Mar. Fish. Rev.,* February 21–25, 1980.

180. Edstrom, R., High resolution gas chromatographic-Hall electrolytic detector analysis of PCB congeners in bluefish (*Pomatomus saltatrux*) and the grey trout (*Cynoscion regalis*) in Virginia, *J. Environ. Toxicol. Chem.,* in press.

181. Bieri, R., Hein, C., Huggett, R., Shou, P., Slone, H., Smith, C., and Su, C., Toxic Organic Compounds in Surface Sediments from the Elizabeth and Patapsco Rivers and Estuaries, Final Report to the U.S. Environmental Protection Agency, EPA-600/53-83-012, EPA, Washington, D.C., 1983.

182. Wright, D. A. and Phillips, D. J. H., Chesapeake and San Francisco Bays: a study in contrasts and parallels, *Mar. Pollut. Bull.,* 19, 405, 1988.

183. Phillips, D. J. H. and Spies, R. B., Chlorinated hydrocarbons in the San Francisco estuarine ecosystem, *Mar. Pollut. Bull.,* 19, 445, 1988.

184. Law, L. M. and Goerlitz, D. F., Selected chlorinated hydrocarbons in bottom material from streams tributary to San Francisco Bay, *Pestic. Monit. J.,* 8, 33, 1974.

185. Chapman, P. M., Dexter, R. N., Cross, S. F., and Mitchell, D. G., A Field Trial of the Sediment Quality Triad in San Francisco Bay, NOAA Tech. Memo. NOS OMA 25, National Oceanic and Atmospheric Administration, Rockville, MD, 1986.

186. Hayes, S. P., Phillips, P. T., Martin, M., Stephenson, M., Smith, D., and Linfield, J., California State Mussel Watch Marine Water Quality Monitoring Program 1983-84, Wat. Qual. Monit. Rep. 85-2WQ, California Water Research Control Board, Sacramento, 1985.

187. Hayes, S. P. and Phillips, P. T., California State Mussel Watch Marine Water Quality Monitoring Program 1965-86, Wat. Qual. Monit. Rep. 87-2WQ, California Water Research Control Board, Sacramento, 1986.

188. Girvin, D.C., Hodgson, A. T., and Panietz, M. H., Assessment of Trace Metal and Chlorinated Hydrocarbon Contamination in Selected San Francisco Bay Estuarine Shellfish, Final Rep., 74-51291 to the San Francisco Bay Regional Water Quality Control Board, Oakland, CA, 1975.

189. Risebrough, R. W., Chapman, J. W., Okazaki, R. K., and Schmidt, T. T., Toxicants in San Francisco Bay and Estuary, Report of the Association of Bay Area Governments, Berkeley, CA, 1978.

190. Linn, J. D., Reiner, C., Crane, D., Smith, L., and Seto, W., Toxic Substances Monitoring Program — 1986 Data Report, Submitted by the California Department of Fish and Game to the State Water Resources Control Board, Sacramento, 1987.

191. Earnest, R. D. and Benville, P. E., Correlation of DDT and lipid levels for certain San Francisco Bay fish, *Pestic. Monit. J.,* 5, 235, 1971.

192. Crosby, D. G., Hogan, K., and Bowes, G. W., The potential impact of chemical residues on the California striped bass fishery, in Third Progress Report of the Cooperative Striped Bass Study, Appendix B, Toxic Substances Control Program, Special Projects Report No. 83-3, State Water Resources Control Board, Sacramento, 1983.

193. Crosby, D. G., Hogan, K., Bowes, G. W., and Foster, G. L., The potential impact of chlorinated hydrocarbon residues on California striped bass, in Report of the Cooperative Striped Bass Study, Tech. Suppl. 1, Sec. 1, State Water Resources Control Board, Sacramento, 1986.

194. Whipple, J. A., Crosby, D. G., and Jung, M., Third Progress Report, Cooperative Striped Bass Study, Special Projects Rep. No. 83-3 sp, State Water Resources Control Board, Sacramento, 1983.

195. Spies, R. B., Rice, D. W., Montagna, P. A., Ireland, R. R., Felton, J. S., Healy, S. K., and Lewis, P. R., Pollutant Body Burdens and Reproduction in *Platichthys stellatus* from San Francisco Bay, Tech. Rep. UCID 19993-84, Lawrence Livermore National Laboratory (for the National Oceanic and Atmospheric Administration) Livermore, CA, 1985.

196. Ohlendorf, H. M. and Miller, M. R., Organochlorine contaminants in California waterfowl, *J. Wildl. Manage.,* 48, 867, 1984.

197. Custer, T. W., Hensler, G. L., and Kaiser, T. E., Clutch size, reproductive success, and organochlorine contaminants in Atlantic Coast black-crowned night-herons, *Auk,* 100, 699, 1983.

198. Henny, C. J., Blus, L. J., Krynitsky, A. J., and Bunck, C. M., Current impact of DDE on black-crowned night-herons on the Intermountain West, *J. Wildl. Manage.,* 48, 1, 1984.

199. Crecelius, E. A. and Bloom, N., Temporal trends of contamination in Puget Sound, in *Oceanic Processes in Marine Pollution,* Vol. 5, *Urban Wastes in Coastal Marine Environments,* Wolfe, D. A. and O'Connor, T. P., Eds., Robert E. Krieger Publishing, Malabar, FL, 1988, 149.

200. Calambokidis, J., Peard, T., Steiger, G. H., and Cubbage, J.C., Chemical Contaminants in Marine Mammals from Washington State, Tech. Memo. NOS OMS 6, U.S. National Oceanic and Atmospheric Administration, Seattle, 1984.

201. Chapman, P. M., Summary of biological effects in Puget Sound — past and present, in *Oceanic Processes in Marine Pollution,* Vol. 5, *Urban Wastes in Coastal Marine Environments,* Wolfe, D. A. and O'Connor, T. P., Eds., Robert E. Krieger Publishing, Malabar, FL, 1988, 169.

202. Arndt, D. P., DDT and PCB Levels in Three Washington State Harbor Seal (*Phoca vitulina richardsii*) Populations, M.S. thesis, University of Washington, Seattle, 1973.

203. Calambokidis, J., Bowman, K., Carter, S., Cubbage, J., Dawson, P., Fleischner, T., Schuett-Hames, J., Skidmore, J., and Taylor, B., Chlorinated Hydrocarbon Concentrations and the Ecology and Behavior of Harbor Seals in Washington State Waters, Unpubl. Tech. Rep., Evergreen State College, Olympia, WA, 1978.

204. Riley, R., Crecelius, E. A., Fitzner, R. E., Thomas, B. L., Gurtisen, J. M., and Bloom, N. S., Organic and Inorganic Toxicants in Sediments and Birds from Puget Sound, Tech. Memo. NOS OMS-1, U.S. National Oceanic and Atmospheric Administration, Rockville, MD, 1984.

205. Phillips, D. J. H., Trace metals and organochlorines in the coastal waters of Hong Kong, *Mar. Pollut. Bull.,* 20, 319, 1989.

206. Tanabe, S., Kannan, N., Fukushima, M., Okamoto, T., Wakimoto, T., and Tatsukawa, R., Persistent organochlorines in Japanese coastal waters: an introspective summary from a far east developed nation, *Mar. Pollut. Bull.,* 20, 344, 1989.

5 Heavy Metals

INTRODUCTION

The continuous increase in heavy metal contamination of estuarine and coastal marine waters is directly attributable to industrialization and development in the coastal zone. The contamination by most heavy metals in coastal environments reflects localized impacts from point or multipoint discharges from municipal and industrial sources.[1] Because of their persistence in the environment, their toxicity at high concentrations, and their tendency to accumulate in the tissues of biota, heavy metals pose potentially hazardous conditions for man. Hence, they have been the subject of ever-expanding research activities to control their concentrations in estuarine and coastal marine habitats.

Heavy metals, as defined by Viarengo,[2] are a group of elements with atomic weights ranging from 63.546 to 200.590 and are characterized by similar electronic distribution in the external shell (e.g., Cu, Zn, Cd, and Hg). These exclude the alkaline earth metals, alkali metals, lanthanides, and actinides.[3] Although these elements are toxic to estuarine and marine organisms above a threshold availability, many of them are essential to metabolism at lower concentrations.[4] Necessary trace elements for life processes include (but are not limited to) cobalt, copper, iron, manganese, molybdenum, vanadium, strontium, and zinc;[5,6] however, any of these elements can be toxic to organisms when present in high concentrations.[7] Of great concern as potential environmental contaminants are cadmium, chromium, mercury, lead, selenium, arsenic, and antimony, which have contributed to severe insidious pollution problems in various estuarine and coastal ecosystems of the U.S. Some heavy metals, such as cadmium and lead, have no known biological function,[3] and may greatly affect biotic communities. Concepcion Bay, Washington, and Newark Bay, New Jersey provide examples of two systems impacted by heavy metal contamination.

A significant fraction of heavy metals in many estuaries derives from domestic and industrial wastes. Substantial amounts of copper, lead, and zinc released from pipes and tanks in domestic systems, for example, enter natural waters that discharge to estuaries. In addition, industrial sources — the mining and processing of metal ores, finishing and plating of metals, and manufacture of metal objects, dyes, paints, and textiles — add large amounts of heavy metals to coastal ecosystems.[3]

Heavy metal toxicity in aquatic organisms is of immense interest to toxicologists. Abel[3] lists heavy metals in an approximate order of decreasing toxicity: Hg, Cd, Cu, Zn, Ni, Pb, Cr, Al, and Co. He emphasizes, however, that this sequence is tentative and subject to change, depending on physical-chemical conditions in the environment that affect chemical speciation of the metals. Moreover, the toxicity of a given metal will vary with the species investigated. The biochemistry of an organism plays a vital role in its susceptibility to metal toxicity. Subsequent to being assimilated by aquatic organisms, heavy metals may be sequestered by metallothioneins and lysosomes, thereby enhancing cellular detoxification.[8] Metallothionein, a low molecular weight, sulfur-containing, metal-binding protein has been shown to control elevated levels of trace metals in marine organisms (e.g., *Callinectes sapidus, Crassostrea virginica,* and *Homarus americanus*) even though their principal function is in regulating normal metal metabolism.[9] Thus, heavy metals may be either maintained in a metabolically available form or may be detoxified subsequent to their accumulation by estuarine and marine organisms.[4]

At elevated levels, heavy metals act as enzyme inhibitors in organisms. For example, the metals copper, cadmium, lead, mercury, and silver, when present in large concentrations, inhibit enzyme activity by forming mercaptides with the sulfhydryl groups.[10] For this reason, as well as others, toxicity levels have been established for heavy metals in aquatic environments. Water quality standards, in turn, have been developed on the basis of toxicity data, and extensive tables of heavy metals for freshwater fish, invertebrates, and marine organisms are available.[11]

The concentrations of trace metals in estuarine and coastal marine waters are controlled by many factors. Advective transport, mixing, and differential settling of sediment-sorbed metals, for instance, give rise to substantial variations in trace metal composition in different parts of an estuary.[12] Many metals complex with organic compounds which influence their chemical speciation.[5] The specific physicochemical form of the element rather than its total concentration determines how it will behave in the environment.[13]

Estuarine seafloor sediments and adjoining marsh surfaces serve as reservoirs of trace metals. The flux of trace metals in these parts of estuarine systems depends on a number of processes, including adsorption-desorption reactions, flocculation, and sedimentation. A considerable amount of heavy metal input by man ultimately accumulates in bottom sediments. For example, Turekian et al.[14] disclosed that specific anthropogenic sources accounted for high concentrations of copper, lead, and zinc in sediments of Long Island Sound. Heavy metals entering estuaries via natural weathering processes also ultimately concentrate in seafloor sediments.

SOURCES OF HEAVY METALS

Most of the heavy metals in estuaries are derived from freshwater runoff and the atmosphere.[15] It is possible to discriminate between sources of metals to estuaries and to the deep sea. Bruland[16] identified two major external sources of heavy metals to the oceans: (1) weathering products of continental rocks transported to the sea by atmospheric and riverine action and (2) the interaction of seawater with newly formed oceanic crustal basalt at ridge-crest spreading centers via hydrothermal activity. Both low and high temperature seawater interaction with the upper crustal layer has been documented in the oceans. The Galapagos hydrothermal vent fields exemplify a low temperature system, and the East Pacific Rise at 21°N, a high temperature system. Table 1 lists the concentrations and speciations of common heavy metals and other trace elements in the sea.

ESTUARINE SYSTEMS
River Input

Excluding anthropogenic origins of heavy metals in estuaries, rivers are the most important source of these elements. Heavy metals exist in natural waters in dissolved, colloidal, and particulate phases, although the concentration of dissolved forms is low. When trans-

TABLE 1
Concentration and Speciation of Elements in Ocean Water

Element	Probable main species in oxygenated seawater	Range and average concentration at 35‰ salinity[a]
Li	Li^+	25 μmol/kg
Be	$BeOH^+$, $Be(OH)^0_2$	4–30 pmol/kg; 20 pmol/kg
B	H_3BO_3	0.416 mmol/kg
C	HCO^-_3, CO^{2-}_3	2.0–2.5 mmol/kg; 2.3 mmol/kg
N	NO^-_3; (also as N_2)	<0.1–45 μmol/kg; 30 μmol/kg
O	O^2 (also as H_2O)	0–300 μmol/kg
F	F^-, MgF^+	68 μmol/kg
Na	Na^+	0.468 mol/kg
Mg	Mg^{2+}	53.2 mmol/kg
Al	$Al(OH)^-_4$, $Al(OH)^0_3$	(5–40 mmol/kg; 20 mmol/kg)
Si	H_4SiO_4	<1–180 μmol/kg; 100 μmol/kg
P	HPO^{2-}_4, $NaHPO^-_4$, $MgHPO^0_4$	<1–3.5 μmol/kg; 2.3 μmol/kg
S	SO^{2-}_4, $NaSO^-_4$, $MgSO^0_4$	28.2 mmol/kg
Cl	Cl^-	0.546 mol/kg
K	K^+	10.2 mmol/kg
Ca	Ca^{2+}	10.3 mmol/kg
Sc	$Sc(OH)^0_3$	8–20 pmol/kg; 15 pmol/kg
Ti	$Ti(OH)^0_4$	(<20 nmol/kg)
V	HVO^{2-}_4, $H_2VO^-_4$, $NaHVO_4$	20–35 nmol/kg; 30 nmol/kg
Cr	CrO^{2-}_4, $NaCrO^-_4$	2–5 nmol/kg; 4 nmol/kg
Mn	Mn^{2+}, $MnCl^+$	0.2–3 nmol/kg; 0.5 nmol/kg
Fe	$Fe(OH)^0_3$	0.1–2.5 nmol/kg; 1 nmol/kg
Co	Co^{2+}, $CoCO^0_3$, $CoCl^+$	(0.01–0.1 nmol/kg; 0.02 nmol/kg)
Ni	Ni^{2+}, $NiCO^0_3$, $NiCl^+$	2–12 nmol/kg; 8 nmol/kg
Cu	$CuCO^0_3$, $CuOH^+$, Cu^{2+}	0.5-6 nmol/kg; 4 nmol/kg
Zn	Zn^{2+}, $ZnOH^+$, $ZnCO^0_3$, $ZnCl^+$	0.05-9 nmol/kg; 6 nmol/kg
Ga	$Ga(OH)^-_4$	(0.3 nmol/kg)
Ge	H_4GeO_4, $H_3GeO^-_4$	≤7–115 pmol/kg; 70 pmol/kg
As	$HAsO^{2-}_4$	15–25 nmol/kg; 23 nmol/kg
Se	SeO^{2-}_4, SeO^{2-}_3, $HSeO^-_3$	0.5–2.3 nmol/kg; 1.7 nmol/kg
Br	Br^-	0.84 mmol/kg
Rb	Rb^+	1.4 μmol/kg
Sr	Sr^{2+}	90 μmol/kg
Y	YCO^+_3, YOH^{2+}, Y^{3+}	(0.15 nmol/kg)
Zr	$Zr(OH)^0_4$, $Zr(OH)^-_5$	(0.3 nmol/kg)
Nb	$Nb(OH)^-_6$, $Nb(OH)^0_5$	(≤50 pmol/kg)
Mo	MoO^{2-}_4	0.11 μmol/kg
(Tc)	TcO^-_4	No stable isotope
Ru	?	?
Rh	?	?
Pd	?	?

TABLE 1 (CONTINUED)
Concentration and Speciation of Elements in Ocean Water

Element	Probable main species in oxygenated seawater	Range and average concentration at 35‰ salinity[a]
Ag	$AgCl_2^-$	(0.5–35 pmol/kg; 25 pmol/kg)
Cd	$CdCl_2^0$	0.001–1.1 nmol/kg; 0.7 nmol/kg
In	$In(OH)_3^0$	(1 pmol/kg)
Sn	$SnO(OH)_3^-$	(1–12, ~4 pmol/kg)
Sb	$Sb(OH)_6^-$	(1.2 nmol/kg)
Te	TeO_3^{2-}, $HTeO_3^-$	?
I	IO_3^-	0.2–0.5 μmol/kg; 0.4 μmol/kg
Cs	Cs^+	2.2 nmol/kg
Ba	Ba^{2+}	32–150 nmol/kg; 100 nmol/kg
La	La^{3+}, $LaCO_3^+$, $LaCl^{2+}$	13–37 pmol/kg; 30 pmol/kg
Ce	$CeCO_3^+$, Ce^{3+}, $CeCl^{2+}$	16–26 pmol/kg; 20 pmol/kg
Pr	$PrCO_3^+$, Pr^{3+}, $PrSO_4^+$	(4 pmol/kg)
Nd	$NdCO_3^+$, Nd^{3+}, $NdSO_4^+$	12–25 pmol/kg; 20 pmol/kg
Sm	$SmCO_3^+$, Sm^{3+}, $SmSO_4^+$	2.7–4.8 pmol/kg; 4 pmol/kg
Eu	$EuCO_3^+$, Eu^{3+}, $EuOH^{2+}$	0.6–1.0 pmol/kg; 0.9 pmol/kg
Gd	$GdCO_3^+$, Gd^{3+}	3.4–7.2 pmol/kg; 6 pmol/kg
Tb	$TbCO_3^+$, Tb^{3+}, $TbOH^{2+}$	(0.9 pmol/kg)
Dy	$DyCO_3^+$, Dy^{3+}, $DyOH^{2+}$	(4.8–6.1 pmol/kg; 6 pmol/kg)
Ho	$HoCO_3^+$, Ho^{3+}, $HoOH^{2+}$	(1.9 pmol/kg)
Er	$ErCO_3^+$, $ErOH^{2+}$, Er^{3+}	4.1–5.8 pmol/kg; 5 pmol/kg
Tm	$TmCO_3^+$, $TmOH^{2+}$, Tm^{3+}	(0.8 pmol/kg)
Yb	$YbCO_3^+$, $YbOH^{2+}$	3.5–5.4 pmol/kg; 5 pmol/kg
Lu	$LuCO_3^+$, $LuOH^{2+}$	(0.9 pmol/kg)
Hf	$Hf(OH)_4^0$, $Hf(OH)_5^-$	(<40 pmol/kg)
Ta	$Ta(OH)_5^0$	(<14 pmol/kg)
W	WO_4^{2-}	0.5 nmol/kg
Re	ReO_4^-	(14–30 pmol/kg; 20 pmol/kg)
Os	?	?
Lr	?	?
Pt	?	?
Au	$AuCl_2^-$	(25 pmol/kg)
Hg	$HgCl_4^{2-}$	(2–10 pmol/kg; 5 pmol/kg)
Tl	Tl^+, $TlCl^0$; or $Tl(OH)_3^0$	60 pmol/kg
Pb	$PbCO_3^0$, $Pb(CO_3)_2^{2-}$, $PbCl^+$	5–175 pmol/kg; 10 pmol/kg
Bi	BiO^+, $Bi(OH)_2^+$	≤0.015–0.24 pmol/kg

[a] Parentheses indicate uncertainty about the accuracy or range of concentration given.

With permission from Bruland, K. W., in *Chemical Oceanography,* Vol. 8, Riley, J. P. and Chester, R., Eds., Academic Press, London, 1983, 157. Copyright: 1983 Academic Press Inc. (London) Ltd.

ported in river water, a trace metal may occur in six possible chemical forms, as recounted by Duinker:[17] (1) in solution as inorganic ion and both inorganic and organic complexes; (2) adsorbed onto surfaces; (3) in solid organic particles; (4) in coatings on detrital particles after coprecipitation with and sorption onto mainly iron and manganese oxides; (5) in lattice positions of detrital crystalline material; and (6) precipitated as pure phases, possibly on detrital particles. However, reactions during estuarine mixing have a significant effect on the partitioning of an element between dissolved and particulate phases.[16] Trace metal removal during estuarine mixing is partly coupled to the relative affinities of trace metals for anions in seawater, and for humic acids and hydrous iron oxides, in the presence of seawater cations.[16,18] Sholkovitz[18] advocated that upon contact with seawater in estuaries, dissolved trace metals present as a colloid in physicochemical association with humic acids and hydrous iron oxides flocculate and settle to the bottom. While the uptake of dissolved trace metals onto solid phases already present in estuarine water is important to the overall concentration of the elements in solution, so is the release of material into solution from particulate phases by dissolution, desorption, and autolytic respiratory biological processes.[19] The desorption of certain elements (e.g., Ba) from suspended particulate matter owes to increasing salinity and the concomitant rise in the concentrations of the major seawater cations. Hence, the exchange of trace metals between dissolved and particulate phases appears to be a regular phenomenon in estuarine systems.

Dissolved trace elements undergo varying degrees of recycling, although most of them are ultimately removed to seafloor sediments.[16] This recycling involves removal of elements from solution by coprecipitation with particulate matter or adsorption onto particle surfaces and their subsequent deposition. In addition, some trace elements associated with particulate phases are desorbed from suspended matter under suitable conditions and remobilized in the estuary. As particulate carrier phases experience oxidation and/or dissolution, for example, trace elements are regenerated, either in the water column or from surficial bottom sediments.[16] Bryan[20] reported that approximately 60% of some heavy metals are mobilized from Rhine River sediments as organometallic complexes by the decomposition of organic matter prior to reaching the sea. However, elsewhere (e.g., the Tamar River estuary) much less mobilization of certain elements, such as Cu, Pb, and Zn, has been recorded when compared to the Rhine. The fate of heavy metals in regard to physicochemical processes is contingent upon conditions in the individual estuarine system.[20] Kennish[6] has reviewed the amount of trace metal removal in various estuaries worldwide.

The biota of estuaries are also factors in the absorption and redistribution of heavy metals. "Active" uptake of essential trace elements (e.g., Cu, Co, Cr, Fe, Mo, Ni, V, Zn) can be significant, especially during blooms. Phytoplankton utilize these elements in metal-requiring and metal-activated enzyme systems which catalyze major steps in glycolysis, the tricarboxylic acid cycle, photosynthesis, and protein metabolism.[16] Zooplankton facilitate the removal of heavy metals by consolidating them in fecal material which sinks to the estuarine seafloor. Heavy metals bound to fecal pellets, crustacean molts, and dead plants and animals generally are responsible for more than 90% of their vertical transport in the water column.[20] After being deposited on the seafloor, the heavy metal-bound materials are redistributed due to the bioturbating activities of the benthos. Seagrasses,[21] deposit-feeding invertebrates (e.g., polychaete worms),[22] filter-feeding invertebrates (e.g., mussels, oysters, and scallops),[23-25] and fishes[26-27] accumulate trace metals in their tissues, thereby influencing biogeochemical cycling of the elements.

Removal of certain riverborne trace metals via flocculation and sedimentation leads to their depletion in oceanic environments. Bottom sediments, because of their large concentrations of heavy metals, exert strong control on the biogeochemical cycling of the elements. The estuarine seafloor serves as both a sink for riverborne trace metals as well as a source of the metals for the overlying water.[28] The greatest fraction of riverborne heavy metals accumulate in the upper reaches of estuaries due to flocculation and settlement of finer sediments and in the middle reaches because of the occurrence of the turbidity maximum. Desorption and diagenetic remobilization of particle-bound trace elements downestuary, together with the influx of clean marine sediments through the estuarine mouth, account for the gradual seaward decline in the heavy metal content of estuarine sediments.

Not only are riverborne particulate phases of trace metals subject to settlement to the seafloor, but dissolved forms transported by

rivers tend to be deposited as well due to adsorption and coprecipitation. Moreover, biotic interaction, notably by bacteria, can play a
paramount role in the deposition of heavy metals.[29] Within the
benthic boundary layer, where primary linkage occurs between
water and sediment movement, advective processes result in bed-
shear stresses that mix sediments in the upper layer of the sediment
column and affect the intensity and direction of particulate transport,
and consequently, the movement of trace metals to different places
in the estuary.[30] Hence, even though the largest amounts of heavy
metals initially accumulate in middle and upper estuary regions, erosion and transport often displace them to the outer estuary or to areas
on the continental shelf beyond the estuarine mouth.[16] Estuarine circulation, therefore, is critical to the distribution of heavy metals in
estuarine environments.[31]

Trace metals appear to vary considerably with depth in the
sediment column.[5] The oxidation state of a trace metal depends on
its location relative to oxidized and reduced sedimentary zones.
Changes in the physicochemical properties of the sediments can
contribute to the vertical movement of the metals, making them more
readily available for release to the overlying water column. For
instance, the decomposition of organic detritus alters the pH, dissolved O_2 concentrations, and sulfides in bottom sediments and can
shift heavy metals into positions where they are more likely to be
exchanged with the bottom waters. Clearly, changes in redox conditions frequently remobilize trace metals to interstitial waters. Moreover, microbes, specifically certain bacteria, solubilize metals from
precipitated sulfides. In organic-rich sediments, diagenetic processes prompt the release of dissolved trace metals in interstitial
waters, and, particularly in the upper layers of reduced sediments,
interstitially dissolved trace metals generally reach substantially
higher levels than in overlying waters. Some trace metals form
soluble complexes with organic matter, and during the decomposition process, diffuse out of the bottom sediments. In addition to the
contributions of soluble heavy metals from interstitial water to the
estuarine water column subsequent to diagenetic alteration of sediments, direct desorption of trace metals from bottom sediments to
estuarine waters adds to the pool of dissolved trace metals in
estuarine waters. Remobilization of the trace metals can be facilitated by natural processes such as wave and current action, compac-

tion, biological activity (e.g., microbial processes and bioturbation), and anthropogenic effects (e.g., dredging).[17,32]

In sum, riverborne trace metals carried in solution may be removed from solution upon contact with saline estuarine water.[33,34] As salinity rises, particle-bound heavy metals sorbed to particulate organic matter,[35,36] oxide coatings,[37] and clays[38] can be desorbed. Organic complexation affects the speciation of the metals;[39] important complexing agents in estuaries include marine humic acids and degraded plant material.[40-42] Knowledge of estuarine trace element chemical speciation is limited, although a new generation of sophisticated methods of trace element analysis is now available.[43] Particle flocculation and sedimentation concentrates heavy metals in the upper and middle reaches of estuaries. The estuarine bottom serves as both a sink and source of the metals for the system. The transport and distribution of particulate trace metals are modulated by physical and geochemical processes, along with biotic activity.[44]

Atmospheric Input

The significance of the atmosphere as a source of trace elements to estuaries and coastal marine waters has not been dealt with in sufficient detail. Excluding the investigations on lead ascribable to anthropogenic mobilization, the influence of the atmospheric flux on the trace element chemistry of estuaries has not been resolved. In the ocean basins, only approximate estimates have been proffered for the atmospheric input of most of the trace elements.[16]

Anthropogenic Input

In terms of environmental pollution imposed by man, heavy metals may be classified as: (1) noncritical; (2) toxic but very insoluble or very rare; and (3) very toxic and relatively accessible. Based on this classification scheme developed by Wood,[45] many of the common heavy metals in estuarine systems (i.e., Cu, Zn, Pb, and Ni) are highly toxic and accessible as potential contaminants. Table 2 summarizes Wood's classification of metals in the environment.

Heavy metals can also be classified according to their potential supply and toxicity. By comparing the rates of addition of the metals to the ocean with the concentration considered to pose minimal risk of deleterious environmental effects, a relative critical index has been formulated for heavy metals of importance in marine pollution. Table 3 lists the heavy metals in order of decreasing

TABLE 2
Classification of Elements According to Toxicity and Availability

Noncritical			Toxic but very insoluble or very rare		Very toxic and relatively accessible		
Na	C	F	Ti	Ga	Be	As	Au
K	P	Li	Hf	La	Co	Se	Hg
Mg	Fe	Rb	Zr	Os	Ni	Te	Tl
Ca	S	Sr	W	Rh	Cu	Pd	Pb
H	Cl	Al	Nb	Ir	Zn	Ag	Sb
O	Br	Si	Ta	Ru	Sn	Cd	Bi
N			Re	Ba		Pt	

From Wood, J. M., *Science*, 183, 1049, 1974. With permission.

TABLE 3
Toxic Elements of Importance in Marine Pollution Based on Potential Supply and Toxicity, Listed in Order of Decreasing Toxicity

	Rate of mobilization (10⁹ g/year)			Toxicity	Relative critical index (km³/year)	
Element	A (man) fossil fuels	B (natural) river flow	C total	D (μg/l)	A/D	C/D
Mercury	1.6	2.5	4.1	0.1	16,000	41,000
Cadmium	0.35	?	3.0	0.2	1,750	15,000
Silver	0.07	11	11.1	1	70	11,100
Nickel	3.7	169	164	2	1,350	82,000
Selenium	0.45	7.2	7.7	5	90	1,540
Lead	3.6	110	113.6	10	360	11,360
Copper	2.1	250	252.1	10	210	25,210
Chromium	1.5	200	201.5	10	150	20,150
Arsenic	0.7	72	72.7	10	70	7,270
Zinc	7	720	727	20	330	36,350
Manganese	7.0	250	257	20	350	12,850

[a] Equals the volume of seawater that would be contaminated annually to the indicated level of toxicity by the specified rates of addition, both by natural processes and anthropogenic activities.

From Ketchum, B. H., Marine industrial pollution, in *Oceanography: The Past*, Sears, M. and Merriman, D., Eds., Springer-Verlag, New York, 1980, 397. With permission.

toxicity based on this classification scheme. As is evident from this list, mercury and cadmium represent the two most troublesome heavy metals with respect to toxicity in estuarine and marine environments.

Heavy metals reach the estuarine habitat via a wide range of human activities, including the burning of fossil fuels, antifouling paints, smelting, power station corrosion products (Cu, Cr, and Zn), sewage-sludge disposal, ash disposal, dredged-material disposal, seed dressings and slimicides (Hg), automobile emissions (Pb), oil refinery effluent, and other industrial processes. Among the major anthropogenic sources of heavy metals in aquatic environments mentioned by Wittman and Förstner,[46] the production of cement and bricks, leaching of metals from garbage and solid waste dumps, and industrial processing of ores and metals have impacted many coastal areas. The following sections address some of these anthropogenic sources.

Antifouling Paints

The use of antifouling paints to protect the hulls of ships from boring and fouling organisms has been responsible for the slow release of large amounts of toxic elements to estuarine and marine waters. When paints were substituted for sheathing to protect ocean-going vessels, copper(I)oxide proved to be a most effective antifouling agent. Antifouling paints containing copper(I)oxide provide out-of-dock protection for up to 3 years.[47] Antifouling paints and coatings may consist of organometallic compounds other than copper oxides, although most are incorporated into vinyl-chlorinated rubber, coal-tar epoxy, and several other kinds of paint bases.[48] Effective action of copper oxide antifouling paints involves slow release of copper ions. Gradual leaching of the copper limits the life expectancy of the paint because the paint is only effective as long as the copper has not been exhausted. In addition, the copper represents a significant environmental contaminant.[49] Due to these two drawbacks, antifouling paints containing organotin compounds have rapidly gained favor in recent years as an active agent in antifouling paints.

Triorganotin compounds, notably tributyltin, have been utilized in a variety of paint formulations either solely or in combination with metal in compounds.[49] Highly effective as an antifouling agent, tributyltin-treated surfaces may be devoid of biofouling impacts for years. However, these compounds, similar to metal ion leachates from antifouling paints, are highly toxic, and nontarget organisms may be at great risk when subjected to them.[50] Consequently, they have become the subject of considerable scrutiny.[51-53] "Self-polishing" paints employing tributyltin as an active agent, together with cuprous oxide which diffuses out of the paint matrix, yield exceptional results in regard to surface protection from biofouling. As the layers of the "self polishing" paint wear off, along with the fouling organisms, they are continually replaced by a new layer of toxic paint containing new tributyltin groups.

More information has been accumulated on the environmental effects of copper ion leachates than on those for organotin. The literature contains numerous references to the toxic effects of copper ions on algae. In the case of phytoplankton, for instance, copper has been shown to restrict the uptake and assimilation of nitrate and the uptake of silicate; furthermore, it inhibits photosynthesis, growth (i.e., cell division), and amino acid synthesis in these microscopic plants.[49] Thomas and Robinson,[49] recounting the exposure of the diatoms *Amphora coffeaeformis* and *Nitzochia ovalis* to self-polishing copolymer (SPC)-painted surfaces under static laboratory conditions, detailed the inhibitory effects of SPC toxins on nitrate and silicate uptake, photosynthesis, and survival. Beaumont and Newman[54] discerned reduced growth in *Dunaliella tertiolecta, Pavlova lutheri,* and *Skeletomena costatum* cultured in tributyltin.

Because of the potential threat of organotin antifouling paints on estuarine and marine life, research on the effects of organotins on aquatic organisms has accelerated in recent years. Various mollusks, which tend to sequester heavy metals, have received considerable attention in laboratory investigations. Lee[55] demonstrated the ability of blue crabs (*Callinectes sapidus*), spider crabs (*Libinia emarginata*), oysters (*Crassostrea virginica*), and fish (spot, *Leiostomus xanthurus*) to metabolize tributyltin oxide. Davies et al.[56] observed that the Pacific oyster (*Crassostrea gigas*) accumulated up to 1.41 mg/kg of tin and 0.87 mg/kg tributyltin from antifouling paints over a period of 41 weeks, but lost 90% of this amount during depuration. His and Robert,[57] assessing the effects of two antifouling paints on *C. gigas*, recorded reduced growth rate (i.e., length, width, and weight) and decreased condition factor and shell density in the bivalve. The treatment of fiddler crabs (*Uca pugilator*) with tributyltin levels as low as 0.5 µg/l for 1 to 3 weeks caused acceleration of the righting reflex in females, indicative of hyperactivity.[58] In addition, burrowing activity declined, regenerative growth was retarded, and ecdysis was delayed.[59] Minchin et al.[60] detected tributyltin levels of 0.7 µg/g wet weight in adult populations of the scallop (*Pecten maximus*) in Mulroy Bay on the northern coast of Ireland, and related this contamination with the introduction and subsequent use of organotin net-dips on salmonid farms in the bay. In a study of the cellular and biochemical responses of the American oyster (*Crassostrea virginica*) and blue mussel (*Mytilus edulis*) to tributyltin concentrations of 0.7 µg/l during a 90-d test period (60-d exposure, 30-d recovery), Pickwell and Steinert[61] discovered that both species of shellfish accumulated tin in their digestive glands to comparable levels in flowing seawater of the test tanks. Whereas the oysters suffered virtually no deaths during the 60-d test period, approximately 50% of the mussels died within this interval of time. Batley et al.[62] determined tributyltin accumulation in the Sydney rock oyster, *Saccostrea commercialis*. New research techniques have been developed that promise to advance our understanding of organotin contaminants in shellfish.[63]

Other work has focused on the risk posed by organotin compounds on communities and ecosystems. For example, Waldock[64] raised concern over the wider environmental impact of tributyltin in the estuaries of England. Quevauviller et al.[65] investigated organotin levels in sediments and organisms of the Sado estuary in Portugal. Seligman et al.[66] followed degradation rates and degradation products of tributyltin in ambient waters of San Diego Bay and Skidaway estuary (U.S.). Stang and Seligman[67] ascertained butyltin compounds in the sediment of San Diego Bay. Chesapeake Bay has been the target of a host of organotin studies during the 1980s,[68-71] most dealing with the use of tributyltin as a biocide in antifouling paint.

While some of the most toxic organotin compounds are used as biocides (e.g., tributyltin, triphenyltin, and tricyclohexyltin), other organtin compounds have practical value in a broader range of agricultural, industrial, and recreational applications. Hence, organotin compounds have also been utilized in the production of polyurethane foam. In addition, they are important in the textile industry as preservatives for mildew control. Furthermore, organotins have been successfully employed as stabilizers in polyvinyl chloride.[72] Despite their domestic and industrial usefulness, organotin compounds are toxic to a large number of estuarine organisms. Toxicity tests must continue to be made, therefore, to accurately evaluate both the acute and chronic effects of the compounds to estuarine biota.

Waste Disposal

A significant source of heavy metals to the estuarine environment in the past has been waste disposal, particularly sewage-sludge and dredged-material disposal. Because the major proportion of heavy metals in disposed wastes is associated with solid phases that accumulate on the estuarine bottom, the biological effects are manifested most conspicuously among the benthos. Many commercially and recreationally important fin- and shellfish species, either inhabiting the seafloor or feeding on bottom-dwelling organisms, tend to absorb and accumulate the metals. The contamination of these species may have deleterious effects on humans that consume them.[73] Consequently, programs have been undertaken to delineate the sedimentary records of heavy metals in impacted habitats.[74] Table 4 reveals heavy metal concentrations in contaminated and uncontaminated sediments of several heavily impacted systems.

The concentrations of heavy metals in bottom sediments of estuaries relative to those in the overlying water underscores the severity of the contamination problem incurred by the dumping of millions of metric tons of contaminated materials. For instance, the ratio of the concentrations of Cd in bottom sediments of estuaries to the overlying waters equals 10^3 and that for Zn and Cu amounts to 10^4 and 10^5, respectively. Many hectares of estuaries and coastal marine waters of the mid-Atlantic Bight have been impacted over the years by the dumping of huge volumes of waste solids containing heavy metals. Sewage-sludge and dredged-spoil disposal sites in areas to the east and west of the Hudson Shelf Valley, for example, contained high concentrations of Pb, Cr, Cu, Ni, and Zn — as much as ten times greater than the values registered in uncontaminated neighboring sediment.[75] The Chesapeake Bay ecosystem represents the second largest source area of waste materials, contributing to significant levels of heavy metals in bottom sediments. As stressed by Helz and Huggett,[76] industrial and municipal point sources of trace metal contaminants are scattered along the shores of Chesap-

TABLE 4
Metal Concentrations ($\mu g/g$) in Contaminated and Uncontaminated Sediments[a]

Sediment	Ag	As	Cd	Co	Cr	Cu	Fe	Hg	Mn	Ni	Pb	Sn	V	Zn	References[73]
Firth of Clyde, Scotland															
Control area (mean)	<0.2	8	3.4	34	64	37	5.3×10^4	0.1	1,100	50	86	19	250	160	Mackay et al., 1972
Sewage-sludge dumpsite (max.)	5.	24	7	40	310	210	6.1×10^4	2.2	1,000	70	320	100	400	830	Halcrow et al., 1973
California coast															
Control area (median)	0.2	—	0.33	—	22	8.3	—	0.043	—	9.7	6.1	—	—	43	Hershelman et al., 1981
Los Angeles wastewater outfall (max.)	27	—	66	—	1,500[b]	940	—	5.4	—	130	580	—	—	2,900	
Eastern England															
Humber Estuary TiO$_2$ and smelting (max.)	—	—	—	30	200	160	9.2×10^4	—	1,100	63	220	—	2,000[b]	430	Jaffe and Walters, 1977
Southwestern England estuaries															
Control (Avon) (typical)	0.1	13	0.3	10	37	19	1.9×10^4	0.12	420	28	39	28	—	98	Bryan et al., 1980
Restronguet Creek acid mine waste (max.)	4.1	2,500[b]	1.2	22	37	2,500	5.8×10^4	0.22	560	32	290	1,700[b]	—	3,500	

TABLE 4 (CONTINUED)
Metal Concentrations (μg/g) in Contaminated and Uncontaminated Sediments[a]

Sediment	Ag	As	Cd	Co	Cr	Cu	Fe	Hg	Mn	Ni	Pb	Sn	V	Zn	References
Tasmania															
Derwent Estuary Zn refinery and chlor-alkali (Hg) (max.)	—	—	860[b]	140	200	>400	16 × 10⁴	100[b]	8,900	42	1,000	—	—	>10,000	Bloom and Ayling, 1977
Norway															
Sorfjord smelting (max.)	190[b]	—	850	—	—	12,000[b]	—	—	—	—	31,000[b]	1,350	—	12,000[b]	Skei et al., 1972
Pacific Ocean															
Pelagic clay (>4000 m depth) (max.)	—	—	—	150[b]	—	1,200	7.5 × 10⁴	—	38,000[b]	1,300[b]	—	—	—	390	Förstner and Stoffers, 1981

[a] Reported values rounded to two significant digits.
[b] Highest value for each element.

From Bryan, G. W., in *Wastes in the Ocean*, Vol. 6, *Nearshore Waste Disposal*, Ketchum, B. H., Capuzzo, J. M., Burt, W. V., Duedall, I. W., Park, P. K., and Kester, D. R., Eds., John Wiley & Sons, New York, 1985, 41.

eake Bay, reaching highest densities at Norfolk (Virginia) and Baltimore (Maryland). Environmental contamination in the Patapsco and Elizabeth Rivers, tributaries to the Chesapeake Bay, is serious (see the "Chesapeake Bay" section under "Heavy Metals in Estuarine Systems"). A compilation of trace metals identified in the sediments of these rivers, as well as other areas of the system, are given in Table 5. In the highly industrialized areas of Elizabeth River and Baltimore Harbor, evidence of an adverse impact of heavy metals upon aquatic organisms is gradually mounting.[76]

Elsewhere on the East Coast of the U.S., Boston Harbor receives large volumes of industrial and sewage wastes, with more than 50% of the freshwater inflow into the harbor being primary-treated sewage effluent.[77] Two primary-treatment plants at Deer Island and Nut Island, respectively, discharge sewage effluent at a rate of 20 m³/s into the harbor. In addition, untreated (raw) sewage enters the harbor directly due to malfunctions at the treatment plants and the existence of combined sewer overflows. The plants also discharge sludge on the outgoing tide. The inner harbor exhibits the greatest impact of heavy metals; although the aforementioned sewage influx accounts for much of these pollutants, industrial and nonpoint sources of heavy metals, which are poorly characterized qualitatively and quantitatively, likewise represent important sources.

Heavy metal concentrations in Boston Harbor generally exceed those of ambient New England coastal water and the surface water of the open ocean based on reliable source data.[77,78] Thus, the concentrations of Cu, Cd, Ni, Pb, and Zn are one to four orders of magnitude greater than those of metals in the surface water of the open ocean, and one to two orders of magnitude greater than the concentrations of heavy metals in adjacent coastal waters. The peak concentrations of most metals occur in the inner harbor, especially in marginal shallow areas, typified by poor tidal flushing and possibly remobilization of the metals from sediments, and in waters overlying sediments containing high metal concentrations.

Estuarine and coastal marine waters along the West Coast of the U.S. also have been impacted by heavy metal pollutants derived from sewage and industrial waste disposal.[75] Tons of gross heavy metals have been dumped into San Francisco Bay during the past several decades. Commencement Bay and Elliott Bay, two estuaries in proximity to major metropolitan centers (Tacoma and Seattle)

TABLE 5
Summary of Data Sources Concerning the Elemental Composition of Chesapeake Bay Sediments

Area	Elements	References[76]
Main stem	Al, C, Ca, Co, Cr, Cu, Fe, H, K, Mg, Mn, Na, Ni, Pb, S, Si, Ti, V	Sommer and Pyzik, 1974
	Cr, Cu, Ni Pb	Schubel and Hirschberg, 1977
	Cd, Cu, Fe, Mn, Ni, Zn	Cronin et al., 1974
	Ag, Al, Cd, Co, Cr, Cu, Fe, Mn, Ni, Pb, V, Zn	Goldberg et al., 1978
	Al, Co, Cr, Cu, Fe, Ga, Mn, Ni, Org C, Org N, Pb, Si, Ti, V, Zn	Sinex, 1981
	As, Cd, Cu, Fe, Hg, Mn, Ni, Pb, Sn, Zn	Harris et al., 1980
Back River	Cd, Cu, Pb, Zn	Helz et al., 1975
Patapsco	Cd, Cr, Cu, Hg, Mn, Ni, Pb, Zn	Villa and Johnson, 1974
	As, Cd, Cr, Cu, Hg, Mn, Ni, Pb, Zn	Tsai et al., 1979
	Al, Co, Cr, Fe, Mn, Ni, Si, Ti, V, Zn	Sinex and Helz, 1982
Rhode River	Cd, Cr, Fe, Mn, Zn	Frazier, 1976
Patuxent	Cd, Co, Cr, Cu, Fe, Mn, Ni, Pb, Zn	Ferri, 1977
Potomac	Ag, Ba, Cd, Co, Cr, Cu, Fe, Li, Mn, Ni, Pb, Sr, V, Zn	Pheiffer, 1972
	Ba, Co, Cr, Cu, Fe, Mn, Pb, Sr, Ti, V, Zi, Zn	Mielke, 1974
Rappahannock, York	Cu, Zn	Huggett et al., 1975
Elizabeth	Al, Cd, Cr, Cu, Fe, Hg, Pb, Zn	Johnson and Villa, 1976
	Al, Co, Cr, Fe, Mn, Ni, Si, Ti, V, Zn	Helz et al., 1983
	Cd, Co, Cr, Cu, Fe, Mn, Ni, Pb, Zn	Rule, 1986

From Helz, G. R. and Huggett, R. J., in *Contaminant Problems and Management of Living Chesapeake Bay Resources,* Majumdar, S. K., Hall, L. W., Jr., and Austin, H. M., Eds., Pennsylvania Academy of Science, Easton, 1987, 270. With permission.

along Puget Sound, receive heavy metals from various anthropogenic sources. Near major sewage outfalls of Seattle, Cr, Pb, and Zn concentrations in mussels (*Mytilus edulis*) were higher than in control areas.[79] Likewise, the concentrations of Cu, Cr, and Pb concentrations in dungeness crabs purportedly are two to five times greater in Commencement Bay than in control sites such as Discovery Bay.

Filter-feeding bivalves tend to accumulate heavy metals and, therefore, have been utilized in the assessment of water and sediment contamination.[80,81] This is so because the bivalves filter particulate matter from the water, have considerable water-tissue contact for the uptake of dissolved heavy metals, and cannot regulate the concentrations of the absorbed metals as well as other organisms (e.g., fish). Table 6 presents heavy metal concentrations in *Mytilus edulis* from various regions of Puget Sound. These data reveal that elevated levels of heavy metals occur in mussels inhabiting areas in proximity to major pollutant sources. For example, large amounts of Cu and Hg in the mussels of Commencement Bay reflect bioaccumulation of Cu and Hg ascribable to input from the American Smelting and Refining Company (ASARCO) refinery.

Sewage sludge discharged through pipelines in the Southern California Bight strongly enrich Santa Monica Bay bottom sediments in trace metals. In the mid-1970s, Schafer[82] reported that the combined annual mass emission rates of the five largest municipal effluents in Southern California equalled about 0.9×10^6 metric tons of suspended particulates, which contributed 1500 metric tons of zinc, 650 metric tons of chromium, 620 metric tons of copper, 200 metric tons of lead, 50 metric tons of cadmium, 25 metric tons of silver, and 2 to 3 metric tons of mercury to the coastal marine environment of Southern California.[75] The Hyperion Joint Water Pollution Control Plant near Los Angeles accounts for the bulk of this material. Near the outfall of this plant, trace metal enrichment contaminates the seafloor of Santa Monica Bay. Here, cadmium levels were as much as 300 times higher than those of uncontaminated sediments in the region. Similarly, the concentrations of mercury, lead, copper, silver, and nickel and chromium were 140, 85, 80, 60, and 15 times greater, respectively, than in uncontaminated sediments.[75,83]

Estuaries in northern Europe also are plagued by heavy metal contamination attributable to waste disposal practices. Significant amounts of heavy metal pollution associated with waste disposal in the past have been documented in Liverpool Bay, the Thames estuary, the Rhine-Waal/Meuse/Scheldt estuaries, the German Bight, the Baltic Sea, and elsewhere. Sewage-sludge disposal has been a major source of heavy metal enrichment over the years in the Firth of Clyde, Liverpool Bay, the Thames estuary, and the Scheldt estuary.[84-87] At the outer Thames estuary, deposits at Barrow Deep have yielded extremely high values of copper (as much as 2500 mg/kg dry wt) and cadmium (30 to 70 mg/kg dry wt).[75] Between 1977

TABLE 6
Average Concentrations (mg kg⁻¹ wet wt) of Selected Metals in Tissue from Mussels (*Mytilus edulis*) from Puget Sound

Location	Hg	Pb	Cd	Cu
Puget Sound (Central Basin)	0.109	6.6	3.7	7.4
Commencement Bay (Near refinery)	0.836	—	3.6	75.0
Puget Sound (U.S. Mussel Watch)	—	1.2	2.0	4.3
Willapa Bay, WA (U.S. Mussel Watch)	—	2.2	3.0	5.9
Sequim Bay	0.102	—	5.2	5.8

From Ginn, T. C. and Barrick, R. C., in *Oceanic Processes in Marine Pollution*, Vol. 5, *Urban Wastes in Coastal Marine Environments*, Wolfe, D. A. and O'Connor, T. P., Eds., Robert E. Krieger Publishing, Malabar, FL, 1988, 157. With permission.

and 1987, the quantities of sewage sludge disposed at the Roughs Tower dumpsite in the outer Thames estuary amounted to about 1.59 × 10⁵ metric tons/year. Additionally, approximately 0.23 × 10⁶ metric tons of dredged spoils were dumped at Roughs Tower each year over this 10-year period. This disposal has resulted in thick accumulations of clay and silt at the dumpsite. Concomitantly, the concentrations of trace metals have become elevated as well.[86] The dumping of about 3 × 10⁶ metric tons of dredged spoils from the Mersey estuary and its approaches at site Z in Liverpool Bay not only smothers the benthos but also raises the heavy metal content of the sediments.[85] Meanwhile, the benthos of the Mersey estuary are not spared from trace metal pollutants either; recent surveys of macro-algae, gastropods, polychaetes, and suspension- and deposit-feeding bivalves disclose that they can accumulate Ag, As, Cd, Cr, Cu, Fe, Hg, Mn, Ni, Pb, Se, Sn, and Zn from these waters.[88] The metal contamination problems in the Clyde estuary and Firth of Clyde originate from the large volume of effluents — approximately half of the population and industries of Scotland discharge effluents to these habitats — released to natural waters.

Smelting

ASARCO, a lead-copper smelter on Commencement Bay, Washington, provides a specific example of the potential environmental damage associated with siting a major smelting operation along the shores of an estuary. ASARCO began operations near Tacoma, WA, in 1889 and the smelting and refining of metals since that time have contributed to significant accumulations of As, Sb, Cu, Pb, and Zn in the southern part of Puget Sound. Schell and Nevissi[89] computed the atmospheric input of lead ($\cong$2477 metric tons/year), zinc ($\cong$744 metric tons/year), and copper ($\cong$408 metric tons/year) to Puget Sound via smokestack emissions from the smelter. The bioaccumulation of Cu and Zn has been verified in mussels collected in proximity to the ASARCO refinery.

The temporal trends in heavy metal contamination of Puget Sound can be inferred from the chemical composition of age-dated sediment cores. Crecelius and Bloom[90] recounted the sedimentary histories of the metals Pb, Hg, Ag, Cu, and Cd and related them to the urbanization of the Seattle-Tacoma area. In brief, contaminant concentrations of Pb, Cu, Ag, and Hg in the sound commenced during the years 1880 to 1890, reached maximum levels in sediments deposited between 1945 and 1965, and, in the case of Pb, Ag, and Hg, declined subsequent to 1965. Copper did not undergo any significant change in concentration during the past 30 years. Moreover, Cd displayed no significant change over the entire length of the cores representing nearly 1 century. While the ASARCO lead-copper smelter unequivocally contributed to heavy metal contamination of Commencement Bay, other anthropogenic pollution sources (e.g., municipal sewage, industrial wastewater, coal combustion waste, and automobile emissions) cannot be discounted as factors in heavy metal contaminant loading in Puget Sound.

BIOACCUMULATION
OF HEAVY METALS

Because estuarine and marine invertebrates, especially mollusks, concentrate heavy metals, they are generally regarded as excellent indicators of heavy metal pollution in their environment.[91] Some species (e.g., the blue mussel, *Mytilus edulis*) are unusually adept at accumulating the metals in certain tissues. Furthermore, their ability to integrate variations in heavy metal concentrations through time makes them ideally suited as biomonitors of contaminated habitats.[92] The exceptional powers of mussels to accumulate heavy metals, together with their widespread distribution, led Goldberg[93] to propose them as biomonitors of coastal pollution. The common limpet, *Patella* sp., has also been suggested as a biological monitor of heavy metals because it is highly efficient at concentrating the metals.[94] It has become fashionable in environmental investigations to implement transplantation experiments, by placing indicator organisms along a suspected pollution gradient and examining their responses to contaminants in question.[95,96] When transplanted from uncontaminated to contaminated sites, the experimental organisms typically exhibit bioaccumulation.[97,98] Most studies, however, have used species native to an impacted locale to monitor the habitat. In some cases, the uptake of contaminants by transplants has been compared to that of the natives in an effort to devise a remedial course of action.[99]

Studies of bioaccumulation of metals in estuarine organisms have also involved vertebrates (mainly fish),[100] albeit at a lower frequency than investigations of invertebrates. For example, during their research on the bioaccumulation of metals in Puget Sound organisms, Ginn and Barrick[79] reported on metal uptake by demersal fish (e.g., starry flounder, Dover sole, English sole, and rock sole), marine birds (especially from the two industrialized areas of Puget Sound — Commencement Bay and Elliott Bay), and harbor seals from the southern segment of the sound. The ability of marine fish to regulate metal concentrations in their muscle tissue is reflected in the lack of substantial bioaccumulation of metals by Puget Sound fish. This finding contrasts with the greater metal uptake by filter-feeding bivalve mollusks in the sound that have less ability than the fish to regulate the metals.

Whereas bioaccumulation of trace metals has been established among various estuarine organisms, little biological magnification has been demonstrated in this environment for most metals, an exception being methyl mercury.[5,20,79] However, some species (e.g., oysters) exhibit selective enrichment of certain metals (e.g., copper, cadmium, and zinc), despite the absence of data in support of a generalized accumulation of metals in the higher trophic levels.[75] The highest concentrations of metals often occur in individuals near sources of pollution and appear to be unrelated to any biomagnification effect. Assessment of the amplification of metals at higher trophic levels has been limited by the relatively few detailed investigations involving the collection of a statistically meaningful quantity of specimens from several trophic levels and from a single location (i.e., with similar metal exposure levels).[79]

Metal speciation and bioavailability regulate bioaccumulation of trace metals by estuarine organisms.[101,102] Metal toxicity, in turn, may be a function of the free metal ionic activity in seawater rather than the total concentration of the metal.[2,103] A wide range of environmental factors exists that can influence trace metal uptake and toxicity in these organisms.[101,104] The presence of naturally occurring organic compounds, for instance, may play a critical role in chemical speciation and bioavailability of trace metals. Since approximately 90% of the dissolved organic compounds in estuarine waters remain uncharacterized, much uncertainty surrounds the coupling of the compounds to trace metal availability.[104] The physiological condition of an organism and its ability to regulate metal uptake further obfuscate bioaccumulation of the metals in the estuarine environment.[105]

As expressed by Bryan,[106] metal uptake by marine fauna takes place via ingestion of food organisms and particulate matter and, more directly, via diffusion of dissolved metals. The molecular mechanisms by which the soluble form of heavy metals penetrate into the cells of marine invertebrates is not completely known. Although a passive-transport process appears to be involved in the uptake of dissolved metals, other concomitant uptake strategies cannot be discounted, and the quantitative significance of each one is contingent upon the chemical composition of the seawater in relation to the heavy metal and the organism in question.[2] Subsequent to cellular penetration, the metals react with structural and enzymatic components of membranes and successively with soluble enzymes, metabolites, and organelles.[107] Heavy metal cation homeostasis is closely linked to complexation, primarily by sulfhydryl-containing molecules, such as amino acids, peptides, and proteins.[2] In particular, the sulfhydryl-rich proteins, metallothioneins, have a high affinity for heavy metals. Moreover, lysosomes (cellular structures involved in intracellular digestion and transport)[102] accumulate trace metals, and are a factor in heavy metal homeostasis.[108] Metallothioneins have been found in crustaceans, echinoderms, mollusks, and algae; metal-rich lysosomes have been detected in annelids, crustaceans, hydroids, mollusks, and algae.[2]

The capacity of estuarine organisms to store, remove, or detoxify metal contaminants varies considerably.[102] The metals may be stored in their skeletal structure, concretions, or intracellular matrices.[105] The release of feces, eggs, and molting products, in turn, removes heavy metals, tending to counter storage effects.

Among the faunal groups inhabiting estuaries, crustaceans and fish are the best regulators of heavy metals.[102] In both groups, metallothionein proteins sequester free metal ions and prevent their toxic action on specific tissue sites.[109,110] The detoxification of heavy metals in these marine organisms — often ascribable to metallothioneins and similar metal-binding proteins — may likewise arise through compartmentalization within extracellular or particulate intracellular structures or blood cells.[102,105] Despite their potential for regulating heavy metal concentrations in organisms, metallothioneins and other heavy metal-binding proteins have a finite binding capacity for the metal contaminants, and once exceeded, toxic effects of the metals may become evident.[111]

The responses of estuarine organisms to the toxic effects of heavy metals are manifested in a variety of ways. In general, changes occur in organismal physiology, reproduction, and development. Hence, growth inhibition due to pollutant exposure has been commonly discerned among crustaceans,[112] mollusks,[113] echinoderms,[114] hydroids,[115] protozoans,[116] and algae.[117] Another conspicuous physiological alteration is that of oxygen consumption.[118,119] Feeding behavior, respiratory metabolism, and digestive efficiency may also be compromised. Nevertheless, at low concentrations, heavy metals can actually enhance the rate of growth, oxygen consumption, and other physiological processes, a condition termed "hormesis".[120] The continued exposure of estuarine organisms to elevated levels of metal contaminants often yields unfavorable pathological responses, including tissue inflammation or degeneration, lack of repair and regeneration of damaged tissue, neoplasm formation, and genetic derangement.[102]

Table 7 compiles trace metal concentrations in groups of marine organisms monitored primarily in estuarine areas. Arsenic, cadmium, copper, lead, mercury, nickel, silver, and zinc values are listed for phytoplankton, other algae, mussels, oysters, gastropods, crustaceans, fish, mammals, and seals. Highest readings correspond to highly impacted estuarine systems, particularly those affected by

TABLE 7
Metal Concentrations in Marine Fauna (ppm) (B = brown algae; G = green algae; M = mammals; s = seals; + = ppm wet wt; other values in ppm dry wt)

Metal		Phyto-plankton	Algae	Mussels	Oysters	Gastropods	Crus-taceans	Fish	Seals, mammals
Arsenic	Geometric mean (1)	—	20	15	10	20	30	10	—
	Newfoundland (2)	—	9.8–17 (b)	1.6–5	—	4.0–11.5	3.8–7.6	0.4–0.8	2.2–11.6+
	England (3)	—	26.0–54 (b)	1.8–15	2.6–10	8.1–38	16	1.7–8.7	—
	Greenland (4)	—	36 (b)	14–17	—	—	63–80	14.7–307	—
Cadmium	Geometric mean (1)	2	0.5	2	10	6	1	0.2	—
	Spain (5)	—	0.8–4 (b)	0.5–8	2.9–3.5	1.1–9	0.7–32	<0.4–4.3	—
	England (3) (6–9)	—	0.2–53 (b)	3.7–65	6–54	3.5–1,120	2.8–33	0.06–3.96+	—
	Australia (10–12)	—	—	4.2–83	9–174	2.8–30	—	0.05–0.4+	—
	Norway (13)	—	1.0–13 (b)	1.9–140	—	0–51	1.9–7	<0.01–0.03+	—
Copper	Geometric mean (1)	7	15	10	100	60	70	3	—
	Spain (5)	—	5–26 (g)	6–14	120–435	5–50	110–435	<0.6–10	—
	England (6–8,14–15)	—	4–141 (b)	7–15	20–6,480	0–1,750	6–64+	0.5–14.6+	—
	California (16–21)	—	—	7–77	10–2,100	3–177	(4–150)	(16–29.3)	14.5–386(m)
	Norway (13)	—	9–170 (g)	3–120	—	17–190	2–90	—	—
Lead	Geometric mean (1)	4	4	5	3	5	1	3	—
	Spain (5)	—	4–20 (g)	2–15	4–11	10–27	<1.2–11	<1.2–2.2	—
	England (6,8,15,21,22)	—	16–66 (g)	7–19	5–17	0.2–0.8	8	14–28	0–4+
	California (18,24–26)	—	3–1,200 (g)	0.3–42	—	0.6–21	8.3	<0.001–5.3	0.3–34.2 (s)
	Norway (13)	—	—	2–3,100	—	0–39	—	—	—
Mercury	Geometric mean (1)	0.17	0.15	0.4	0.4	0.2	0.4	0.4	0.6–103 (m)
	Hawaii (27,28)	—	—	—	—	0–0.03	0.03–0.12	0.02–23	0.1–700 (s)
	California (21,29)	—	—	—	—	<0.01–0.07+	0.02–0.04+	0.02–0.2	—
	Atlantic (30–35)	0.2–5.3	<0.01–0.07+	<0.01–0.13+	0.02–180+	—	<0.05–0.6+	0.1–9.0	—
	Mediterranean (36,37)	—	<0.5–07+	0.25–0.4	—	0.1–3.5+	0.3–4.5+	0.1–29.8+	—
	Australia (11,12)	—	—	0.05–0.23+	1.5–8.2	0.32–0.65	—	0.3–16.5	0.1–106 (m)
	Norway (38–40)	0.5–25.2	—	0.24–0.84	—	0.61	0.31–0.39	0.14–7.3	0.4–225 (s)
	England (3,8,9,22,41–43)	—	<0.01–25.5 (g)	0.64–1.86	0.56–1.2	0.02–1.84	–0.98	0.02–1.8	—

TABLE 7 (Continued)
Metal Concentrations in Marine Fauna (ppm) (B = brown algae; G = green algae; M = mammals; s = seals; + = ppm wet wt; other values in ppm dry wt)

Metal		Phytoplankton	Algae	Mussels	Oysters	Gastropods	Crustaceans	Fish	Seals, mammals
Nickel	Geometric mean (1)	3	3	3	1	2	1	1	
	England (3,6,9)	—	4–33 (g,b)	5–12	2–174	8.8–12.3	1.1–12.3	0.5–10.6	
	California (18,21)		—	3.3–20	—	1.8–18.5			
Silver	Geometric mean (1)	0.2	0.2	0.3		1	0.4	0.1	
	California (18,44,21)	—	—	0.7–46	—	0.4–10.7		0.1–1.2 (m)	
Zinc	Geometric mean (1)	38	90	100	1,700	200	80	80	
	South Africa (45,46)	0.6–710	5.6+ (g)	73–113	400–886	12+	17+	3.2–7.2+	
	California (17,18,44,47)	0.1–725	46–244	70–8,430	1.7–288	—	80	78–875	
	Spain (5)	—	63–345 (g)	190–370	310–920	60–120	79–330	21–220	
	Australia (11,12)	—	—	170–1,350	3,740–38,700	56–1,050	—	4–375	
	England (3,6,8,9)	—	28–1,240 (g)	12–779	1,830–99,200	9.7–4,500	36–82	2–342	
	Norway (13)	—	20–2,310 (g)	105–2,370	—	87–2,900	12–32		

(1) Bryan, 1976; (2) Penrose et al., 1975; (3) Leatherland & Burton, 1974; (4) Bohn, 1975; (5) Stenner and Nickless, 1975; (6) Boyden, 1975; (7) Stenner and Nickless, 1974a; (8) Peden et al., 1973; (9) Wright, 1976; (10) Talbot et al., 1976a,b; (11) Bloom and Ayling, 1977; (12) Mackay et al., 1975; (13) Stenner and Nickless, 1974b; (14) Bradfield et al., 1976; (15) Boyden and Romeril, 1974; (16) Young and McDermott, 1975; (17) Ruddell and Rains, 1975; (18) Schwiner, 1973; (19) Parsons et al., 1973; (20) Fletcher et al., 1975; (21) Martin et al., 1976; (22) Hardisty et al., 1974; (23) Roberts et al., 1976; (24) Chow et al., 1976; (25) Chow et al., 1974; (26) Braham, 1973; (27) Klemmer et al., 1976; (28) Schultz et al., 1976; (29) Eganhouse and Young, 1976; (30) Windom et al., 1973; (31) Johnson and Braman, 1975; (32) Freeman et al., 1974; (33) Koper, 1974; (34) Greig et al., 1976; (35) Gibbs et al., 1974; (36) Renzoni et al., 1974; (37) Vucetic et al., 1974; (38) Skei et al., 1976; (39) Andersen and Neelakantan, 1974; (40) Havre et al., 1973; (41) Jones et al., 1972; (42) Raymont, 1972; (43) Pentreath, 1976; (44) Alexander and Young, 1976; (45) As et al., 1975; (46) Watling and Watling, 1976a,b; (47) Knauer and Martin, 1972.

[a] Other values in pm dry weight

From Förstner, U., in *Chemistry and Biogeochemistry of Estuaries*, Olausson E. and Cato, I., Eds., John Wiley & Sons, Chichester, U.K., 1980, 307. With permission.

waste disposal mines and smelters. For example, in the Sörfjord, Norway, lead and zinc concentrations in biota were much greater than normal due to effluents from a Pb-Zn smelter. Green algae had lead values 300 times greater, and mussels contained cadmium in amounts 4 to 70 times higher, than normal.[75]

Although estuarine and marine organisms show a range of abilities to concentrate trace metals, some generalizations have been advanced concerning organismal uptake of the metals.[23] According to Riley and Chester,[121] the affinity of cations for organisms is as follows: 4+ and 3+ elements > 2+ transition metals > 2+ Group IIA metals > 1+ Group I metals. Furthermore, these workers remark that the organisms take up lighter elements in a particular group less strongly than the heavier elements. "Higher" organisms are less likely to concentrate trace elements than "lower" organisms.

HEAVY METALS IN ESTUARINE SYSTEMS

Case Studies
Chesapeake Bay

During the past several decades, Chesapeake Bay has been subjected to increased levels of industrial and municipal point sources of contaminants along its shores.[122] In addition, the influx of pollutants from the Susquehanna River has exacerbated an already tenuous situation attributable to the shoreline input of these materials. Both point and nonpoint sources of heavy metals in the bay have contributed to the degradation in the overall quality of Chesapeake Bay, perhaps reflected most conspicuously in declining recreational and commercial resources. Decreases in oyster and striped bass populations in recent years may be coupled to this problem. The gradual retreat of aquatic vegetation, such as eelgrass, throughout the estuarine ecosystem possibly offers clues to the regional impact of chemical contaminants or some other type of pollutant. Clearly, certain contaminants are regionally dispersed in the bay, for example, polychlorinated biphenyls, phthalate esters, anthropogenic trace metals (Cu, Zn, Pb), polycyclic aromatic hydrocarbons (PAHs), herbicides, and weapon derived radionuclides. However, no link has been established between these contaminants and the regional changes in biotic communities.[76]

Trace metals enter Chesapeake Bay primarily via the Susquehanna River; other sources include shoreline erosion, atmospheric deposition (particularly for lead and zinc), municipal wastewaters, and industry.[123] The largest point source of heavy metals in Chesapeake Bay is Baltimore Harbor, which contributes little quantity of metals to the bay, and secondarily, the Hampton Roads complex near Norfolk, VA. Contamination of the Baltimore Harbor and Hampton Roads areas derives from heavy industrial activity in Baltimore and Norfolk, which has resulted in the spread of heavy metal contamination to the Patapsco and Elizabeth Rivers, respectively.

The Susquehanna River, the only river that empties directly into the main body of Chesapeake Bay, delivers $3.46 \times 10^{10}\,m^3$ of water per year and 0.9×10^6 metric tons of sediment per year to the northern bay, with most of its fine-grained, metal-rich sediment (predominantly clay and silt with small amounts of fine sand) being deposited north of Baltimore Harbor.[123-125] Freshwater flow and sediment influx from the Susquehanna greatly exceed other sources for the estuary. More than 90% of the freshwater input into the bay is via this river. The volume of fluid flow and sediments transported by the Susquehanna River is highly irregular; the spring fishet, occurring during <10% of the year, and major storm-generated floods carry most of the sediment to the estuary where it is largely deposited in the upper 30 km of the bay. Hence, the spring freshet generally accounts for more than 70% of the total annual fluvial input of suspended sediment, which averages about 1×10^6 metric tons, and 75% of these sediments are deposited in the upper 25 to 30 km of the estuary.[126,127] Here, the average sedimentation rate is about 0.7 cm/year.[128] Subsequent to their initial deposition, tidal currents and waves redistribute them.[128] Shoreline erosion may be the largest source of sediment in the middle and lower reaches of the bay,[129] while relict sand enters the estuary from the continental shelf.[128] Schubel and Carter[129] estimate that the bay receives 0.6×10^6 tons of clay and silt annually from shoreline erosion.

The annual cycle of sedimentation is punctuated at times by major storms and ensuing floods. These episodic events can have a devastating effect on the sediment regime of the estuary. The Great Flood of March 1936 deposited 30 cm of sediment in the northern bay, and Tropical Storm Agnes added another 15 cm of sediment in this area

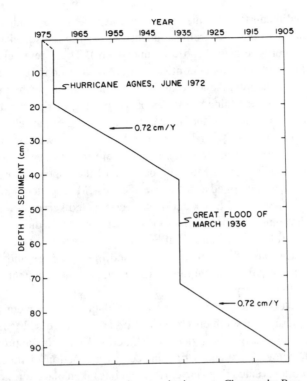

FIGURE 1. Sedimentation at a site in upper Chesapeake Bay (39°23′N, 76°05′W) showing thick accumulations of sediments due to the Great Flood of March 1936 and Hurricane Agnes in June 1972. These two events accounted for approximately 50% of the sediment accumulation at the site between 1900 and 1975. (From Schubel, J. R. and Hirschberg, D. J., in *Estuarine Interactions*, Wiley, M. L., Ed., Academic Press, New York, 1978, 285. With permission.)

during June 1972 (Figure 1). Tropical Storm Agnes was responsible for the discharge of more than 31×10^6 metric tons of sediment through the Susquehanna River to the upper bay over a 10-d period, an amount >30 times the average annual input.[127] The Great Flood of 1936 and Tropical Storm Agnes together accounted for more than half of all the sediment deposited in the upper bay between 1900 and 1975. In addition, floods associated with two major storms (Agnes and Eloise) delivered about 80% of the suspended sediment to the bay by the Susquehanna River during the decade 1966 to 1976.[130]

Helz et al.[131] compared the composition of heavy metals in surface sediments (top 10 cm) in northern Chesapeake Bay with the composition of material from sediment sources (i.e., Susquehanna River, shoreline erosion, and the atmosphere). These researchers ascertained the processes controlling Fe, Mn, and Zn in sediments of the bay. They concluded that the Fe content of upper bay sediments could be explained in terms of the mixing of riverine and shore-derived materials and hydrodynamic sorting of particles. The high Zn/Fe ratio observed in the mid-bay, in turn, was ascribed to the deposition of anthropogenic zinc from the atmosphere. Chemical remobilization of Mn from deposits of the upper bay or transition zone was invoked as the cause of excess Mn registered in mid-bay.

Sinex[132] and Sinex and Helz[133] conducted surveys of the spatial distribution of heavy metals in Chesapeake Bay sediments, whereas Cantillo[134] and Helz et al.[135] performed surveys of the temporal variations of the metals in the sediments. Sinex and Wright[123] reviewed the distribution of trace metals in the sediments and biota of the estuary.

Helz et al.[136] generated baseline data on trace metal levels in Chesapeake sediments by using Si:Al ratios regressed against metal concentrations from deeper sediments. Using this approach, they formulated contamination indices as a function of time.[122] Their work revealed that zinc enrichment has taken place for more than 1 century in the middle and upper bay. Similarly, cadmium and lead enrichment has occurred in the upper bay; atmospheric deposition appears to be partly responsible for this enrichment.[137]

General patterns of trace metal distributions are evident from the data collected by the aforementioned investigators. First, trace metal concentrations decline downestuary, reflecting the diminishing influence of the Susquehanna River in a seaward direction. Second, "hot spots" of heavy metal contamination exist in Baltimore Harbor and Hampton Roads due to industrial sources flanking the shoreline in these areas. Three-layered circulation in the Patapsco subestuary characterized by landward flow of surface and bottom waters into the harbor and seaward flow at mid-depth enables sediments and heavy metals to be trapped in the harbor.[138] This circulation pattern causes the harbor basin to behave as a sink for the metals. Third, the composition of trace metals in the lower bay parallels that of the continental shelf, which is the sediment source for the lower bay.[139]

A contamination index devised by the Environmental Protection Agency (EPA) to determine the degree of metal contamination in bay sediments showed high values for nearly the entire upper bay (0 to 75 km), Baltimore Harbor, and part of the Hampton Roads area. Those channel sites >75 km downestuary that also have elevated readings of heavy metals are most likely the result of material derived from the Susquehanna River and released from the upper bay,[140] most frequently during floods produced by major storms.[123]

Sinex and Helz[133] compared different geographical regions of the estuary by employing an enrichment factor, EF, as follows

$$EF = \frac{(X / Fe)\,sediment}{(X / Fe)\,Earth's\ crust} \qquad (1)$$

where X/Fe represents the ratio of the concentration of element X to iron, the element of normalization. An EF value of unity signifies neither enrichment nor depletion in relationship to the average Earth's crust. Table 8 gives enrichment factors for Chesapeake Bay, based on Equation 1, and also for Delaware and San Francisco Bays. The enrichment factors calculated for areas of Chesapeake Bay are consistent with the discussion presented above regarding the spatial

---

TABLE 8
Enrichment Factors (average or typical range) for Chesapeake Bay and Other Areas

Area	Cr	Mn	Ni	Cu	Zn	Pb	References
Chesapeake Bay (whole)	1	2	1	1	5	5	See Ref. 123
Upper (0—75 km)	1	3–6	–	2	6–8	4–7	Sinex, 1981
Middle (75—200 km)	1	1	–	1	4–6	3–4	Sinex & Helz, 1981
Lower (200—300 km)	1	<1	–	1	2–4	2–4	
Baltimore Harbor	4	1	1	–	8	–	Sinex & Helz, 1982
Hampton Roads	1	1	1	2–4	5–10	7–21	Rule, 1986
Susquehanna River	0.6	6	3	2	8	7	Sinex & Helz, 1981
Offshore[a]	0.7	0.8	0.6	0.2	2	3	Rule, 1986
Delaware Bay	3	–	12	2	10	16	Bopp & Biggs, 1973
San Francisco Bay	–	<1	2	2	2	–	Eaton, 1979

[a] Proposed disposal site on the inner Virginian continental shelf approximately 27 km off the Chesapeake Bay mouth.

From Sinex, S. A. and Wright, D. A., *Mar. Pollut. Bull.,* 19, 425, 1988. With permission.

distribution of trace metals baywide, i.e., highest values correspond to the upper bay, Baltimore Harbor, and Hampton Roads. The upper bay readings are comparable to those of suspended material in the Susquehanna River. Some of the highest EFs (>10) have been computed for Delaware Bay, where industrial sources heavily abut the shoreline.

Analysis of heavy metals in sediment cores from the main bay, Elizabeth River, and the Baltimore Harbor indicates surface enrichment of zinc at all sites. Because zinc enrichment in sediments has taken place over the past century, enrichment factors for zinc approach unity at the base of sediment cores (about 1 m in length) extracted from the main bay.[135] This pattern of surface sediment enrichment is replicated in cores from the Elizabeth River.[76] Zinc contaminates a thicker wedge of sediment in Baltimore Harbor; here, contamination can be identified in sediments at depths of 3 m.[141] The wedge of contamination thickens in an upestuary direction at Baltimore Harbor.[123]

While much data have been gathered on the spatial and temporal distributions and concentrations of trace metals in Chesapeake Bay sediments, less information has been obtained on trace metal concentrations in the water body and biota of the estuary. Research on the effects of the metals on the biota, for instance, is less advanced than that performed in other major estuaries such as San Francisco Bay.[123] Prompted by a 17-year decline in the commercially important striped bass fishery and the growing body of knowledge related to toxicity in spawning grounds and the possible role of trace metals and acid precipitation in the demise of this finfish,[122,142] effects of trace metals on Chesapeake Bay biota have only recently been focused upon. Mehrle et al.[143] reported on the toxicity of As, Cd, Cu, Pb, and Se, as well as a group of organic toxicants, on striped bass larvae. Chemical toxicity of striped bass larvae have been investigated in the Choptank,[143] Nanticoke,[144] and Potomac Rivers.[145] Aluminum has been implicated as a factor contributing to toxicity of the larvae in the Choptank and Nanticoke Rivers, whereas high levels of cadmium and copper may be responsible for increased larval mortality in the Potomac River.[123] Comparisons of the body burdens of aluminum, cadmium, and copper in striped bass larvae from the Choptank, Nanticoke, and Potomac Rivers can be found elsewhere.[146]

Bender and Huggett[147] detailed contaminant effects on Chesapeake Bay shellfish, including those attributable to trace metals. Certain metals, for example, Cd, Cu, and Zn, are higher in oyster (*Crassostrea virginica*) tissues near urbanized areas. Huggett et al.[148] found that residues of Cd, Cu, and Zn in *C. virginica* were contingent upon both the source of the contaminant as well as the position of the organism in the estuary. Salinity may be a controlling factor in the body burdens of Ag, Cd, Cu, and Zn in *C. virginica*.[149] The U.S. EPA[150] summarizes the distribution of the trace metals As, Cd, Cr, Cu, Hg, Pb, and Zn in oysters tissues from the upper and lower reaches of Chesapeake Bay.

The toxicity of trace metals to Chesapeake Bay biota should be studied on a tributary-by-tributary basis as well as in the main body of the estuary. The Choptank and Nanticoke Rivers, which are relatively minor tributaries, represent important spawning grounds for certain species, such as the striped bass. Other larger influent systems, for example the Potomac River, exert even greater biotic influence on Chesapeake Bay. The Susquehanna River supplies about 50% of the freshwater flow and most of the trace metal input to the main body of the estuary. Hence, most future investigations concerning the interaction of trace metals and biota in the bay probably will deal in some way with the effects of this riverine system.

San Francisco Bay

Among the major estuaries in the U.S., perhaps the one most strongly modified by anthropogenic activity is San Francisco Bay. The water quality and biotic communities of this estuary have been greatly impacted by more than a 50% reduction of freshwater inflow over the years, much of which is now diverted for irrigation. Diking and filing of wetlands bordering the bay have destroyed habitat for fish and waterfowl, thereby exacerbating the impact of low freshwater inflow. Other significant modifications include the introduction of exotic plants and animals, some of which are pest species, that have altered the composition of the communities. The disposal of sewage waste, toxic chemicals, and heavy metals creates an array of hazardous conditions for living resources in the estuary. The possible synergistic effect of various human activities may further modify an already disturbed environment,[151] attesting to the potential dangers of urban and industrial development.

San Francisco Bay is a three-component system consisting of
Suisun and San Pablo Bays in the northern reach, Central Bay, and
South Bay.[152] North of San Francisco the estuary can be characterized as partially mixed, and south of San Francisco, typically well
mixed.[153] Approximately 90% of the freshwater input to the estuarine system occurs via the Sacramento and San Joaquin Rivers at the
eastern end of Suisun Bay after passing through the Delta zone
(Figure 2). The Sacramento and San Joaquin River system carries
runoff from about 40% of California's surface area. Not all of the
runoff enters the bay. Diversions, storage, and consumption reduce
some of the supply, with the remaining water flowing into the system
at the eastern end of Suisun Bay.

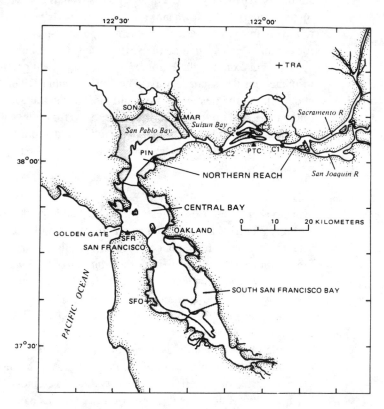

FIGURE 2.   Map of San Francisco Bay showing the south, central, and northern
segments as well as the Delta region and the influent systems of the San Joaquin
and Sacramento Rivers. (From Walters, R. A. and Gartner, J. W., *Est. Coastal Shelf
Sci.,* 21, 17, 1985. With permission.)

Riverine inflow from the Delta region is highly variable and seasonally dependent. During the summer and fall minimum inflows of 100 to 300 m³/s correspond to low runoff periods. In contrast, high runoff in winter and spring yield maximum inflows of 12,000 m³/s.[153] Changes in flow rates and, concomitantly, suspended sediment concentrations may dramatically shift trace metal abundances in the water column. For example, the suspended sediment of the Susquehanna River entering Chesapeake Bay is depleted in Cu, Mn, and Zn during high flow conditions compared to low flow conditions.[154] Similarly, the depletions of Cu, Mn, and Zn on suspended matter of the Mississippi River during periods of high flow amounts to 30 to 40% relative to that during periods of low flow.[155] The loss of freshwater input to San Francisco Bay, mostly diverted for agricultural use, leads to hydrodynamic changes that modulate trace metal influx and residence times in the estuary.[156,157]

Although San Francisco Bay has undergone radical environmental changes for more than 1 century due to human activities, quantitative investigations of trace metals in the estuary only commenced in the early 1970s.[157] Eaton[158] and Gordon[159] conducted surveys of dissolved trace metal behavior and distribution in the northern reach, and Girvin et al.[160] and Kinkel et al.[161] described the concentration and distribution of trace metals in South San Francisco Bay. Kinkel et al.[161] also assessed the mechanisms responsible for controlling trace metal distribution in the water column. Luoma and Phillips[157] provide a comprehensive overview of trace element composition, distribution, and impacts in the bay.

As in the case of Chesapeake Bay, the source of most trace metals in San Francisco Bay is via major river systems (Table 9). In contrast to Chesapeake Bay, however, San Francisco Bay receives most of its anthropogenic inputs of trace metals from sources on the estuary itself and not from incoming rivers.[157] Luoma and Phillips[157] note that 50 municipal wastewater treatment plants discharge effluent into the bay; between 1984 and 1986, they released 2.9×10^9 l/d of effluent into the bay and its delta. Eighteen industrial facilities, including 6 petroleum refineries, discharge an additional 4×10^5 l/d of waste to the bay. Compounding these inputs is metal enrichment derived from mine wastes, some dating back to the late 1800s when mining operations on the Sacramento and San Joaquin Rivers concentrated on removing copper, gold, mercury, zinc, and silver, accounting for their releases to the northern bay.[161] Today, trace metals of greatest concern because of their frequency or severity of contamination in sediments, the water column, and biota appear to be copper, cadmium, mercury, selenium, and silver, and they exhibit complex temporal and spatial trends which, in some cases, are strongly coupled to anthropogenic element input.[157]

In terms of point sources of contaminants, nickel (29 kg/d), copper (31 kg/d), and zinc (74 kg/d) have the highest absolute values. Relative to riverine inputs, however, nickel and selenium are released in largest amounts from point sources. In regard to urban runoff, lead and zinc loadings far outdistance the other trace metals (Table 9). Of specific anthropogenic concern, three dredged-spoil disposal sites in Central Bay and the northern reach of the estuary receive approximately 6×10^6 m^3 of metal-enriched sediments from harbor and marina locales. In addition, 40 hazardous waste disposal

TABLE 9
Sources of Trace Metals in San Francisco Bay, Including Municipal and Industrial Point Sources, Urban Runoff, and the Sacramento-San Joaquin River System[a]

Metal	Point source	Urban runoff	Rivers	Point source riverine[b]	Anthropogenic riverine[c]
Ag	7.5	?	26	0.28	>0.28
As	5.7	9	37	0.15	0.39
Cd	4.0	3	27	0.15	0.26
Cr	14	15	92	0.15	0.32
Cu	31	59	203	0.15	0.44
Hg	0.8	0.2	3	0.26	0.33
Ni	29	?	82	0.35	>0.35
Pb	17	250	66	0.26	4.0
Se	2.5	?	7.4	0.33	>0.33
Zn	74	268	288	0.25	1.19

[a] Concentrations in kg/dry wt.
[b] Ratios of point source inputs and total anthropogenic input (sum of point source and urban runoff) relative to riverine input. Data from Walters, R. A. and Gartner, J. W., *Est. Coastal Shelf Sci.*, 21, 17, 1985.
[c] Data from Gunther, A. J., Davis, J. A., and Phillips, D. J. H., An Assessment of the Loading of Toxic Contaminants to the San Francisco Bay-Delta, Tech. Rep., Aquatic Habitat Institute, Richmond, CA, 1987.

locations rimming the shoreline are purportedly leaching elements into the estuary,[162] and the roiling of sediments at historic waste disposal dumps in the bay may contribute to trace metal contamination in the water column. Aside from the local anthropogenic trace metal input, a large quantity of trace metals originates from riverine influx as emphasized above, and this function provides background concentrations in the estuary.

The problem trace metals, copper, mercury, selenium, and silver, are enriched in areas of specific anthropogenic activity. Thus, in the waters of South Bay, in particular, copper concentrations exceed the U.S. EPA water quality recommendation of 2.9 µg/l. Mercury enters the bay principally from the Sacramento and San Joaquin Rivers, being derived from upstream sites of historic placer gold mining operations. An additional fraction leaches into the bay from natural cinnabar deposits in the Coast Range of the Central Valley. As a result, many larger striped bass in the delta and bay contain mercury in their axial muscle at levels >0.5 µg/g wet wt. Similar concentrations are seen in several other species of fish in the upstream catchment draining to the bay. Selenium originates mainly from the leaching of seleniferous soils on agricultural land in the San Joaquin River catchment and from the discharges of six major oil refineries, mostly as selenite, in the northern reach of the bay. The biotic impact of the selenium releases has not been firmly established. Elevated levels of silver are especially apparent in the water and biota of South Bay.[163,164] Municipal effluents containing industrially derived silver are largely responsible for the silver input. Despite these impacts, water quality in the estuary has been enhanced somewhat by improvements in municipal and industrial effluent treatment in the 1960s and early 1970s.[156]

Luoma and Phillips[157] have provided a detailed account of the characteristics of trace metal contamination in the estuary and the biological impacts of the contamination. The following discussion draws heavily from their work. The physical and geochemical processes that control the distribution and transport of dissolved and particulate trace metals in San Francisco Bay change both in space and time, thereby contributing to the variability of trace metals in the estuary. Seasonal shifts in freshwater inflow, mentioned above, as well as seasonal cycles in wind and tidal velocity regimes foster changes in trace metal concentrations either directly by modulating the influx of the elements or indirectly, for example, by altering sediment properties. Hence, fluctuations in the amount of organic material, iron, and manganese accompany changes in grain size distribution of sediments which can have a marked effect on trace metal concentrations. Luoma et al.[165] stated that, although trace metal concentrations in the surface sediments of intertidal and shallow subtidal habitats generally fluctuate by two- to fourfold, at times they may vary by as much as tenfold. In the case of copper, the quantity of the element correlates strongly with total organic carbon. Furthermore, copper speciation appears to be controlled by metal-humic material complexation in at least part of the estuary,[42] consequently, dissolved organic matter concentrations, which can also be highly variable, play a vital role in regulating trace metal distribution in the water column.

Trace metals in San Francisco Bay have a heterogeneous spatial distribution owing to both localized anthropogenic input and variability in physical and geochemical processes. Despite few analytically reliable studies on some of the trace metals, three generalizations have been formulated to delineate trace metal distributions in the estuary: (1) the concentrations of many metals peak in the urbanized/industrialized region of the estuary; (2) patches of extreme enrichment characterize most of the metals and are ascribable to anthropogenic sources such as municipal discharges, industrial effluents, and marine operations; and (3) the South Bay is enriched in some metals compared to the northern reach. Enrichment of South Bay in cadmium, copper, selenium, and silver may be partially explained by basin geomorphology and hydrology. South Bay is a semi-enclosed system with only slight freshwater inflow relative to the northern reach. Most of the freshwater entering South Bay consists of discharges from five large waste treatment plants. Certain long-term changes in trace metal concentrations are evident in South Bay. Between 1976 and 1985, for instance, the amount of cadmium in solution doubled in South Bay. Moreover, silver concentrations in sediments of South Bay equalled 0.5 to 0.7 μg/g in 1970 and 1.02 ± 0.18 μg/g in 1986. The increase in cadmium and silver during the past 2 decades has occurred concomitantly with an expanding human population which has doubled in the bay area since the early

1970s. Despite the rising contamination of cadmium and silver in sediments and biota of the bay, available data collected since the early 1970s indicate little change in enrichment of other trace metals (e.g., Cu, Hg, Pb, and Zn).

Metal-induced stress in aquatic communities in the bay is most evident among the benthos inhabiting areas of localized contamination. While average trace metal enrichment figures baywide may not differ significantly from other impacted estuaries,[166] some trace metals in sediments and biota attain levels comparable to those of the most contaminated systems known. Exceptionally high levels are most frequently registered for cadmium, mercury, lead, and silver.[156,165,167] At several sites in South Bay, copper in bivalves reaches very high levels as well, even though sediment enrichment seldom surpasses 100 $\mu g/g$.[165,168] This suggests that bioavailability of Cu may be enhanced by geochemical processes in the bay. In contrast, lead may be enriched in sediments but not in benthic biota; geochemical processes may reduce the bioavailability of lead.

Physiological stress and ecological change in aquatic communities due to heavy metals have been examined most thoroughly at a contamination site at Palo Alto in South Bay. Here, evidence of metal stress exists at several levels of biological organization, specifically at the cellular, organism, and population levels. Metal stress at the community and ecosystem levels, while possibly being important, has not been established at Palo Alto in part because of the difficulty of separating its impact from changes imposed by natural variability.

Luoma and Phillips (p. 423)[157] summarize the biological impacts of trace metal contamination in San Francisco Bay in the following way:

> "Stress in benthic communities is evident in extremely contaminated localities. Metal-induced stress on one benthic species is suggested by converging lines of evidence at a Cu-Ag enriched locality in South Bay. In general, however, studies to date have lacked the systematic approach, comprehensive nature and continuity of support necessary to establish rigorous scientific proof of trace element impacts on species in the Bay. There is a particular dearth of knowledge about upper trophic level species."

SUMMARY AND CONCLUSIONS

Heavy metals are a group of elements that are potentially toxic to estuarine and marine organisms above a threshold availability. However, many of these elements also remain essential for proper metabolism of biota at lower concentrations. Cobalt, copper, iron, manganese, molybdenum, vanadium, strontium, and zinc include some of the trace metals necessary for normal life processes; they can be toxic to organisms at elevated concentrations. Trace elements commonly of greatest concern in estuarine and coastal marine waters are arsenic, cadmium, chromium, copper, lead, mercury, selenium, and silver, which at times contribute to severe contamination problems.[169] At higher concentrations, heavy metals act as enzyme inhibitors and can result in the demise of large numbers of susceptible organisms.

Heavy metals enter estuaries from three principal sources: (1) freshwater input; (2) the atmosphere; and (3) anthropogenic activity. Excluding the anthropogenic origin of the trace metals, which frequently is locally significant, the primary source of these elements in estuaries is riverine influx. Anthropogenic inputs superimpose substantial amounts of heavy metals upon background concentrations in some systems. These inputs are derived from a multitude of activities such as smelting operations, ash disposal, sewage-sludge disposal, dredged-spoil dumping, and the burning of fossil fuels. Municipal and industrial discharges in urbanized/industrialized regions generally account for heavy, localized contamination in impacted systems. Examples are Commencement Bay and Elliott Bay on Puget Sound, Boston Harbor, and Newark Bay. Historically, sewage-sludge and dredged-spoil disposal have played a key role in the contamination of bottom sediments with heavy metals. Disposal sites in the coastal waters of the New York Bight and Southern California Bight contain tons of heavy metals derived from man. In Europe, Liverpool Bay, the Thames estuary, the Rhine-Waal/Meuse/Scheldt estuaries, the German Bight, and the Baltic Sea have been subjected to heavy metal enrichment by the disposal of various wastes.

Estuarine organisms, particularly mollusks, concentrate heavy metals and, consequently, have been employed as bioindicators of heavy metal pollution. Mussels are especially useful in this regard and have been utilized in many investigations of suspected contamination of estuarine waters. While bioaccumulation of trace metals is conspicuous among estuarine organisms, evidence of a general biomagnification effect is lacking.

The free metal ionic activity may be more important in metal toxicity of an organism than the total concentration of a metal. Metal uptake by estuarine organisms occurs via ingestion of food and particulate inorganic matter and via diffusion of dissolved metals. The metals may be stored in the skeletal structure, concretions, or intracellular matrices of an organism, and they are released in feces, eggs, and molting products. Metallothioneins and other heavy metal-binding proteins bind metal contaminants within the organism, thereby helping to control heavy metal concentrations.

Chesapeake Bay and San Francisco Bay are two major estuaries in the U.S. exhibiting heavy metal enrichment in some areas. Most heavy metals enter these two estuaries in large influent systems. However, anthropogenic input of metals accounts for significant localized contamination of areas in proximity to urbanized/industrialized zones. San Francisco Bay may be more heavily impacted by anthropogenic activity because of the long history of mining and urban and industrial development taking place along its shores and in watershed areas upstream of the mouth of influent rivers. Metal-induced stress in aquatic communities of both estuaries is most conspicuous within benthic habitats subjected to high trace metal enrichment. Determining the significance of trace metal contamination at the ecosystem level is more problematic and remains particularly challenging for future investigations.

REFERENCES

1. Sunda, W. G., Neritic-oceanic trends in trace-metal toxicity to phytoplankton communities, in *Oceanic Processes in Marine Pollution*, Vol. 1, *Biological Processes and Wastes in the Ocean*, Capuzzo, J. M. and Kester, D. R., Eds., Robert E. Krieger Publishing, Malabar, FL, 1987, 19.
2. Viarengo, A., Heavy metals in marine invertebrates: mechanisms of regulation and toxicity at the cellular level, *Rev. Aquat. Sci.*, 1, 295, 1989.
3. Abel, P. D., *Water Pollution Biology*, Ellis Horwood, Chichester, U.K., 1989.

4. Rainbow, P. S., The biology of heavy metals in the sea, *Int. J. Environ. Stand.,* 25, 195, 1985.

5. Stickney, R. R., *Estuarine Ecology of the Southeastern United States and Gulf of Mexico,* Texas A&M University Press, College Station, 1984.

6. Kennish, M. J., *Ecology of Estuaries,* Vol. 1, CRC Press, Boca Raton, FL, 1986.

7. Nemerow, N. L., *Stream, Lake, Estuary, and Ocean Pollution,* Van Nostrand Reinhold, New York, 1985.

8. Viarengo, A., Moore, M. N., Mancinelli, G., Mazzucotelli, A., and Pipe, R. K., Significance of metallothioneins and lysosomes in cadmium toxicity and homeostasis in the digestive gland cells of mussels exposed to the metal in presence or absence of phenanthrene, *Mar. Environ. Res.,* 17, 184, 1985.

9. Engel, D. W. and Brouwer, M., Trace metal-binding proteins in marine molluscs and crustaceans, *Mar. Environ. Res.,* 13, 177, 1984.

10. Johnston, R., Ed., *Marine Pollution,* Academic Press, London, 1976.

11. Mance, G., *Pollution Threat of Heavy Metals in Aquatic Environments,* Elsevier, New York, 1987.

12. Armannsson, H., Burton, J. D., Jones, G. B., and Knap, A. H., Trace metals and hydrocarbons in sediments from the Southampton water region, with particular reference to the influence of oil refinery effluent, *Mar. Environ. Res.,* 15, 31, 1985.

13. Landner, L., Ed., *Speciation of Metals in Water, Sediment, and Soil Systems,* Springer-Verlag, Berlin, 1987.

14. Turekian, K. K., Cochran, J. K., Benninger, L. K., and Aller, R. C., The sources and sinks of nuclides in Long Island Sound, in *Estuarine Physics and Chemistry: Studies in Long Island Sound,* Saltzman, B., Ed., Academic Press, New York, 1980, 129.

15. Nixon, S. W., Between coastal marshes and coastal waters: a review of twenty years of speculation on the role of salt marshes in estuarine productivity and water chemistry, in *Estuarine and Wetland Processes — With Emphasis on Modeling,* Hamilton, P. and Macdonald, K. B., Eds., Plenum Press, New York, 1980.

16. Bruland, K. W., Trace elements in seawater, in *Chemical Oceanography,* Riley, J. P. and Chester, R., Eds., Academic Press, London, 1983, 157.

17. Duinker, J. C., Suspended matter in estuaries: adsorption and desorption processes, in *Chemistry and Biogeochemistry of Estuaries,* Olausson, E. and Cato, I., Eds., John Wiley & Sons, Chichester, U.K., 1980, 121.

18. Sholkovitz, E. R., The flocculation of dissolved Fe, Mn, Al, Cu, Ni, Co, and Cd during estuarine mixing, *Earth Planet. Sci. Lett.,* 41, 77, 1978.

19. Burton, J. D., Basic properties and processes in estuarine chemistry, in *Estuarine Chemistry,* Burton, J. D. and Liss, P. S., Eds., Academic Press, London, 1976, 1.

20. Bryan, G. W., Heavy metal contamination in the sea, in *Marine Pollution*, Johnston, R., Ed., Academic Press, London, 1976.

21. Pulich, W., Barnes, S., and Parker, P., Trace metal cycles in seagrass communities, in *Estuarine Processes*, Vol. 1, Wiley, M., Ed., Academic Press, New York, 1976, 493.

22. Bryan, G. W. and Gibbs, P. E., Polychaetes as indicators of heavy-metal availability in marine deposits, in *Oceanic Processes in Marine Pollution*, Vol. 1, *Biological Processes and Wastes in the Ocean*, Capuzzo, J. M. and Kester, D. R., Eds., Robert E. Krieger Publishing, Malabar, FL, 1987, 37.

23. Day, J. H., The chemistry and fertility of estuaries, in *Estuarine Ecology: With Particular Reference to Southern Africa*, Day, J. H., Ed., A. A. Balkema, Rotterdam, 1981, 57.

24. Koepp, S. J. and Santoro, E. D., Bioaccumulation of Hg, Cd, and Pb in *Mytilus edulis* transplanted to a dredged-material dumpsite, in *Oceanic Processes in Marine Pollution*, Vol. 1, *Biological Processes and Wastes in the Ocean*, Capuzzo, J. M. and Kester, D. R., Eds., Robert E. Krieger Publishing, Malabar, FL, 1987, 50.

25. Nelson, D. A., Miller, J. E., and Calabrese, A., Effect of heavy metals on bay scallops, surf clams, and blue mussels in acute and long-term exposures, *Arch. Environ. Contam. Toxicol.*, 17, 595, 1988.

26. Cross, F. A., Willis, J. N., Hardy, L. H., Jones, N. Y., and Lewis, J. M., Role of juvenile fish in cycling of Mn, Fe, Cu, and Zn in a coastal plain estuary, in *Estuarine Research*, Vol. 1, Cronin, L. E., Ed., Academic Press, New York, 1975, 45.

27. Weis, J. S., Renna, M., and Vaidya, S., Mercury tolerance in killifish: a stage-specific phenomenon, in *Oceanic Processes in Marine Pollution*, Vol. 1, *Biological Processes and Wastes in the Ocean*, Capuzzo, J. M. and Kester, D. R., Eds., Robert E. Krieger Publishing, Malabar, FL, 1987, 31.

28. Förstner, U., Sources and sediment associations of heavy metals in polluted coastal regions, in *Proc. 2nd Symp. The Origin and Distribution of the Elements*, Ahrens, L. H., Ed., Pergamon Press, Oxford, 1978, 849.

29. Brinckman, F. E. and Iverson, W. P., Chemical and bacterial cycling of heavy metals in the estuarine system, in *Marine Chemistry in the Coastal Environment*, Church, T. M., Ed., ACS Symp. Ser. Vol. 18, American Chemical Society, Washington, D.C., 1975, 319.

30. Wright, L. D., Benthic boundary layers of estuarine and coastal environments, *Rev. Aquat. Sci.*, 1, 75, 1989.

31. Neilson, B. J., Brubaker, J., and Kuo, A., Eds., *Estuarine Circulation*, Humana Press, Clifton, NJ, 1989.

32. Officer, C. B. and Lynch, D. R., Bioturbation, sedimentation, and sediment-water exchanges, *Est. Coastal Shelf Sci.*, 28, 1, 1989.

33. Li, Y. H., Burkhardt, M., and Teraoka, H., Desorption and coagulation of trace elements during estuarine mixing, *Geochim. Cosmochim. Acta,* 48, 2011, 1984.
34. Ackroyd, D. R., Bale, A. J., Howland, R. J. M., Knox, S., Millward, G. E., and Morris, A. W., Distribution and behavior of dissolved Cu, Zn, and Mn in the Tamar estuary, *Est. Coastal Shelf Sci.,* 23, 621, 1986.
35. Davis, J. A., Complexation of trace metals by adsorbed natural organic matter, *Geochim. Cosmochim. Acta,* 48, 679, 1984.
36. Wangersky, P. J., Biological control of trace metal residence time and speciation, a review and synthesis, *Mar. Chem.,* 18, 269, 1986.
37. Balls, P. W., Composition of suspended particulate matter from Scottish coastal waters — geochemical implications for the transport of trace metal contaminants, *Sci. Tot. Environ.,* 57, 171, 1986.
38. Gibbs, R. J., Segregation of metals by coagulation in estuaries, *Mar. Chem.,* 18, 149, 1986.
39. van den Berg, C. M. G., Merks, A. G. A., and Duursma, E. K., Organic complexation and its control of dissolved concentrations of copper and zinc in the Scheldt estuary, *Est. Coastal Shelf Sci.,* 24, 785, 1987.
40. Delfino, J. J. and Otto, R. G., Trace metal transport in two tributaries of the upper Chesapeake Bay: the Susquehanna and Bush Rivers, *Mar. Chem.,* 20, 29, 1986.
41. Jones, G. B., Thomas, F. J., and Burdon-Jones, C., Influence of *Trichodesmium* blooms on cadmium and iron speciation in Great Barrier Reef lagoon waters, *Est. Coastal Shelf Sci.,* 23, 378, 1986.
42. Kuwabara, J. S., Chang, C. C. Y., Cloern, J. E., Fries, T. L., Davis, J. A., and Luoma, S. N., Trace metal associations in the water column of south San Francisco Bay, California, *Est. Coastal Shelf Sci.,* 28, 307, 1989.
43. Aston, S. R., Trace element analysis, in *Practical Estuarine Chemistry: A Handbook,* Head, P. C., Ed., Cambridge University Press, Cambridge, 1985, 126.
44. Paulson, A. J., Feely, R. A., Curl, H. C., Jr., and Tennant, D. A., Estuarine transport of trace metals in a buoyant riverine plume, *Est. Coastal Shelf Sci.,* 28, 231, 1989.
45. Wood, J. M., Biological cycles for toxic elements in the environment, *Science,* 183, 1049, 1974.
46. Wittmann, G. T. W. and Förstner, U., Metal enrichment of sediments in inland waters — the Hartbeespoort Dam, *Water SA,* 1, 76, 1975.
47. Houghton, D. R., Toxicity testing of candidate antifouling agents and accelerated antifouling paint testing, in *Marine Biodeterioration: An Interdisciplinary Study,* Costlow, J. D. and Tipper, R. C., Eds., Naval Institute Press, Annapolis, MD, 1984, 255.

48.   Burton, D. T., Biofouling control procedures for power plant cooling water systems, in *Condenser Biofouling Control,* Garey, J. R., Jorden, R. M., Aitken, A. H., Burton, D. T., and Gray, R. H., Eds., Ann Arbor Science, Ann Arbor, MI, 1980, 251.

49.   Thomas, T. E. and Robinson, M. G., The physiological effects of the leachates from a self-polishing organotin antifouling paint on marine diatoms, *Mar. Environ. Res.,* 18, 215, 1986.

50.   Pinkney, A. E., Wright, D. A., and Hughes, G. M., A morphometric study of the effects of tributyltin compounds on the gills of the mummichog, *Fundulus heteroclitus, J. Fish. Biol.,* 34, 665, 1989.

51.   Simmonds, M., The case against tributyltin, *Oryx,* 20, 217, 1986.

52.   Laughlin, R. B., Jr. and Linden, O., Tributyltin — contemporary environmental issues, *Ambio,* 26, 252, 1987.

53.   Batley, G. E., Mann, K. J., Brockbank, C. I., and Maltz, A., Tributyltin in Sydney Harbor and Georges River waters, *Aust. J. Mar. Freshwater Res.,* 40, 39, 1989.

54.   Beaumont, A. R. and Newman, P. B., Low levels of tributyltin reduce growth of marine microalgae, *Mar. Pollut. Bull.,* 17, 457, 1986.

55.   Lee, R. R., Metabolism of tributyltin oxide by crabs, oysters, and fish, *Mar. Environ. Res.,* 17, 145, 1985.

56.   Davies, I. M., McKie, J. C., and Paul, J. D., Accumulation of tin and tributyltin from antifouling paint by cultivated scallops (*Pecten maximus*) and Pacific oysters (*Crassostrea gigas*), *Aquaculture,* 55, 103, 1986.

57.   His, E. and Robert, R., Comparative effects of two antifouling paints on the oyster *Crassostrea gigas, Mar. Biol.,* 95, 83, 1987.

58.   Weis, J. S. and Perlmutter, J., Effects of tributyltin on activity and burrowing behavior of the fiddler crab, *Uca pugilator, Estuaries,* 10, 342, 1987.

59.   Weis, J. S., Gottlieb, J., and Kwiatkowski, J., Tributyltin retards regeneration and produces deformities of limbs in the fiddler crab, *Uca pugilator, Arch. Environ. Contam. Toxicol.,* 16, 321, 1987.

60.   Minchin, D., Duggan, C. B., and King, W., Possible effects of organotins on scallop recruitment, *Mar. Pollut. Bull.,* 18, 604, 1987.

61.   Pickwell, G. W. and Steinert, S. A., Accumulation and effects of organotin compounds in oysters and mussels: correlation with serum biochemical and cytological factors and tissue burdens, *Mar. Environ. Res.,* 24, 215, 1988.

62.   Batley, G. E., Fuhua, C., Brockbank, C. I., and Flegg, K. J., Accumulation of tributyltin by the Sydney rock oyster, *Saccostrea commercialis, Aust. J. Mar. Freshwater Res.,* 40, 49, 1989.

63.   Page, D. S., An analytical method for butyltin species in shellfish, *Mar. Pollut. Bull.,* 20, 129, 1989.

64. Waldock, M. J., TBT in UK estuaries, 1982-1986: evaluation of the environmental problem, in *Oceans '86 Conference Record: Science-Engineering-Adventure,* Vol. 4, *Organotin Symposium,* IEEE Publishing, New York, 1986, 1324.

65. Quevauviller, P., Lavigne, R., Pinel, R., and Astruc, M., Organo-tins in sediments and mussels from the Sado estuarine system (Portugal), *Environ. Pollut.,* 57, 149, 1989.

66. Seligman, P. F., Grovhoug, J. G., and Richter, K. E., Measurement of butyltins in San Diego, California: a monitoring strategy, in *Oceans '86 Conference Record: Science-Engineering-Adventure,* Vol. 4, *Organotin Symposium,* IEEE Publishing, New York, 1986, 1289.

67. Stang, P. M. and Seligman, P. F., Distribution and fate of butyltin compounds in the sediment of San Diego Bay, in *Oceans '86 Conference Record: Science-Engineering-Adventure,* Vol. 4, *Organotin Symposium,* IEEE Publishing, New York, 1986, 1256.

68. Olson, G. J. and Brinckman, F. E., Biodegradation of tributyltin by Chesapeake Bay microorganisms, in *Oceans '86 Conference Record: Science-Engineering-Adventure,* Vol. 4, *Organotin Symposium,* IEEE Publishing, New York, 1986, 1196.

69. Hall, L. W., Jr., Lenkevich, M. J., Hall, W. S., Pinkney, A. E., and Bushong, S. J., Monitoring organotin concentrations in Maryland Waters of Chesapeake Bay, in *Oceans '86 Conference Record: Science-Engineering-Adventure,* Vol. 4, *Organotin Symposium,* IEEE Publishing, New York, 1986, 1275.

70. Huggett, R. J., Unger, M. A., and Westbrook, D. J., Organotin concentrations in the southern Chesapeake Bay, in *Oceans '86 Conference Record: Science-Engineering-Adventure,* Vol. 4, *Organotin Symposium,* IEEE Publishing, New York, 1986, 1262.

71. Klauda, R. J. and Bender, M. E., Contaminant effects on Chesapeake Bay finfishes, in *Contaminant Problems and Management of Living Chesapeake Bay Resources,* Majumdar, S. K., Hall, L. W., Jr., and Austin, H. M., Eds., Pennsylvania Academy of Science, Easton, 1987, 321.

72. Hall, L. W., Jr., Tributyltin environmental studies in Chesapeake Bay, *Mar. Pollut. Bull.,* 19, 431, 1988.

73. Bryan, G. W., Bioavailability and effects of heavy metals in marine deposits, in *Wastes in the Ocean,* Vol. 6, *Nearshore Waste Disposal,* Ketchum, B. H., Capuzzo, J. M., Burt, W. V., Duedall, I. W., Park, P. K., and Kester, D. R., Eds., John Wiley & Sons, New York, 1985, 41.

74. Finney, B. and Huh, C.-A., High resolution sedimentary records of heavy metals from the Santa Monica and San Pedro Basins, *Mar. Pollut. Bull.,* 20, 181, 1989.

75. Förstner, U., Inorganic pollutants, particularly heavy metals in estuaries, in *Chemistry and Biogeochemistry of Estuaries,* Olausson, E. and Cato, I., John Wiley & Sons, Chichester, U.K., 1980, 307.

76. Helz, G. R. and Huggett, R. J., Contaminants in Chesapeake Bay: the regional perspective, in *Contaminant Problems and Management of Living Chesapeake Bay Resources,* Majumdar, S. K., Hall, L. W., Jr., and Austin, H. M., Eds., Pennsylvania Academy of Science, Easton, 1987, 270.

77. Wallace, G. T., Jr., Waugh, J. H., and Garner, K. A., Metal distribution in a major urban estuary (Boston Harbor) impacted by ocean disposal, in *Oceanic Processes in Marine Pollution,* Vol. 5, *Urban Wastes in Coastal Marine Environments,* Wolfe, D. A. and O'Connor, T. P., Eds., Robert E. Krieger Publishing, Malabar, FL, 1988, 157.

78. Bruland, K. W. and Franks, R. P., Mn, Ni, Cu, Zn, and Cd in the western North Atlantic, in *Trace Metals in Sea Water,* Wong, C. S., Boyle, E., Bruland, K. W., Burton, J. D., and Goldberg, E. D., Eds., Plenum Press, New York, 1983, 395.

79. Ginn, T. C. and Barrick, R. C., Bioaccumulation of toxic substances in Puget Sound organisms, in *Oceanic Processes in Marine Pollution,* Vol. 5, *Urban Wastes in Coastal Marine Environments,* Wolfe, D. A. and O'Connor, T. P., Eds., Robert E. Krieger Publishing, Malabar, FL, 1988, 157.

80. Goldberg, E. D., Koide, M., Hodge, V., Flegal, A. R., and Martin, J., U.S. mussel watch: 1977-1978 results on trace metals and radionuclides, *Est. Coastal Shelf Sci.,* 16, 69, 1983.

81. Martin, M. and Castle, W., Petrowatch: petroleum hydrocarbons, synthetic organic compounds, and heavy metals in mussels from the Monterey Bay area of central California, *Mar. Pollut. Bull.,* 15, 259, 1984.

82. Schafer, H. A., Characteristics of Municipal Wastewater Discharges, Southern California Coastal Water Research Project, Project El Segundo, annual report, 1975, 57.

83. Schafer, H. A. and Bascom, W., Sludge in Santa Monica Bay, Southern California Coastal Water Research Project, El Segundo, annual report, 1976, 77.

84. Haig, A. J. N., Use of the Clyde estuary and Firth for the disposal of effluents, in *The Environment of the Estuary and Firth of Clyde,* Allen, J. A., Barnett, P. R. O., Boyd, J. M., Kirkwood, R. C., Mackay, D. W., and Smyth, J. C., Eds., *Proc. R. Soc. Edinburgh, Sect. B,* 90(Spec. iss.), 393, 1986.

85. Rowlatt, S. M., Rees, H. L., and Rees, I. S., Changes in sediments following the dumping of dredged materials in Liverpool Bay, Proc. Counc. Meet. International Council for the Exploration of the Sea, ICES, Copenhagen, 1986.

86. Rowlatt, S. M. and Limpenny, D. S., The effects on the sea bed of dumping dredged material and sewage sludge at Roughs Tower in the outer Thames estuary, Proc. Counc. Meet. International Council for the Exploration of the Sea, ICES, Copenhagen, 1987.

87. Araùjo, M. F. D., Bernard, P. C., and Van Grieken, R. E., Heavy metal contamination in sediments from the Belgian coast and Scheldt estuary, *Mar. Pollut. Bull.,* 19, 269, 1988.

88. Langston, W. J., A Survey of Trace Metals in Biota from the Mersey Estuary, Tech. Rep., Marine Biology Association, Plymouth, U.K., 1988.

89. Schell, W. R. and Nevissi, A., Heavy metals from waste disposal in central Puget Sound, *Environ. Sci. Technol.,* 11, 887, 1977.

90. Crecelius, E. A. and Bloom, N., Temporal trends of contamination in Puget Sound, in *Oceanic Processes in Marine Pollution,* Vol. 5, *Urban Wastes in Coastal Marine Environments,* Wolfe, D. A. and O'Connor, T. P., Eds., Robert E. Krieger Publishing, Malabar, FL, 1988, 149.

91. Phillips, D. J. H. and Rainbow, P. S., Barnacles and mussels as biomonitors of trace elements: a comparative study, *Mar. Ecol. Prog. Ser.,* 49, 83, 1988.

92. White, S. L. and Rainbow, P. S., On the metabolic requirements for copper and zinc in molluscs and crustaceans, *Mar. Environ. Res.,* 16, 215, 1985.

93. Goldberg, E. D., The mussel watch, a first step in global marine monitoring, *Mar. Pollut. Bull.,* 6, 111, 1975.

94. Navrot, J., Amiel, A. J., and Kronfeld, J., *Patella vulgata:* a biological monitor of coastal marine pollution — a preliminary study, *Environ. Pollut.,* 7, 303, 1974.

95. Bayne, B. L., Moore, M. N., Widdows, J., Livingstone, D. R., and Salkeld, P., Measurement of the responses of individuals to environmental stress and pollution: studies with bivalve molluscs, *Phil. Trans. R. Soc. London B,* 286, 563, 1979.

96. Simpson, R. D., Uptake and loss of zinc and lead by mussels (*Mytilus edulis*) and relationships with body weight and reproductive cycle, *Mar. Pollut. Bull.,* 10, 74, 1979.

97. Bryan, G. S. and Hummerstone, L. G., Heavy metals in the burrowing bivalve *Scrobicularia plana* from contaminated and uncontaminated estuaries, *J. Mar. Biol. Assoc. U. K.,* 58, 401, 1978.

98. Cain, D. J. and Luoma, S. N., Copper and silver accumulation in transplanted and resident clams (*Macoma balthica*) in south San Francisco Bay, *Mar. Environ. Res.,* 15, 115, 1985.

99. Bryan, G. W. and Gibbs, P. E., Heavy metals in the Fal estuary, Cornwall: a study of long-term contamination by mining waste and its effects on estuarine organisms, *Occasional Publ. Mar. Biol. Assoc. U.K.,* 2, 1, 1983.

100. Atchison, G. J., Henry, M. G., and Sandheinrich, M. B., Effects of metals on fish behavior: a review, *Environ. Biol. Fish.,* 18, 11, 1987.

101. Batley, G. E., Ed., T*race Element Speciation: Analytical Methods and Problems,* CRC Press, Boca Raton, FL, 1989.

102. Capuzzo, J. M., Burt, W. V., Duedall, I. W., Park, P. K., and Kester, D. R., The impact of waste disposal in nearshore environments, in *Wastes in the Ocean,* Vol. 6, *Nearshore Waste Disposal,* Ketchum, B. H., Capuzzo, J. M., Burt, W. V., Duedall, I. W., Park, P. K., and Kester, D. R., Eds., John Wiley & Sons, New York, 1985, 3.

103. Sunda, W., Engel, D. W., and Thuotte, R. M., Effect of chemical speciation on toxicity of cadmium to grass shrimp *Palaemonetes pugio:* importance of free cadmium ion, *Environ. Sci. Technol.,* 12, 409, 1978.

104. Zamuda, C. D., Wright, D. A., and Smucker, R. A., The importance of dissolved organic compounds in the accumulation of copper by the American oyster, *Crassostrea virginica, Mar. Environ. Res.,* 16, 1, 1985.

105. George, S. G., Subcellular accumulation and detoxication of metals in aquatic animals, in *Physiological Mechanisms of Marine Pollutant Toxicity,* Vernberg, W. B., Calabrese, A., Thurberg, F. P., and Vernberg, F. J., Eds., Academic Press, New York, 1982, 3.

106. Bryan, G. W., Recent trends in research on heavy metal contamination in the sea, *Helgol. Wiss. Meeresunters.,* 33, 6, 1980.

107. Viarengo, A., Biochemical effects of trace metals, *Mar. Pollut. Bull.,* 16, 153, 1985.

108. Moore, M. N., Lysosomal responses to environmental chemicals in some marine invertebrates, in *Pollutant Effects on Marine Organisms,* Giam, C. S., Ed., D. C. Heath & Co., Lexington, MA, 1977, 143.

109. Jenkins, K. D., Brown, D. A., Hershelman, G. P., and Meyer, W. C., Contaminants in white croakers *Genyonemus lineatus* (Ayres, 1855) from the southern California Bight. I. Trace metal detoxification/toxification, in *Physiological Mechanisms of Marine Pollutant Toxicity,* Vernberg, W. B., Calabrese, A., Thurberg, F. P., and Vernberg, F. J., Eds., Academic Press, New York, 1982, 177.

110. Roesijadi, G., The significance of low molecular weight, metallothionein-like proteins in marine invertebrates: current status, *Mar. Environ. Res.,* 4, 167, 1981.

111. Lee, R. F., Davies, J. M., Freeman, H. C., Ivanovici, A., Moore, M. N., Stegeman, J., and Uthe, J. F., Biochemical techniques for monitoring biological effects of pollution in the sea, in *Biological Effects of Pollution and the Problems of Monitoring,* Vol. 179, Rapports et Procés-Verbaux des Réunions, McIntyre, A. D. and Pearce, J. B., Eds., Conseil International pour l'Exploration de la Mer, Copenhagen, 1980, 48.

112. Leland, H. V. and Kuwabara, J. J., Trace metals in *Fundamentals of Aquatic Toxicology,* Rand, G. M. and Petrocelli, S. R., Eds., Hemisphere Publishing, Washington, D.C., 1984, 134.

113. Redpath, K. J., Growth inhibition and recovery in mussels (Mytilus edulis) exposed to low copper concentrations, *J. Mar. Biol. Assoc. U.K.,* 65, 421, 1985.

114. Fujisawa, H. and Amemiya, S., Effects of zinc and lithium ions on the strengthening cell adhesion in sea urchin blastulae, *Experientia,* 38, 852, 1982.

115. Stebbing, A. R. D., The effects of low metal levels on a clonal hydroid, *J. Mar. Biol. Assoc. U.K.,* 56, 977, 1976.

116. Irato, P. and Piccini, E., Effetti del cadmio in *Tetrahymena pyriformis, Boll. Soc. Ital. Ecol.,* 1987, 207.

117. Lustingman, B. K., Enhancement of pigment concentrations in *Dunaliella tertiolecta,* as a result of copper toxicity, *Bull. Environ. Contam. Toxicol.,* 37, 710, 1986.

118. MacInnes, J. and Thurberg, F., Effects of metals on the behaviour and oxygen consumption of the mud snail, *Mar. Pollut. Bull.,* 4, 185, 1973.

119. Thurberg, F. P., Dawson, M. A., and Collier, R. S., Effects of copper and cadmium on osmoregulation and oxygen consumption in two species of estuarine crabs, *Mar. Biol.,* 23, 171, 1973.

120. Stebbing, A. R. D., Hormesis-stimulation of colony growth in *Campanularia flexuosa* (Hydrosoa) by copper, cadmium and other toxicants, *Aquat. Toxicol.,* 1, 227, 1981.

121. Riley, J. P. and Chester, R., *Introduction to Marine Chemistry,* Academic Press, London, 1971.

122. Wright, D. A. and Phillips, D. J. H., Chesapeake and San Francisco Bays: a study in contrasts and parallels, *Mar. Pollut. Bull.,* 19, 405, 1988.

123. Sinex, S. A. and Wright, D. A., Distribution of trace metals in the sediments and biota of Chesapeake Bay, *Mar. Pollut. Bull.,* 19, 425, 1988.

124. Schubel, J. R., Suspended Sediment of the northern Chesapeake Bay, Chesapeake Bay Institute, Tech. Rep. 35, Johns Hopkins University, Baltimore, 1968.

125. Officer, C. B., Lynch, D. R., Setlock, G. H., and Helz, G. R., Recent sedimentation in Chesapeake Bay, in *The Estuary as a Filter,* Kennedy, V. S., Ed., Academic Press, Orlando, FL, 1984, 131.

126. Biggs, R. B., Sources and distribution of suspended sediment in northernChesapeake Bay, *Mar. Geol.,* 9, 187, 1970.

127. Schubel, J. R. and Hirschberg, D. J., Estuarine graveyards, climatic change, and the importance of the estuarine environment, in *Estuarine Interactions,* Wiley, M. L., Ed., Academic Press, New York, 1978, 285.

128. Schubel, J. R., Turbidity maximum of the northern Chesapeake Bay, *Science,* 161, 1013, 1968.

129. Schubel, J. R. and Carter, H. H., Suspended sediment budget for Chesapeake Bay, in *Estuarine Process,* Vol. 2, *Circulation, Sediments and Transfer of Material in the Estuary,* Wiley, M., Ed., Academic Press, New York, 1976, 48.

130. Gross, M. G., Karweit, M., Cronin, W. B., and Schubel, J. R., Suspended sediment discharge of the Susquehanna River to northern Chesapeake Bay, 1966-1976, *Estuaries,* 1, 106, 1978.

131. Helz, G. R., Sinex, S. A., Ferri, K. L., and Nichols, M. Processes controlling Fe, Mn, and Zn in sediments of northern Chesapeake Bay, *Est. Coastal Shelf Sci.,* 21, 1, 1985.

132. Sinex, S. A., Trace Element Geochemistry of Modern Sediments from Chesapeake Bay, Ph.D. thesis, University of Maryland, College Park, 1981.

133. Sinex, S. A. and Helz, G. R., Regional geochemistry of trace elements in Chesapeake Bay sediments, *Environ. Geol.,* 3, 315, 1981.

134. Cantillo, A. Y., Trace Element Deposition Histories in the Chesapeake Bay, Ph.D. thesis, University of Maryland, College Park, 1982.

135. Helz, G. R., Setlock, G. H., Cantillo, A. Y., and Moore, W. S., Processes controlling the regional distribution of Pb-210, Ra-226, and anthropogenic zinc in estuarine sediments, *Earth Planet. Sci. Lett.,* 76, 23, 1986.

136. Helz, G. R., Sinex, S. A., Setlock, G. H., and Cantillo, A. Y., Chesapeake Bay Trace Elements, EPA-600/53-83-012, U.S. Environmental Protection Agency, Washington, D.C., 1983.

137. Muhlbaier, J., The Chemistry of Precipitation Near the Chalk Point Power Plant, Ph.D. thesis, University of Maryland, College Park, 1978.

138. Schubel, J. R. and Pritchard, D. W., Responses of upper Chesapeake Bay to variations in discharge of the Susquehanna River, *Estuaries,* 9(4A), 236, 1986.

139. Officer, C. B., Lynch, D. R., Setlock, G. H., and Helz, G. R., Recent sedimentation rates in Chesapeake Bay, in *The Estuary as a Filter,* Kennedy, V., Ed., Academic Press, New York, 1984, 131.

140. Eaton, A., Grant, V., and Gross, M. G., Chemical tracers for particle transport in Chesapeake Bay, *Est. Coastal Mar. Sci.,* 10, 75, 1980.

141. Sinex, S. A. and Helz, G. R., Entrapment of zinc and other trace elements in a rapidly flushed, industrial harbor, *Environ. Sci. Technol.,* 16, 820, 1982.

142. Hall, L. W., Jr., Studies of striped bass in three Chesapeake Bay spawning habitats, *Mar. Pollut. Bull.,* 19, 478, 1988.

143. Mehrle, P. M., Buckler, D., Finger, S., and Ludke, L., Impact of Contaminants on Striped Bass, Interim Tech. Rep., U.S. Fish and Wildlife Service, Columbia National Fisheries Research Laboratory, Columbia, MO, 1984.

144. Hall, L. W., Jr., Pinkney, A. E., Horseman, L. O., and Finger, S. E., Mortality of striped bass larvae in relation to contaminants and water quality in a Chesapeake Bay tributary, *Trans. Am. Fish. Soc.,* 114, 861, 1985.

145. Hall, L. W., Jr., Hall, W. S., Bushong, S. J., and Herman, R. L., *In situ* striped bass (*Morone saxatilis*) contaminant and water quality studies in the Potomac River, *Aquat. Toxicol.,* 10, 73, 1987.

146. Wright, D. A., Dose related toxicity of copper and cadmium in striped bass larvae from the Chesapeake Bay: field considerations, *Water Sci. Technol.,* in press.

147. Bender, M. E. and Huggett, R. J., Contaminant effects on Chesapeake Bay shellfish, in *Contaminant Problems and Management of Living Chesapeake Bay Resources,* Majumdar, S. K., Hall, L. W., Jr., and Austin, H. M., Eds., Pennsylvania Academy of Science, Easton, 1987, 373.

148. Huggett, R. J., Bender, M. E., and Slone, H. D., Utilizing metal concentration relationships in the eastern oyster (*Crassostrea virginica*) to detect heavy metal pollution, *Water Res.,* 7, 451, 1973.

149. Phelps, H. L., Wright, D. A., and Mihursky, J. A., Factors affecting trace metal accumulation by estuarine oysters *Crassostrea virginica, Mar. Ecol. Prog. Ser.,* 22, 187, 1985.

150. U.S. Environmental Protection Agency, Chesapeake Bay: A Profile of Environmental Change, Appendix 1, U.S. Environmental Protection Agency, Region 3, Philadelphia, 1983.

151. Nichols, F. H., Cloern, J. E., Luoma, S. N., and Peterson, D. H., The modification of an estuary, *Science,* 231, 567, 1984.

152. Conomos, T. J., Properties and circulation of San Francisco Bay waters, in *San Francisco Bay: The Urbanized Estuary,* Conomos, T. J., Ed., Pacific Division of the American Association for the Advancement of Science, San Francisco, 1979, 47.

153. Walters, R. A. and Gartner, J. W., Subtidal sea level and current variations in the northern reach of San Francisco Bay, *Est. Coastal Shelf Sci.,* 21, 17, 1985.

154. Helz, G. R. and Sinex, S. A., Influence of infrequent floods on the trace metal composition of estuarine sediments, *Mar. Chem.,* 21, 1, 1986.

155. Trefry, J. H. and Presley, B. J., Heavy metal transport from the Mississippi River to the Gulf of Mexico, in *Marine Pollutant Transfer,* Windom, H. L. and Duce, R. A., Eds., Lexington Books, Lexington, MA, 1976, 39.

156. Luoma, S. N. and Cloern, J. E., The impact of waste-water discharge on biological communities in San Francisco Bay, in *San Francisco Bay: Use and Protection,* Kockelman, W. J., Conomos, T. J., and Leviton, A. E., Eds., Pacific Division of the American Association for the Advancement of Science, San Francisco, 1982, 137.

157. Luoma, S. N. and Phillips, D. J. H., Distribution, variability, and impacts of trace elements in San Francisco Bay, *Mar. Pollut. Bull.,* 19, 413, 1988.

158. Eaton, A., Observations on the geochemistry of soluble copper, iron, nickel, and zinc in the San Francisco Bay estuary, *Environ. Sci. Technol.,* 13, 425, 1979.

159. Gordon, R. M., Trace Element Concentrations in Seawater and Suspended Particulate Matter from San Francisco Bay and Adjacent Coastal Waters, M.S. thesis, San Jose State University, San Jose, CA, 1980.

160. Girvin, D.C., Hodgson, A. T., Tatro, M. E., and Anaclerio, R. N., Spatial and Seasonal Variations of Silver, Cadmium, Copper, Nickel, Lead, and Zinc in South San Francisco Bay Water During Two Consecutive Drought Years, Lawrence Berkeley Publ. #UCID 8008, University of California, Berkeley, 1978.

161. Kinkel, A. R., Jr., Hall, W. E., and Albers, J. P., Geology and Base-Metal Deposits of West Shasta Copper-Zinc District, Shasta County, California, Geol. Surv. Prof. Pap. 285, U.S. Government Printing Office, Washington, D.C., 1956.

162. Gunther, A. J., Davis, J. A., and Phillips, D. J. H., An Assessment of the Loading of Toxic Contaminants to the San Francisco Bay-Delta, Tech. Rep., Aquatic Habitat Institute, Richmond, CA, 1987.

163. Luoma, S. N., Cain, D., and Johansson, C., Temporal fluctuations of silver, copper, and zinc in the bivalve *Macoma balthica* at five stations in south San Francisco Bay, *Hydrobiologia,* 129, 109, 1985.

164. Smith, D. R., Stephenson, M. D., and Flegal, A. R., Trace metals in mussels transplanted to San Francisco Bay, *Environ. Toxicol. Chem.,* 5, 129, 1986.

165. Luoma, S. N., Dagovitz, R., and Axtmann, E., Trace elements in sediments and the bivalve *Corbicula* sp. from the Suisun Bay/San Joaquin Delta of San Francisco Bay, *Mar. Environ. Res.,* in press.

166. Bradford, W. L. and Luoma, S. N., Some perspectives on heavy metal concentrations in shellfish and sediment in San Francisco Bay, California, in *Contaminants and Sediments,* Vol. 2, Baker, R. A., Ed., Ann Arbor Science, Ann Arbor, MI, 1980, 501.

167. Katz, A. and Kaplan, I. R., Heavy metals behavior in coastal sediments of southern California: a critical review and synthesis, *Mar. Chem.,* 10, 261, 1981.

168. Luoma, S. N. and Cain, D. J., Fluctuations of copper, zinc, and silver in tellinid clams as related to freshwater discharge — south San Francisco Bay, in *San Francisco Bay: the Urbanized Estuary,* Conomos, T. J., Ed., Pacific Division of the American Association for the Advancement of Science, San Francisco, 1979, 231.

169. Cutter, G. A., Trace elements in estuarine and coastal waters, Rev. Geophys., (Supplement), Contrib. Oceanogr., 1991, 639.

6 Radioactivity

INTRODUCTION

Since the discovery of nuclear fission in 1939, large quantities of radioactive waste have accumulated throughout the world. Radioactive waste is produced whenever radioactive materials are employed, such as in nuclear weapons testing and in nuclear reactors for the generation of electricity. The waste exists in liquid, gaseous, or solid form and based on its origin and potential hazard, may be classified into four categories: (1) high-level waste; (2) transuranic waste; (3) low-level waste; and (4) uranium mill tailings.[1] In addition to differing in physical form, radioactive wastes can be distinguished by their chemical form and, consequently, their potential environmental impact, as well as the nature of the radiation they emit.[2]

The management of radioactive waste became a major concern during the 1960s and 1970s as countries began detonating more nuclear devices and commercialized operation of nuclear power plants escalated. Prior to 1970, the long-term management of these wastes in the U.S. was not given adequate consideration by either the scientific community or the federal government. The 1972 International Convention on the Prevention of Marine Pollution by Dumping of Wastes and Other Matter and the U.S. Marine Protection, Research, and Sanctuaries Act of 1972 (commonly called the Ocean Dumping Act) prohibited the dumping of high-level radioactive wastes in the sea.[3] Furthermore, the passage of the Low-Level Radioactive Waste Policy Act of 1980 and the Nuclear Waste Policy Act of 1982 provided a framework for resolving many of the management problems associated with low- and high-level radioactive waste. The Low-Level Radioactive Waste Policy Act established two major national policies: (1) each state must assure adequate disposal capacity for the low-level radioactive waste generated within its boundaries, except for the waste produced by federal defense or research and development; and (2) a regional grouping of states can be allied through interstate agreements called compacts to provide the required disposal facilities. (Congress must approve a compact ratified by a group of states and can rescind its consent to a compact every 5 years.) The Nuclear Waste Policy Act establishes the following: (1) a schedule for the siting, construction, and operation of high-level radioactive waste repositories; (2) the working and decision-making relationships between the federal government, state governments, and Indian tribes; (3) the federal policy and responsibility for nuclear waste management; and (4) a fund to cover the costs of nuclear waste disposal.[2] Despite these comprehensive radioactive waste legislations, many scientific and technical questions remain regarding the proper waste disposal design required to safeguard the public and the environment in future years. Marine radiobiology appears to be a science still in its infancy.[4]

This chapter investigates the sources, forms, and fates of radioactive wastes in estuarine and marine waters. It also details the effects of ionizing radiation on organisms inhabiting these environments. Finally, the radioactive waste management strategies employed in the marine realm are assessed, together with the engineering practices designed to ensure the safe disposal of radioactive materials in these sensitive areas.

RADIOACTIVITY AND RADIATION

RADIOACTIVITY DEFINED

The term "radioactivity" denotes the process in which a parent radionuclide undergoes spontaneous disintegration of its nucleus with the emission of one or more radiations and the formation of a daughter nuclide.[5] Regardless of the parent material, whether it be a natural or artificial radioactive substance, the instability of the nucleus results in the release of corpuscular or electromagnetic radiation. It is the propagation of energy through space by this ionizing radiation — α, β, and neutron particles and γ-rays — that poses a hazard to living organisms and, at high intensities, causes severe damage to biological tissue. Table 1 gives the basic characteristics of these radiation types.

Through the process of radioactive decay, radioactive substances gradually become stable. Radioactive decay may be expressed as follows:

$$\frac{dN}{dt} = -\lambda N \tag{1}$$

where N equals the number of radioactive atoms, t refers to time, and λ is the decay constant. dn/dt represents the derivative of the number of radioactive atoms with respect to time. The decay constant, λ, serves as a direct measure of radionuclide instability and indicates the fraction of N expected to decay per unit time. Equation 1 can be integrated and written as:

$$\int \frac{dN}{N} = -\lambda \int dt \tag{2}$$

$$\ell nN = -\lambda t + C \tag{3}$$

As recounted by Whicker and Schultz,[5] one can set $N = N_0$ at $t = 0$ to evaluate the constant of integrations, C

$$\ell n \, N_0 = -\lambda(0) + C \tag{4}$$

$$\ell n \, N_0 = C \tag{5}$$

and by substituting Equation 5 into Equation 3, one obtains the symbolic representation

$$\ell n \, N - \ell n \, N_0 = -\lambda t \tag{6}$$

which is equivalent to

$$N = N_0 \, e^{-\lambda t} \tag{7}$$

The half-life $(T_{1/2})$ of a radioactive substance relates to that value of t in which the number of radioactive atoms, N, or the disintegration rate, dN/dt, declines by a factor of 2. This is given by

TABLE 1

Characteristics of Common Nuclear Radiations

Radiation	Rest mass	Charge	Typical energy range	Path length (order of magnitude)		General comments
				Air	Solid	
α	4.00 amu	2+	4–10 MeV	5–10 cm	25–40 μm	Identical to ionized He nucleus
β (negatron)	5.48×10^{-4} amu 0.51 meV	–	0–4 MeV	0–1 m	0–1 cm	Identical to electron
Positron (β positive)	5.48×10^{-4} amu 0.51 meV	+	–	0–1 m	0–1 cm	Identical to electron except for charge
Neutron	1.0086 amu 939.55 MeV	0	0–15 MeV	0–100 m	0–100 cm	Free half-life: 16 min
γ (e.m. photon)	–	0	10 keV–3 MeV	0.1–10 m[a]	1 mm–1 m	Photons from nuclear transitions

[a] Exponential attenuation in the case of electromagnetic radiation.

From Eicholz, G. G., *Environmental Aspects of Nuclear Power*, Ann Arbor Science Publishers, Ann Arbor, MI, 1976. With permission.

$$N = \tfrac{1}{2} N_0 \qquad (8)$$

and

$$t = T_{\tfrac{1}{2}} \qquad (9)$$

Substituting into Equation 6 yields

$$\ell n \frac{N_0}{2} - \ell n\, N_0 = -\lambda T_{\tfrac{1}{2}} \qquad (10)$$

$$\ell n\, 2 = \lambda T_{\tfrac{1}{2}} \qquad (11)$$

Because the natural logarithm of 2 equals about 0.693,

$$T_{\tfrac{1}{2}} = \frac{0.693}{\lambda}, \qquad (12)$$

thereby establishing the relationship between the half-life of a radionuclide and the decay constant.

The half-life of a radionuclide sheds some light on its potential long-term hazard. For example, a small quantity of radionuclide with a half-life greater than several years will persist long enough to pose a long-term hazard to aquatic organisms exposed to it. In contrast, the same quantity of a radionuclide with a half-life of only minutes will pose much less of long-term hazard.[5]

It is customary to express the amount of a radioactive substance in terms of its mass or activity, the latter defined as the disintegrations per unit time of radioisotope per gram. The activity per unit weight or volume of a radioactive substance, known as specific activity, gives insight into its biological effects as well as its environmental and physiological behavior.[5] Specific activity is reported in such units as millicuries per gram, disintegrations per second per milligram, or counts per minute per milligram.[6]

COMMON TYPES OF IONIZING RADIATION
α Particle

Consisting of two protons and two neutrons, the α particle is essentially a helium nucleus devoid of orbiting electrons. The large, relatively slow moving, positively charged α particle is comparatively heavy and intensely ionizing in the tissue through which it passes. Although an α particle loses its energy in a short distance, being stopped by only a few centimeters of air or only 40 μm of tissue, it can cause more damage to tissues than ionizing particles having a longer path because larger quantities of energy are released on a very small amount of tissue, thereby leading to concentrated damage.[7] Consequently, an α particle poses a hazard to an organism when deposited internally.

β Particle

A more penetrating type of ionizing radiation (up to several centimeters) is the β particle, a small negatively or positively charged particle with the same mass as an electron (i.e., 5.49×10^{-4} amu). The negatively charged β, called the negatron, occurs more frequently than the positively charged one, termed the positron. However, radionuclides with a neutron/proton ratio less than optimum (e.g., ^{65}Zn) tend to emit positrons. Negatrons and positrons have identical properties except for electrical charge. As in the case of the α particle, β radiation raises risks to an organism when the nuclei emitting source material is taken into the body.

Neutron

A particle with much greater range than either α or β radiation is the neutron. Generated in the atmosphere via cosmic-ray interactions, the neutron particle, through a process of neutron activation, reacts with elements (e.g., Ni) to form other radionuclides (e.g., ^{14}C and tritium). It is also produced anthropogenically during fission reactions in nuclear reactors and at sites of nuclear detonations. Because of its long range, a neutron particle represents an external danger to the whole body of an organism. By the process of moderation, neutrons adversely affect living tissue; a high energy neutron penetrating biological material, upon colliding with a proton, can dislodge it from a molecule, whereby it essentially becomes an internal projectile. With sufficient energy, the proton can travel some distance through the tissue to cause secondary damage by ionization and excitation of molecules.[5] Although less common than α and β radiations, neutron particles are encountered in high quantities around the core of nuclear reactors and at sites of nuclear explosions.

γ-Rays

Characterized as highly penetrating electromagnetic energy waves, γ-rays are strongly ionizing to the tissue through which they pass. The emission of γ photons commonly accompanies the release of α and β particles during the decay of many radionuclides. β-Emitters, in particular, usually release energy in the form of γ-rays. Daughter nuclides typically achieve stability by emission of γ-rays, a process known as isomeric transition.

The overall damage of radiation to living tissue depends on the type of radiation and the dose, which is measured in rems or rads. It is contingent, therefore, upon the transmission of energy to the cell and its constituents and upon the number of cells struck by the radiations.[2] The next section focuses on the natural and anthropogenic sources of ionizing radiation that yield particles or rays of energy either by the spontaneous decay of a radionuclide or by the external stimulation of certain materials.[5] It also deals with the relative significance of natural vs. anthropogenic sources of radiation on aquatic organisms.

SOURCES OF RADIATION
Natural Sources
Cosmic Radiation

A continuous source of radiation enters the Earth's atmosphere from outer space and is termed cosmic or galactic radiation. Cosmic radiation largely exists in the form of high-energy protons that collide with atoms in the upper atmosphere (e.g., nitrogen, oxygen, and argon) to produce secondary particles, principally neutrons and protons, together with some pions, kaons, and electromagnetic radiations.[8] Primary cosmic radiation — that not yet interacting with matter in the Earth's atmosphere, lithosphere, or hydrosphere — mainly consists of protons (approximately 85%) and alpha particles (about 14%) with small amounts of heavier nuclei (<1%). Secondary cosmic radiation accounts for nearly all cosmic rays recorded at sea level; here muons, electrons, and protons are responsible for nearly 70, 30, and <1%, respectively, of the flux of cosmic rays.[5]

Cosmogenic radionuclides formed by the interaction of primary cosmic rays with matter in the atmosphere and on the surface of the Earth contribute to the total radiation dose from natural background, albeit in small quantities. Table 2 registers radionuclides generated

TABLE 2
Natural Radionuclides Produced by Cosmic Rays

Radionuclide	Half-life	Average atmospheric production rate (atoms/cm² s)	Tropospheric concentration (pCi/kg air)	Principal radiations and energies (MeV)	Observed average concentration in rainwater (pCi/l)
^{3}H	12.3 years	0.25	3.2×1.0^{-2}	β^- 0.0186	—
^{7}Be	53.6 d	8.1×10^{-3}	0.28	γ0.477	18.0
^{10}Be	1.5×10^6 years	3.6×10^{-2}	3.2×1.0^{-8}	β^- 0.555	—
^{14}C	5730 years	2.2	3.4	β^- 0.156	—
^{22}Na	2.6 years	5.6×10^{-5}	3.0×1.0^{-5}	β^+ 0.545, 1.28	7.6×10^{-3}
^{24}Na	15.0 h	—	—	β^- 1.4, 1.37; 2.75	0.08–0.16
^{32}Si	~650 years	1.6×10^{-4}	5.4×1.0^{-7}	β^- 0.210	—
^{32}P	14.3 d	8.1×10^{-4}	6.3×1.0^{-3}	β^- 1.71	"A few"
^{33}P	24.4 d	6.8×10^{-4}	3.4×1.0^{-3}	β^- 0.246	"A few"
^{35}S	88 d	1.4×10^{-3}	3.5×1.0^{-3}	β^- 0.167	0.2–2.9
^{36}Cl	3.1×10^5 years	1.1×10^{-3}	6.8×1.0^{-9}	β^- 0.714	—
^{38}S	2.87 h	—	—	β^- 1.1, 1.88	1.8–5.9
^{38}Cl	37.3 min	—	—	β^- 4.91, 1.60, 2.17	4.1–67.6
^{39}Cl	55.5 min	1.6×10^{-3}	—	β^- 1.91, 0.25, 1.27, 1.52	4.5–22.5

From Eisenbud, M., *Environmental Radioactivity*, 3rd ed., McGraw-Hill, New York, 1987. With permission.

by this process. Most of these nuclides occur in minute concentrations, the two exceptions being tritium (^{3}H) and ^{14}C. After forming in the atmosphere, tritium mixes with water on Earth as well as H_2 gas reservoirs, whereas ^{14}C enters the atmospheric CO_2 pool subsequent to combining with oxygen.[5] Cosmogenic radionuclides, in general, may settle to the Earth's surface with precipitation, either in solution or sorbed on particulates, or they may behave as a gas and enter into equilibrium reactions at the surface.[9]

Primordial Radionuclides

These radionuclides largely found in the Earth's crust were generated at the time of formation of the Earth. Detectable primordial radionuclides of significance include ^{40}K, ^{238}U, and ^{232}Th, all with a half-life in excess of 10^9 years. Other primordial, long-lived radionuclides are shown in Table 3. In estuaries and oceans, primordial radionuclides occur throughout the water column and in sediments, being derived from the weathering and erosion of rocks and leaching of soils. While most of the primordial radionuclides exist

TABLE 3
Singly Occurring Primordial Radionuclides

Radionuclide	Half-life (year)	Radiation
^{40}K	1.26×10^9	β,
^{50}V	6×10^{15}	β,
^{87}Rb	4.8×10^{10}	β
^{115}In	6×10^{14}	β
^{123}Te	1.2×10^{13}	EC[a]
^{138}La	1.1×10^{11}	β,
^{142}Ce	$>5 \times 10^{16}$	α
^{144}Nd	2.4×10^{15}	α
^{147}Sm	1.1×10^{11}	α
^{149}Sm	$>1 \times 10^{15}$	α
^{152}Gd	1.1×10^{14}	α
^{174}Hf	2×10^{15}	α
^{176}Lu	2.2×10^{10}	β,
^{180}Ta	$>1 \times 10^{12}$	β
^{187}Re	4.3×10^{10}	β
^{190}Pt	6.9×10^{11}	α

[a] Electron capture.

From Eisenbud, M., *Environmental Radioactivity,* 3rd ed., Academic Press, New York, 1987. With permission.

in the lithosphere, considerable spatial variation in their concentrations has been demonstrated. Among the primordial radionuclides, ^{40}K represents the main, naturally occurring source of internal radiation in organisms.

Anthropogenic Sources

The major sources of human-generated radioactivity in estuarine and marine environments over the past 40 years have been the nuclear fuel cycle and the detonation of nuclear explosives. Agricultural, industrial, medical, and scientific uses of radioisotopes also have contributed to the pool of artificial radionuclides. While the radioactivity from all natural radionuclides in seawater amounts to about 750 dpm/l (97% derived from ^{40}K), the oceanic input from human activity has varied significantly, owing to reductions in fallout from nuclear explosions after the second nuclear test ban treaty in 1963, the increase in the number of nuclear power plants on line after 1970, and accelerated usage of radionuclides in agriculture, industry, and medicine since 1975.

The ocean has been receiving artificial radionuclides since late 1944 when the Hanford atomic plant first discharged effluent into the Pacific Ocean via the Columbia River.[10] A year later, atomic blasts at Hiroshima and Nagasaki in Japan released radioactivity to the atmosphere. Between 1945 and 1980, more than 1200 nuclear weapons tests were conducted throughout the world. The global stock of nuclear power plants grew from 66 to 398 from 1970 to 1986, and these became a greater source of anthropogenic radioactivity in estuaries and oceans than nuclear detonations after the termination of atmospheric nuclear testing in 1980. The Chernobyl (U.S.S.R.) nuclear power plant disaster in April 1986 released about 50 million curies of radioactivity products to the atmosphere, far in excess of contributions from previous detonations of nuclear explosives in the atmosphere. Radioecological studies in estuarine and marine environments have been performed since 1946 when research was initiated on the impact of radionuclides on populations of invertebrates and fish at Bikini Atoll in the Pacific Ocean.

Nuclear Fuel Cycle

Nuclear power plants generate large amounts of radioactive waste. However, the actual operation of a nuclear power plant is only one stage in the typical nuclear fuel cycle which consists of uranium mining, milling, conversion, isotopic enrichment, fuel element fabrication, reactor operation, and fuel reprocessing. Some workers also enlist waste disposal as a component of the cycle. Weber and Wiltshire (p. 10)[2] describe the radioactive waste produced during the nuclear fuel cycle:

1. *Uranium mining:* Routine ventilation of mines results in the release of radon gas and uranium-bearing dust.

2. *Milling:* Uranium ore is crushed, ground, and chemically processed to produce a compound (U_3O_8) known as "yellow-cake". This operation releases a small amount of radon gas and uranium dust. After the refining process, the tailings are pumped in slurry form to a settling pond. The water gradually dissipates through seepage and evaporation, eventually leaving behind huge piles of relatively dry, finely ground tailings that contain radium (which decays into radon) and other long-lived radioisotopes, especially ^{230}Th.

3. *Conversion:* Yellowcake is converted to uranium hexafluoride (UF_6). Depending on the technique used, the process produces wastes that are either mostly solid or a sludge, with a small part discharged as gas. These wastes contain mainly radium and some uranium and thorium.

4. *Enrichment:* With the application of heat, UF_6 becomes a gas that permits the concentration (enrichment) of ^{235}U, the uranium isotope required for reactor fuel. In this process, small quantities of radioactive gas are vented directly into the atmosphere and some liquid waste from cleanup operations is diluted and discharged to the environment.

5. *Fuel fabrication:* Enriched UF_6 gas is converted chemically to solid uranium dioxide (UO_2), which is formed into ceramic pellets that are placed in zircalloy cladding to make fuel rods. These are bundled together into fuel assemblies containing 50 to 300 rods. The radioactive waste resulting from these operations include gases and liquid waste containing very small quantities of uranium and thorium.

6. *Power plant operation:* As the ^{235}U fuel in the nuclear reactor fissions and generates heat for electric power production, the fission fragments (products) accumulate and gradually reduce the efficiency of the chain reaction. After an average use of 3 to 4 years, spent fuel rods are then removed from the reactor.

Currently almost all spent fuel rods are being stored underwater in large pools at reactor sites. Other radioactive wastes generated at nuclear power plants include fission product gases, such as krypton and xenon; filter media left over from treating contaminated cooling and cleaning water; and miscellaneous solid wastes, such as protective clothing and cleaning paper.

7. *Reprocessing:* During this stage unconsumed uranium and plutonium are chemically separated from the fission products in the spent fuel so they can be used again. However, commercial spent fuel is not being reprocessed in the U.S. at the present time.

Figure 1 shows the basic elements and options within the nuclear fuel cycle and depicts releases of radioactivity to the environment.

Nuclear Explosions

In addition to the man-generated radioactivity produced in the nuclear fuel cycle, radioactive products generated by atmospheric nuclear detonations prior to 1980 accounted for substantial elevations of activity above background. Before the Limited Test Ban Treaty was signed by the U.S. and U.S.S.R. in 1963, most of the radioactive material released to the atmosphere by nuclear tests was attributable to these two countries (Table 4). All of the atmospheric nuclear tests ultimately resulted in the introduction of artificial radioactivity into estuaries and oceans.[9]

Biologically important constituents of fallout from atmospheric nuclear explosions have included ^{89}Sr and ^{90}Sr, which tend to concentrate in the bones and exoskeletons of aquatic organisms, ^{137}Cs, and [131]I. Other fission products with potential biological significance are given in Table 5. The bulk of the fallout from explosions has derived from the stratosphere extending from about 9000 to 18,000 m to about 50,000 m above sea level. By downward mixing and diffusion, fallout radionuclides enter the troposphere, mainly in the middle and upper latitudes in late winter and early spring in both the northern and southern hemispheres.[9] The residence time of radioactive debris in the stratosphere ranges from less than 1 year to more than 5 years. In general, the radioactivity injected at low latitudes will fallout more rapidly than that injected at high latitudes. Moreover, winter fallout proceeds more quickly than summer fallout.[5] As asserted by Joseph et al.,[9] the probable latitudinal distribution of the release of strato-

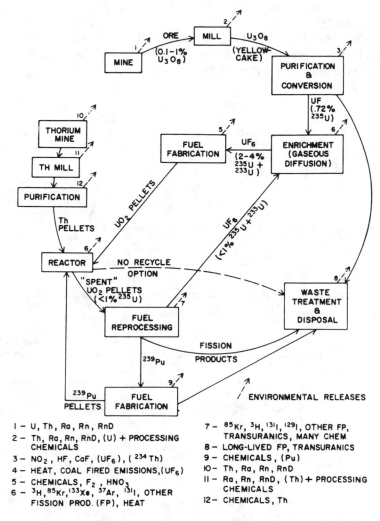

FIGURE 1. Basic elements and options within the nuclear fuel cycle, illustrating radioactive releases to the environment. (From Whicker, F. W. and Schultz, V., *Radioecology: Nuclear Energy and the Environment,* Vol. 1, CRC Press, Boca Raton, FL, 1982. With permission.)

spheric debris into the troposphere and the precipitation patterns on the ground have combined to yield a spring maximum in fallout. Nonetheless, at times fallout peaks in the autumn rather than the spring because of greater autumnal precipitation.[5] Highest deposition of fallout radionuclides has taken place in the mid latitudes; for

TABLE 4
Number of Nuclear Tests Conducted by Major Nuclear Powers for the 10-Year Period Following the Limited Nuclear Test Ban Treaty of 1963, and for the Period 1945–1973

Period	Environment	U.S.	U.S.S.R	U.K.	France	China	Total
1963-1973	Atmosphere	0	0	0	29	14	43
	Subsurface	261	121	2	9	1	394
	Total	261	121	2	38	15	437
1945-1973	Atmosphere	193	161	21	33	14	422
	Subsurface	372	124	4	13	1	514
	Total	565	285	25	46	15	963

From Barnaby, F., *IAEA Bull.*, 15(4), 13, 1973. With permission.

TABLE 5
Some Fission Product Radionuclides of Potential Biological Importance

Radionuclide	Fission yield (%)[a]	Radiation	Half-life	Important element analogs
^{3}H	0.01	β	12 years	H
^{85}Kr	0.29	β,γ	10 years	—
^{90}Sr	5.77	β	28 years	Ca
^{89}Sr	4.79	β	51 d	Ca
^{137}Cs	6.15	β,γ	27 years	K
^{131}I	3.1	β,γ	8.1 d	I
^{129}I	0.9	β,γ	1.7×10^{7} years	I
$^{144}Ce^{b}$	6.0	β,γ	285 d	—
$^{103}Ru^{b}$	3.0	β,γ	40 d	—
$^{106}Ru^{h}$	0.38	β,γ	1.0 years	—
$^{95}Zr^{b}$	6.2	β,γ	65 d	—
$^{140}Ba^{b}$	6.32	β,γ	12.8 d	Ca
^{91}Y	5.4	β,γ	58 d	—
$^{143}Ce^{b}$	5.7	β,γ	33 h	—
$^{147}Nd^{b}$	2.7	β,γ	11 d	—

[a] Based upon thermal neutron fission of ^{235}U.
[b] Decay to radioactive daughters.

From Whicker, F. W. and Schultz, V., *Radioecology: Nuclear Energy and the Environment*, Vol. 1, CRC Press, Boca Raton, FL, 1982. With permission.

example, the deposition of ^{90}Sr in the past has been highest between 40 and 50° in both hemispheres.[5] Table 6 lists average values of fallout radioactivity in the North Atlantic Ocean in the early 1970s. The longer-lived radionuclides from this source enable their detection in estuarine and marine systems.

Over the years, the principal emphasis in seawater analysis of fallout radionuclides has been on ^{90}Sr and ^{137}Cs. The focus on ^{90}Sr in biological as well as environmental samples is due to its unique combination of relatively long 28-year half-life, highly energetic β particle of its ^{90}Y daughter, and the resemblance of this fission product to calcium in metabolic processes.[11] For these reasons, a large database on ^{90}Sr distribution was developed between 1955 and 1980.

Sediment, plankton, and benthic algae rapidly adsorb fallout radionuclides as they enter an aquatic ecosystem.[5] The distribution and abundance of these system components, therefore, determine the fate of a significant fraction of fallout radionuclides in estuarine and marine waters. Since plankton populations, in particular, support primary consumers in these waters, they are an important pathway for the movement of radionuclides to upper trophic level organisms in food chains. Furthermore, physical processes, such as water currents, assist in the redistribution of the radionuclides in these environments. Hence, both biotic and abiotic factors play critical roles in the distribution of fallout radioactivity within aquatic ecosystems.

TABLE 6
Average and Range of Radioactivity in the North Atlantic Ocean in the Early 1970s Due to Radioactive Fallout from Nuclear Weapons Tests

Radioisotope	Radioactivity (pCi/l)	
Tritium (^{3}H)	48	(31–74)
^{137}Cs	0.21	(0.03–0.80)
^{90}Sr	0.13	(0.02–0.50)
^{14}C	0.02	(0.01–0.04)
^{239}Pu		(0.0003–0.0012)

From Clark, R. B., *Marine Pollution,* 2nd ed., Clarendon Press, Oxford, 1989. With permission.

TYPES OF RADIOACTIVE WASTES

The U.S. Interagency Review Group on Nuclear Waste Management[12,13] published the following definitions for the various categories of radioactive waste:

1. High-level wastes. These wastes are either fuel assemblies that are discarded after having served their useful life in a nuclear reactor (spent fuel) or the portion of the wastes generated in the reprocessing that contain virtually all of the fission products and most of the actinides not separated out during reprocessing. The wastes are being considered for disposal in geologic repositories or by other technical options designed to provide long-term isolation of the wastes from the biosphere.

2. Transuranic wastes. These wastes are produced primarily from the reprocessing of defense spent reactor fuels, the fabrication of plutonium to produce nuclear weapons, and, if it should occur, plutonium fuel fabrication for use in nuclear power reactors. Transuranic wastes contain low levels of radioactivity but varying amounts of long-lived elements above uranium in the periodic table of elements, mainly plutonium. They are currently defined as materials containing >370 Bq g^{-1} of transuranic activity.

3. Low-level wastes. These wastes contain <370 Bq g^{-1} of transuranic contaminants. Although low-level wastes require little or no shielding, they have low, but potentially hazardous concentrations of radionuclides, and consequently require management. Low-level wastes are generated in almost all activities involving radioactive materials and are presently being disposed of by shallow land burial.

4. Uranium mine and mill tailings. These wastes are the residues from uranium mining and milling operations. They are hazardous because they contain low concentrations of radioactive materials which, although naturally occurring, contain long-lived radionuclides. The tailings, with a consistency similar to sand, are generated in large volumes, about 10^{10} kg year^{-1} in the U.S., and are presently stored in waste piles at the site of mining and milling operations. A program is underway either to immobilize or bury uranium mine and mill tailings to prevent them from being dispersed by wind or water erosion.

5. Decontamination and decommissioning wastes. As defense and civilian reactors and other nuclear facilities reach the end of their productive lifetimes, parts of them will have to be handled as either high- or low-level wastes, and disposed of accordingly. Decontamination and decommissioning activities will generate significant quantities of wastes in the future.

6. Gaseous effluents. These wastes are produced in many defense and commercial nuclear activities, such as reactors, fuel fabrication facilities, uranium enrichment plants, and weapons manufacturing facilities. They are released into the atmosphere in a controlled manner after passing through successive stages of filtration and mixing with air where they are diluted and dispersed.

While low- and intermediate-level radioactive wastes are still dumped at sea, high-level wastes are not, owing to the agreement of the 1972 London Dumping Convention which regulates all dumping of radioactive wastes in marine waters. The International Atomic Energy Agency[14] drafted an operational definition of high-level radioactive wastes:

> "For the purpose of Annex I to the [London Dumping] Convention, high-level radioactive wastes or other high-level radioactive matter unsuitable for dumping at sea means any waste or other matter with an activity per unit gross mass exceeding: (a) 3.7×10^{10} Bg t^{-1} for α-emitters but limited to 3.7×10^{9} Bg t^{-1} for ^{226}Ra and supported ^{210}Po; (b) 3.7×10^{12} Bg t^{-1} for β- and α-emitters of unknown half-lives; and (c) 3.7×10^{16} Bg t^{-1} for ^{3}H and β- and α-emitters with half-lives of less than 0.5 y. The above activity concentrations shall be averaged over a gross mass not exceeding 1000 t."

The disposal of low-level wastes at sea has been accomplished by packaging the products in steel drums, encasing them in concrete, and disposing the contents at approved sites. Although the contents are expected to corrode and leach through the containment at some time in the future, the delay should provide enough time to ensure the loss of radioactivity via radionuclide decay. The slow release of the pollutants should also result in great dilution, thereby minimizing impact on the marine environment.[7]

RADIOACTIVITY AND ESTUARINE ORGANISMS

GENERAL

One of the principal concerns regarding radioactivity in estuarine and marine environments is the uptake of the pollutants by organisms and their potential biomagnification through food chains to man. As a consequence, organisms of direct dietary importance to man (e.g., crabs, clams, mussels, oysters, and fish) have been monitored continuously for radionuclides in many regions of the world. However, the ingestion of seafood does not represent the major source of mankind's exposure to radioactivity. Most of the annual average doses of radiation originates from natural radiation sources on land, with inhalation being the most important pathway followed by external irradiation and ingestion of radioactive substances.[15] The inhalation of radon, a radioactive noble gas present in relatively high concentrations in indoor air in certain regions, is responsible for a large fraction of the effective dose equivalent from inhalation.[16] Coal-fired power plants emit an additional airborne dose of radon, and so some people receive an even greater dose because they inhabit areas in close proximity to coal-fueled industries.

People living in the U.S. receive only about 0.1 to 0.2% of anthropogenic radioactivity from seafood.[17] This is so because of the oceanic processes of physical dilution, isotopic dilution, and chemical competition which limit source concentrations and because seafood only comprises approximately 1% of the U.S. diet.[16,18] Hence, the ocean effectively removes radionuclides from pathways to man.[16]

Many estuarine and marine organisms concentrate ^{210}Po.[19] Naturally occurring ^{210}Po represents the major radiation dose to these organisms,[20] and it ultimately returns to man via seafood.[16,20] Natural levels of ^{210}Po in surface seawater range from 0.6 to 4.2×10^{-2} pCi/l, which are greater than the concentrations of thorium nuclides, but less than those of ^{40}K, tritium, isotopes of uranium, and a number of other radionuclides (Table 7). The total background radioactivity in surface seawater ($\cong$340 pCi/l) is less than that in marine muds (20,000 to 30,000 pCi/kg) and sands (5,000 to 10,000 pCi/kg). Since

TABLE 7
Natural Levels of Radioactivity in Surface Seawater

Radionuclide	Concentration (pCi/l)
^{40}K	320
Tritium (^{3}H)	0.6–3.0
^{87}Rb	2.9
^{234}U	1.3
^{238}U	1.2
^{14}C	0.2
^{228}Ra	$(0.1–10.0) \times 10^{-2}$
^{210}Pb	$(1.0–6.8) \times 10^{-2}$
^{235}U	5×10^{-2}
^{226}Ra	$(4.0–4.5) \times 10^{-2}$
^{210}Po	$(0.6–4.2) \times 10^{-2}$
^{222}Rn	2×10^{-2}
^{228}Th	$(0.2–3.1) \times 10^{-3}$
^{230}Th	$(0.6–14.0) \times 10^{-4}$
^{232}Th	$(0.1–7.8) \times 10^{-4}$

From Clark, R. B., *Marine Pollution,* 2nd ed., Clarendon Press, Oxford, 1989. With permission.

fine sediments, with their large surface area, adsorb more radioactivity than coarse sediments and have higher concentrations of organic matter and benthic populations, they may provide a more important pathway for the movement of radionuclides, particularly heavier species, through food chains.

Differences in background radiation dose in estuarine organisms derive, in part, from their position in an estuary. For instance, cosmic radiation supplies a dose rate of about 4×10^{-8} Sv/h at the water surface, which declines to approximately 5×10^{-9} Sv/h at 20 cm depth. The cosmic dose rate drops even further at greater depths, being negligible below 100 m.[21] The background dose for phytoplankton, zooplankton, and pelagic fish principally results from ^{210}Po and a few other α-emitting isotopes.[1,20] While the benthos may be spared the effects of dose rate from cosmic radiation, radiation from seafloor sediments can be an even more significant source as noted above.

Once radioactive substances enter the estuarine environment, various biological, chemical, and physical processes act to concen-

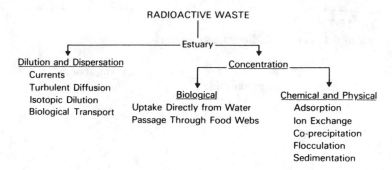

FIGURE 2.   Processes tending to concentrate, disperse, and dilute radioactive materials in estuaries. (From Langford, T. E., *Electricity Generation and the Ecology of Natural Waters,* Liverpool University Press, Liverpool, U.K., 1983. With permission.)

trate, disperse, and dilute it (Figure 2). Whicker and Schultz[5] encapsulate some of these processes. Upon entering an estuary, radionuclides are subject to turbulent and molecular diffusion, causing dispersion which lowers their concentration away from an effluent source. In time, gravitational settling, precipitation scavenging, impaction, and chemical adsorption or exchange gradually promote the deposition of the radionuclides. Radioactive particles tend to attach to particulate materials, with the strength of the attachment contingent upon various kinds of physical and chemical forces related to the properties of the nuclide and the particulate surfaces. The interaction of radionuclides with particles in estuaries and oceans has been described most succinctly in terms of simple ion exchange or adsorption equilibria.[22] While suspended sediment (e.g., clay minerals) and organic detritus create ideal surfaces for the sorption of the pollutants (e.g., ^{137}Cs), radioactive particles also attach to live organisms such as bacterioplankton, phytoplankton, and zooplankton, if they come in close proximity. Particulates generally accumulate radionuclides to equilibrium concentrations much greater than in the water medium.

The seafloor is the ultimate repository for most radionuclides in estuarine systems.[23-25] Once deposited, the nuclides can be resuspended along with sediments by turbulence from bottom currents or storms which roil grains. Furthermore, the bioturbating activity of benthic fauna displaces the sediment and alters the erodibility of the substrate, thereby enhancing particle resuspension. Because bottom sediments serve as the major reservoir of radioactive materials, they are extremely important to estuarine contamination. As radionuclides migrate downward from surface sediment layers, they eventually become isolated from contact with biological processes. At this point, the estuary effectively becomes a sink for the removal of the substances from biotic compartments in the system.

Radionuclides enter biotic compartments of estuaries by organismal uptake from water, sediments, or other organisms via ingestion. The uptake of radionuclides by plants occurs directly from the water medium as well as from sediments. Some of these contaminants may even be assimilated from surface-deposited material that enters plant tissue through stomates or epidermal tissues.

In grazing food webs, herbivores assimilate radioactive material largely from consumption of primary producers. One direct pathway of radionuclides to man involves the ingestion of marine algae containing natural or synthetic radioactivity. Studies of marine algae hold promise in assessing transport pathways of radioactive-waste material in aquatic systems.[26-29]

Because of the accumulation of radionuclides in bottom sediments and associated organic detritus, the manipulation and processing of radionuclide-sorbed organic detritus by detritus feeders is a key linkage to the recycling of radioactive material through biotic compartments. The radionuclides directly attach to the detritus from air or water via deposition and sorption phenomena, and subsequent mineralization of the detritus by microbes releases the radioactive substances to sediments or interstitial waters. Alternatively, radionuclides enter the organic detritus reservoir after the death of plants and animals that sequestered radionuclides in their tissues when alive. The solubilization or resuspension of the nuclides in the detritus enables them to reenter estuarine food chains or to remobilize to other areas of the system. Detritus food chains represent a source of radionuclides for grazing food chains and vice versa.

Carnivores accumulate radionuclides by eating herbivores and other prey that store radionuclides. They also receive radionuclides from environmental compartments such as water and sediments. The propensity of these organisms to accumulate radioactive materials from the environment is related to their behavioral patterns, physical attributes, and other inherent characteristics. The external morphology of the animals, for example, influences surface adsorp-

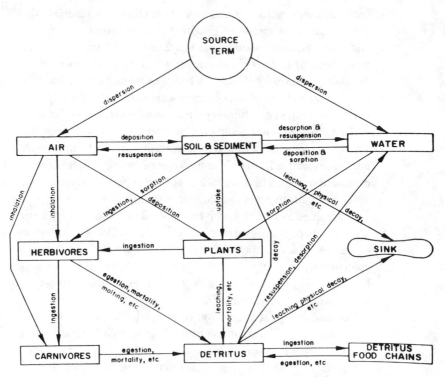

FIGURE 3.   General transport processes operating on radionuclides in ecosystems. Boxes
define ecosystem components, and arrows depict the flow of the materials through functional
processes. (From Whicker, F. W. and Schultz, V., *Radioecology: Nuclear Energy and the
Environment,* Vol. 1, CRC Press, Boca Raton, FL, 1982. With permission.)

tion of radionuclides, with smaller individuals generally adsorbing
more of the substances per unit mass than larger forms. In addition,
benthic feeders coming in contact with contaminated sediments
might be expected to have higher radionuclide concentrations than
pelagic feeders inhabiting pristine waters. Finally, the organism's
physiology and metabolism greatly affects the intake, assimilation,
tissue distribution, and retention by the carnivores.

Figure 3 illustrates the general transport processes acting on
radionuclides released to the environment and shows the movement
of the material through various biotic and abiotic compartments. The
arrows depict the flow of the pollutants among the components of the
system. The quantitative importance of competing transport processes determines the concentrations of radionuclides accumulating
in each compartment.

Impact of Radioactivity on Estuarine Organisms
Bioaccumulation

The accumulation of radioactive substances in estuarine organisms is of direct interest to man because of the transference of the substances through food chains and their potential radiation hazard.[30] Thus, it is essential to monitor the concentrations of the contaminants in order to properly manage the estuarine ecosystem.[31] Although radionuclides behave chemically in the same manner as naturally occurring, nonradioactive isotopes, fear of bioaccumulation and biomagnification in food chains has precipitated numerous investigations of the accumulation of radionuclides by estuarine and marine organisms.[7,32-34]

Many estuarine organisms quickly accumulate radionuclides. Most conspicuous in this regard are smaller organisms (e.g., phytoplankton and zooplankton) which have large surface area to volume ratios. The degree of uptake by larger biota (e.g., macroalgae, macroinvertebrates, and fish), while usually much less than that of smaller organisms, nevertheless can be significant in those forms that ineffectively regulate the radionuclides in their tissues.[32]

For large numbers of marine species, the uptake of radionuclides is proportional, or nearly proportional, to the elemental concentration in the seawater medium.[32] Small plankton and macroalgae exemplify this phenomenon. Elemental concentration factors among classes of marine organisms vary widely (Table 8), due largely to species differences. When the radionuclide concentration is assessed along food chains, little evidence exists for biomagnification at high trophic levels, although exceptions can be found, such as ^{137}Cs, which preferentially accumulates in higher trophic level fish.[35] Bioaccumulation of radionuclides by estuarine and marine organisms appears to be manifested most clearly in areas near nuclear fuel processing plants and industries producing nuclear weapons.[36,37]

Effects of Radiation on Organisms
Somatic Effects

Many studies have been conducted on estuarine and marine organisms to ascertain the effects of ionizing radiation on individuals. A component of these studies has dealt with estimates of dose rates from natural background sources (Table 9) in an effort to

TABLE 8
Concentration Factors (C.F.) for Different Classes of Marine Organisms

Element	Group	C.F. range	Mean C.F.
Cs	Plants	17–240	51
	Mollusca	3–28	15
	Crustacea	0.5–26	18
	Fish	5–244	48
Sr	Plants	0.2–82	21
	Mollusca	0.1–10	1.7
	Crustacea	0.1–1.1	0.6
	Fish	0.1–1.5	0.43
Mn	Plants	2,000–20,000	5,230
	Mollusca	170–150,000	22,080
	Crustacea	600–7,500	2,270
	Fish	35–1,800	363
Co	Plants	60–1,400	553
	Mollusca	1–210	166
	Crustacea	300–4,000	1,700
	Fish	20–5,000	650
Zn	Plants	80–2,500	900
	Mollusca	2,100–33,0000	47,000
	Crustacea	1,700–15,000	5,300
	Fish	280–15,500	3,400
Fe	Plants	300–6,000	2,260
	Mollusca	1,000–13,000	7,600
	Crustacea	1,000–4,000	2,000
	Fish	600–3,000	1,800
I	Plants	30–6,800	1.065
	Mollusca	20–20,000	5,010
	Crustacea	20–48	31
	Fish	3–15	10
Ce	Plants	120–4,500	1,610
	Mollusca	100–350	240
	Crustacea	5–220	88
	Fish	0.3–538	99
K	Plants	4–31	13
	Mollusca	3.5–10	8
	Crustacea	8–19	12
	Fish	6.7–34	16
Ca	Plants	1.8–31	10
	Mollusca	0.2–112	16.5
	Crustacea	0.5–250	40
	Fish	0.5–7.6	1.9

TABLE 8 (CONTINUED)
Concentration Factors (C.F.) for Different Classes of Marine Organisms

Element	Group	C. F. range	Mean C.F.
Cu	Plants	—	1,000
	Mollusca	—	286
	Fish	0.1–5	2.55
Mo	Plants	12–42	23
	Mollusca	11–27	17
	Crustacea	8.9–17.3	13
	Fish	7.6–23.8	17
Mn	Plants	15–2,000	448
	Mollusca	1–3.6	2.2
	Crustacea	1–100	38
	Fish	0.4–26	6.6
Ar–Nb	Plants	170–2,900	1,119
	Mollusca	8–165	81
	Crustacea	1–100	51
	Fish	0.05–247	86

From Eisenbud, M., *Environmental Radioactivity,* 3rd ed., Academic Press, New York, 1987. With permission.

yield baseline data to compare dose-rate increments and effects contributed by anthropogenic activities.[38] In general, higher organisms are much less tolerant than primitive organisms (Figure 4); sensitivity to ionizing radiation among categories of aquatic organisms is as follows: bacteria < protozoa < algae < Mollusca < Crustacea < fishes < mammals.[1,39] The lethal doses of ionizing radiation on these organisms vary by nearly three orders of magnitude.[40] Even within groups of similar organisms, the sensitivity to ionizing radiation differs substantially among some species. For example, the $LD_{50/40}$ (the lethal dose killing 50% of the organisms within 40 d) value for the blue crab, *Callinectes sapidus,* is two orders of magnitude lower than that of the grass shrimp, *Palaemonetes pugio.*[40] Mortality from irradiation not only depends on the intensity and period of exposure, but also on other factors, including temperature.[37] Hence, sensitivity to ionizing radiation rises with increasing temperature.[40]

A single large dose of radiation may culminate in death immediately subsequent to irradiation or some time after the event. Death can ensue days, weeks, or even months after irradiation.[41] In addition

TABLE 9
Estimates of Annual Doses (mrad/year) Received by Marine Organisms from Natural Sources of Radiation

Source	Taxonomic group	Marine (20 m depth)
Cosmic	Phytoplankton	4.4
	Zooplankton	4.4
	Mollusca	4.4
	Crustacea	4.4
	Fish	4.4
Water	Phytoplankton	3.5
	Zooplankton	1.8
	Mollusca	0.9
	Crustacea	0.9
	Fish	0.9
Sediment (β^+)	Phytoplankton	0
	Zooplankton	0
	Mollusca	27–324
	Crustacea	27–324
	Fish	0–324
Internal	Phytoplankton	17–64
	Zooplankton	23–138
	Mollusca	65–131
	Crustacea	69–188
	Fish	24–37
Sum of natural sources	Phytoplankton	25–72
	Zooplankton	29–168
	Mollusca	97–460
	Crustacea	101–517
	Fish	29–366

From Whicker, F. W. and Schultz, V., *Radioecology: Nuclear Energy and the Environment,* Vol. 1, CRC Press, Boca Raton, FL, 1982. With permission.

to the possible delay in organismal damage from acute lethal doses of radioactivity, sublethal genetic damage cannot be detected until the next or a later generation.[7]

While acute and chronic effects of radiation exposure have been repeatedly demonstrated in the laboratory, only one episode of radiation mortality of aquatic organisms has been confirmed in the natural environment. The exposure of carnivorous fish in the Marshall Islands to ^{131}I from fallout during nuclear weapons testing ultimately caused thyroid damage and death.[37] Present radiation

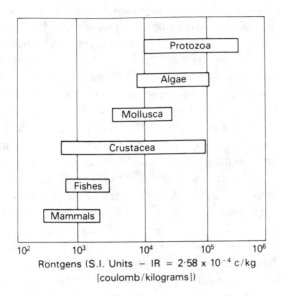

FIGURE 4. LD$_{50}$ ranges for various groups of organisms irradiated by X- or γ-rays. (From Langford, T. E., *Electricity Generation and the Ecology of Natural Waters,* Liverpool University Press, Liverpool, U.K., 1983. With permission.)

conditions in the sea have not resulted in any measurable environmental impact on estuarine and marine organisms or their habitats.[7]

It is difficult to extrapolate findings of acute radiation exposure to low-level chronic exposures. The total dose required for injury and death of an organism, compared to an acute single large dose, increases if the organism is irradiated continuously at low dose rates.[42] Chronic exposures do not necessarily lead to ill effects on the biota because tissue repair processes may be adequate. In essence, competition takes place between injury to the organismal tissue and repair at low dose rates.[1] The probability of injury, as expected, rises at higher doses. According to the International Atomic Energy Agency,[21] therefore, experimental research to evaluate the possible effects of irradiation in contaminated environments should be conducted by employing the lowest dose rates practicable to minimize the extent of extrapolation of effects from high to low dose rates.

Probably the most in-depth investigations of the somatic effects of irradiation have targeted commercially exploited populations, particularly finfish. The lethal radiation dose for fishes ranges from about 10 to 50 Gy, with embryos, juveniles, and preadults most susceptible, presumably because of the higher metabolic and proliferative state of their tissues.[43] The LD_{50} for eggs, larvae, and juvenile fish is less than that for adults.[44] Damage both to cell-renewal systems and to the CNS accounts for most radiation injuries responsible for the death of fishes. Woodhead and Pond (p. 159)[43] expound on this subject by stating that, "The causes of acute radiation death are damage to: (1) the central nervous system at 88 Gy or more; (2) the gastrointestinal tract at 35 to 280 Gy; and (3) the hematopoietic system at 8.8 to 18 Gy. Latent effects, such as shortened lifespan, cancer induction, and reduced growth may occur at 8.8 Gy or less. In contrast to the volume of information gathered on acute radiation injury in fishes, less data have been collected on chronic low-level radiation effects." Most of the studies of chronic radiation exposure of fish have been unable to uncover detrimental effects either on individuals or populations. Moreover, fish appear to be capable of repairing and recovering from radiation damage. With typically lower concentrations of radionuclides in the water column than in the seafloor sediments, pelagic fishes usually are exposed to less radiation dose than benthic species.[7]

Genetic Effects

Susceptible marine organisms may be expected to experience genetic disturbances as a consequence of radiation exposure. Genetic research over the years on marine organisms has attempted to ascertain mutations and chromosomal aberrations subsequent to acute and chronic exposures to radiation. Fishes have long been utilized in basic genetics research, being a focal point of modeling for more than 50 years.[43] These organisms exhibit a similar spectrum of chromosomal damage after radiation exposure as mammals.[45] Low-level radiation potentially can damage DNA in fish as well as invertebrates, causing mutation; there is much less information available on the genetic effects of exposure than on the somatic effects.[43] Templeton[46] gives an overview of the genetic effects of ionizing radiation on aquatic populations.

RADIOACTIVE WASTE DISPOSAL

GENERAL

Between 1946 and 1970, many countries, including the U.S., used the oceans for the dumping of low-level radioactive wastes. In addition to the packaging of these wastes in steel drums and their dumping at sea, low-level wastes were buried in shallow trenches at sites owned and operated by the U.S. Government. The last disposal of radioactive waste at sea by the U.S. occurred in 1970, but by that time, approximately 107,000 containers of waste (4.3×10^{15} Bq) had been dumped at U.S. Atomic Energy Commission-licensed sites in the northwestern Atlantic and the Pacific Oceans. Nearly all of the waste (95 to 98%) was dumped at four sites. European countries disposed of an additional 1×10^5 metric tons of radioactive material (3×10^{16} Bq) from 1949 to 1980, mainly in the northeast Atlantic.[47]

In regard to the release of unpackaged, low-level radioactive wastes from nuclear power plants, discharges are extremely low with little activity that contributes a very small amount to the total radioactivity introduced into the marine environment. The principal radionuclide input into shallow coastal waters has been attributed to nuclear fuel reprocessing facilities, such as at Windscale and Downreay in England and La Hague in France. The annual discharge of α emitters in liquid effluents from these nuclear fuel reprocessing facilities from 1972 to 1977 ranged from about 1.1×10^{11} to 1.7×10^{14} Bq, and that of β and γ emitters approximately 2.0×10^{13} to 7.7×10^{15} Bq.[47] Much of these effluents passed through the North Sea into the Norwegian Sea in surface currents with a minor component carried directly into the North Atlantic.[47]

The London Dumping Convention, which entered into force on 30 August 1975, prohibits the dumping of high-level radioactive wastes in the ocean. At present, no single solution to high-level radioactive waste disposal has been proffered by national or international agencies. The favored planning strategy adopted for the disposal of these wastes is the use of mined geologic repositories.[1,48] Alternatively, the utilization of subseabed disposal has been gaining favor.[49]

Until proven to be technically feasible and environmentally acceptable, however, these disposal alternatives will not be implemented
and high-level nuclear wastes will continue to be stored in temporary
surface storage facilities. For example, defense wastes are currently
stored in double-walled steel tanks on federal reservations, while
spent fuel assemblies from commercial nuclear power plants have
been placed in pools of water at the plant sites.[50] In contrast to the
dilemma involving high-level radioactive waste, the permanent
disposal of low-level radioactive waste in the U.S. now exists in
shallow land-burial facilities — commercial and government-
owned.[2]

STORAGE OR DISPOSAL SYSTEMS

Radioactive waste disposal can be interim (retrievable) or permanent, depending on the planning strategy. Interim storage may be
opted for when (1) the radionuclides comprising the waste are
sufficiently short-lived as to make storage practical; (2) no decision
has been reached regarding the ultimate method of disposal; and (3)
a later safe use of the material is contemplated.[8] Retrievable storage
alternatives include tank storage, near-surface storage vaults, water-
filled canals, air-cooled vaults, sealed-cask storage, and tunnel
storage. Permanent disposal systems, aside from the most highly
recommended alternatives of deep crustal burial on land and subseabed placement in the oceans, entail: (1) ice sheet disposal, whereby
the radioactive waste would be embedded in continental ice sheets
(e.g., in Antarctica or Greenland); and (2) extraterrestrial disposal,
whereby the radioactive waste would be jettisoned from the Earth
entirely after being packaged in special flight containers and placed
onboard space shuttles destined for the sun, for Earth or solar orbit,
or for solar system escape.

Among the permanent underground disposal options are the
following: (1) deep-well waste disposal involving the injection of
liquid chemical wastes into wells to depths of 1000 to 5000 m into
strata sufficiently isolated from the biosphere by impermeable
overburden; (2) deep-hole waste disposal characterized by the
deposit of nuclear waste in holes drilled deep into geologic formations up to 10,000 m underground; (3) mine storage in bedrock

consisting of the construction of deep-seated, waste storage vaults at depths of 600 m or more underground; (4) bedded salt formations in which essentially impermeable domed salt deposits serve as nuclear waste repositories; (5) rock melt waste disposal, whereby natural heat produced during radioactive decay of liquid or slurry waste would melt crystalline rock, enabling the waste to slowly descend under gravity until it resolidifies in a relatively insoluble matrix deep underground, thereby trapping the radioactive material; and (6) hydraulic fracturing of host rock, such as impermeable shale, followed by the injection of a radioactive waste solution with a cement base mix into the 300-m deep formation which quickly sets and fixes the radionuclides permanently in the fractured rock. Geologic disposal must consider the adaption of multiple barriers, either natural geologic or engineered, to isolate and contain nuclear waste for at least 10,000 years. The selection of a suitable permanent land repository for the waste is dependent upon the host rock itself and the geologic surroundings. Eichholz (p. 611)[8] itemizes the geologic criteria that must be met before a prospective disposal site can be chosen:

1. Burial site devoid of surface water and stable geomorphically
2. Groundwater flow paths that do not lead to surface flow
3. Predicted residence time of radionuclides in the order of hundreds of years (hydrologic system must be simple enough to make possible reliable residence-time predictions)
4. The highest water table several meters below the burial zone

He also lists the basic data required for site evaluation:

1. Depth to water table
2. Location and distance to points of water use
3. Minimum of 2 years precipitation and land pan evaporation records
4. Water table contour map for different seasons of the year
5. Magnitude of annual water table fluctuations
6. Detailed stratigraphic and structural data to base of shallowest confining aquifer
7. Base-flow data on nearby perennial streams

8. Chemistry of water in aquifer, confining beds, and of leachate from burial trenches
9. Laboratory measurements of porosity, permeability, mineralogy, and ion exchange capacity of each lithology in saturated and unsaturated zones
10. A record of at least 2 years of moisture content and *in situ* soil moisture-tension in the upper 10 to 15 m of unsaturated zone at burial site
11. Three-dimensional distribution of heat to base of shallowest confining aquifer
12. Field test determination of storage coefficient and transmissivity
13. Definition of recharge and discharge areas
14. Field measurements of dispersion coefficient
15. Laboratory and field determination of the distribution coefficient
16. Rates of denudation and slope retreat

Apart from information on the hydrologic regime, other data needed to accurately assess potential disposal sites are the physical, chemical, and thermal properties of the host rock; the tectonic stability of the repository area and region; the depth of the planned repository below the land surface (assumed to range from 600 to 1000 m); and the natural multibarrier safety features provided by the location.[50]

The U.S. Department of Energy (DOE) in February 1983 identified nine sites in five geohydrologic settings as possibly being acceptable for a mined geologic repository for spent nuclear fuel and high-level radioactive waste.[2] In December 1984, the DOE proposed the nomination of five sites for the repository, i.e., sites at Hanford, Washington, Yucca Mountain, Nevada, Deaf Smith County, Texas, Davis Canyon, Utah, and Richton Dome, Mississippi. The DOE also searched for a second high-level waste repository, concentrating on granite formations in 17 eastern and midwestern states: Connecticut, Georgia, Maine, Maryland, Massachusetts, Michigan, Minnesota, New Hampshire, New Jersey, New York, North Carolina, Pennsylvania, Rhode Island, South Carolina, Vermont, Virginia, and Wisconsin.

While controversy has raged over the selection of primary and secondary repository sites on land, subseabed disposal of radioactive wastes presents an attractive alternative. Advantages of the disposal of these wastes at subseabed sites in mid-gyre, mid-plate regions are their seismic and tectonic stability, ability of thick layers of fine sediments to retain the wastes if displaced, and the availability of large volumes of water for dilution of any wastes that escape through the seafloor. Major disadvantages of the subseabed concept include the need for special port facilities, considerable transportation, and precise placement of the wastes. Furthermore, any accidents at sea during transport may result in substantial biological accumulation of radionuclides in biota that can pose a hazard to man. Finally, international political uncertainties and lack of global cooperation threaten to offset the technically attractive features associated with subseabed disposal.[8,50]

WASTE MANAGEMENT STRATEGIES

The safe management of radioactive wastes generally incorporates three basic strategies: dilution and dispersion, isolation and containment, and isotopic dilution with stable isotopes. In the case of estuaries, which most commonly receive artificial radionuclides in low-level wastes discharged in liquid effluents from nuclear power plants, the preferred management option is one of dilution and dispersion. The release of some radioactive materials into estuarine and marine environments appears to be the inevitable consequence of the use of nuclear power, but it is absolutely necessary to strictly regulate the amount of the artificial radioactivity discharged to these environments. In fact, the disposal of certain categories of radioactive wastes in power plant effluents offers some advantages over any storage approach. Preston (p. 108)[51] explains why disposal rather than storage may be the best option in the management of radioactive waste.

> "...Storage, however, is expensive, and it may not in the long run offer any better chance than does prompt disposal in optimizing the choice between cost of protection and the value of the radiation detriment thus avoided. Storage is thus, except in a very few cases, no substitute for disposal, and disposal should be the preferred management option as soon as it offers a reasonable chance of optimizing with respect to the resulting radiation detriment."

The packaging and dumping of radioactive wastes in the deep ocean, a once routine practice, exemplifies two waste management strategies. For instance, the use of 200-l oil drums to contain the waste and their placement in a deep-sea environment represents a practice of isolation and containment. If the radioactive waste releases from the drums on the seafloor, it will be subject to dilution and dispersion. Today, dumping at sea must be in compliance with all articles of the London Dumping Convention designed to preclude harm to the marine environment.[52]

Isolation and containment, practices deemed to be critical in the handling of high-level radioactive wastes, may not necessarily be adapted as a strategy in managing certain types of radioactive wastes, such as the low-level effluents discharged by nuclear power plants, as mentioned above. In June 1990, the U.S. Nuclear Regulatory Commission (NRC) even approved a policy allowing low-level radioactive waste (e.g., clothing, equipment, and parts from nuclear power plants) to be dumped in ordinary landfills and recycled into consumer products (e.g., frying pans and jewelry). While isolation and containment, together with dilution and dispersion, are methods of dealing with radioactive wastes away from their source, isotopic dilution serves as a novel approach of controlling waste toxicity (radiation dose) at the source, consequently avoiding further treatment. According to this practice, certain radioactive wastes can be maintained below the maximum permissible concentrations by adding stable isotopes of the same chemical forms and equilibrating them sufficiently before the wastes are discharged to the environment. This management scheme may prove to be the most economical means of controlling certain radioactive wastes.[1]

SUMMARY AND CONCLUSIONS

Estuarine organisms receive radiation from both natural (e.g., cosmic radiation) and anthropogenic (e.g., low-level radioactive wastes from nuclear power plants) sources. Between 1944 and 1980, human-generated radioactivity in the form of fallout from the detonation of nuclear explosives in the atmosphere also contributed significant amounts of radionuclides to estuarine and marine ecosystems. For example, biologically important constituents of fallout from atmospheric nuclear explosions included ^{89}Sr, ^{90}Sr, ^{137}Cs, and other radionuclides. The uptake, biomagnification, and transference of radionuclides originating from human activities and their passsage through marine food chains back to man have been a major societal concern. Consequently, detailed management strategies and monitoring programs involving radioactive wastes have been devised for the estuarine and marine biosphere.

Radioactive waste is classified under six categories: (1) high-level wastes; (2) transuranic wastes; (3) low-level wastes; (4) uranium mine and mill tailings; (5) decontamination and decommissioning wastes from nuclear reactors; and (6) gaseous effluents. Several waste management strategies exist to minimize the environmental impact of disposal of these wastes. These are dilution and dispersion, isolation and containment, and isotopic dilution with stable isotopes. The preferred management option for the release of low-level radioactive wastes in estuaries appears to be dilution and dispersion, as is evident in the discharge of radionuclides in liquid effluents of nuclear power plants.

Estuarine organisms accumulate radionuclides from water, sediments, and other organisms via ingestion. Certain radionuclides that enter biotic compartments (e.g., ^{210}Po) tend to be concentrated. Phytoplankton and zooplankton, in particular, quickly assimilate radionuclides. The uptake of these contaminants by larger estuarine organisms, although generally less than that of smaller organisms, nevertheless can be significant especially in those forms that cannot effectively regulate radionuclide concentrations in their tissues. Most evidence indicates, however, the lack of biomagnification of radionuclides in estuarine and marine food chains. Radiation impacts on estuarine and marine populations are manifested as somatic and genetic disturbances.

Various radioactive waste disposal schemes have been formulated to deal with anthropogenic nuclear wastes. A number of these disposal options are interim (retrievable) and others are permanent. The two most favorable permanent disposal systems consist of the deep crustal burial of radioactive waste on land and subseabed placement of the contaminants in mid-gyre, mid-plate regions of the oceans. Retrievable storage alternatives are tank storage, near-surface storage vaults, water-filled canals, air-cooled vaults, sealed cask storage, and tunnel storage. Careful planning and execution of these options will minimize the risk of radioactive contamination of terrestrial and marine ecosystems.

REFERENCES

1. Park, P. K., Kester, D. R., Duedall, I. W., and Ketchum, B. H., Radioactive wastes and the ocean: an overview, in *Wastes in the Ocean,* Vol. 3, *Radioactive Wastes and the Ocean,* Park, P. K., Kester, D. R., Duedall, I. W., and Ketchum, B. H., Eds., John Wiley & Sons, New York, 1983, 3.
2. Weber, I. P. and Wiltshire, S. D., *The Nuclear Waste Primer,* Nick Lyons Books, New York, 1985.
3. Schell, W. R. and Nevissi, A. E., Radionuclides at the Hudson Canyon disposal site, in *Wastes in the Ocean,* Vol. 3, *Radioactive Wastes and the Ocean,* Park, P. K., Kester, D. R., Duedall, I. W., and Ketchum, B. H., Eds., John Wiley & Sons, New York, 1983, 183.
4. Cognetti, G., Radioecology in the Mediterranean, *Mar. Pollut. Bull.,* 21, 1, 1990.
5. Whicker, F. W. and Schultz, V., *Radioecology: Nuclear Energy and the Environment,* Vol. 1, CRC Press, Boca Raton, FL, 1982.
6. Lapedes, D. N., Ed., *Dictionary of Scientific and Technical Terms,* McGraw-Hill, New York, 1974.
7. Clark, R. B., *Marine Pollution,* 2nd ed., Clarendon Press, Oxford, 1989.
8. Eichholz, G. G., *Environmental Aspects of Nuclear Power,* Ann Arbor Science Publishers, Ann Arbor, MI, 1976.
9. Joseph, A. B., Gustafson, P. F., Russell, I. R., Schuert, E. A., Volchok, H. L., and Tamplin, A., Sources of radioactivity and their characteristics, in *Radioactivity in the Marine Environment,* National Academy of Sciences, Washington, D.C., 1971, 6.
10. Beasley, T. M., Ball, L. A., Andrews, J. E., III, and Halverson, J. E., Hanford-derived plutonium in Columbia River sediments, *Science,* 214, 913, 1981.
11. Volchok, H. L., Bowen, V. T., Folsom, T. R., Broecker, W. S., Schuert, E. A., and Bien, G. S., Oceanic distributions of radionuclides from nuclear explosions, in *Radioactivity in the Marine Environment,* National Academy of Sciences, Washington, D.C., 1971.

12. U.S. Interagency Review Group on Nuclear Waste Management, Report to the President by IRG, TID-29442, March 1979, National Technical Information Service, U.S. Department of Commerce, Washington, D.C., 1979.

13. U.S. Interagency Review Group on Nuclear Waste Management, Subgroup Report on Alternative Technology Strategies for the Isolation of Nuclear Waste, TID-28318, October 1979, National Technical Information Service, U.S. Department of Commerce, Washington, D.C., 1979.

14. International Atomic Energy Agency, The IAEA Revised Definition and Recommendations of 1978 Concerning Radioactive Wastes and Other Radioactive Matter Referred to in Annex I and II to the Convention on the Prevention of Marine Pollution by Dumping of Wastes and Other Matter (London Dumping Convention), Info. Circ. No. 205/Add. 1/Rev. 1, August 1978, International Atomic Energy Agency, Vienna, 1978.

15. United Nations Scientific Committee on the Effects of Atomic Radiation, Ionizing Radiation: Sources and Biological Effects, Tech. Rep., United Nations, New York, 1982.

16. Osterberg, C. L., Nuclear power wastes and the ocean, in *Wastes in the Ocean,* Vol. 4, *Energy Wastes in the Ocean,* Duedall, I. W., Kester, D. R., Park, P. K., and Ketchum, B. H., Eds., John Wiley & Sons, New York, 1985, 127.

17. Eisenbud, M., The status of radioactive waste management: needs for reassessment, *Health Phys.,* 40, 429, 1981.

18. Roels, O., Mariculture fertilized by upwelling, *Sea Technol.,* 23, 63, 1982.

19. Cherry, R. D. and Shannon, L. V., The alpha radioactivity of marine organisms, *Atomic Energy Rev.,*12, 3, 1974.

20. Cherry, R. D. and Heyraud, M., Evidence of high natural radiation dose domains in mid-water oceanic organisms, *Science,* 218, 54, 1982.

21. International Atomic Energy Agency, Effects of Ionizing Radiation on Aquatic Organisms and Ecosystems, Safety Ser., No. 11, International Atomic Energy Agency, Vienna, 1976.

22. Edgington, D. N. and Nelson, D. M., The chemical behavior of long-lived radionuclides in the marine environment, in Behavior of Long-Lived Radionuclides Associated with Deep-Sea Disposal of Radioactive Wastes: Report of a Co-ordinated Research Programme Organized by the International Atomic Energy Agency, Tech. Rep., International Atomic Energy Agency, Vienna, 1986, 41.

23. Hayes, D. W. and Sackett, W. M., Plutonium and cesium radionuclides in sediments of the Savannah River estuary, *Est. Coastal Shelf Sci.,* 25, 169, 1987.

24. Donoghue, J. F., Bricker, O. P., and Olsen, C. R., Particle-borne radionuclides as tracers for sediment in the Susquehanna River and Chesapeake Bay, *Est. Coastal Shelf Sci.,* 29, 341, 1989.

25. Hamilton, E. I., Radionuclides and large particles in estuarine sediments, *Mar. Pollut. Bull.,* 20, 603, 1989.

26. Bonatto, S., Carraro, G., Strack, S., Lüttke, A., Colard, J., Koch, G., and Kirchmann, R., Ten years of investigation on radioactive contamination of the marine environment: incorporation, by marine algae and animals, of hydrogen-3 and other radionuclides present in effluents of nuclear or industrial origin, in Impacts of Radionuclide Releases into the Marine Environment, International Atomic Energy Agency, Vienna, 1981, 649.

27. Bonatto, S., Bossus, A., Nuyts, G., Kirchmann, R., Mathot, P., Colard, J., and Cinelli, F., Experimental uptake study of ^{60}Co, ^{137}Cs, ^{125}Sb, and ^{65}Zn in four marine algae, in *Wastes in the Ocean,* Vol. 3, *Radioactive Wastes and the Ocean,* Park, P. K., Kester, D. R., Duedall, I. W., and Ketchum, B. H., Eds., John Wiley & Sons, New York, 1983, 287.

28. Arapis, G., Bonatto, S., Bossus, A., Nuyts, G., Gerber, G. B., Kirchmann, R., Colard, J., Mathot, P., and Cogneau, M., Tritium and technetium uptake and distribution in three marine algae, in *Oceanic Processes in Marine Pollution,* Vol. 1, *Biological Processes and Wastes in the Ocean,* Capuzzo, J. M. and Kester, D. R., Eds., Robert E. Krieger Publishing, Malabar, FL, 1987, 133.

29. Van der Ben, D., Cogneau, M., Robbrecht, V., Nuyts, G., Bossus, A., Hurtgen, C., and Bonotto, S., Factors influencing the uptake of technetium by the brown alga, *Fucus serratus, Mar. Pollut. Bull.,* 21, 84, 1990.

30. Kiefer, J., *Biological Radiation Effects,* Springer-Verlag, New York, 1990.

31. Michaelis, W., Ed., *Estuarine Water Quality Management,* Springer-Verlag, New York, 1990.

32. Fowler, S. W., Biological transfer and transport processes, in *Pollutant Transfer and Transport in the Sea,* Kullenberg, G., Ed., CRC Press, Boca Raton, FL, 1982, 1.

33. Tateda, Y. and Koyanagi, T., Accumulation of radionuclides by common mussel *Mytilus edulis* and purple bifurcate mussel *Septifer virgatus, Bull. Jpn. Soc. Sci. Fish.,* 52, 2019, 1986.

34. Martin, J. M. and Thomas, A. J., Origins, concentrations, and distributions of artificial radionuclides discharged by the Rhone River to the Mediterranean Sea, *J. Environ. Radioact.,* 11, 105, 1990.

35. Pentreath, R. J., Radionuclides in marine fish, *Oceanogr. Mar. Biol. Annu. Rev.,* 15, 365, 1977.

36. Eisenbud, M., *Environmental Radioactivity,* 3rd ed., Academic Press, New York, 1987.

37. Langford, T. E., *Electricity Generation and the Ecology of Natural Waters,* Liverpool University Press, Liverpool, U.K., 1983.

38. Blaylock, B. G. and Trabalka, J. R., Evaluating the effects of ionizing radiation on aquatic organisms, in *Advances in Radiation Biology,* Vol. 7, Lett, J. T. and Adler, H., Eds., Academic Press, New York, 1978, 103.

39. Seymour, A. M., The impact on ocean ecosystems (nuclear war), *Ambio,* 11, 132, 1982.

40. Rice, T. R. and Baptist, J. P., Ecological effects of radioactive emissions from nuclear power plants, in *Human and Ecological Effects of Nuclear Power Plants,* Sagan, L. A., Ed., Charles C Thomas, Springfield, IL, 1974, 373.

41. Polikarpov, G. G., *Radioecology of Aquatic Organisms: The Accumulation and Biological Effect of Radioactive Substances,* North Holland, Amsterdam, 1966.

42. Cosgrove, G. E. and Blaylock, B. G., Acute and chronic irradiation effects in mosquito fish at 15 or 25°C, in Radionuclides in Ecosystems (Proc. 3rd Symp. Radioecology), Nelson, D. J., Ed., Conf-710501, U.S. Atomic Energy Commission Technical Information Center, Oak Ridge, TN, 1973, 579.

43. Woodhead, A. D. and Pond, V., Effects of radiation exposure on fishes, in *Oceanic Processes in Marine Pollution,* Vol. 1, *Biological Processes and Wastes in the Ocean,* Capuzzo, J. M. and Kester, D. R., Eds., Robert E. Krieger Publishing, Malabar, FL, 1987, 157.

44. Egami, N. and Ijiri, K.-I., Effects of irradiation on germ cells and embryonic development in teleosts, *Int. Rev. Cytol.,* 59, 195, 1979.

45. Woodhead, D. S., Influence of acute irradiation on induction of chromosome aberrations in cultured cells of the fish, *Ameca splendens,* in Biological and Environmental Effects of Low-Level Radiation, Vol. 1, International Atomic Energy Agency, Vienna, 1976, 67.

46. Templeton, W. L., Effects of Irradiation on aquatic populations, in *Environmental Toxicity of Aquatic Radionuclides: Models and Mechanisms,* Miller, M. W. and Stannard, J. N., Ann Arbor Science Publishers, Ann Arbor, MI, 1976, 287.

47. Hagen, A. A., History of low-level radioactive waste disposal in the sea, in *Wastes in the Ocean,* Vol. 3, *Radioactive Wastes and the Ocean,* Park, P. K., Kester, D. R., Duedall, I. W., and Ketchum, B. H., Eds., John Wiley & Sons, New York, 1983, 47.

48. U.S. Department of Energy, Excerpts from the President's Policy Statement on Comprehensive Radioactive Waste Management Program, Nuclear Waste Isolation Activities Report, May 1980, U.S. Department of Energy, Division of Waste Isolation, Washington, D.C., 1980.

49. Heath, G. R., Hollister, C. D., Anderson, D. R., and Leinen, M., Why consider subseabed disposal of high-level nuclear wastes?, in *Wastes in the Ocean,* Vol. 3, *Radioactive Wastes and the Ocean,* Park, P. K., Kester, D. R., Duedall, I. W., and Ketchum, B. H., Eds., John Wiley & Sons, New York, 1983, 303.

50. Hoskins, E. R. and Russell, J. E., Geologic and engineering dimensions of nuclear waste storage, in *Nuclear Waste: Socioeconomic Dimensions of Long-Term Storage,* Murdock, S. H., Leistritz, F. L., and Hamm, R. R.; Eds., Westview Press, Boulder, CO, 1983, 19.

51. Preston, A., Deep-sea disposal of radioactive wastes, in *Wastes in the Ocean,* Vol. 3, *Radioactive Wastes and the Ocean,* Park, P. K., Kester, D. R., Duedall, I. W., and Ketchum, B. H., Eds., John Wiley & Sons, New York, 1983, 107.
52. Duedal, I. W., A brief history of ocean disposal, *Oceanus,* 33, 29, 1990.

7 Dredging and Dredged-Spoil Disposal

INTRODUCTION

Major physical, chemical, and biological changes in estuaries, rivers, harbors, and ports are associated with dredging and dredged-spoil disposal. Recreational and commercial usage of coastal waterways justifies the need for dredging to maintain sufficient water depth. Nonmaintenance pursuits, such as the construction of power plants and port facilities, also require dredging activities that transfer significant volumes of sediment. Dredging in estuaries causes damage directly by the physical disturbance involved in removing and relocating sediment.[1] The resuspension of bottom sediments releases nutrients and remobilizes contaminants, thereby affecting water quality and overall estuarine chemistry. The removal of bottom sediments destroys the benthic habitat, and operation of the dredge increases mortality of the benthos by mechanical injury during dredging, as well as by smothering of individuals with sediments after the organisms are picked up.

Some workers contend that the physical impacts at dredged sites, owing to sediment excavation, and at dredged-spoil disposal locations, due to sediment dumping, are not great.[2] Others, however, underscore the sedimentologic and hydraulic consequences of dredging and disposal and their economic and ecologic ramifications.[3] As conveyed by Nichols,[3] dredging and disposal of spoils upset the hydraulic and sediment regime of an estuary. The deepening of an estuarine channel, for instance, alters the stability it has attained by its adjustment of bed geometry and hydraulic regime

over time. This modification reduces tidal currents which accelerates sediment deposition in the channel. Hence, dredging is partly a
self-perpetrating activity. The disposal of sediment also impairs
natural processes. The disposal of dredged material usually results
in misplaced sediment in a foreign environment of deposition.[3] An
anomalous bed topography is often created by the uneven or heterogeneous distribution of the sediment, typically in disequilibrium
with the prevailing energy regime. If disposed alongside a dredged
channel, therefore, the misplaced sediment gradually returns to the
channel and eventually must be dredged again.

Probably of greater concern are the potential environmental
impacts attributable to the dispersal of pollutants from dredged
materials. The release of toxic substances into the water poses a
threat to the aquatic biota. The adverse effects of these contaminants
may be exacerbated by increased nutrient and turbidity loads that can
be detrimental, in particular, to planktonic organisms. Capuzzo et
al.[4] identifies four general categories of chemical contamination
associated with dredging and disposal of sediments: (1) high concentrations of organic matter fostering anoxia and the presence of
hydrogen sulfide; (2) transition and heavy metal contamination; (3)
petroleum hydrocarbons; and (4) synthetic organic chemicals.

The effects of dredged material on the water quality of an estuary
must be considered on a case-by-case basis since changes in water
quality cannot be accurately predicted from the examination of the
material to be dredged.[5,6] While it is likely that dredged material from
urban estuaries and harbors contain elevated levels of at least some
contaminants ascribable to industrial and municipal sewer discharges as well as nonpoint source discharges,[4] the sediments often
do not release the contaminants during dredging.[6] Thus, Windom,[7]
assessing the water quality at dredged sites in Georgia estuaries,
discerned no detrimental changes in dissolved oxygen, chemical
oxygen demand, biochemical oxygen demand, pH, and iron concentration. Levels of phosphate dropped and those of mercury increased, presumably due to the dredging. At diked-spoil disposal
areas, however, Windom[7] noted significantly higher concentrations
of copper, iron, and zinc, and elevated oxygen and pH readings.

The amount of chemical contaminants in dredged spoils varies
widely among estuaries, being highest in systems bordered by heavy
industry and intense shipping activity. Clark[8] compares some of the

pollutant loads in dredged spoils derived from the estuaries and harbors of England. According to Clark,[8] dredged spoils from the Tees estuary contain 7 ppm of mercury, 320 to 460 ppm of lead, 1500 ppm of chromium (one sample), and 3000 ppm of zinc. Dredged spoils from the docks of the Manchester Ship Canal, in contrast, have 20.7 ppm of mercury and 5080 ppm of lead. At Swansea, large quantities of cadmium (18 ppm) exist in dredged spoils, while at Newhaven, low levels of chromium (1 ppm) occur in dredged spoils relative to other locales (e.g., the Tees estuary).

The most acute biotic impacts of dredging and dredged-spoil disposal involve the benthos. The immediate effect of dredging is the physical removal of benthic organisms and their translocation to spoil disposal areas. Mortality results from mechanical damage or smothering and approaches 100% when the organisms are dumped at terrestrial disposal sites. Less mortality originates from subaqueous disposal, but injuries and death can still be quite high depending on the types of organisms affected and their new habitat. Through recolonization and succession, the benthic communities slowly recover from dredging effects.[9]

This chapter examines the physical, chemical, and biological changes in estuaries ascribable to dredging and dredged-material disposal. An overview is presented on the types of dredging operations and equipment employed in estuaries. The information focuses on the potential impacts of dredging and dredged-spoil disposal on aquatic communities.

DREDGING DEVICES

The disposal of dredged spoils in estuarine and marine waters accounts for most of the materials currently dumped in these environments. In the nearshore waters of the U.S., dredged materials comprise about 80% by weight of all materials dumped there.[4,10] An average of 300 million m³ of sediment are removed each year for the maintenance of existing waterways in the U.S., with an additional 80 million m³ extracted for new dredging projects.[6] A substantial volume of this material is disposed of in the ocean, although incomplete estimates have been compiled.[11] More than 100 coastal disposal sites in the U.S. receive dredged material. Nearly all open ocean dumping of dredged spoils takes place on the continental shelf. A total of 28 million metric tons of dredged spoils is dumped annually at 60 licensed offshore sites in England and Wales.[8]

A major environmental concern relates to the disposal of contaminated dredged material in estuarine and coastal marine habitats. Consequently, a number of alternate methods for the disposal of this material has been proposed, such as sub-bottom containment, artificial island creation, land containment, and the removal of contaminants by predumping treatment.[2,4,12-14] Capuzzo et al.[4] briefly summarize each of these alternatives to conventional disposal. Sub-bottom containment involves the capping of contaminated dredged spoils with clean sediments to minimize the reentry of chemical pollutants into the water column. The most contaminated dredged material is targeted for artificial island creation; in this case, isolation of contaminated sediments in confined areas protects estuarine and coastal marine waters. Upland disposal of dredged sediments entails the use of soils with high absorption capacity to preclude the infiltration of pollutants into groundwater or transport to wetlands. The addition of substances that bind contaminants either chemically or physically within deposited dredged material serves as a potentially effective method for the predumping treatment to remove contaminants. Unfortunately, this method is costly, and its long-term effectiveness has been questioned.[2] In general, the disposal of contaminated sediments at sea or in upland habitats poses problems for biotic communities and groundwater quality, respectively.[15] For these reasons, a variety of bioassays have been developed to assess the impacts of the dredged spoils on organisms.[16] Engler[17] details two primary bioassays applied to the assessment of dredged material: (1) bioassays dealing with liquid-phase or water-column effects; and (2) bioassays treating solid-phase effects.

Dredges are either mechanical or hydraulic devices. Mechanical dredges lift sediment from the seafloor and transport it to a disposal site. Hydraulic dredges, however, produce a slurry, suspending the sediment in the water and pumping it to a discharge site. Various types of mechanical and hydraulic dredges exist;[11] Table 1 lists these dredges and their relationship to sediment type and disposal method. The following is a description of some of them, adopted from Stickney[6] and Machemehl:[18]

TABLE 1
Types of Dredging Devices and Their Relationship to Sediment Type and Disposal Method

Dredge type	Sediment type	Disposal conveyance
Mechanical devices		
Dipper dredge	Blasted rock	Vessel
Bucket dredge	Coarse grain size	Vessel
Ladder dredge	Fine grain size	Vessel
Agitation dredge	Mud, clay	Prevailing current
Hydraulic devices		
Agitation dredge	Mud, clay	Prevailing current
Hopper dredge	Fine grain size	Vessel
Suction dredge	Soft mud, clay	Pipeline
Cutterhead dredge	Consolidated, coarse grain size	Pipeline
Dustpan dredge	Sand	Pipeline
Sidecasting dredge	Fine grain size	Short pipe

From Kester, D. R., Ketchum, B. H., Duedall, I. W., and Park, P. K., in *Wastes in the Ocean,* Vol. 2, *Dredged Material Disposed in the Ocean,* Kester, D. R., Ketchum, B. H., Duedall, I. W., and Park, P. K., Eds., John Wiley & Sons, New York, 1983, 3. With permission.

Dipper dredge: This dredge is basically a power shovel mounted on a barge. The barge has three spuds (two forward and one at the stern) for stability. Spuds are large-diameter pipes that can be raised and lowered to help hold the barge in position or allow it to pivot around a given point. Dipper dredges can handle up to nearly 10 m³ of hard material per cycle. They are commonly used to remove blasted rock or loose boulders. Dredged material is discharged within reach of the dipper boom. This kind of dredge is usually not used in water deeper than about 20 m.

Ladder dredge: A ladder dredge is essentially an endless chain of buckets mounted on a barge. The barge is maintained in position by side cables. The buckets can each carry between 1 and 2 m³ of material. The excavated material is dumped into chutes or onto belts and is discharged over the side of the barge. Ladder dredges are generally not used in water over about 30 m.

Bucket dredge: A bucket dredge is basically a crane mounted on a barge. The bucket (a clamshell, orange peel, or dragline type) can be changed to suit the job. The barge uses either spuds or anchor lines for stability. This kind of dredge can handle moderately stiff material in confined areas. It is generally used on small-scale dredging projects. The material is dumped within the length of the boom.

Hydraulic pipeline dredge: This is the most versatile of the modern dredges and is also the most widely used. Using a cutter head, a hydraulic pipeline dredge can excavate material ranging from light silts to heavy rock. Dredged material is pumped through floating pipes and pipes laid onshore and can be discharged in areas remote from the dredge itself. Hydraulic pipeline dredges range in size and can work in depths of up to 20 m.

Hopper dredge: A hopper dredge is a self-propelled vessel designed to dredge, load, and retain spoil in hoppers and haul it to aquatic disposal areas. Capacities for hopper dredges can exceed 10,000 m³. Such dredges can work in depths over 20 m.

Hydraulic dredges and hopper dredges process most of dredged material in coastal areas, being responsible for about 69 and 24%, respectively, of the total material handled.

The agitation dredge incorporates some of the qualities of both mechanical and hydraulic dredges.[19] In this case, an object dragged along the seafloor resuspends bottom sediments that are transported away from the dredged site by a flowing current of water.[11] Factors considered in dredge selection include the availability of the device, its maneuverability, the sediment type, conflicts with prevailing maritime traffic, and historical precedent for a specific region.[11,19]

Studies performed during the past 15 years support the view that spoils released from dredges at a dumpsite are deposited quickly.[20] Less than 1% of dredged silt released at a dumpsite in Long Island Sound, according to Gordon,[21] remained in suspension after dumping and was dispersed by the tides. Morton[77] drew a similar conclusion for dredged material dropped on three disposal mounds in Long Island Sound. Subsequent to the discharge of the spoils, 95% of the sediment was found at two of the sites and 90% at the third locale, indicating a minor loss beyond the boundaries of the mounds. Sustar and Wakeman[22] reported the loss of only 1 to 5% of dredged material as suspended mud in San Francisco Bay. Tavolaro[23] likewise chronicled a loss of approximately 4% of dredged material during transport and discharge at an open-water site in the New York Bight. Barge or hopper-dredged discharges have proven to be very effective in dumping dredged spoils onto the seafloor, with <5% of the sediment widely dispersed in the water column by this practice.[20]

ENVIRONMENTAL EFFECTS OF DREDGING AND DREDGED-SPOIL DISPOSAL

Most investigations of dredging and dredged-spoil disposal operations focus on environmental impacts while ignoring or downplaying ecological benefits. The adverse effects of dredging can be categorized into three groups: (1) primary impacts on habitat and organisms; (2) changes in water quality; and (3) tertiary effects. The beneficial effects are (1) increased recreational and commercial usage of a waterway; (2) increased circulation in estuaries and embayments; (3) increased nutrients; and (4) sediment supply for beach nourishment, landfill projects, and soil improvement.[24]

ADVERSE EFFECTS
Primary Impacts on Habitat and Organisms

The immediate impact of dredging in an estuary is the destruction of habitat by the removal of sediments and the increase in mortality of benthic organisms due to the mechanical action of the dredge and smothering by sediments. The operation of the dredge and dumping of dredged spoils constitute environmental perturbations, particularly for the benthos.[25] Recovery of a dredged site occurs slowly at times, with opportunistic pioneering forms, such as polychaete worms (e.g., *Capitella* spp.) initially recolonizing the site and later supplanted by equilibrium assemblages of organisms in a successional pattern.[9,26] The recovery process may require 1 or more years.

Hard[27] commented that benthic fauna tend to repopulate dredged-spoil dumping grounds at reasonably predictable rates, densities, and diversities. The rate of recolonization is a function of the locality, sediment composition, and type of organisms comprising the benthic community. In general, the same species inhabiting a dumping locale or a dredged site usually return to the disturbed area after a period of time. The more motile forms have lower mortality rates than the sedentary varieties when dumped upon, but even the burrowing types may succumb to the deposition of spoils if they are buried deeply in sediment.[6] However, dredging or dredged-spoil disposal activities often do not adversely affect motile populations.[5]

The species composition of a benthic community at a dredged site or a spoil disposal location is most likely to be altered by the release of toxic contaminants from the sediments or by acute changes in sediment type.[6,28] Regardless of the changes in species composition, the earliest colonizers usually consist of pioneering members of the community that exhibit greater productivity than the more stable equilibrium populations which become established at a later time, as previously mentioned.[26] The rate of recolonization may be rapid; as discovered by Bybee,[29] the penaeid shrimp of Tijuana Slough repopulated a dredged area only 3 to 4 weeks subsequent to dredging. The abundance of the shrimp population returned to levels of commercial production in about 6 months. Bonvicini Pagliai et al.[30] also found rapid recolonization by macrobenthos at a dredged site of the Thyrrenian Sea in the Gulf of Cagliari, Sardinia, Italy. Here, the main structural parameters of the benthic community 6 months after dredging paralleled those of the predredged period and at neighboring nondredged areas. Stickney and Perlmutter[31] observed the recolonization of predominantly short-lived, motile benthic infauna within days after dredging. Work by Connor and Simon[32] on a benthic community in Tampa Bay, Florida, corroborate these findings, showing that complete recovery of the benthos occurred at an impacted area of the bay 6 months following dredging. Other investigators have documented much slower rates of recovery among the benthos, such as at a dredged site in a brackish water bay of southern Finland.[33] Similarly, Groot[34] estimated a recovery time of 3 years for benthic fauna of dredged sandy banks in the North Sea. Hence, the recolonization and recovery time of benthic fauna after dredging is quite variable and dependent on site-specific environmental conditions.[30]

Floral communities may be more easily impacted by dredging activities. Substrate modifications resulting from direct dredging affect benthic macroflora, at least locally, at the site of dredging and not uncommonly beyond the limits of the dredged site. For example, Godcharles[35] did not ascertain any recolonization of seagrasses (i.e.,

Thalassia testudinum and *Syringodium filiforme*) for at least 1 year following clam dredging of a grassflat habitat in Florida. While the action of a dredge directly harms benthic macroflora, secondary effects (e.g., turbidity) can influence the floral community as well. Increased turbidity in the vicinity of a dredged site frequently attenuates light sufficiently to lower primary production by phytoplankton or bottom-dwelling plants, although this is not always the case.[36]

Equally devastating to plant communities is the disposal of dredged spoils at upland sites which eradicates significant biomasses of terrestrial vegetation. In the past, salt marshes and other wetland habitats have received large volumes of dredged spoils that destroyed vegetation and altered their surfaces by modifying the soil fertility and the morphology of the marshland. Dredging, filling, diking, and ditching of these tidal wetlands — in some cases to reclaim land from the shallow continental shelf — place constraints on developing floral and faunal communities.

Because of increasing societal demands for land in coastal regions, especially in densely populated areas, salt marshes and other tidal wetlands have been reclaimed in the past at the expense of natural communities of organisms. Thus, approximately 20% of Manhattan Island in New York City has been constructed on reclaimed tidal marsh and shallow harbor areas.[37] Over the last 8 centuries, about 6300 km² of coastal habitat has been reclaimed in The Netherlands alone.[38] The dredging and filling of salt marshes make them attractive sites for industrial development, owing to their proximity to water transport.[37] Unfortunately, industrial and housing development on wetlands have destroyed many hectares of important wildlife habitat and decimated aquatic and terrestrial communities of organisms.

Cahoon and Cowan[39] refer to the conversion of wetlands to either open water or upland (spoil bank) habitat as wetland loss. The Louisiana coastal zone provides an excellent example of wetland loss, in which dredging for site access, navigation, and pipeline canals has produced elevated spoil banks alongside excavated canals. In essence, the dredged spoils form a new habitat for birds and wildlife,[40] but the canals and spoils also directly impact the wetlands by: (1) converting marsh habitat to open-water and upland (spoil bank) habitat; and (2) altering the local hydrologic regime (i.e., sheetflow over the marsh, subsurface water flow, sediment dispersal, and saltwater intrusion), thereby influencing the health of the marsh.[39,41,42] Lindstedt and Nunn[43] estimated that between 1900 and 1985, 77,760 ha of Louisiana's coastal wetlands were converted to open-water habitat by dredging. Furthermore, indirect impacts ascribable to canal dredging, spoil banking, and levee building were responsible for as much as 60% of the coastal wetland loss due to changes in regular overbank flooding.[42] The use of high-pressure spray disposal possibly offers a suitable alternative to traditional dredging technology for the minimization of wetland conversion to nonwetland habitat.

Substantial loss of wetlands is not restricted to the Louisiana coast. Dredging and other activities have eliminated two thirds of the cordgrass marshes in the southern California range of the light-footed clapper rail.[44,45] The dredging of inlets and construction of jetties in southeastern North Carolina lagoons have at times eliminated the transport of marine sediments to marsh surfaces, periodically contributing to their historical demise.[46]

A trend has developed over the years toward land disposal of dredged spoils rather than their disposal at aquatic sites, due primarily to concerns over the dumping of contaminated sediments in open water.[47] Unequivocally, the uncertainties associated with open-water disposal of dredged material fostered restrictions on the use of aquatic disposal sites in the late 1960s and early 1970s.[48] As a result of additional environmental constraints on open-water disposal of polluted sediments, confined dredged-material disposal areas have been constructed on shore to accept contaminated sediments dredged from estuaries, harbors, and other impacted coastal locales.[49] These disposal areas serve as effective sedimentation basins for the storage of solids, with the settled material being overlain by clarified supernatant that is typically discharged after a retention time varying from hours to weeks. The effluent contains predominantly particle-associated (i.e., adsorbed, coprecipitated, or ion exchanged) contaminants as well as dissolved concentrations of contaminants.[50,51] With the passage of the Water Resources Development Act of 1986, the dredging of potentially contaminated sediments from areas not recently maintained is expected to increase during the 1990s.[47] This dredging should lead to the greater utilization of confined dredged-material disposal sites on land.

Since the 1960s, the disposal of dredged material on coastal marshes and wetlands in general has been discouraged because of the recognition that these wetland systems represent major nursery and breeding grounds for a multitude of organisms and zones of extremely high biological productivity for estuaries and nearshore regions. Nevertheless, wetland disposal of dredged spoils has been the only practical method in some situations. Concern over this disposal, as stated earlier, has centered on direct impacts on biota and indirect effects resulting from altered physical and chemical characteristics of the substrate which causes undesirable changes in indigenous flora and fauna. A more attractive alternative to using marshes and other wetlands as dredged-spoil disposal sites is the creation of marshes or spoil islands from dredged material. This management scheme, therefore, would promote environmentally compatible disposal operations.[47]

In open-water spoil disposal, two types of sites are considered — retentive and dispersive sites.[52] Retentive sites ensure that disposed sediments remain within each site; low-energy hydrodynamic environments generally characterize these disposal localities. However, when a low-energy hydrodynamic environment is lacking, an inerodible cap may be placed over the disposed sediments to prevent their movement and the mobilization of contaminants from the dredged material. Despite the design effectiveness of a sheltered site or a protective cap of clean sediments for retentive purposes, both approaches can fail. Major storms roil bottom sediments even in the most protected estuarine areas, and the construction of an impenetrable protective cap over disposed sediments may not be achievable. Still the monitoring of capped disposal sites in ocean waters for extended periods of time has demonstrated that capping is technically feasible under normal tidal and wave conditions.[53,54] The use of a 50-cm cap of clay, silt, and sand in laboratory experiments has been shown to preclude the transfer of contaminants from dredged sediments into biota in overlying water, despite the existence of large numbers of bioturbating infauna.[55] Cap materials mainly comprised of clay and silt appear to afford greater protection against contaminant transfer to biota than those consisting of larger amounts of sand. The New England Division and the New York District of the U.S. Army Corps of Engineers have experimented with the use of clean sediments to cap contaminated dredged material in order to mitigate

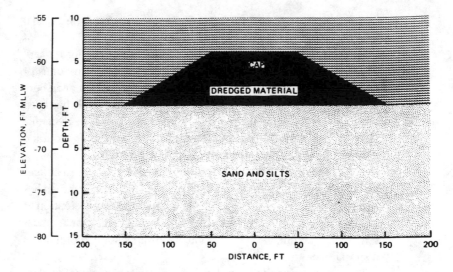

FIGURE 1. Idealized profile of the Duwamish Waterway capping mound in Seattle, WA. (From Brannon, J. M. and Poindexter-Rollings, M.E., *Sci. Total Environ.*, 91, 115, 1990. With permission.)

ecological impact of estuarine and marine waters.[56] Figure 1 illustrates an idealized profile of a sediment cap overlying dredged material in the Duwamish Waterway in Seattle, WA.

Dispersive sites are chosen in the expectation that the disposed sediments will be transported away from the site by ambient currents, enabling additional disposal to take place. Selection of dispersive sites typically is based on their high-energy hydrodynamics which prevent the settlement of disposed sediments to the seafloor or enhance their erosion once they deposit. Factors evaluated in the selection of an open-water disposal site include not only hydrodynamics, but also the character of the bottom, the bathymetry of the area, the volume of sediment to be disposed, and the method of disposal. Field observations, laboratory tests, and numerical modeling studies all are performed in the design of disposal management plans for both retentive and dispersive open-water disposal sites.[52] These preliminary investigations enable critical appraisal of the dumping locale to minimize environmental impacts.

In the past, open-water disposal of dredged sediments in the estuarine environment has been primarily conducted to minimize dredging costs and to preserve contained disposal site capacity.[52]

However, the increasing awareness of the public concerning aquatic pollution and habitat alteration associated with anthropogenic activities, such as dredging and dredged-spoil disposal, has led to a reassessment of priorities. In the case of dredging operations, much research is now directed toward environmental impacts, either directly or indirectly, on the structure and function of biotic communities. In this respect, spoil disposal may be of even greater concern than dredging, with high levels of suspended solids, sediment buildup, and oxygen depletion coupled to the dumping of dredged material accounting for a variety of adverse effects on the communities. Moreover, the presence of biostimulants and toxins, often chemically or physically sorbed within the sediment matrix, poses potentially long-term hazards to the water quality and constituent flora and fauna of an impacted site. To understand the impact of dredging and disposal on the environment, the U.S. Army Corps of Engineers has performed a significant amount of dredged material research (Table 2). This comprehensive program of research, study, and experimentation on dredged material has yielded much information regarding the environmental effects of dredging and disposal in estuaries and coastal marine waters.[47]

Effects on Water Quality

The U.S. Army Corps of Engineers dredges over 250 million m^3 of sediment each year to maintain more than 30,000 km of waterways and about 1000 harbor projects. Of this volume of sediment, approximately 60% is dredged in the estuarine and coastal zone.[52] The principal effect of dredging on the water quality of estuaries concerns chemical exchanges between the dispersed sediment and water. As the dredged sediments disperse, several events occur that affect water quality. Initially, heavy metals in the water sorb to the sedimentary particles in suspension, thereby lowering their concentrations in the water. Later, the sediments may release some of their heavy metal load, raising the metal concentration in the water. Second, estuarine sediments release ammonia when dispersed which stimulates phytoplankton production and is followed by higher values of pH, dissolved oxygen, and biological oxygen demand (BOD).[57]

Turbidity

Increases in the quantity of suspended sediments during dredging

TABLE 2
Dredged Material Research Program[a]

Research area	Research task
1. Environmental Impact of Open Water Disposal	A. Evaluation of Disposal Sites
	B. Fate of Dredged Materials
	C. Effects of Dredging and Disposal on Water Quality
	D. Effects of Dredging and Disposal on Aquatic Organisms
	E. Pollution Evaluation
2. Environmental Impact of Land Disposal	A. Environmental Impact Studies
	B. Marsh Disposal Research
	C. Containment Area Operation Research
3. New Disposal Concepts	A. Open-Water Disposal Research
	B. Inland Disposal Research
	C. Coastal Erosion Control Studies
4. Productive Uses of Dredged Material	A. Artificial Habitat Creation Research
	B. Habitat Enhancement Research
	C. Land Improvement Research
	D. Products Research
5. Multiple Utilization Concepts	A. Dredged-Material Drainage/Quality Improvement Research
	B. Wildlife Habitat Program Studies
	C. Disposal Area Reuse Research
	D. Disposal Area Subsequent Use Research
	E. Disposal Area Enhancement

6. Treatment Techniques and Equipment
   A. Dredged Material Dewatering and Related Research
   B. Pollutant Constituent Removal Research
   C. Turbidity Control Research

7. Dredging/Disposal Equipment and Techniques
   A. Dredge Plant Related Studies
   B. Accessory Equipment Research
   C. Dredged-Material Transport Concept Research

[a] U. S. Army Corps of Engineers.

From Kirby, C. J., Keeley, J. W., and Harrison, J., in *Estuarine Research*, Vol. 2, *Geology and Engineering*, Cronin, L. E., Ed., Academic Press, New York, 1975, 523. With permission.

and dredged-spoil disposal operations precipitate an obvious change
in the water quality of estuaries, although the impacts appear to be
greatest in those systems having low ambient turbidity levels (e.g.,
southern Florida). In estuaries characterized by high natural turbidity, such as upper Chesapeake Bay and other systems along the
Atlantic and Gulf Coasts receiving large riverine sediment input,
effects on primary productivity are less conspicuous. By raising
light-extinction coefficients, suspended sediments limit primary
production of phytoplankton as well as benthic micro- and macroflora.[57] Consequently, the availability of food for primary consumers declines. Furthermore, the reduction of light penetration into the
water column alters the chemistry and temperature of the water,
fostering additional changes in the environment. Nonetheless, the
high turbidity ascribable to dredging and dredged-material disposal
may be relatively brief because of the movement of large volumes of
sediment in a small area over a short time interval.[47] Stickney (p.
256),[6] borrowing from the work of Windom,[57] identified four general
patterns of change in water quality in estuarine environments
through dredging: "...(1) turbidity increases, which are most significant where natural turbidity is low; (2) relatively rapid return to
normal turbidity levels after dredging; (3) the release of ammonia
during dredging, which may increase nitrate levels, thus increasing
algal production; and (4) initial decreases in concentrations of heavy
metals near the dredge, but eventual increases in iron, copper, zinc,
cadmium, and lead."

Truitt[48] discussed the transport of sediment as suspended solids
into the water column during dredged material disposal by barge
and hopper at open-water locations. Three distinct transport phases
or stages define the behavior of dredged material released from a
vessel (Figure 2). Upon release from a barge or hopper, the dredged
material descends as a dense, fluid-like jet through the water column.
Some of this material consists of solid blocks or clods of very dense
cohesive material. Site water is entrained in the jet as the sediment
descends rapidly toward the estuarine seafloor. Not all of the
sediment drops to the seafloor; a fraction of it remains in the upper
part of the water column and cannot be accounted for in a mass
balance. Currents transport this low density material away from the
disposal site. Since some of this suspended material may be contaminated, it is a cause of concern in estuaries and can degrade water

quality of uncontaminated areas removed from the impacted disposal location. The descending jet and its core of cohesive material tend to collapse, most frequently when they strike the seafloor, and much less commonly when the jet encounters a layer in the water column of equal or greater density than the jet.

The potential for large-scale contamination of clean areas in an estuary generated by the transport of contaminated sediment away from a descending jet of disposed material is not great. Mass balance calculations of dredged material released from barges and hoppers at open-water sites, for example, indicate that typically <5% of the volume of sediment discharged from the vessels cannot be accounted for at the disposal site. Field measurements of the losses of dredged sediment exiting barges or hoppers in Long Island Sound, New York Bight, San Francisco Bay, and Duwamish Waterway, for instance, amount to 1, 3.7, 1 to 5, and 2 to 4%, respectively. Table 3 summarizes these findings as well as the loss of material resulting from open-water disposal of dredged sediment by barge or hopper operations elsewhere. Hence, the principal adverse effects of

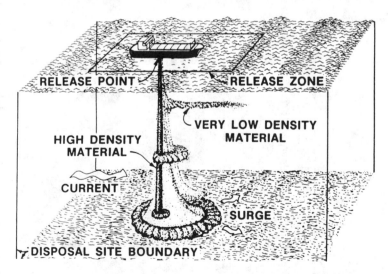

FIGURE 2. Sediment transport processes operating during open-water dredge disposal. (From Pequegnat, W. E., Pequegnat, L. H., James, B. M., Kennedy, E. A., Fay, R. R., and Fredericks, A. D., Procedural Guide for Designation Surveys of Ocean Dredged-Material Disposal Sites, Tech. Rep. EL-81-1, U.S. Army Corps of Engineers, Washington, D. C., 1981.)

TABLE 3
Summary of Field Studies of Dredged Material Behavior during Open-Water Disposal

Data source[48]	Site	Site characteristics			Dredging/disposal characteristics				Sediment in upper water column (% of original)
		Water depth (m)	Bottom currents (cm/s)	Dredged sediment	Dredge type	Disposal type	Typical volume (cu/m)	Monitoring technique/ device	
Gordon, 1974	Long Island Sound	18–20	6–30	Silt-clay	Clamshell	Scow	900–2300	Transmissometer	1
Sustar and Wakeman, 1977	Carquinez[a]	14	9–24	Silt-clay	Trailing Suction hopper	Hopper	1000	Transmissometer and gravimetric	1–5
Bokuniewicz et al., 1978	Ashtabula (Lake Erie)	15–18	0–21	Sandy silt	Trailing Suction hopper	Hopper	690	Transmissometer and gravimetric	1[b]
	New York Bight	26	6–24	Marine silt	Trailing Suction hopper	Hopper	6000	Transmissometer and gravimetric	1[b]
	Saybrook (Long Island	52	21–70	Marine silt	Clamshell	Scow	1100	Transmissometer and	1[b]

(Sound) Elliot Bay	67	0–21	Sandy silt	Clamshell	Scow	380–535	gravimetric Transmissometer and gravimetric	1[b]
Rochester (Lake Ontario)	17–45	0–21	Riverine silt	Trailing Suction hopper	Hopper	690	Transmissometer and gravimetric	1[b]
Tavolaro (1982) New York Bight	15–25	N/R	Silt-clay	Clamshell	Scow	1375–3000	Mass balance	3.7
Truitt (1986) Duwamish Waterway	20–21	6	Silt-clay	Clamshell	Scow	840	Gravimetric and mass balance	2–4

[a] Limited data at two additional sites included.
[b] Synthesis of all sites reported.

From Truitt, C. L., *J. Coastal Res.*, 4, 489, 1988. With permission.

dredged-spoil disposal at open-water sites on the overall water quality of an estuary are mainly confined to the location of the descending jet of material.

Dredging action, in addition to the open-water disposal of dredged sediment, impinges on the water quality of estuaries via the generation of turbidity, and the redeposition of the dredge-induced suspended sediment can impact benthic communities. Nichols et al.[25] monitored turbidity and suspended sediment concentrations in central Chesapeake Bay associated with hopper dredging, specifically the resuspension of sediment as the dredge dragheads move across the bottom and the high force discharge of dredged material as the hopper overflows. Two separate turbidity plumes arise from this activity: (1) a near-bottom plume produced by the agitation, cutting, and turbulence of the draghead as it moves through seafloor sediments; and (2) an upper plume in near-surface and mid-depth water generated by the discharge of hopper overflow slurries (Figure 3). Suspended sediment concentrations in the near field (about 300 m from the dredge) reached 840 to 7200 mg/l or 50 to 400 times the normal background level. These levels did not persist very long, occurring for <3 min at a fixed point. In the far field (>300 m from the dredge), the amount of suspended sediment during the period of overflow discharge (40 to 60 min) averaged 81 mg/l in the upper water layer or approximately five times the normal background. During an entire dredge-loading cycle (1.5 to 2.0 h), near-bottom water contained an average suspended sediment concentration of 137 mg/l or eight times the normal background.

Three phases of transport and dispersion typify the overflow discharge plume, i.e., convective descent, dynamic collapse, and long-term passive diffusion (Figure 3). During convective descent, the dredged material released from the vessel settles as a sinking jet, with the bulk of the overflow discharge dropping rapidly to the bottom. Downstream from the discharge port, at a point where the discharge velocity and density approach that of the ambient water, dynamic collapse takes place. This phase is manifested by the exponential decline of far-field suspended sediment concentrations at a distance of 300 to 400 m from the overflow discharge point. Beyond this distance, long-term passive diffusion develops, in which the plume is affected, not by its own dynamics, but by the ambient current and turbulence. Internal waves disperse some of this suspended material.

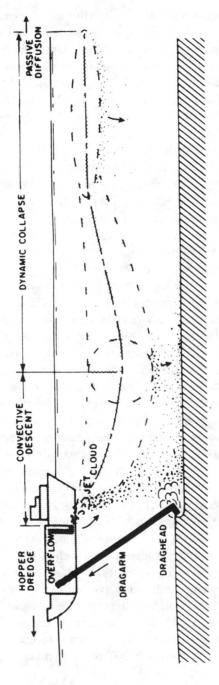

FIGURE 3.   Hopper dredging operations in Chesapeake Bay showing an upper turbidity plume produced by overflow discharge and a near-bottom turbidity plume produced by draghead agitation and settling of particulates from the upper plume. The conceptual model depicts three transport phases for hopper overflow discharge: convective descent, dynamic collapse, and passive diffusion. (From Nichols, M., Diaz, R. J., and Schaffner, L. C., *Environ. Geol. Water Sci.*, 15, 31, 1990. With permission.)

The consequence of the elevated turbidity and suspended sediment concentrations in the upper and near-bottom plumes is the exceedance of certain water quality standards. However, the hopper dredging in the estuary violated water quality standards of turbidity for recreation, aquatic life, and shellfishing established in Maryland only in the near-field zone, <800 m behind the dredge. Despite the high suspended sediment levels in the near field, benthic communities in Chesapeake Bay appear to be minimally impacted. Four possible reasons explain why macrobenthic assemblages are largely unaffected by the deposition of the dredged material: (1) anthropogenic pollutants do not contaminate the dredged sediments; (2) the grain size of the dredged material is the same as that of the natural background sediments; (3) the rate of sediment deposition from the plumes is not excessive for the survival of the biota; and (4) the fauna comprising the benthic communities have high motilities, flexible life-history strategies, and short lives.[58]

The role of anthropogenic pollutants associated with the dredged material is difficult to assess since sediments containing high levels of certain pollutants (e.g., heavy metals) do not necessarily release them into the water column during dredging or disposal.[6] However, pollutants introduced into estuarine waters are partitioned into, and concentrated in, bottom sediments where they warrant a great deal of attention, especially in urban estuaries.[59] Estuarine and nearshore biota exposed to contaminated sediments can transfer them up food chains to ultimately threaten human health.[60] Lunsford et al.,[61] analyzing the uptake of the organochlorine insecticide kepone *in situ* in the wedge clam (*Rangia cuneata*) during the dredging of contaminated sediments in the James River estuary, found significantly elevated residue levels above background in clams along a disposal area 2 weeks after dredging commenced. Hall[62] researched the effects of dredging and reclamation on the concentrations of heavy metals (i.e., Cd, Cr, Cu, Fe, Ni, Mn, and Pb) in the water column and sediments of an estuary and suggested that both activities released additional quantities of the metals into the environment, with current and wave action enhancing their distribution.

Future work on the ecological impact of dredging in estuaries must address the large variation of contaminant concentrations in sediments which requires prior knowledge of sediment composition and history to properly interpret.[63] Differences in mineralogy, grain size, organic matter, water content, and sources of anthropogenic inputs in estuarine and nearshore oceanic sediments contribute to wide variations in contaminant concentrations, in some cases (e.g., Cd, Hg, and Pb) by two to three orders of magnitude within a fairly restricted area.[64,65] The highest concentrations of contaminants in sediments are documented for major river estuaries in proximity to highly populated and industrialized regions. Even in some of these cases, the concentrations of contaminants in edible organisms do not give rise to alarm. However, in many others, the concentrations of contaminants surpass national and international concentration limits and pose a hazard to human health. In the past, "hot spots" such as the Hudson-Raritan estuary, Los Angeles Bight, Minamata Bay, and Ems estuary fell into this category.[63] These impacted systems must be monitored vigilantly to reduce acute or chronic contamination and return them to a revitalized state.

The suspension of contaminated sediments during dredging or dredged-spoil disposal operations presents a difficult problem when evaluating potential impacts on the environment. When these operations are executed in heavily utilized harbors and waterways containing large quantities of inorganic and organic pollutants, it is imperative to chemically characterize the dredged material. This goal can be achieved most expeditiously via four approaches: (1) bulk chemical analysis; (2) the elutriate test; (3) selective chemical leaching; and (4) bioassay tests.[11] Bulk chemical analysis provides a basis for estimating mass loads of wastes to estuarine and marine environments and, therefore, a basis for assessing the amount of substances that possibly can alter the biology and chemistry of a disposal site. The elutriate test measures the quantity of a substance exchanged between the sediment and the aqueous phase during dredging and disposal. The chemical state of pollutants associated with dredged sediments is determined via selective chemical leaching. Bioassay tests, in turn, yield an operational measure of the toxicity of dredged material on estuarine organisms.

To accurately delineate the effects of suspended dredged material on water quality, quantitative chemical analyses of the sediments should be pursued. Examples of specific chemical analyses of the sediments include the determination of sediment sorption capacities, ionic and cationic sediment exchange capacities, and sediment elemental partitioning.[47] The partitioning of contaminants between solid and solution phases, the persistence of the pollutants sorbed to sediments in the estuarine environment, and their effects on biota all are important in the assessment of impact of contaminants contained in dredged material. Processes such as dissolution, diagenesis, resuspension, and microbial degradation influence the flux and residence times of contaminants released with sediments into the water column during dredging and dredged-spoil disposal operations and, consequently, their distribution and fate in the estuary (Figure 4).

Nutrients

High concentrations of soluble phosphorus and nitrogen commonly occur in polluted as well as unpolluted estuarine sediments. These nutrients can be released when the sediments are suspended. Thus, Biggs[66] observed that phosphorus and nitrogen increased from 50 to 100 times in the immediate vicinity of the site of overboard dredging and spoil disposal in Chesapeake Bay. The addition of these nutrients to the water column promotes eutrophication at certain times in susceptible systems. For example, estuaries already high in natural nutrient levels may become eutrophic during the summer months when temperature and light conditions are favorable for plant growth.

Windom[57] cataloged the release of nutrients (nitrogen and phosphorus) from dispersed sediments during dredging in Charleston Harbor, SC, Cape Fear River, NC, and Terry Creek, GA. Initially, ammonia increased markedly in the water and nitrate levels were largely unaffected. Subsequently, ammonia concentrations declined due to uptake by microflora.

Phosphate fluxes were less predictable. Concentrations of phosphate initially increased in some cases but decreased in others. Adsorption of phosphate onto suspended sediments no doubt contributed to the amelioration of phosphate levels through time at some locations. Despite the variations in phosphate levels, the release of nitrogen compounds from dredged sediments had a greater effect on plant production.

BENEFICIAL EFFECTS
Increased Circulation

The deepening of waterways in estuaries results in improved circulation, leading to a more even distribution of temperature,

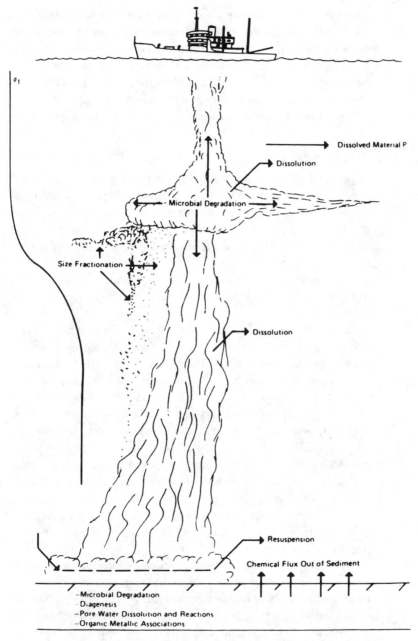

FIGURE 4. The processes influencing the distribution and fate of organic pollutants associated with dredged material. (From U.S. Army Corps of Engineers, Dredged-Material Research Program, Vicksburg, MS, 1973-1978.)

salinity, dissolved oxygen, and other physical-chemical conditions. These changes enable some species of organisms to spread and colonize broader habitat areas. The greater depths of waterways may facilitate fin- and shellfish migrations (e.g., flounders and shrimp), and for some species, create sanctuaries from low or high temperatures during winter and summer months, respectively. They may also generate additional spawning grounds for certain populations.

Sediment Supply and
Habitat Restoration

Dredged spoils, when environmentally compatible, have been used in salt marsh creation, spoil-island development, beach nourishment, and substrate enhancement. The utilization of dredged material in habitat creation serves as an attractive alternative to either the subaerial or land disposal of spoils commonly practiced because of the definite need for these types of habitats.[47] The consequential damage to wetlands and open-water sites attributable to the disposal of dredged material makes this alternative particularly enticing.

In some cases, the disposal of dredged spoils in water-bottom habitats can actually improve substrates and enhance the development of benthic communities. The upgrading of seafloor sediments by dredged-spoil disposal, while altering the natural substrate, may facilitate the growth of recreationally or commercially important shellfish (e.g., clams and oysters). The potential value of this method appears to be greatest where uncontaminated spoils cover sediments of poor quality that hinder the establishment of dense shellfish populations or preclude their harvest. However, since dredged spoils frequently are contaminated with heavy metals, chlorinated hydrocarbons, petroleum hydrocarbons, and other chemical compounds due to dredging of urban estuaries and other heavily utilized navigable waterways, they generally are of little or no value for substrate improvement of the benthic habitat.

The deposition of sandy dredged spoils on beaches depleted of sand as a result of rising sea level and coastal erosion may be beneficial not only to humans but to birds and other organisms as well. Coastal birds (e.g., black skimmers and terns) nest on bare sand and would benefit from the disposal of these sediments. Because coastal birds utilizing beaches play a significant role in marine food webs, the revitalization of these key habitats has important implications in marine ecology.

Nutrients

Dredged materials commonly contain high levels of nitrogen and phosphorus as stipulated previously and, therefore, may have value when dumped in subaerial environments. While the release of nutrients during dredging or dredged-spoil disposal can effectively enhance primary productivity in subaqueous environments, in many estuarine systems the release of additional nutrients has the adverse effect of promoting eutrophic conditions because of naturally high concentrations of nitrogen and phosphorus in bottom sediments or the water column.[67] Many shallow estuaries in the U.S., for example, receive large inputs of organic nutrients, exhibit high biological production, and are not in need of nutrient replenishment.

CASE STUDIES

CHESAPEAKE BAY

Dredging of Chesapeake Bay maintains the economic viability of major port cities, such as Baltimore and Norfolk.[25] Each year, more than 6000 ships carry approximately 103 million metric tons of cargo valued at more than $27,000 million to these two cities.[68] Consequently, it is necessary to periodically deepen the main shipping channel between Baltimore and the sea and the approach channels to Hampton Roads and Norfolk. The U.S. Army Corps of Engineers plans to dredge an estimated 39.5×10^6 m^3 of sediment from the Baltimore channel, deepening it from 12.8 to 15.2 m along 90-km segments of the bay. This project will take place over a 3.5-year period and cost more than $250 million to complete. Concurrently, dredging of the approach channels to Hampton Roads and Norfolk is designed to deepen the waterways from 13.7 to 16.8 m. With the advent of larger cargo ships and the expansion of port facilities in the decades ahead, future dredging of these waterways should extract considerably more sediment ($>210 \times 10^6$ m^3) from Chesapeake Bay.[69,70]

Nichols et al.[25] assessed the effects of dredging operations in central Chesapeake Bay by focusing on the responses of benthic

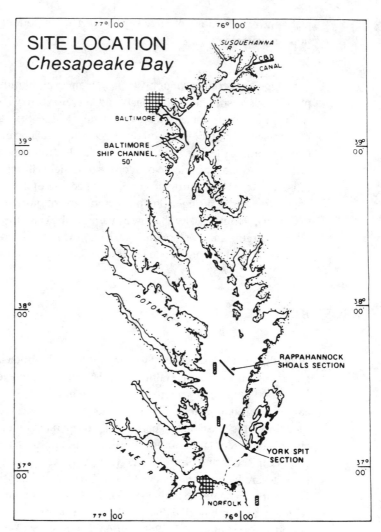

FIGURE 5.   Map of Chesapeake Bay showing the location of the Baltimore shipping channel, dredged-spoil disposal sites (hachured), and the benthic study area at Rappahannock Shoals in the central bay. (From Nichols, M., Diaz, R. J., and Schaffner, L. C., *Environ. Geol. Water Sci.,* 15, 31, 1990. With permission.)

macrofauna at Rappahannock Shoals (Figure 5) to the redeposition of dredged-induced suspended sediment. In 1987, the dominant benthic macrofaunal species in this area included the burrowing polychaetes, *Nephytes* cf. *cryptomma, Pseudeurythae paucibranchiata,* and *Sigambra tentaculata,* and the smaller tubi-

TABLE 4
Dominant Benthic Species around Rappahannock Shoals Channel, May 1987

Species	Average abundance/0.06 m2	
	Mean	SD
P[a] *Paraprionospio pinnata*	55.6	24.9
G *Cyclostremiscus pentagona*	18.9	57.9
P *Pseudeurythoe paucibranchiata*	17.7	17.3
P *Streblospio benedicti*	11.9	17.5
P *Mediomastus ambiseta*	11.5	18.2
N *Tubulanus pellucidus*	7.1	4.1
P *Bhawania heteroseta*	7.0	9.8
G *Anachis lafresnayi*	6.7	4.3
P *Nephtys* cf. *cryptomma*	4.4	3.0
P *Sigambra tentaculata*	3.8	3.6

[a] P = polychaete, N = Nemertean, G = Gastropod.

From Nichols, M., Diaz, R. J., and Schaffner, L. C., *Environ. Geol. Water Sci.,* 15, 31, 1990. With permission.

colous spionid polychaete, *Paraprionospio pinnata*. Somewhat less abundant was the deposit-feeding polychaete, *Clymenella torquata*.[71] A reevaluation of the benthic community in Rappahannock Shoals Channel in 1987 uncovered very similar dominance patterns among the macrofauna, with the four polychaete species, *P. pinnata, P. paucibranchiata, Streblospio benedicti,* and *Mediomastus ambiseta,* and one gastropod species, being most abundant (Table 4). Subsequent to dredging, polychaetes (67% of all individuals) and gastropods (17%) dominated the community followed by nemerteans (5%), bivalves (4%), amphipods (2%), and oligochaetes (1%).[72] No consistent pattern of increasing or decreasing abundance could be discerned with thickness of dredged material or distance from the channel. Community structure patterns also exhibited no clear pattern relative to dredged-material thickness. The total wet weight biomass of the benthic macrofauna likewise showed no pattern with depth of dredged material or distance from the channel, although a trend toward lower annelid biomass was evident near the channel.[25]

Results of this study indicate no significant impacts of the dredging operations on the benthic macrofauna of Rappahannock Shoals. The estuarine infaunal assemblage inhabiting this region of Chesapeake Bay is typically resilient. Dredging and dredged-material disposal clearly create an environmental perturbation, and both activities have the potential to impact the benthic community. However, the life history strategies and population dynamics of the benthos in this estuary enable them to cope quite successfully with the environmental changes.

Alberni Inlet, British Columbia

Levings et al.[73] studied the benthic communities of Alberni Inlet, concentrating on the spatial and temporal distribution of the constituent populations and their responses to dredged-material disposal. Located on the southwest coast of Vancouver Island, Alberni Inlet is a long (60 km), narrow (average width 1.3 km) fjord partially isolated from continental shelf waters by sills (Figure 6). In the past, Alberni Inlet has been the site of disposal of dredged material derived from Port Alberni, a harbor at the head of the inlet. As at other dredged-disposal sites in British Columbia, material dumped in the inlet is rich in wood fibers, bark, and chips originating from forest industries. Dredging concentrates the wood debris in bottom sediments, and dredging of harbor sediments removes them to disposal locations.

Crustaceans, mollusks, and polychaetes dominated the benthic infaunal communities of Alberni Inlet during the survey period. Based on cluster analysis of the benthic invertebrate community data, four provinces (i.e., central, wood debris, dumpsite, and dumpsite fringe) can be differentiated in the study area. *Axinopsida serricata*, *Aricidea lopezi* (Polychaeta), *Scoloplos pugettensis* (Polychaeta), *Eudorella pacifica* (Cumacea), *Heterophoxus oculatus* (Amphipoda), and *Macoma carlottensis* (bivalve) numerically dominated the central province. In the wood debris province, which is strongly influenced by waste wood from log storage and sorting operations, invertebrates that utilize wood for food (e.g., woodborers, *Limnoria lignorum* and *Xylophaga washingtoni*) or for habitat (e.g., *Nebalia pugettensis*) were predominant. At the dumpsite province, a zone of faunal obliteration existed; only the oppor-

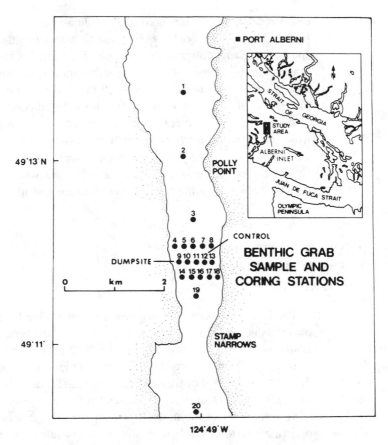

FIGURE 6. Dredged-material disposal study area in Alberni Inlet. (From Levings, C. D., Anderson, E. P., and O'Connell, G. W., in *Wastes in the Ocean,* Vol. 6, *Nearshore Waste Disposal,* Ketchum, B. H., Capuzzo, J. M., Burt, W. V., Duedall, I. W., Park, P. K., and Kester, D. R., Eds., John Wiley & Sons, New York, 1985, 131. With permission.)

tunistic polychaetes, *Capitella* sp., *Schistomeringos japonica,* and *S. rudolphi,* maintained populations there. The polychaetes, *Schistomeringos* spp. and *Trochochaeta multisetosa,* and the bivalve *Axinopsida serricata,* characterized the dumpsite fringe province.

Crustaceans appeared to be nearly totally excluded from the dumpsite. Only one bivalve, *A. serricata* inhabited the modified dumpsite sediments. The opportunistic polychaetes, *Capitella* spp., colonized dumpsite sediments during the dumping season, but other polychaete species probably were adversely affected by the dredged spoils. For example, eight polychaete species, relatively abundant at other sampling stations, were absent at the dumpsite. In conclusion, Levings et al.[73] deemed the disposal of dredged material to have a localized biological impact in Alberni Inlet. The dredged material disposed of in the inlet had a very restricted areal distribution, and it did not spread toward the center of the inlet. Nevertheless, in a fjord already stressed by pulp-mill effluent and log storage, the dredged spoils rich in wood debris represented an additional source of excess organic carbon and a physical disruption to the system.

REGULATION OF DREDGED MATERIAL DISPOSAL

The U.S. Army Corps of Engineers is responsible for dredged-material disposal in U.S. waters. Section 404 of the Federal Water Pollution Control Act Amendments of 1972 established a permit program to be administered by the Corps of Engineers to regulate the discharge of dredged or fill material in navigable waters of the U.S. The U.S. District Court in 1975 extended the responsibility of the U.S. Army Corps of Engineers to regulate the discharge of dredged or fill material to wetlands, both adjacent and isolated. Passage of the Clean Water Act of 1977 further refined the authority of the Corps of Engineers to regulate the disposal of dredged materials.[74]

Section 404 of the Clean Water Act requires authorization from the Department of the Army to discharge dredged or fill material into U.S. waters during construction of bridges, causeways, dams, and dikes. Beyond the construction of these structures, a Corps permit is required for the placement of fill needed for: (1) property protection or reclamation devices such as riprap, groins, seawalls, breakwaters, and revetments; (2) artificial islands; (3) beach nourishment; (4) levees; (5) creation of ponds; (6) intake and outfall pipes and subaqueous utility lines; (7) impoundments; (8) site-development for recreational, industrial, commercial, and other uses; (9) access roadways; and (10) any other work involving the discharge of fill or dredged material into waters or wetlands in the U.S. Moreover, the transportation of dredged material with the intention of dumping it into territorial seas requires authorization from the Secretary of the Army, acting through the Corps of Engineers, as noted in Section 103 of the Marine Protection, Research, and Sanctuaries Act of 1972, also known as the Ocean Dumping Act.[14,74,75]

Section 10 of the River and Harbor Act of 1899 also mandates the U.S. Army Corps of Engineers to regulate certain activities in U.S. waters. Specifically, Section 10 requires authorization from the Secretary of the Army, acting through the Corps of Engineers, for the excavation from or deposition of material in any navigable water of the U.S., the construction of any structure in such waters, and any obstruction or alteration therein. Not only does the law apply to dredging and dredged-spoil disposal activities, but also to excavation, filling, rechannelization, construction, and any other modification of navigable waters (defined as those subject to the ebb and flow of the tide shoreward to the mean high water mark and/or are presently used, or have been used in the past, or may be susceptible for use to transport interstate or foreign commerce). The construction of any structure in or over navigable waters, for example, jetties, groins, wharfs, weirs, bank protection (e.g., bulkhead, revetment, and riprap), moored structures (e.g., pilings), moored floating vessels, boat ramps, and recreational docks, requires compliance with Section 10.[74]

The primary intent of these federal laws is to regulate and limit adverse environmental effects of dredging, dredged-spoil disposal, construction, or any other modification of navigable waters and wetlands. In connection with this goal, the U.S. Army Corps of Engineers' permit program provides a mechanism to ensure that water resources in the U.S. are safeguarded and that they are used in the best interest of the people. To this end, the environmental, social, and economic concerns of the public are considered in the permitting program. The regulatory process, therefore, protects the nation's water resources, and controls their proper development and utilization for future generations.

SUMMARY AND CONCLUSIONS

Dredging involves the removal of sediment from underwater locations, whereas dredged-spoil disposal deals with the discharge, release, or dumping of this sediment at subaerial or subaqueous sites. The principal objective of dredging is to maintain the navigability of harbors and waterways. Depending on the type and volume of sediment, the water depth, and the cost of the sediment removal, various types of dredges may be used to dredge an area. They fall into two main categories, however, namely mechanical and hydraulic devices. Mechanical dredges include bucket, dipper, and ladder dredges; hydraulic dredges encompass agitation, dustpan, cutterhead, hopper, sidecasting, and suction types. During mechanical dredging, sediment is lifted from the seafloor and transported to a disposal location. During hydraulic dredging, agitation by water jets or the action of cutterheads loosens sediment creating a slurry that can be easily transported via a pipeline to a dredged-material disposal area. Alternately, the slurry is stored in hoppers, moved to disposal locales, and dumped. Hydraulic dredges offer a number of advantages over mechanical dredges. Perhaps most importantly, they are more efficient and economical to operate.[76]

The environmental impacts of dredging and dredged-spoil disposal in estuaries mainly concern direct effects on habitat and organisms or indirect effects attributable to changes in water quality. The destruction of habitat by the removal of sediment and the increase in mortality of benthic organisms owing to the mechanical action of the dredge and smothering by sediment account for most of the dredging impacts. Recovery of a dredged site by benthic communities is temporally and spatially variable and site specific, but usually requires at least 6 months to 1 year or more to complete. Opportunistic, pioneering invertebrate species typically comprise the initial colonizers of a dredged-spoil dumpsite. Equilibrium assemblages of benthos succeed these opportunistic forms as time passes. Gradually, the same species that inhabited the dredged-spoil site prior to the disturbance return to repopulate the area. The more motile invertebrates generally incur less mortality than the sedentary types. Two factors that can increase the probability of death of benthic fauna are the release of large concentrations of toxic contaminants from the sediments or acute changes in sediment type during dredging and dredged-spoil disposal operations.

Benthic flora seem to be more easily impacted by dredging activities than benthic fauna. They are directly killed by the action of the dredge itself and secondary effects, such as high turbidity levels, can limit their production. The dumping of spoils over plant communities inhabiting subaqueous or subaerial environments destroys many hectares of plant communities each year.

In the past, coastal marshes have been heavily utilized as waste-disposal repositories. However, the use of coastal marshes and other wetlands as dredged-spoil disposal sites has been curtailed in recent years because of the recognition that these areas are important nursery and breeding grounds for many estuarine organisms and zones of high biological productivity for estuaries and nearshore oceanic regions. Hence, depending on the conditions, open-water release of dredged spoils may be the preferred method of disposal. Indeed, it is necessary to distinguish uncontaminated and contaminated sediment when selecting a particular method of disposal. Coastal revitalization often entails the use of uncontaminated sediment as landfill for shoreline modification or land improvement, such as in the filling of wetlands, sanitary landfill cover, and beach and agricultural soil replenishment. This sediment can be utilized whenever possible to develop habitats (e.g., artificial islands, marshes, and reefs) or improve existing shoreline areas. Contaminated sediment, on the other hand, poses more pressing problems for disposal in estuaries and adjacent biotopes. Open-water disposal of contaminated spoils may be the most practical alternative when the degree of contamination does not contribute to any ecological harm of an area. For most contaminated spoils, however, the disposal alternative usually selected consists of some type of containment of the material (e.g., a protective cap). In the case of open-water disposal, two types of sites — retentive and dispersive — are normally considered. Retentive sites confine the spoils to an isolated area, whereas dispersive sites promote the transportation of disposed sediments away from the site.

Changes in water quality in estuaries associated with dredging and dredged-spoil disposal commonly relate to increases in turbidity, nutrients, and contaminants. These changes can be detrimental to estuarine communities, hindering the establishment of constituent populations, especially among the benthos. However, the deepening of waterways in estuaries often improves circulation substantially, resulting in a more even distribution of temperature, salinity, dissolved oxygen, and other physical-chemical factors which may enable organisms to spread and colonize a greater habitat area.

Dredging and dredged-spoil disposal constitute global problems in estuaries and fringing wetlands. The movement of large volumes of sediment alters sensitive habitats, impacts organisms, and modifies the water quality of estuarine systems. The remobilization of contaminants stored in bottom sediments can raise the contaminant body burdens of organisms, particularly those located in dredged mounds not adequately designed. These contaminants at times enter food chains where they pose a potential threat to estuarine organisms as well as to humans. When the contaminated spoils are properly disposed of in capped mound deposits above the seafloor, in depressions below capped deposits, and in subaqueous depressions, the probability of food-chain contamination and modification declines dramatically. Even for uncontaminated sediments, dredging and dredged-spoil disposal activities influence estuarine communities, most noticeably in the benthos, as reflected in case studies of Chesapeake Bay and Alberni Inlet.

REFERENCES

1. Wilson, J. G., *The Biology of Estuarine Management,* Croom Helm, London, 1988.
2. Gordon, R. B., Bohlen, W. F., Bokuniewicz, H. J., de Picciotto, M., Johnson, J., Kamlet, K. S., McKinney, T. F., Schubel, J. R., Suszkowski, D. J., and Wright, T. D., Management of dredged material, in *Ecological Stress and the New York Bight: Science and Management,* Mayer, G. F., Ed., Estuarine Research Federation, Columbia, SC, 1982, 113.
3. Nichols, M. M., The problem of misplaced sediment, in *Ocean Dumping and Marine Pollution,* Palmer, H. D. and Gross, M. G., Eds., Dowden, Hutchinson & Ross, Stroudsburg, PA, 1979, 147.
4. Capuzzo, J. M., Burt, W. V., Duedall, I. W., Park, P. K., and Kester, D. R., The impact of waste disposal in nearshore environments, in *Wastes in the Ocean,* Vol. 6, *Nearshore Waste Disposal,* Ketchum, B. H., Capuzzo, J. M., Burt, W. V., Duedall, I. W., Park, P. K., and Kester, D. R., Eds., John Wiley & Sons, New York, 1985, 3.

5. Windom, H. L., Environmental aspects of dredging in the coastal zone, *Crit. Rev. Environ. Cont.,* 6, 91, 1976.
6. Stickney, R. R., *Estuarine Ecology of the Southeastern United States and Gulf of Mexico,* Texas A & M University Press, College Station, 1984.
7. Windom, H. L., Environmental aspects of dredging in estuaries, *J. Waterways, Harbors, Coastal Eng. Div. ASCE,* 98(WW4), 475, 1972.
8. Clark, R. B., *Marine Pollution,* 2nd ed., Clarendon Press, Oxford, 1989.
9. Rhoads, D.C., McCall, P. L., and Yingst, J. Y., Disturbance and production on the estuarine seafloor, *Am. Sci.,* 66, 577, 1978.
10. Goldberg, E. D., Proceedings of a Workshop on Assimilative Capacity of U.S. Coastal Waters for Pollutants, Crystal Mountain, Washington, Special Report, Environmental Research Laboratories, U.S. National Oceanic and Atmospheric Administration, Boulder, CO, 1979.
11. Kester, D. R., Ketchum, B. H., Duedall, I. W., and Park, P. K., The problem of dredged-material disposal, in *Wastes in the Ocean,* Vol. 2, *Dredged-Material Disposal in the Ocean,* Kester, D. R., Ketchum, B. H., Duedall, I. W., and Park, P. K., Eds., John Wiley & Sons, New York, 1983, 3.
12. Bokuniewicz, H. J., Submarine borrow pits as containment sites for dredged sediment, in *Wastes in the Ocean,* Vol. 2, *Dredged-Material Disposal in the Ocean,* Kester, D. R., Ketchum, B. H., Duedall, I. W., and Park, P. K., Eds., John Wiley & Sons, New York, 1983, 215.
13. Bokuniewicz, H. J., Energetics of dredged-material dispersal, in *Wastes in the Ocean,* Vol. 6, *Nearshore Waste Disposal,* Ketchum, B. H., Capuzzo, J. M., Burt, W. V., Duedall, I. W., Park, P. K., and Kester, D. R., Eds., John Wiley & Sons, New York, 1985, 305.
14. Kamlet, K. S., Dredged-material ocean dumping: perspectives on legal and environmental impacts, in *Wastes in the Ocean,* Vol. 2, *Dredged-Material Disposal in the Ocean,* Kester, D. R., Ketchum, B. H., Duedall, I. W., and Park, P. K., Eds., John Wiley & Sons, New York, 1983, 29.
15. van Driel, W. and Nijssen, J. P. J., Development of dredged-material disposal sites: implications for soil, flora, and food quality, in *Chemistry and Biology of Solid Waste: Dredged Material and Mine Tailings,* Salomons, W. and Förstner, U., Eds., Springer-Verlag, Berlin, 1988, 101.
16. Ahlf, W. and Munawar, M., Biological assessment of environmental impact of dredged material, in *Chemistry and Biology of Solid Waste: Dredged Material and Mine Tailings,* Salomons, W. and Förstner, U., Eds., Springer-Verlag, Berlin, 1988, 127.
17. Engler, R. M., Prediction of polluted potential through geochemical and biological procedures: development of regulatory guidelines and criteria for the discharge of dredged-fill material, in *Contaminants and Sediments,* Baker, R. A., Ed., Ann Arbor Science Publishers, Ann Arbor, MI, 1980, 143.
18. Machemehl, J. L., Mechanics of dredging and filling, in Proceedings of the Seminar on Planning and Engineering in the Coastal Zone, Coastal Plains Center for Marine Development Service, Washington, D.C., 1972.

19. Mohr, A. W., Mechanical dredges, in *Dredging and Its Environmental Effects,* Krenkel, P. A., Harrison, J., and Burdick, J. C., III, Eds., American Society of Civil Engineers, New York, 1976, 125.

20. Bokuniewicz, H., Behavior of sand caps on subaqueous dredged-sediment disposal sites, in *Oceanic Processes in Marine Pollution,* Vol. 4, *Scientific Monitoring Strategies for Ocean Waste Disposal,* Hood, D. W., Schoener, A., and Park, P. K., Eds., Robert E. Krieger Publishing, Malabar, FL, 1989, 221.

21. Gordon, R. B., Dispersion of dredged spoil dumped in nearshore waters, *Est. Coastal Mar. Sci.,* 2, 349, 1974.

22. Sustar, C. and Wakeman, T., Dredged-Disposal Study: San Francisco Bay and Estuary, Main Report, U.S. Army Corps of Engineers, San Francisco, 1977.

23. Tavolaro, J., A sediment budget study of the clamshell dredging and ocean disposal activities in the New York Bight, *Environ. Geol. Water Sci.,* 6, 133, 1984.

24. U.S. Army Corps of Engineers, Final Environmental Statement: Operation and Maintenance of the New Jersey Intracoastal Waterway and Manasquan, Barnegat, Absecon, and Cold Spring Inlets, New Jersey, Unpubl. Tech. Rep., U.S. Army Corps of Engineers, Philadelphia, 1975.

25. Nichols, M., Diaz, R. J., and Schaffner, L. C., Effects of hopper dredging and sediment dispersion, Chesapeake Bay, *Environ. Geol. Water Sci.,* 15, 31, 1990.

26. Rhoads, D. C. and Boyer, L. F., The effects of marine benthos on physical properties of sediments: a successional perspective, in *Animal-Sediment Relations: The Biogenic Alteration of Sediments,* McCall, P. L. and Tevesz, M. J. S., Eds., Plenum Press, New York, 1982, 3.

27. Hard, C. G., Aspects of dredged-material research in New England, in *Estuarine Research,* Vol. 2, *Geology and Engineering,* Cronin, L. E., Ed., Academic Press, New York, 1975, 537.

28. Maurer, D., Keck, R. T., Tinsman, J. C., and Leathem, W. A., Vertical migration and mortality of benthos in dredged material. I. Mollusca, *Mar. Environ. Res.,* 4, 299, 1981.

29. Bybee, J. R., Effects of hydraulic pumping operations on the fauna of Tijuana Slough, *Calif. Fish Game,* 55, 213, 1969.

30. Bonvicini Pagliai, A. M., Cognetti Varriale, A. M., Crema, R., Curini Galletti, M., and Vandini Zunarelli, R., Environmental impact of extensive dredging in a coastal marine area, *Mar. Pollut. Bull.,* 16, 483, 1985.

31. Stickney, R. R. and Perlmutter, D., Impact of intracoastal waterway dredging on a mud bottom benthos community, *Biol. Conserv.,* 7, 211, 1975.

32. Conner, W. G. and Simon, J. L., The effects of oyster shell dredging on an estuarine benthic community, *Est. Coastal Mar. Sci.,* 9, 749, 1979.

33. Bornsdorff, E. J., Macrozoobenthic recolonization of a dredged brackish water bay in SW Finland, *Ophelia,* Suppl. 1, 145, 1980.

34. Groot, S. J., An assessment of the potential environmental impact of large scale sand dredging for the building of artificial islands in the North Sea, *Ocean. Manage.,* 5, 211, 1979.

35. Godcharles, M. F., A Study of the Effects of a Commercial Hydraulic Clam Dredge on Benthic Communities in Estuarine Areas, Marine Research Laboratory Tech. Ser. No. 64, Florida Department of Natural Resources, St. Petersburg, 1971.

36. Corliss, J. and Trent, L., Comparison of phytoplankton production between natural and altered areas in West Bay, Texas, *Fish. Bull. U.S.,* 69, 829, 1971.

37. Gross, M. G., *Oceanography: A View of the Earth,* 3rd ed., Prentice-Hall, Englewood Cliffs, NJ, 1982.

38. Gross, M. G., Human impact, in *The Encyclopedia of Beaches and Coastal Environments,* Schwartz, M. L., Ed., Hutchinson & Ross, Stroudsburg, PA, 1982, 465.

39. Cahoon, D. R. and Cowan, J. H., Jr., Environmental impacts and regulatory policy implications of spray disposal of dredged material in Louisiana wetlands, *Coastal Manage.,* 16, 341, 1988.

40. Bettinger, K. M. and Hamilton, R. B., Avian use of levee habitat types, Rockefeller Wildlife Refuge, Louisiana, in *Proc. 4th Coastal Marsh and Estuary Management Symp.,* Bryan, C. F., Zwank, P. J., and Chabreck, R. H., Eds., Louisiana State University Printing Office, Baton Rouge, 1985, 165.

41. Swenson, E. M. and Turner, R. E., Spoil banks: effects on a coastal marsh water level regime, *Est. Coastal Shelf Sci.,* 24, 599, 1987.

42. Turner, R. E. and Cahoon, D. R., Eds., Causes of Wetland Loss in the Coastal Central Gulf of Mexico, Vol. 2, Technical Narrative, Final Tech. Rep., Minerals Management Service, New Orleans, Contract No. 14-13-0001-30252, OCS Study/MMS 87-0120, New Orleans, 1987.

43. Lindstedt, D. and Nunn, L., Petroleum development in Louisiana's coastal zone, in *Proc. of the 4th Symposium on Coastal and Ocean Management,* July 1985, Baltimore, Md., (Coastal Zone '85), Vol. 2, Magoon, O. T., Miner, D., Clark, D., and Tobin, L. T., Eds., American Society of Civil Engineers, New York, 1985, 1410.

44. Zedler, J. B., The Ecology of Southern California Coastal Marshes: A Community Profile, FWS/OS-81/54, U.S. Fish and Wildlife Service, Washington, D.C., 1982.

45. Day, J. W., Jr., Hall, C. A. S., Kemp, W. M., and Yàñez-Arancibia, A., *Estuarine Ecology,* John Wiley & Sons, New York, 1989.

46. Hackney, C. T. and Cleary, W. J., Saltmarsh loss in southeastern North Carolina lagoons: importance of sea level rise and inlet dredging, *J. Coastal Res.,* 3, 93, 1987.

47. Kirby, C. J., Keeley, J. W., and Harrison, J., An overview of the technical aspects of the corps of engineers national dredged-material research program, in *Estuarine Research,* Vol. 2, *Geology and Engineering,* Cronin, L. E., Ed., Academic Press, New York, 1975, 523.

48. Truitt, C. L., Dredged-material behavior during open-water disposal, *J. Coastal Res.*, 4, 489, 1988.

49. Palermo, M. R. and Thackston, E. L., Flocculent settling above zone settling interface, *J. Environ. Eng.*, 114, 770, 1988.

50. Palermo, M. R. and Thackston, E. L., Test for dredged-material effluent quality, *J. Environ. Eng.*, 114, 1295, 1988.

51. Palermo, M. R. and Thackston, E. L., Verification of predictions of dredged-material effluent quality, *J. Environ. Eng.*, 114, 1310, 1988.

52. McAnally, W. H. and Adamec, S. A., Jr., Designing Open Water Disposal for Dredged Muddy Sediments, Unpubl. Tech. Rep., U.S. Army Engineer Waterways Experiment Station, Vicksburg, MS, 1987.

53. O'Connor, J. M. and O'Connor, S. G., Evaluation of the 1980 Capping Operations at the Experimental Mud Dump Site, New York Bight Apex, Tech. Rep. D-83-3, U.S. Army Engineer Waterways Experiment Station, Vicksburg, MS, 1983.

54. Parker, J. H. and Valente, R. M., Long-term Sand Cap Stability: New York Dredged-Material Disposal Site, Contract Rep. CERC-88-2, U.S. Army Engineer Waterways Experiment Station, Vicksburg, MS, 1988.

55. Brannon, J. M., Hoeppel, R. E., and Gunnison, D., Capping contaminated dredged material, *Mar. Pollut. Bull.*, 18, 175, 1987.

56. Brannon, J. M. and Poindexter-Rollings, M. E., Consolidation and contaminant migration in a capped dredged-material deposit, *Sci. Total Environ.*, 91, 115, 1990.

57. Windom, H. L., Water-quality aspects of dredging and dredge-spoil disposal in estuarine environments, in *Estuarine Research,* Vol. 2, *Geology and Engineering,* Cronin, L. E., Ed., Academic Press, New York, 1975, 559.

58. Schaffner, L. C., Diaz, R. J., Olsen, C. R., and Larsen, I. L., Faunal characteristics and sediment accumulation processes in the James River estuary, Virginia, *Est. Coastal Shelf Sci.*, 25, 211, 1987.

59. Alden, R. W., III, Butt, A. J., and Young, R. J., Jr., Toxicity testing of sublethal effects of dredged materials, *Arch. Environ. Contam. Toxicol.*, 17, 381, 1989.

60. Furness, R. W. and Rainbow, P. S., *Heavy Metals in the Marine Environment,* CRC Press, Boca Raton, FL, 1990.

61. Lunsford, C. A., Weinstein, M. P., and Scott, L., Uptake of Kepone® by the estuarine bivalve, *Rangia cuneata,* during the dredging of contaminated sediments in the James River, Virginia, *Water Res.*, 21, 411, 1987.

62. Hall, L. A., The effects of dredging and reclamation on metal levels in water and sediments from an estuarine environment of Trinidad, West Indies, *Environ. Pollut.*, 56, 189, 1989.

63. Fowler, S. W., Critical review of selected heavy metal and chlorinated hydrocarbon concentrations in the marine environment, *Mar. Environ. Res.*, 29, 1, 1990.

64. Balls, P. W., Copper, lead, and cadmium in coastal waters of the western North Sea, *Mar. Chem.*, 15, 363, 1985.

65. Figueres, G., Martin, J. M., Meybeck, M., and Seyler, P., A comparative study of mercury contamination in the Tagus estuary (Portugal) and major French estuaries (Gironde, Loire, Rhône), *Est. Coastal Shelf Sci.,* 20, 183, 1985.

66. Biggs, R. B., Environmental effects of overboard spoil disposal, *J. Sanitary Eng. Div.,* 94 (SA3), 477, 1968.

67. Jaworski, N. A., Retrospective study of the water quality issues of the upper Potomac estuary, *Rev. Aquat. Sci.,* 3, 11, 1990.

68. The Port of Greater Hampton Roads, Annu. Tech. Rep., Hampton Roads Maritime Association, Hampton Roads, VA, 1985.

69. Gross, M. G. and Cronin, W. B., Dredging and disposal in Chesapeake Bay, 1975 to 2025, in *Ocean Dumping and Marine Pollution,* Palmer, H. D. and Gross, M. G., Eds., Dowden, Hutchinson & Ross, Stroudsburg, PA, 1979, 131.

70. Nichols, M., What are the best guidelines for dredging and placement of dredged material?, in *Ten Critical Questions for Chesapeake Bay in Research and Related Matters,* Cronin, L. E., Ed., Chesapeake Research Consortium, Baltimore, 1983, 113.

71. Diaz, R. J., Schaffner, L. C., Byrne, R. J., and Gammisch, D. R., Baltimore Harbor and Channels Aquatic Benthos Investigations, Final Report to the U.S. Army Corps of Engineers Baltimore District, Baltimore, 1985.

72. Nichols, M. and Diaz, R., Plume Monitoring of Rappahannock and York Spit Channels, Baltimore Harbor and Channels (Phase 1), Contract Rep. to the U.S. Army Corps of Engineers Baltimore District, Baltimore, 1987.

73. Levings, C. D., Anderson, E. P., and O'Connell, G. W., Biological effects of dredged-material disposal in Alberni Inlet, in *Wastes in the Ocean,* Vol. 6, *Nearshore Waste Disposal,* Ketchum, B. H., Capuzzo, J. M., Burt, W. V., Duedall, I. W., Park, P. K., and Kester, D. R., John Wiley & Sons, New York, 1985, 131.

74. U.S. Army Corps of Engineers Philadelphia District, Are You Planning Work in a Waterway or Wetland?, Unpubl. Tech. Rep., U.S. Army Corps of Engineers, Philadelphia, 1990.

75. Peddicord, R. K. and Hansen, J. C., Technical implementation of the regulations governing ocean disposal of dredged material, in *Wastes in the Ocean,* Vol. 6, *Nearshore Waste Disposal,* Ketchum, B. H., Capuzzo, J. M., Burt, W. V., Duedall, I. W., Park, P. K., and Kester, D. R., John Wiley & Sons, New York, 1985, 71.

76. Herbich, J. B. and Haney, J. P., Dredging, in *The Encyclopedia of Beaches and Coastal Environments,* Schwartz, M. L., Ed., Hutchinson & Ross, Stroudsburg, PA, 1982, 379.

77. Morton, R. W., Precision bathymetric study of dredged-material capping experiment in Long Island Sound, in *Wastes in the Ocean,* Vol. 2, *Dredged-Material Disposal in the Ocean,* Kester, D. R., Ketchum, B. H., Duedall, I. W., and Park, P. K., Eds., John Wiley & Sons, New York, 1983, 99.

8 Effects of Electric Generating Stations

INTRODUCTION

The construction and operation of electric generating stations in the coastal zone cause a wide range of ecological impacts on aquatic communities. The excavation of sediment at the site of power stations, together with the dredging of intake and discharge canals, docks, and other structures during construction, can produce significant changes in natural sedimentation patterns of nearby waters manifested in increased sediment loads potentially dangerous to benthic organisms. Other adverse effects of the construction of power plants involve alterations in the local hydrography and water quality of receiving waters. Calefaction or thermal loading related to cooling water discharges directly interferes with physiological processes of biota, such as enzyme activity, feeding, reproduction, respiration, and photosynthesis. Behavioral changes (e.g., attraction or avoidance responses) are commonly observed in organisms subjected to thermal discharges as well. Less conspicuous effects include indirect stresses on individuals due to their greater vulnerability to disease, to changing gaseous solubilities, and to chemical toxicants associated with thermal enrichment.[1] In the case of open-cycle, once-through cooling systems, the temperature of the heated discharge released from condensers is quite high, amounting to an average of approximately 12°C above ambient levels for 100 MWe oil or coal-fired power stations and about 15°C above ambient for 1000 MWe nuclear power facilities.[2]

Of greater potential impact on the biotic communities of estuaries than waste heat discharges are the losses of various life history stages of invertebrates and fishes impinged on intake screens or entrained in condenser cooling systems. It is not uncommon for millions of fish and invertebrates to become impinged on power plant intake screens each year. Many of these organisms die from asphyxiation resulting from constrained gill movements or from internal and external mechanical damage from operation of the traveling screens and screen wash sprays. Mortality of these fauna may be immediate or latent.[1]

Many more eggs and larvae of invertebrates and fishes are killed because of entrainment. The absolute numbers exceed 10^{10} individuals annually at some power plants for susceptible forms. The following biotic factors influence the overall impact of entrainment on estuarine populations: (1) the species composition and abundance of affected organisms; (2) their abundance in the adjacent waters; (3) their survival rates during entrainment relative to natural survival; and (4) the ecological roles of entrained organisms and their reproductive strategies.[3] As noted by Marcy et al.,[3] abiotic factors affecting entrainment are the site of the power plant, the cooling system design and operation, the volume of cooling water withdrawn, and the ambient conditions of the cooling water.

In addition to power plant impacts ascribable to thermal discharges, impingement, and entrainment, other adverse effects on aquatic organisms may arise from the use of biocides to control fouling organisms on heat exchanger surfaces and the leaching of copper or other heavy metals from condenser tubes. Nuclear power plants also release low-level radioactive waste in liquid discharges. The concentrations of these radionuclides do not pose a hazard to estuarine organisms or humans. Finally, stack emissions from fossil-fueled power plants contain polycyclic aromatic hydrocarbons (PAHs) that accumulate in the bottom sediments of estuarine and marine environments where they can contaminate benthic communities.[1]

This chapter describes the effects of electric generating stations on estuarine organisms. It focuses on three principal areas of impact: thermal discharges, impingement, and entrainment. Also discussed are the potential problems of biocides used to reduce biofouling in the cooling water circuit of these facilities. The overall impact of a power plant on the estuarine environment depends greatly on its size and, therefore, its cooling water demands, as well as its siting. Hence, it is necessary to assess electric generating stations on a case-by-case basis to accurately determine their impact on the estuarine environment.

HISTORICAL DEVELOPMENT

In the 1950s and 1960s, much of the environmental concern of power plant operation surrounded the biological effects of thermal effluents. A series of papers published in the late 1950s highlight the nature of the problem of power plants being addressed at that time.[4-8] The work of Markowski,[8] in particular, ushered in a decade of ecological studies on the calefaction of riverine, estuarine, and coastal marine waters by power plants. This period of investigation culminated with a national symposium on thermal pollution co-sponsored by Vanderbilt University (Nashville, TN) and the Federal Water Pollution Control Administration of the U.S. Department of the Interior held at Portland, OR from June 3 to 5, 1968. The book, *Biological Aspects of Thermal Pollution,*[9] contains the papers and discussions presented at the symposium.

Environmental impact assessments of power plants shifted some-what in the 1970s to emphasize the entire cooling system and to examine in detail the possible impacts of entrainment and impinge-ment on aquatic communities.[10] With increasing societal awareness of these environmental issues in the 1960s and the subsequent passage of federal legislation, such as the National Environmental Policy Act (NEPA) of 1969, the Federal Water Pollution Control Act Amendments (Public Law 92-500) of 1972, and the Clean Water Act of 1977 (Public Law 95-217), efforts were made to evaluate overall plant impacts on estuarine environments. The legal provisions of Public Law 92-500 dealing with thermal discharge effects and entrainment and impingement effects are Sections 316a and 316b, respectively. Section 316b requires the use of the best available cooling system technology with regard to the design location, construction, and capacity of intake structures.[11] For the large electric generating stations in operation today, this usually is inter-preted to mean dry or wet cooling towers.[1]

As regulatory attention suddenly turned in the 1970s to entrain-ment and impingement impacts, five entrainment and impingement workshops were held over a 10-year period (1971 to 1980), bringing together scientists from industry, government, academia, and consulting agencies. The purpose of these workshops was to gain a better understanding of entrainment and impingement problems plaguing power plants, the regulatory issues surrounding them, and the new approaches and techniques of data assessment that could lead to a mitigation of impacts. A list of these workshops follows:

1. First National Workshop on Entrainment, Impingement, and Intake Screening held at the Johns Hopkins University in Baltimore, MD, in 1971.
2. Second National Workshop on Entrainment and Impingement held at The Johns Hopkins University in Baltimore, MD, in 1973.
3. Third National Workshop on Entrainment and Impingement held in New York City in 1975.
4. Fourth National Workshop on Entrainment and Impingement held in Chicago in 1977.
5. Fifth National Workshop on Entrainment and Impingement held in San Francisco in 1980.[12]

Efforts during the last decade have attempted to place the environmental impacts of power plants — those associated with thermal discharges, entrainment, and impingement — into a reasoned perspective whereby the most serious and insidious pollutant problems are addressed in a cost-effective, prioritized manner.[12] These efforts entail both qualitative and quantitative impact assessments. Only then can marine scientists accurately delineate the importance of thermal discharges, entrainment, and impingement relative to other estuarine impacts.

EFFECTS OF POWER PLANT OPERATION

THERMAL DISCHARGES

The calefaction or thermal loading of estuaries is most commonly coupled to waste heat releases from power plants. The electric utility industry alone accounts for more than two thirds of all water used by industry for cooling purposes.[11,13] Power plants, whether they be conventional (coal, oil, or gas fired) or nuclear, steam electric generating facilities, require large volumes of water for cooling because much of the heat energy produced by them is rejected in the condensing process. For example, fossil-fueled power plants operate at about 40% efficiency, meaning that approximately 60% of the heat energy generated must be rejected to the environment. The picture for nuclear power stations is even more bleak, with their thermal efficiency placed at only about 33%.[14] Thus, the average heat rejection of fossil-fueled generating plants exceeds 5500 Btu/kWh of electricity produced, and that of nuclear facilities about 50% greater.[15,16] Since the cooling water requirements of a 1000 MW nuclear power plant exceed 3.0×10^6 l/min and approach 5 billion l/d,[17] these electric generating facilities have been located at sites where adequate quantities of cooling water can be ensured, and where they are protected. Consequently, many of them have been built along estuarine waters during the last 25 years.

Problems related to the calefaction of estuarine waters due to waste heat discharges of electric generating plants were less acute prior to 1960. This is because the size of power stations in the U.S. increased substantially during the last 30 years. For instance, in 1955 the average size of electric generating plants in the U.S. amounted to 35 MW; the largest plant equalled 300 MW.[16] A trend toward larger units first occurred in the 1960s. Most electric generating facilities constructed in the 1970s were 500 MW or more, and of those built in the 1980s, an increasing number of them surpassed 1000 MW. Between 1970 and 1986, the number of nuclear power plants worldwide grew from 66 to 398, with many being >1000 MW.

As Tebo and Little[16] invoke, the average quantity of condenser cooling water required for operation of fossil-fueled and nuclear power plants combined approximates 3.4 m³/MW, yielding a temperature rise of 8 to 9°C (15 to 16°F). Two 500-MW units, therefore, use about 57 m³/s of condenser cooling water and discharge the effluent at temperatures 8 to 9°C (15 to 16°F) above ambient levels. Table 1 shows cooling water requirements for a 1000 MW steam-electric power plant.

All waste heat discharged from a power station is ultimately dissipated in the environment. Potentially detrimental increases in the temperature of a receiving body take place when the volume of

TABLE 1
Estimated Cooling Water Requirements for a 1000 MWe Steam-Electric Plant Operating at Full Load

	Type of plant			
	Fossil		Nuclear	
	(1980)	(1990)	(1980)	(1990)
Plant heat rate,[a] Btu/kWh	9500	9000	10,500	10,000
Condenser flows-cms, for various temperature rises across the condenser				
10°F (5.5°C)	58.9	53.5	82.6	76.7
15°F (8.3°C)	39.3	35.7	55.2	51.2
20°F (11.1°C)	29.4	26.9	41.2	38.5
Consumptive use, cms for various types of cooling				
Once through	0.34	0.28	0.48	0.42
Cooling ponds[b]	0.45	0.40	0.62	0.57
Cooling towers[c]	0.79	0.74	1.13	0.99

[a] For fossil-fueled plants in operation in 1970, a heat rate of 10,000 Btu/kWh and a temperature rise of 13°F (7.2°C) were assumed, except where reported heat rate data were available.
[b] Where appropriate, an additional allowance was made for natural evaporation from the pond surface.
[c] Evaporative towers; includes blowdown and drift.

From Eicholz, G. G., *Environmental Aspects of Nuclear Power*, Ann Arbor Science Publishers, Ann Arbor, MI, 1976. With permission.

cooling water discharges is high relative to the dissipative capacity of the receiving body. Thermal impact is most likely to develop in shallow, enclosed, and/or poorly mixed estuaries, and in systems vulnerable to the siting of clusters of power plants that strain available water resources. Examples of concentrations of power plants on receiving bodies include a series of five generating facilities on the mid-Hudson River and a series of stations sited along the subestuaries of the Chesapeake Bay estuarine system. These power plants have been the subject of ongoing ecological investigations for years.

Though it is possible that thermal discharges can selectively eliminate large components of a healthy aquatic ecosystem, most impacts appear to be less pronounced.[18] The major adverse effects of waste heat occur in the near-field region closest to a power station.[19] Here, events such as heat shock and cold shock mortality of biota are more probable than in areas farther removed from the plant site. However, subtle effects on aquatic life may be evident even in these far-field areas where temperature changes of only 1 or 2°C elicit physiological or behavioral responses in organisms. The ability of fish to thermoregulate behaviorally, (i.e., to avoid or select environmental temperature),[20] for example, is apparent in estuaries receiving thermal discharges. Certain finfish species will avoid or be attracted to a thermal plume, depending on the season of the year and other factors.

Water Quality

Temperature strongly influences the physical properties of water (Table 2). As a result, the calefaction of estuarine waters via power plants can alter water quality either directly or indirectly. Oxygen solubility in estuaries is a function of temperature, with values declining from 14.6 mg/l at 0°C to 6.6 mg/l at 40°C. While the oxygen demand of biota rises at higher temperatures, less dissolved oxygen exists in the water, creating life-threatening conditions under certain circumstances. The synergistic effect of elevated water temperatures and organic loading together with an accelerated rate of bacterial respiration in the summer may promote anoxia or hypoxia in susceptible systems, thereby contributing to the elimination of many forms of aquatic life. The occurrence of summer anoxia in the deep waters of the mesohaline portion of Chesapeake Bay provides an example. Summer oxygen depletion in the deeper waters of this bay has been an annual event throughout the historical record.[21] This type of problem can be exacerbated by anthropogenic activities in some estuaries, especially the input of thermal effluent, which may further depress dissolved oxygen levels.

Other water properties — density, viscosity, surface tension, and nitrogen solubility — also decrease with increasing temperature (Table 2). These alterations likewise have far-reaching implications for the estuarine ecosystem. Density differences, of course, foster stratification of water bodies, although circulation and depth affect stratification stability as well. Hence, temperature-induced changes in density owing to power plant thermal effluents have potentially

TABLE 2
Water Properties as a Function of Temperature

Temp. (°C)	Vapor Pressure (torr)	Viscosity (centipoise)	Density (g/ml)	Surface tension (dynes/cm)	Oxygen solubility (mg/l)	Nitrogen solubility (mg/l)
0	4.579	1.787	0.99984	75.6	14.6	23.1
5	6.543	1.519	0.99997	74.9	12.8	20.4
10	9.209	1.307	0.99970	74.2	11.3	18.1
15	12.788	1.139	0.99910	73.5	10.2	16.3
20	17.535	1.002	0.99820	72.8	9.2	14.9
25	23.756	0.890	0.99704	72.0	8.4	13.7
30	31.824	0.798	0.99565	71.2	7.6	12.7
35	42.175	0.719	0.99406	70.4	7.1	11.6
40	55.324	0.653	0.99224	69.6	6.6	10.8

From Eicholz, G. G., *Environmental Aspects of Nuclear Power*, Ann Arbor Science Publishers, Ann Arbor, MI, 1976. With permission.

significant consequences on the hydrography of estuaries. Diminishing viscosity at higher temperatures leads to increased settling rates; thus, it influences sedimentation. The diminution in nitrogen solubility, in turn, can contribute to an alteration of primary production.

The rise in vapor pressure at higher temperatures fosters increases in evaporation. When the water temperature increases from 10 to 32.2°C, the evaporation flux rises from 0.09 kg/m^2/d to 1.42 kg/m^2/d. The calefaction of estuarine waters in the near-field region of a power plant, where temperature changes are controlled principally by the geometry and hydrodynamics of the discharge,[14] can substantially raise the evaporation in the summer months.

Estuarine Organisms

The addition of waste heat to an estuarine water body alters rates of chemical solubility and biochemical reactions that elicit physiological and behavioral responses in organisms. The physiological activity of an estuarine organism is clearly linked to temperature. According to the Arrhenius or van't Hoff law, for example, the physiological activity of an organism increases by two- or threefold with each 10°C rise in temperature. For poikilothermic organisms, which cannot actively regulate their internal heat balance, changes in metabolism accompany shifts in water temperature. The modification of metabolic rates is manifested by restricted growth in some species. In others, altered rates of biochemical reactions attributable to temperature eliminate reproduction. Still other populations have a much greater risk of death when the temperature exceeds some upper limit that may not be much greater than the optimal temperature for the species. Several mechanisms may cause thermal death of higher organisms: the failure of smooth muscle peristalsis, denaturation of proteins in the cells, increased lactic acid in the blood, and oxygen deficit due to increased respiratory activity.[1,22] Although an organism may not succumb to elevated temperatures, it can experience sublethal effects, such as a drop in the hatching success of eggs, an inhibition in larval development, a degeneration of erythrocytes, and a lowering of respiration rates. Many of these harmful effects are avoided through the process of acclimation. In addition, other factors affect the thermal tolerance of an estuarine organism including salinity, food availability, nutrient concentrations, dissolved oxygen levels, and anthropogenic pollutants.[1]

As illustrated by Brett's temperature tolerance polygon (Figure 1),[23] upper and lower temperature thresholds for reproduction, growth, and survival underscore the temperature dependencies of aquatic organisms. This diagram indicates that the thermal requirements for survival are less restrictive than those for growth and reproduction. The thermal requirements for survival partly determine the presence or absence of a species in an impacted area. The productivity levels within a species, however, are contingent upon reproduction and growth requirements.[24]

Odum[25] embellished upon the local detrimental effects of thermal discharges on aquatic biota. He noted that as calefaction of a system proceeds: (1) critical stenothermal periods in life histories may be exceeded; (2) organisms often are more susceptible to toxic substances; and (3) normal algal populations tend to be replaced by less desirable blue-green species. Blue-green algae are nuisances because they cause odor problems and can develop enormous biomasses in impacted systems. Eruptions of blue-green algae commonly place large biological oxygen demands on a system as their dead cells deplete dissolved oxygen resources. Blue-green algae gradually replace diatoms and green algae as temperatures rise from

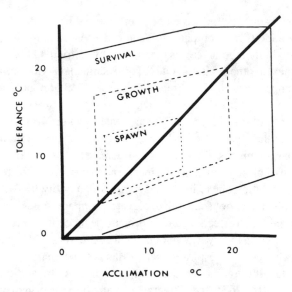

FIGURE 1.   Temperature tolerance polygon for aquatic organisms. (From Brett, J. R., in Water Pollution, Taft, R. A.,
Ed., San. Eng. Center Tech. Rep., unpublished manuscript,
1960, 110. With permission.)

10 to 38°C (Figure 2). Woolcott[26] found blue-green algae to be most
abundant in areas subjected to warm-water effluents of a power
station on the James River. Sampling at thermally impacted stations
revealed fewer attached forms of green and yellow-green algae, as
well as diatoms, than at unheated control locations.

Thermal loading reduces primary production in some locales, and
this reduction combined with other modifications in community
respiration, species composition, nutrient dynamics, and secondary
production no doubt affects the entire aquatic system.[1] Morgan and
Stross[27] documented a depression in rates of phytoplankton photosynthesis when thermal releases at the Chalk Point Power Plant
raised ambient water temperatures above 20°C in the Patuxent River
estuary. Specifically, temperature increases of approximately 8°C
lowered photosynthesis when natural water temperatures were 20°C
or warmer. Thermal discharges in Biscayne Bay, Florida, purportedly lowered the production of seagrasses and decreased the abundance and diversity of phytoplankton.[28] When exposed to tempera-

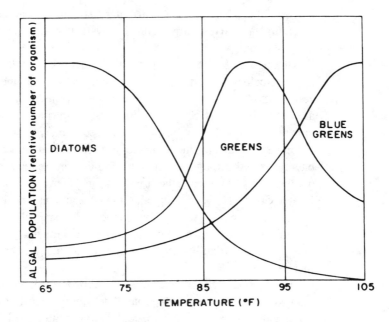

FIGURE 2. Population shifts of algae with changes in temperature. (From Eicholz, G. G., *Environmental Aspects of Nuclear Power,* Ann Arbor Science, Ann Arbor, MI, 1976. With permission.)

tures of 5°C above ambient, 9.3 ha of seagrasses (*Thalassia testudinum*) disappeared from South Biscayne Bay. A 50% loss of seagrass beds occurred in areas exposed to temperature elevations of 3 to 4°C. Macroalgal populations were also depressed in this area, to 30% of their original distribution.[29] Hall et al.[1] and Zieman and Wood[30] suggest that subtropical and tropical ecosystems may be more susceptible to thermal discharges than temperate ecosystems; in these regions, entire communities are likely to experience heavy mortality when temperatures surpass 30 to 31°C.

Zooplankton metabolism depends upon temperature, which strongly affects the overall physiology and ecology of this planktonic group.[31,32] When passing through the condensers of a power plant, estuarine copepods as well as other zooplankters are killed when upper limits of thermal tolerances are exceeded.[31] High temperatures in near-field regions of power plants can also result in sublethal impacts such as diminished growth and lower biomasses. These types of effects have been reported among zooplankton in

Oyster Creek, New Jersey, a tributary stream of Barnegat Bay receiving thermal discharges from the Oyster Creek Nuclear Generating Station.[33]

Although correlated with thermal discharges from the Oyster Creek Nuclear Generating Station, reductions in abundance of zooplankton in Oyster Creek are generated by primary entrainment, as well as thermal effects. Mortality caused by primary (organisms pumped through the station and discharged into Oyster Creek) and secondary (organisms incorporated into the thermal plume without having passed through the station) entrainment reduces the total number of zooplankton in the discharge canal of the power plant and Oyster Creek. However, reproduction of zooplankton in the discharge canal and Oyster Creek produces a greater abundance of some forms. Changes in abundance of zooplankton in Oyster Creek due to natural variability are difficult to assess in light of the above factors.[33]

Fluctuations in abundance of some microzooplankton, macrozooplankton, and ichthyoplankton have been observed in the discharge canal and Oyster Creek. Microzooplankton found to be less abundant at the mouth of Oyster Creek than at the condenser discharge of the station included barnacle larvae (72% less abundant at the mouth of the discharge canal), polychaete larvae (42%), copepod nauplii (19%), unidentified bivalve larvae (25%), *Acartia tonsa* (57%), and *Acartia* spp. (61%). Other microzooplankton were more numerous at the mouth of Oyster Creek than at the condenser discharge of the station; these included rotifers (103% more numerous at the mouth of the discharge canal), cyphonaute larvae (3547%), gastropod larvae (70%), and *Mulinia lateralis* larvae (155%).[33]

Benthic invertebrates, because of their relatively limited mobility, have been the focal point of some power plant impact studies. Hence, Loveland and Vouglitois[34] reported a power plant impact on the benthic invertebrates in Oyster Creek, manifested by a significantly lower species richness of benthic forms in Oyster Creek relative to control sites in Barnegat Bay. While Kennish and Olsson[35] discerned reduced growth of hard clams (*Mercenaria mercenaria*) in close proximity to Oyster Creek in summer due to the calefaction of estuarine water, Kennish[36] detected no significant increases in mortality of adults in Barnegat Bay ascribable to thermal discharges from the Oyster Creek Nuclear Generating Station. In areas of the middle reach of Chesapeake Bay receiving thermal discharges from the Calvert Cliffs Nuclear Power Plant, oysters (*Crassostrea virginica*) grew more rapidly than elsewhere in the system.[37] As in the case of the Barnegat Bay ecosystem, the impact of thermal discharges from the Indian River Power Plant on the benthos of the Indian River estuary, Delaware, is confined to the area within the discharge canal of the power plant.[38]

Probably the most detailed investigations of temperature influences on the behavior of aquatic poikilotherms involve finfish. The calefaction of estuarine waters provides an artificial temperature cue that can affect the occurrence and concentration of fish populations.[1] Fish perceive temperature variations in two ways: (1) cutaneous thermal receptors in the organism detect temperature changes and send nerve impulses to the CNS; and (2) temperature acts on a chemical process in some or all tissues.[39,40] Once they encounter heated effluent, fish typically either avoid the water or are attracted to it. In essence, behavioral responses of the fish to the heated water leads to a modification of their distribution patterns. Alterations in fish distributional patterns indicate at least a reallocation of food resources and can reflect changes in productivity of an estuary. Conceivably, the productivity of fishes in an area exposed to calefaction may increase; however, fish kills caused by cold or heat shock, serve to offset this possible benefit.[24] Not to be forgotten are the physiological responses of the organism to the temperature elevations of heated effluents that seemingly influence many aspects of its life, notably gonad maturation and spawning, egg- and embryogenesis, and growth and ontogenetic development through adult stages.

To properly assess the effects of thermal discharges on fish productivity in an area, detailed long-term sampling programs must be executed. O'Connor and McErlean (p. 499)[24] state the complexities associated with power plant impact assessment of fish productivity:

> "The total impact of a power plant upon productivity in estuarine environments must be measured in terms of changes within the entire nektonic community. Productivity may be altered directly, as demonstrated by changes in total biomass, or indirectly, as demonstrated by changes in the dynamics of the food chain. Both effects must be evaluated; both are difficult to quantify. A description of the food chain

for fish communities requires exhaustive sampling at all trophic levels, stomach-content analyses, and behavioral observations. A study of this nature requires years of gathering data, both before and after a plant starts operating, if alterations attributable to power plants are to be identified..."

Fish have the ability to detect even slight temperature changes — some species as little as 0.05 to 0.20°C.[41] Temperature preference and avoidance studies in the laboratory and field confirm the sensitivity of estuarine and marine fishes to temperature alterations. The methods used to estimate temperature preference and avoidance responses from laboratory and field data have been treated elsewhere,[40] and are not repeated here. Suffice it to say that temperature preference and avoidance data have been generated for many fishes employing a range of laboratory and field techniques, although the most complete set of data exists on freshwater species with more limited information available for estuarine and marine forms.

Olla and Studholme[42,43] and Olla et al.[44,45] have conducted some of the most quantitative and informative investigations on the effects of sublethal temperatures on the activity and behavior of estuarine fishes. They accomplished this work by concentrating on bluefish (*Pomatomus saltatrix*), tautog (*Tautoga onitis*), and Atlantic mackerel (*Scomber scombrus*). They recorded the effects of gradual increases and decreases in temperature on the swimming speed and schooling behavior of the fishes. Increases or decreases in temperature elicited responses of increased swimming speed in bluefish and Atlantic mackerel but decreased activity in tautog, which exhibited a reduction in aggressive behavior and a need to seek shelter from the stressful conditions. Hence, the tautog did not respond to temperature stress by avoiding it, as is typical of most fish.

All finfishes display a range of temperatures below or above which they cannot survive.[20,42] Table 3 lists the temperature ranges for survival of a number of marine, estuarine, and freshwater species. Some of these fishes have broad temperature ranges, and it is difficult to predict how they may be distributed in the field within these thermal limits.[20] This becomes much more evident when taking note of the other (abiotic and biotic) factors that may also influence the thermal responses of fish (e.g., food availability, nutritional state, competition and/or predation, pathological condition, social factors, age of the organism, habitat requirements,

TABLE 3
Lethal Temperature Limits for Adult Marine, Estuarine, and Freshwater Fishes

Species	°C	°F	Acclimation temp.[a]		°C	°F
			°C	°F		
Alewife	—	—	—	—	26.7–32.2	80.0–90.0
Bass, striped	—	—	6.0—7.5	42.8–45.5	25.0–27.0	77.0–80.0
California killifish	14.0–28.0	57.2–82.4	—	—	32.3–36.5	90.1–97.7
Common silverside	7.0–28.0	44.6–82.4	1.5–8.7	34.8–47.8	22.5–32.5	73.3–90.3
Flounder, winter	21.0–28.0	69.8–82.4	1.0–5.4	33.8–41.6	—	—
Herring	7.0–28.0	44.6–82.4	—	—	22.0–29.0	71.6–84.2
	—	—	-1.0	30.2	19.5–21.2	67.1–70.1
Northern swellfish	14.0–18.0	57.2–82.4	8.4–13.0	47.1–55.4	—	—
	10.0–28.0	50.0–82.4	—	—	28.2–33.0	82.9–90.4
Perch, white	4.4	40.0	—	—	27.8	82.0
Salmon (general)	—	—	0.0	32.0	26.7	80.0
Bass, largemouth	20.0	68.0	5.0	41.0	32.0	89.6
	30.0	86.0	11.0	51.8	34.0	93.2
Bluegill	15.0	59.0	3.0	37.4	31.0	87.8
	30.0	86.0	11.0	51.8	34.0	93.2
Catfish, channel	15.0	59.0	0.0	32.0	30.0	86.0
	25.0	77.0	6.0	42.8	34.0	93.2

TABLE 3 (CONTINUED)
Lethal Temperature Limits for Adult Marine, Estuarine, and Freshwater Fishes

Species	Acclimation temp.[a]					
	°C	°F	°C	°F	°C	°F
Perch, yellow	5.0	41.0	—	—	21.0	69.8
(winter)	25.0	77.0	4.0	39.2	30.0	86.0
(summer)	25.0	77.0	9.0	48.2	32.0	89.6
Shad, gizzard	25.0	77.0	11.0	51.8	34.0	93.2
	35.0	95.0	20.0	68.0	37.0	98.6
Shiner, common	5.0	41.0	—	—	27.0	80.6
	25.0	77.0	4.0	39.2	31.0	87.8
	30.0	86.0	8.0	46.4	31.0	87.8
Trout, brook	3.0	37.4	—	—	23.0	73.4
	20.0	68.0	—	—	25.0	77.0

[a] Values are LD_{50} temperature tolerance limits, i.e., water temperatures survived by 50% of the test fish after 1 to 4 d acclimation.

From the Federal Water Pollution Control Administration, Industrial Waste Guide on Thermal Pollution, Federal Water Pollution Control Administration, Corvallis, OR, 1968.

salinity, time of the day, and season of the year).[20] The prior thermal history of the organism (e.g., acclimation temperature) must be considered with respect to thermal tolerance as well and both acute preference and avoidance reactions.

The attraction of fish populations to the heated waters around a power plant is often detrimental because of the effects of cold shock, heat shock, and chronic exposure to extremes of temperature. Direct kills from heat stress are rare because the fish perceive the thermal characteristics of a plume as being incompatible with their preferred temperature range, and, consequently, they move away from the area of stress. However, once a fish has adapted to a thermally impacted area, abrupt startup and shutdown operations at a power plant can generate rapid temperature changes that may be lethal to the organism. In temperate estuaries, winter cold shock mortality of finfish populations in the near-field regions of power plants are not unusual. Some fish populations that normally migrate to the ocean in the fall may become thermally marooned in a discharge plume during the winter months. Unfortunately, when the power plant becomes inoperative, the fish, which have become acclimated to the elevated temperatures, are exposed to sudden temperature reductions resulting in cold shock mortality. Table 4 discloses episodes of fish kills at the Oyster Creek Nuclear Generating Station between January 1972 and December 1982, with most of the events caused by cold shock mortality of fish during the winter. Fish kills of overwintering populations, particularly Atlantic menhaden (*Brevoortia tyrannua*) have been most numerous.

Aside from heat and cold shock mortality, much of the effect of thermal discharges on fish involves avoidance or attraction to impacted areas. Thus, Gallaway and Strawn[46] observed fish attracted to the thermal discharges of a power plant on Galveston Bay, Texas, during spring, autumn, and winter. In Barnegat Bay, New Jersey, effects of thermal discharges on fish populations are confined primarily to Oyster Creek. The distribution of some species has been affected because they may prefer, or may avoid, Oyster Creek at various times of the year in response to the heated water from the Oyster Creek Nuclear Generating Station. From June through September, for example, Atlantic menhaden (*Brevoortia tyrannus*), bluefish (*Pomatomus saltatrix*), and weakfish (*Cynoscion regalis*) avoid some part of Oyster Creek, and young winter flounder

TABLE 4

Fish Mortality in the Discharge Canal of the Oyster Creek Nuclear Generating Station (OCNGS) and Oyster Creek Caused by Operation of the OCNGS

Date	No.	Species	Size range	Probable cause
1/29/72	100,000–1,000,000	Atlantic menhaden	76–127 mm	Thermal shock
1/5/73	18,000–1,200,000	Atlantic menhaden	102–356 mm	Thermal shock
1/8/73	20	Bay anchovy	—	Thermal shock
2/16/73–2/21/73	Several thousand	Atlantic menhaden	—	Thermal shock
8/9/73	2,000–4,000	Atlantic menhaden	127–356 mm	Thermal shock
1/7/74	500	Atlantic menhaden	203–280 mm	Chlorine
1/11/74–1/15/74	9,900–180,000	Atlantic menhaden	102–356 mm	Thermal shock
	100–3,600	Bluefish	228–356 mm	Thermal shock
10/9/74	200	Crevalle jack	—	Thermal shock
2/4/75	100	Atlantic menhaden	—	Thermal shock
	50–100	Bluefish	—	Thermal shock
11/24/75	7–100	Crevalle jack	—	Thermal shock
12/29/75	15–100	Atlantic menhaden	100–250 mm	Thermal shock
	3–200	Bluefish	90–170 mm	Thermal shock
10/21/77	120–200	Blue runner	—	Thermal shock
		Crevalle jack	—	Thermal shock
1/15/79	682	Atlantic menhaden	165–225 mm	Thermal shock
8/2/79	50–100	Striped bass	34–44 mm	Thermal shock
		Northern puffer	Missing parts	
		Goosefish	Missing parts	
		Tautog		
12/17/79	Unknown	Unknown	Unknown	Unknown
12/20/79	12	Bluefish	—	Unknown
	1	Weakfish	—	
	1	Sea robin	—	
	1	Black sea bass	—	
	1	Atlantic menhaden	—	
1/5/80	5,483	Atlantic menhaden	240 mm[a]	Thermal shock
	952	Bluefish	295 mm[a]	
	544	Spot	120 mm[a]	
	43	Weakfish	501 mm[a]	
	5	Scup	200 mm[a]	
	1	Butterfish	—	
	1	Northern kingfish	240 mm[a]	

TABLE 4 (CONTINUED)
Fish Mortality in the Discharge Canal of the Oyster Creek Nuclear Generating Station (OCNGS) and Oyster Creek Caused by Operation of the OCNGS

Date	No.	Species	Size range	Probable cause
11/22/80	3,638	Blue runner	206 mm[a]	Thermal shock
		Crevalle jack	173 mm[a]	
	1,038	Bluefish	267 mm[a]	
	17	Smooth dogfish	601 mm[a]	
	3	Ladyfish	298 mm[a]	
	2	Northern kingfish	—	
	1	Gray snapper	118 mm[a]	
	1	American eel	—	
	1	Mojarra	221 mm[a]	
12/9/82–	166	Crevalle jack	110–204 mm	Thermal shock
12/10/82	80	Blue runner	171–218 mm	
	76	Bluefish	274–476 mm	
	28	Atlantic needlefish	250–661 mm	
	9	Scup	205–247 mm	
	2	American eel	—	
	1	Conger eel	—	
	1	Northern kingfish	185 mm	
	1	Ladyfish	410 mm	

[a] Mean size.

From Kennish, M. J., Roche, M. B., and Tatham, T. R., in *Ecology of Barnegat Bay, New Jersey,* Kennish, M. J. and Lutz, R. A., Eds., Springer-Verlag, New York, 1984, 318. With permission.

(*Pseudopleuronectes americanus*) avoid most of Oyster Creek because of high water temperatures. From autumn through early winter, Atlantic menhaden, bluefish, spot (*Leiostomus xanthurus*), bay anchovy (*Anchoa mitchilli*), weakfish, and several other incidental species are attracted to the discharge canal. Some individuals of these species overwinter in the canal.[33] Similar patterns of avoidance and attraction to discharge canals of power plants have been ascertained at other locations.

WASTE HEAT UTILIZATION

Because of the vast amounts of waste heat rejected by electric generating stations in the form of thermal effluents — only about 38 to 40% of the heat produced by these stations actually generates electricity, the remainder being released to the environment — proposals have been advanced over the years advocating the beneficial utilization of the waste heat. Sonn[47] sets the picture in perspective, noting that an electric generating unit with a 100-MWe capacity releases approximately 200 MWt of waste heat, together with about 275,000 l/min of cooling water effluent. If fully utilized, the waste heat from a unit of this size could heat approximately 120 to 150 ha of greenhouses or produce 1.5 to 2.0 million kg of finfish each year. Two major requirements for waste heat utilization must be met, however; in particular, the uses for the rejected heat should be economical and capable of removing most of the thermal discharges of a power plant.

The beneficial uses of waste heat from electric generating stations fall into three main categories: (1) industrial and residential applications; (2) agriculture; and (3) aquaculture. One of the most promising areas for employing waste heat is in industrial and residential space heating in winter and absorption-type air conditioning in summer.[14] Potential major industrial users of waste heat at temperatures up to 82°C include food processing, textile, furniture, chemical, concrete block manufacturers, pulp and paper, iron and steel foundries, and automobile and truck manufacturers.[48] Effective waste heat utilization by these industries relies on thermal augmentation of condenser discharges. If the waste heat is augmented for process temperatures in the range of 60 to 82°C, possible industrial users are food processing, concrete block manufacturers, iron and steel foundries, and automobile manufacturers. Temperature augmentation up to 60°C probably provides applications largely to the food processing industry but also has value in residential space heating. Kim et al.[48] point out some of the problems associated with the industrial and residential use of thermal effluent from power plants, specifically the low discharge temperature, high utilization costs, the necessity of locating the user near the heat source, the potential interruptable nature of the service, the possibility of chemical contamination, and seasonal variations in source temperature. Additional waste heat utilization research must be conducted by industry, but at least three problems have largely frustrated attempts to do so: (1) the need for meaningful commitments to finance waste heat research and development; (2) the lack of site-specific information regarding the potential of waste heat use; and (3) the need for information concerning the economics and engineering of waste heat delivery systems.[49]

Some beneficial uses of waste heat in agriculture involve the heating and cooling of greenhouses, soil warming, crop drying, irrigation, frost protection, reclamation of livestock wastes, wastewater treatment, and environmental control of livestock and poultry housing.[50] Of these potential uses, applications to greenhouse production may be the most likely candidate to lead to fruition. The large-scale irrigation of farm land appears to be less practical. For example, a 1000 MWe power plant would need 26,000 to 52,000 ha of farm land to effectively dissipate its waste heat. Moreover, the power plant would require the continuous discharge of thermal effluent over a 24-h period, which would greatly exceed the irrigation demands of the crops, especially during periods of rainfall.[14] While the use of power plant waste heat to warm field soils and increase crop yields is technically feasible, this application has not proven to be profitable.[50]

The feasibility of utilizing thermal discharges in fin- and shellfish aquaculture has been investigated for many years.[51-54] The most common organisms employed in marine culture system studies are salmon *(Oncorhynchus* and *Salmo)*, flounder *(Pseudopleuronectes americanus)*, abalone *(Haliotis)*, clams *(Mercenaria mercenaria)*, mussels *(Mytilus)*, oysters *(Crassostrea* and *Ostrea)*, scallops *(Argopecten* and *Patinopecten)*, shrimp *(Penaeus)*, and lobsters *(Homarus americanus)*. Commercial culture in thermal effluents in the U.S. is primarily limited to oyster and salmon smolt operations.[49] Freshwater organisms being studied in waste heat aquaculture projects include catfish *(Ictalurus punctatus)*, eels *(Anguilla)*, tilapia *(Sarotherodon)*, carp *(Cyprinus)*, yellow perch *(Perca flavescens)*, striped bass *(Morone saxatilis)*, rainbow trout *(Salmo gairdneri)*, and prawns *(Macrobrachium)*. Of these animals, the catfish has met with the greatest degree of commercial success in power plant effluents.[55]

BIOCIDES

The biofouling of heat exchanger surfaces in power plants causes a broad spectrum of engineering and environmental problems. Most problems of biofouling in power plants entail microbiological infestation followed by macroinvertebrate buildup (Figure 3).[56] The

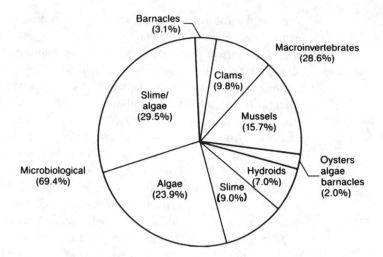

FIGURE 3. Biofouling organisms in cooling-water systems based on a survey of 365 units. (From Chow, W., in Proceedings: Condenser Biofouling Control — State-of-the-Art Symposium, Chow, W. and Massalli, Y. G., Eds., Tech. Rep., Electric Power Research Institute, Palo Alto, CA, 1985, 1-1. With permission.)

micro- and macrofouling of condenser tube walls lower the heat transfer efficiency of the condenser, accelerate the fluid frictional resistance, and increase the corrosion of the metals.[57] It can be extremely costly and labor intensive to control biofouling to maintain performance in both once-through and closed-cycle cooling water systems. The most common and effective method of biofouling control in power plant components is the use of biocides, mainly chlorine, in liquid, gaseous, or hypochlorite form.[58] Although chlorine removes bacterial slimes very efficiently when added at regular intervals to the cooling water, it does not control macrofouling (e.g., mussel or clam control) as well. In effect, mussels and other larger shell-bearing organisms respond to intermittent chlorine dosing by closing their shells until the biocide concentration dissipates. The effectiveness of chlorine in preventing biofilm development on heat exchanger surfaces stems from its ability to rapidly kill microorganisms directly or to hydrolyze the extracellular polymers that hold the biofilm together.[59]

The application of chlorine poses other problems, however, because of its general toxicity to organisms. Chlorine easily kills nontarget organisms entrained in cooling water, such as phytoplankton, meroplankton, holozooplankton, and ichthyoplankton. Organisms in receiving waters outside of the cooling-water system also may be threatened by the capacity of the biocide to form toxic residual organic compounds (i.e., chloramines) and by its enhanced residual toxicity in seawater where bromine is liberated.[58] Consequently, constraints have been placed on power station discharges of chlorine in an effort to minimize the chlorine concentration released to receiving waters while still maintaining effective biofouling control. For example, the U.S. Environmental Protection Agency (EPA) substantially lowered the residual chlorine emissions allowed for steam-electric power plants in its revised effluent guidelines of November 1982. As iterated by Chow (pp. 1 to 4),[56] "...A utility had to demonstrate to EPA its need for chlorination, and at plants permitted to use chlorine, the concentration of total residual chlorine would not be allowed to exceed 0.2 mg per liter. Furthermore, discharge would be permitted for no more than two hours a d from each power plant unit."

An extensive database has been collected on the toxicity of chlorine to freshwater and marine organisms.[60-64] Mattice and Zittel[61] described laboratory bioassays of chlorine toxicity on these organisms, underscoring the concentration and duration of chlorine exposure. Figure 4 shows acute and chronic toxicity thresholds for the biota, with the thresholds being defined as median effect levels on either mortality or sublethal physiological rates. Hence, they formulated a conceptual model for evaluating chlorine releases and predicting acute and chronic effects upon particular types of organisms (i.e., freshwater or marine). Results of this study indicate that the freshwater acute toxicity threshold runs from a chlorine concentration-time point of 1.0 mg/l-1 min to 0.0015 mg/l-7500 min. Similarly for marine organisms, the marine acute toxicity threshold runs from a chlorine concentration-time point of 0.15 mg/l-1 min to 0.02 mg/l-100 min (Figure 4). The chronic toxicity data suggest that relatively low concentrations of chlorine may elicit sublethal responses, including physiological, biochemical, or behavioral (e.g., avoidance) changes in both freshwater and marine forms.[63] These findings have been basically corroborated by Seegert et al.[65] Based on the site-specific assessment of power plant chlorination in the marine environment, acute and chronic toxicity thresholds for saltwater organisms amounted to approximately 0.3 mg/l for short exposure times

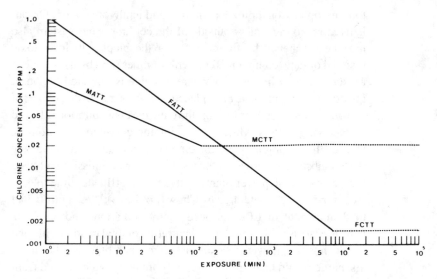

FIGURE 4. Chlorine-exposure diagram illustrating acute and chronic toxicity thresholds for marine and freshwater organisms. (From Mattice, J. S. and Zittel, H. E., J. *Water Pollut. Cont.*, 48, 2284, 1976. With permission.)

and 0.02 mg/l for long-term exposures.[1,61] The acute toxicity thresholds were lower for marine than freshwater organisms, whereas the opposite held for chronic exposures.[61]

Chlorination tends to ameliorate phytoplankton productivity. For instance, Carpenter et al.[66] calculated a 79% reduction of photosynthetic activity in entrained water at the Millstone Point Nuclear Power Station on northeastern Long Island Sound immediately subsequent to the application of chlorine in concentrations from 0.1 to 1.2 mg/l. Eppley et al.[67] discerned a 70 to 80% depression in marine phytoplankton photosynthesis in the thermal plume at the San Onofré nuclear power stations in California which had chlorine levels of 0.02 and 0.04 mg/l.

As in the case of phytoplankton, survival of zooplankton to chlorine exposure is time and dose dependent.[58] At chlorine levels above 0.5 mg/l, prolonged exposure of zooplankton in the receiving waters of power plants can be lethal to many species.[68] Larval and postlarval forms of marine invertebrates and fish are very sensitive to chlorination.[69,70] Capuzzo et al.,[70] for example, documented decreased growth and metabolic activity in larval lobsters *(Homarus americanus)* subjected to prolonged sublethal exposure of chlorine or chloramine. Marked increases in respiration were observed in larvae after 30- and 60-min exposures to 1.0 and 0.1 mg/l concentrations. They found greater sensitivity of lobster larvae to chloramine, with responses taking place even at concentrations below the level of detectability of the chemical.[71] A synergistic effect of chlorine and temperature occurred with both free and combined chlorine products.[58,69]

Macrobenthic invertebrates likewise may be highly sensitive to chlorine exposure in power plant cooling water discharges. Brungs,[60] Mattice and Zittel,[61] and Morgan and Carpenter[63] review the literature on this topic. Much of the data accumulated on this biotic group derive from bioassay or experimental treatment rather than field observations.[58]

A greater volume of information on the effects of chlorine, chlorine derivatives, and their toxicity has been accumulated on ichthyoplankton and juvenile and adult finfish.[60,63,64,72] In general, chlorine begins to be lethally effective on marine fishes at concentrations of about 0.01 mg/l,[61] although temperature, the presence of other compounds (e.g., nitrogenous substances), as well as the physiological condition of the animal may affect its tolerance to the biocide.[58,63] Avoidance thresholds in estuarine fishes range from approximately 0.04 to 0.15 mg/l.[73] Chlorine attacks the gills of fish resulting in the oxidation of hemoglobin to methemoglobin, with the organism succumbing due to anoxia.[64] The gill epithelium tends to slough off.[58]

Ichthyoplankton tolerance (of estuarine fishes) to chlorine appears to be age related, as shown by studies of the eggs and larvae of blueback herring (*Alosa aestivalis*), striped bass (*Morone saxatilis*), and white perch (*Morone americana*). In these species, newly laid eggs are less tolerant to chlorine than older eggs, but newly hatched larvae are more tolerant than older larvae.[74] The larvae of plaice (*Pleuronectes platessa*) and sole (*Solea solea*) exhibit lower tolerance to chlorine than do their eggs.[63]

IMPINGEMENT AND ENTRAINMENT

The mortality of various life stages of estuarine organisms from impingement on trash racks and intake screens and entrainment and passage through condenser cooling systems represents a potentially greater impact on aquatic populations and communities than that caused by thermal discharges and biocides. Not only do impingement and entrainment adversely affect biotic productivity in estuaries directly by killing individuals, they also decrease food supplies and thereby may influence the structure of food webs.[24] The numerical losses of estuarine organisms from impingement and entrainment at power plants can be extremely high. At the P. H. Robinson generating plant in Galveston Bay, Texas, more than 7 million fish were impinged in 1 year (1969 to 1970). Impingement mortality of Atlantic menhaden (*Brevoortia tyrannus*) at the Millstone Nuclear Power Station on Niantic Bay, Connecticut, exceeded 2 million individuals in the late summer and early autumn of 1971.[1] Approximately 13 million fish and macroinvertebrates were impinged at the Oyster Creek Nuclear Generating Station on Barnegat Bay, New Jersey, from September 1975 through August 1977.[75] During the period January to December 1986, impingement of fish and invertebrates at the Pilgrim Nuclear Power Station on Cape Cod Bay, Massachusetts, averaged 1.26 and 1.91 individuals per hour, respectively.[76] Entrainment losses of Atlantic menhaden and river herring at the Brayton Point Power Plant on Mount Hope Bay, Massachusetts, during the summer of 1971 ranged from 7 to more than 160 million individuals. Effects of entrainment at the Millstone Nuclear Power Station killed some 36 million young fish in 16 d in November 1971.[77] An estimated 5.16×10^{13} microzooplankton were entrained in the Oyster Creek Nuclear Generating Station from September 1975 through August 1976 and 4.03×10^{13} microzooplankton from September 1976 through August 1977. Macrozooplankton losses during these intervals amounted to an estimated 4.25×10^{11} and 9.98×10^{10} individuals, respectively, from 1975 to 1976 and 1976 to 1977.[75] The number of fish eggs entrained through the Pilgrim Nuclear Power Station in 1986 equalled 1.70×10^9 and the number of larval fish, 2.76×10^8.[76]

Although the absolute number of estuarine organisms impinged or entrained by power plants commonly ranges into the millions over a yearly cycle, the actual damage of these losses to resident populations of phytoplankton, zooplankton, benthos, and fishes is difficult to determine. Clearly, the need exists for comparison studies of power plant-induced mortality of populations to the resident population sizes in waters not impacted by impingement and entrainment.

For instance, Marcy et al.[78] aptly summarized the percentages of larval fish populations entrained by various power plants in the U.S.; the numbers entrained ranged from about 2 to 12% of the estimated population sizes.

Much speculation and controversy surrounds the total effect of impingement and entrainment mortality. Mathematical population models have been formulated to assess these impacts on resident species.[79-82] Major limitations to the use of the models include inaccurate estimates of population sizes and the inability to accurately account for marked fluctuations in reproductive success and survival of organisms from year to year.[58,79] In addition, biological compensation, a basic tenet of fisheries management, may operate to mollify or eliminate impingement and entrainment effects.[24] For example, high egg and larval mortalities can be compensated for by increases in fecundity and survival of resident populations, together with predator mortality, unless the populations diminish to very low levels.[58,83]

A number of abiotic and biotic factors affect the impingement and entrainment of estuarine organisms, as recounted in the introduction to this chapter. Among the most important abiotic factors are the siting and cooling water system design of the power plant and the volume and ambient conditions of the cooling water used. Biotic factors of significance relate to the abundance, survival, ecological roles, and reproductive strategies of the affected organisms.[3] Considerable variation in impingement or entrainment impact may arise from a multitude of causes such as seasonal fluctuations, organismal migrations, climatic changes, tides, storms, and unscheduled changes in plant operations.[58]

Impingement

The mortality of estuarine organisms from impingement on intake screens at power plants became a pressing problem in the U.S. with the passage of the 1972 Federal Water Pollution Control Act Amendments. Specifically, Section 316(b) of this legislation stresses that cooling water intakes at power plants must reflect the best technology available to minimize adverse environmental impact. As a result, utilities have dealt with the impingement problem by examining methods of impingement mitigation either by backfitting new devices into their intake structures to lower the mortality of impinged organisms or by installing behavioral barriers and guidance systems to direct organisms away from intakes and, therefore, lessen the number of individuals impinged on screens. Consideration is also given, as a last resort, to the complete conversion from once-through cooling in an existing system to closed-cycle operation of cooling towers and cooling ponds.

The intake screening on thermal power stations generally consists of drum or band type screens having mesh apertures between 5 and 9 mm, or approximately 50% of the condenser tube diameter.[58] Drum screens are more common in England than other countries where band screens predominate. The intake screens usually rotate vertically in a screen well with pressurized spray acting to remove the buildup of impinged debris and organisms. Organisms washed off of the screens typically collect in sluiceways or channels, with water jets transporting them to natural waterways. The debris is accumulated in large metal baskets for transport to disposal sites. Many variations to this basic design exist throughout the world. In the past, poor handling of organisms subsequent to their washing off of intake screens has exacerbated impingement mortality.

Following passage of the Federal Water Pollution Control Act Amendments of 1972, a significant volume of impingement data on fish and macroinvertebrates was gathered at operating sites in the U.S.[84-86] Concurrently, power stations in other countries (e.g., England) revealed impingement statistics as well. Table 5 provides an example of impingement data obtained during the 1970s — estimates of the number of white perch (*Morone americana*) impinged at all Hudson River power plants (i.e., Bowline, Lovett, Indian Point, Roseton, Danskammer, and Albany) during the years 1974 to 1977. Using a simple model derived from Ricker's theory of fisheries dynamics, Barnthouse and Van Winkle[87] assessed the impact of the power plant impingement on the 1974 and 1975 year classes of the Hudson River white perch population. Their analysis disclosed that the abundance of the 1974 white perch year class in the Hudson River was lowered by at least 10%, and probably by 20% or more, due to impingement, while the abundance of the 1975 year class dropped by at least 8%, and probably by 15% or more. Given the additional impact attributable to entrainment by the power plants, the authors

TABLE 5
Monthly Estimates of the Number of White Perch Impinged at All Hudson River Power Plants Combined, for 1974 and 1975 Year Classes

		Year class			
		1974		**1975**	
		Number of years of vulnerability		Number of years of vulnerability	
Age (years)	Month	**2**	**3**	**2**	**3**
0	6	0		0	
	7	3,486		8,898	
	8	14,887		97,910	
	9	26,239		83,980	
	10	112,957		93,888	
	11	245,492		239,150	
	12	607,434		348,596	
	1	415,724		589,206	
	2	270,571		182,891	
	3	139,751		130,261	
	4	609,090		111,820	
	5	91,910		40,151	
1	6	37,242	18,621	27,014	13,507
	7	22,126	11,063	13,835	6,918
	8	14,122	7,061	6,770	3,385
	9	19,924	9,962	13,791	6,896
	10	19,534	9,767	25,676	12,838
	11	28,005	14,002	12,552	6,276
	12	7,803	3,902	48,102	24,051
	1	38,078	19,039	143,010	71,505
	2	9,293	4,646	43,558	21,779
	3	12,444	6,222	49,579	24,790
	4	14,103	7,052	38,692	19,346
	5	7,612	3,806	56,365	28,183
2	6		13,057		35,710
	7		6,918		8,805
	8		3,385		12,662
	9		6,896		8,736
	10		12,838		17,362
	11		6,276		19,145
	12		24,051		10,890

TABLE 5 (CONTINUED)
Monthly Estimates of the Number of White Perch Impinged at All
Hudson River Power Plants Combined, for 1974 and 1975 Year
Classes

		Year class			
		1974		**1975**	
		Number of years of vulnerability		Number of years of vulnerability	
Age (years)	**Month**	**2**	**3**	**2**	**3**
	1		71,505		
	2		21,779		
	3		24,790		
	4		19,346		
	5		28,182		

From Van Winkle, W., Barnthouse, L. W., Kirk, B. L., and Vaughan, D. S., Evaluation of Impingement Losses of White Perch at the Indian Point Nuclear Station and Other Hudson River Power Plants, Tech. Rep., ORNL/NUREG/TM-361, Oak Ridge National Laboratory, Oak Ridge, TN, 1980.

suggested the implementation of impingement mitigatory measures to protect the white perch population in the system.

In another study, Lawler et al.[88] evaluated the impact of impingement, along with entrainment, at the Brunswick Steam Electric Plant adjacent to the Cape Fear estuary in North Carolina. Table 6 lists annual impingement and entrainment losses for seven finfish and one shellfish species over the 2-year period from September 1976 through August 1978 and the 4-year interval from January 1974 through August 1978. Atlantic menhaden (*B. tyrannus*) and bay anchovy (*Anchoa mitchilli*) showed the greatest impingement, with both species having mean annual losses exceeding 1 million individuals.

At the Fawley Generating Station in England, Holmes[89] recorded weekly impingement of up to 60,000 fish (Figure 5). Under-yearlings and smaller species comprised the bulk of the impinged fish. The highest impingement occurred during the winter, early spring, and autumn. The lowest abundance and diversity of fish on the intake

TABLE 6
Average Yearly Entrainment and Impingement Losses at the Brunswick Steam Electric Plant[a]

Species	Entrainment[b]	Impingement[c]	Impingement[d]
Spot	186,000	724	350
Croaker	123,000	356	235
Shrimp	171,000	675	760
Flounder	7,200	21	12
Mullet	5,200	34	18
Trout	38,500	169	205
Menhaden	32,500	9,744	4,000
Anchovy	913,000	2,748	1,600

[a] Number of organisms $\times 10^3$.
[b] Computed from weekly average plant flows from September 1976 through August 1978 and 5-year average entrainment densities for the same weekly period. These flows are close to the full plant flow.
[c] Two-year averages of measured impingement losses from September 1976 through August 1978.
[d] Five-year averages of measured impingement losses from January 1974 through August 1978.

From Lawler, J. P., Hogarth, W. T., Copeland, B. J., Weinstein, M. P., Hodson, H. G., and Chen, H. Y., in *Issues Associated With Impact Assessment, Proc. 5th Natl. Workshop on Entrainment and Impingement,* Jensen, L. D., Ed., EA Communications, Sparks, MD, 1981, 159.

screens arose in summer. Various species displayed different seasonal impingement patterns, however, owing to different migration and abundance patterns. For instance, peak impingement of Atlantic herring (*Clupea harengus*) developed in mid-winter, and sand smelt (*Atherina presbyter*), in autumn and winter. Pout (*Trisopterus luscus*) had relatively constant impingement figures year round.[58,89]

Seasonal fluctuations in impingement numbers seem to reflect two primary effects: (1) the seasonal migration or emigration of populations in an estuary which brings organisms into and out of intake areas; and (2) changes in temperature that control the ability of organisms to withstand water currents and, hence, impingement on intake screens.[58] For example, most of the impingement of finfish on intake screens of the Oyster Creek Nuclear Generating Station takes place in the spring and autumn when species migrate into Barnegat Bay (spring) or emigrate from the bay (autumn).[75] Rapidly

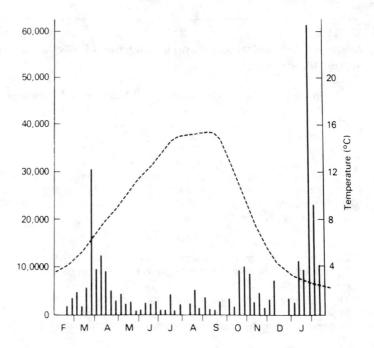

FIGURE 5. Weekly impingement of fish on intake screens of the Fawley Generating Station, England, in 1973. Vertical bars depict impingement. Dashed line gives water temperature. (From Langford, T. E., *Electricity Generation and the Ecology of Natural Waters,* Liverpool University Press, Liverpool, U.K., 1983. With permission.)

changing temperatures in the estuary during these seasons can have a profound effect on the swimming performance of fishes and their ability to avoid intake screens.[90]

Other biological and environmental factors also influence the swimming performance of fishes at power plant intakes. Aside from the type of species and its developmental stage, parasite infestation, nutritional condition, and presence of predators or prey all affect swimming performance. Abiotic factors of note that likewise play a role in this respect are salinity, oxygen concentration, suspended matter, tides, time of day, and season.[91] Langford[58] documented highest impingement of fish at low tide at the Fawley Generating Station (Figure 6).

Methods of mitigating impingement range from the modification of standard vertical traveling screens to the installation of behavioral barriers and guidance systems. White and Brehmer[92] described how

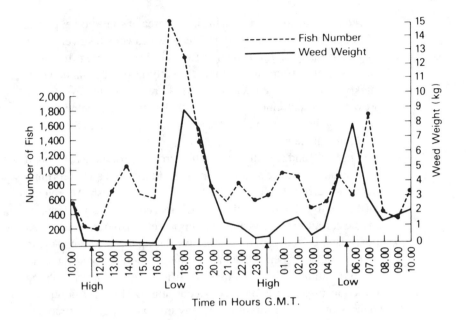

FIGURE 6.   Hourly impingement of fish and weeds on intake screens of the Fawley Generating Station, England, on November 20 and 21, 1975. (From Langford, T. E., *Electricity Generation and the Ecology of Natural Waters,* Liverpool University Press, Liverpool, U.K., 1983. With permission.)

the 1974 retrofitting of Ristroph vertical traveling screens onto the Surry Power Station located along the James River, Virginia, substantially reduced impingement mortality. An 18-month study of the Ristroph screens indicated that an average of 93.3% of all impinged fish survived the impingement process. A similar retrofit of Ristroph screens at the Oyster Creek Nuclear Generating Station in 1978 also reduced fish mortality from 48 to 24%.[33]

Apart from the modification of intake screens, deflection methods and exclusion structures have been utilized to attract or repel fish from intake structures to ameliorate impingement. Among the deflection techniques, the use of electrical barriers, air bubble curtains, artificial lights, acoustic barriers, louvers, and velocity caps has met with only marginal success.[93,94] Of these various methods, louvers and velocity caps appear to offer the most promise of directing fish away from intake structures and guiding them into areas for removal or bypass.[93] Fine screens or clinker bunds surrounding intake ponds from which cooling water is taken have served as exclusion devices at the intake area of some power plants. Although these physical barriers effectively lower intake velocities, they clog easily and thus require frequent backwashing. Impingement mortalities are reduced most greatly at power plants where intake velocities have been substantially reduced. However, it has proven to be extremely expensive to lower intake velocities by increasing the size of intakes, the pumping capacity, and the screening at the intake structure.

Langford[58] and Hocutt[94] explain the intricacies of these behavioral barriers and guidance systems. The use of electrode arrays forming electrical barriers frightens, attracts, repels, or stuns fish. Alternating current tends to stun or repel fish, and direct current, attracts fish. Air bubble curtains yield a visual stimulus to deter fish from entering an area. These curtains are more effective during the day than at night, supporting the evidence that the curtains act more as a visual than as a tactile stimulus. Fish responses to artificial lights vary from strong attraction to strong avoidance. According to Hocutt,[94] photic responses of fishes to illumination may involve hypnotic attraction, positive phototaxis, conditioned response where light is associated with food, optimum preferred light intensity or wavelength, disorientation behavior similar to the disorientation of nocturnally migrating birds, and curiosity and reflex action. Fish exhibit sensitivity to mechanical, infrasonic, sonic, and also some ultrasonic vibrations. However, experiments employing acoustic barriers often are inconclusive or reveal acoustic stimuli to be ineffective. Louver screens, vertical arrays of slats placed at a 90° angle to the direction of flow, show promise in deflecting fish away from intake structures. Velocity caps, in turn, create a horizontal flow that fish perceive and generally avoid.[9] Recently, the utilization of horizontal rather than conventional vertical traveling screens has provided great potential for guiding fish from intakes. They elicit behavioral responses in fish similar to those of lower design systems. In addition to these technologies, other intake designs and screen modifications may be useful in impingement mitigation, including perforated pipe intakes, drum screens, wedgewire screens, rotating disc screens, open setting screens, fine mesh screens, radial well intakes, porous like filters, and rapid filter beds.[95]

Entrainment

Planktonic organisms, microinvertebrates, and small juvenile fish passively drawn into the cooling water condenser systems of power plants suffer the consequences of entrainment, with increased risk of death from thermal, mechanical, and chemical (e.g., biocidal) effects. As conveyed by Marcy (p. 90),[10] entrainment mortality is contingent upon:

> "...the ambient temperature and the quality of the receiving water, (1) the seasonal densities, sizes, life-stages and relative susceptibility to injury of the species involved, (3) the amplitude of the temperature rise (delta T) as the water passes through the condenser cooling system, (4) the duration of exposure to these temperatures, (5) the mechanical abrasions resulting from turbulence (shear forces) and pressure changes in passage through the system, (6) the exposure to biocides used for fouling control, and (7) gas bubble disease or the formation of air embolism caused by pressure and temperature changes in the cooling system."

Thus, the entrainment impact consists of the cumulative effect of all of these stresses. However, no consensus exists regarding the relative significance of thermal, mechanical, and chemical stresses to in-plant mortalities of entrained organisms. Some investigators consider thermal components of entrainment mortality to be most important, whereas others allude to damage associated with the physical shock of in-plant passage, notably pressure changes, abrasion, collision with structures, and shear and acceleration forces. The toxic effects of biocides, anticorrosion agents, and corrosion by-products can completely eradicate biotic populations. Davis and Coughlan,[96] for example, observed nearly complete suppression of entrained plankton activity at the Fawley Generating Station after periods of chlorination. Secondary entrainment in the thermal plume or discharge canal of a power station subject organisms to additional stress and cause delayed mortality, and shocked or stunned individuals are much more susceptible to predation by fishes or large invertebrates in outfall areas. Schubel and Marcy[97] give a comprehensive review of the biological effects of power plant entrainment and detail the physical and chemical stresses that entrained organisms encounter during their inner-plant passage.

Thermal Stress

In entrainment assessment of estuarine organisms, thermal stresses are usually subdivided into two categories: (1) thermal shock in which sudden temperature changes result from primary entrainment through a power plant; and (2) gradually declining temperatures owing to entrainment in thermal plumes or effluent canals after in-plant passage. Some organisms that have not undergone in-plant passage may still be exposed to heat, mechanical shear, and toxic chemicals when secondarily entrained within the discharge plume. Comprehensive treatment of the effects of temperature shock and time-temperature exposures on invertebrates and fishes are presented by Schubel et al.[98,99] Several approaches used to assess thermal and other impacts of power plant entrainment are laboratory simulation studies, laboratory thermal tolerance work, and *in situ* investigations of survival at operating power plants.[100]

For common North American species of aquatic invertebrates and fishes instantaneous mortalities generally occur when the organisms are exposed to a ΔT of 10°C above summer ambient temperatures and a ΔT of 17 to 20°C above winter ambient temperatures.[58,98,99] The literature contains an extensive amount of thermal-effects data relevant to entrainment exposures of the early life stages (egg to young-of-the-year). The early life stages of a wide variety of fishes appear to safely tolerate short-duration exposures (<60 min) to temperatures 10 to 15°C above spring and early summer ambient temperatures in temperate climates and 7 to 10°C above temperatures typical of mid-summer.[100]

Early investigators surmised that the temperature rise across condensers was largely responsible for entrainment mortality of biota. As thermal data accumulated, however, the importance of other factors became more evident. Hoss et al.,[101] for example, ascertained little or no distress in larval flounder, menhaden, pinfish, and spot exposed to thermal shocks up to 12°C. Schubel,[102] subjecting eggs of blueback herring (*Alosa aestivalis*), alewife (*A. pseudoharengus*), American shad (*A. sapidissima*), white perch (*Morone americana*), and striped bass (*M. saxatilis*) in the laboratory to time-excess temperature histories typical of those experienced by the organisms entrained in power plants with a variety of design and operating criteria, could not discern detrimental effects to egg development or hatching success at ΔTs up to 10°C. Marcy[103,104] registered a 34% survival of entrained young fish (2.6 to 40 mm in size) immediately after their passage through the Connecticut Yankee Nuclear Power Station. He assigned approximately 20% of the total mortality of these forms to heat shock and 80% to mechanical abrasion during their 93-s travel through the cooling system. A series of other studies corroborated Marcy's conclusions concerning the importance of mechanical stresses to mortality of organisms during entrainment, as discussed below.

Mechanical Effects

During the early and mid-1970s, a growing body of data began to point to mechanical damage as the primary cause of entrainment mortality of organisms in power plants. Owing to turbulence, changes in pressure, abrasion of suspended organic particles, and impacts against plant piping, mechanical damage to plankton and larval fish is commonly responsible for more than 50% of the total entrainment mortality. Work on phytoplankton,[105] zooplankton,[106,107] and larval fish[102-104,108] revealed the overriding importance of mechanical damage. About 60% of the phytoplankton and zooplankton mortality cited by Beck and Miller[106] was ascribed to mechanical damage. According to Carpenter et al.,[107] 70% of the copepods entrained and killed in Millstone Nuclear Power Station experienced mechanical damage in the cooling water system. Schubel[109] acknowledged the high percentages of ichthyoplankton mortality incurred during in-plant passage, indicating that mechanical stresses exert a greater role in this mortality than does temperature.

Biocidal Effects

The injection of biocide, most commonly chlorine, to control biofouling of heat exchangers in power plants leads to increased mortalities of aquatic organisms (see discussion in the section entitled, "Effects of Power Plant Operation, Biocides", above) that are often difficult to differentiate from those caused by thermal and mechanical effects. Bacteria and phytoplankton, in particular, experience massive mortalities due to in-plant chlorination, but these organisms have short generation times that mollify adverse effects on them in receiving waters. Phytoplankton, for instance, have generation times of hours or days and, although disruptions in phytoplankton productivity may result from power plant entrainment, recovery of phytoplankton populations exposed to chlorinated water during in-plant passage or in the near- or far-field discharge mixing zones, is highly possible. Nonetheless, every effort must be made to limit chlorine impacts on this group because energy shortages can occur within higher trophic levels dependent on the microscopic plants.[110]

Phytoplankton populations have different chlorine sensitivities,[111] as do other organisms.[63] Hence, the phytoplankter, *Chlamydomonas* sp., is much more resistant to chlorine than *Skeletonema costatum*. Chlorine concentrations of 1.5 to 2.3 ppm have been shown to be lethal to *S. costatum* in 5 to 10 min.[111]

The copepod, *Acartia tonsa,* suffers 100% mortality when exposed to chlorine concentrations of more than 1.0 ppm over time intervals between about 8 and 100 min.[58,63] Significantly lower mortalities take place at lesser time exposures (Figure 7). A similar effect has been observed for other invertebrates as well.[58]

Higher water temperatures exacerbate chlorine effects for both invertebrates and larval fish.[63,68] For many marine fauna, tempera-

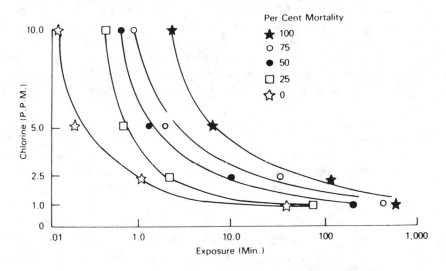

FIGURE 7.   Mortality data (response isopleths) for the calanoid copepod, *Acartia tonsa,* exposed to chlorine. (From Morgan, R. P., II and Carpenter, E. J., in *Power Plant Entrainment: A Biological Assessment,* Schubel, J. R. and Marcy, B. C., Eds., Academic Press, New York, 1978, 95. With permission.)

ture has a synergistic effect on chlorine toxicity.[112] However, the decay of chlorine also occurs at elevated temperatures,[113] and the increased chlorine toxicity at higher temperatures will be offset by a decrease in residual chlorine in power station effluent.[114] These changes may yield corresponding increases in survivorship of biota in the near-field regions of power plants.

Figure 8 depicts the typical changes in chlorine levels, as well as temperature, pressure, and velocity encountered in the cooling water system of electric generating stations. As recounted above, the effects of these factors, individually or synergistically, can significantly raise the mortality of estuarine organisms in transit through the cooling-water circuits of these facilities. It remains the task of future research to accurately evaluate the impact of the losses of these organisms on aquatic communities in the receiving waters of estuaries.

CASE STUDIES

OYSTER CREEK NUCLEAR
GENERATING STATION

Located approximately 3.5 km west of Barnegat Bay, New Jersey, the Oyster Creek Nuclear Generating Station (OCNGS) consists of a single boiling water reactor and a turbine generator that uses steam produced in the reactor to yield 620 MW (summer net) of electric power. Operating with a nominal thermal efficiency of 32%, the station employs a once-through cooling system. Four circulating pumps, each with a design capacity of 435 m^3/min, deliver a total of 1739 m^3 of cooling water to the main condenser. This water is drawn from Barnegat Bay via the intake canal of the station, with waste heat rejected to Barnegat Bay via a discharge canal and Oyster Creek. During its 3.5 km traverse to the bay, much of the waste heat dissipates to the atmosphere by evaporation and conduction.

Biofouling is controlled on the heat exchanger surfaces by injecting chlorine sequentially into six condenser sections for six 20-min intervals during a 24-h period. Maximum chlorination occurs during the summer to compensate for more rapid growth of fouling organisms at that time. Waste chlorinated water (<0.05 mg/l) is rejected to the discharge canal. Biocidal injection, together with thermal dis-

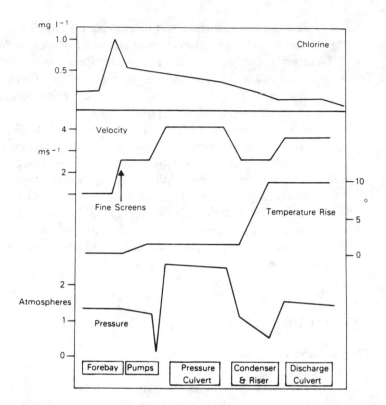

FIGURE 8. Typical changes in chlorine as well as velocity, temperature, and pressure in the cooling-water system of an electric generating station. (From Langford, T. E., *Electricity Generation and the Ecology of Natural Waters*, Liverpool University Press, Liverpool, U.K., 1983. With permission.)

charges, impingement, and entrainment, account for essentially all of the long-term plant impacts on the biota of the Barnegat Bay ecosystem. The following discussion on the effects of the OCNGS on aquatic organisms is derived from the work of Kennish et al.[33]

Impact on Aquatic Biota
Construction Effects

During construction of the OCNGS, the south branch of Forked River and Oyster Creek were dredged and modified extensively. The existing stream beds were widened and deepened; dredging of Oyster Creek and the south branch of Forked River was concluded in December 1966 and April 1967, respectively.[115] Construction of the intake and discharge canals connected the south branch of

Forked River and OCNGS. These modifications destroyed most of the original freshwater and low salinity habitats in the affected portions of the streams. Pumping at the station caused salinity, temperature, and dissolved oxygen in both streams to become similar to those of the bay. The direction of water flow in the south branch of Forked River was reversed such that it became directed to the station during operation. As a consequence of these changes in water quality and flow, estuarine fishes and invertebrates character-istic of Barnegat Bay replaced the typically brackish and freshwater communities in the lower portions of the south branch of Forked River and Oyster Creek. The most acute effects of construction, therefore, occurred in Forked River and Oyster Creek where water quality and aquatic communities changed significantly because of the modification in bathymetry, bottom composition, water flow, and water quality in both streams.

Cooling System Effects

At full power, when all four circulating pumps are operating, the temperature rise across the OCNGS condenser approaches 12.8°C. The temperature change may increase to 18.3°C during a period when one circulating pump is out of service; however, such maxi-mum temperature differentials persist for short periods of time. Thus, the monthly average temperature change recorded across the condenser usually is <10°C.

Waste heat from the OCNGS mainly affects organisms in Oyster Creek with few impacts demonstrated in Barnegat Bay. In Oyster Creek, planktonic and benthic invertebrate communities experience substantial seasonal fluctuations in the composition and abundance of constituent populations compared to those of Barnegat Bay. The greatest change from conditions in Barnegat Bay, however, may be the attraction and avoidance of fishes to the thermal effluent during different seasons of the year. Although the attraction of fishes to thermal discharges has occasionally resulted in fish kills due to heat and cold shock (Table 4), it has also improved the sport fishery catch in the area. Apart from waste heat effects, thermal, chemical, and physical stresses incurred during entrainment or impingement have increased mortality of various ontogenetic stages of estuarine fishes and invertebrates, but do not appear to have affected the structure of aquatic communities of the estuary.

To mitigate the impact of the thermal discharges on aquatic biota, two axial flow dilution pumps, each with a 983 m³/min capacity, divert water from the intake to the discharge canal to control the temperature level. A thermal plume forms in the bay as the thermal effluent flows out of Oyster Creek. The morphology of the plume varies temporally, being affected principally by winds and tides. The plume is confined to roughly a 1.6-km radius about the mouth of Oyster Creek. On calm days the plume often fans out about the mouth of Oyster Creek, its shape being affected by weak tidal currents;[75] however, strong winds from the north or south often drive it as a narrow band abutting the shoreline. At times of peak operation, water temperatures rise 3 to 5°C above ambient levels at the mouth of Oyster Creek. This increase in temperature dissipates rapidly with distance from Oyster Creek and is undetectable along the estuarine substratum beyond 1.6 km.

Phytoplankton exhibit lower primary productivity, biomass, and diversity in Oyster Creek than other areas of the estuarine system. Mountford[116] recorded a maximum decrease in gross productivity of 30.3%, a maximum decline in net productivity of 20.1%, and a maximum drop in biomass of 17.7% at the mouth of Oyster Creek when compared to the mouth of Forked River. Loveland et al.[117] also registered a decrease in gross productivity (35%) at the mouth of Oyster Creek. Mountford[116] and Hein[118] documented lower phytoplankton diversity in the discharge canal and Oyster Creek than in the intake canal and Forked River. These differences were attributed to station operation.

Effects of the thermal discharges on zooplankton have been described previously in the above section, "Effects of Power Plant Operation, Thermal Discharges, Estuarine Organisms". Suffice it to say that the impacts seem to be confined to the OCNGS discharge canal and Oyster Creek. Entrainment losses of zooplankton also contribute to reductions in absolute abundance of zooplankton in Oyster Creek relative to other areas of the system.

The benthic invertebrate community of Oyster Creek is dominated by species reported to be indicators of stress (e.g., *Capitella capitata* and *Corophium* sp.). Benthic sampling during the 1970s showed that the species composition in Forked River paralleled that of Barnegat Bay; however, in Oyster Creek, there existed a scarcity of species that numerically dominated the communities in Forked River and the bay. Furthermore, the species richness in Oyster Creek was significantly less than the species richness in Forked River. These findings reflect a power plant impact on the benthic invertebrates of Oyster Creek.

According to Hoagland and Crocket,[119] who performed studies for the U.S. Nuclear Regulatory Commission (NRC), thermal discharges from the station have extended the breeding season of shipworms, increased their growth rates (and hence the amount of their damage to wooden structures), and reduced their winter mortality rates. The establishment of the tropical-subtropical shipworm, *Teredo bartschi*, in Oyster Creek during the mid-1970s and its subsequent spread into Forked River have also been ascribed to thermal discharges from the station, as well as to changes of salinity and water flow in Oyster Creek and Forked River due to operation of the station.[120] In addition to *T. bartschi*, the occurrence of the tropical-subtropical shipworm, *T. furcifera*, in the Barnegat Bay system during the mid- to late 1970s was related to the presence of thermal effluent in the Oyster Creek area.[121] Maciolek-Blake et al.[122,123] contend that operation of the OCNGS has not affected populations of *T. navalis* and *Bankia gouldii*.

A low total number of fouling species has been recorded in Oyster Creek compared to other areas of the estuary. The low species diversity of the fouling community in Oyster Creek results from thermal discharges of the station.[124] Subtle changes in seasonality of the epibenthos in Oyster Creek also appear to be due to the thermal discharges.[125] The number of epibenthic species, for example, decreases in the discharge canal and Oyster Creek during summer and increases during winter. Although thermal discharges of the station may affect the seasonal abundances and settling locations of fouling organisms in the discharge canal and Oyster Creek, no gross detrimental or beneficial effects of the effluent have been observed on these organisms.[125]

In addition to avoidance and attraction responses of finfishes to the thermal plume and their increased mortality caused by heat and cold shock, as discussed above, some fishes and macroinvertebrates are susceptible to impingement and entrainment at the OCNGS. Between September 1975 and September 1977, the most abundant fishes and macroinvertebrates impinged on intake screens, in numerical order, included the blue crab, sand shrimp, bay anchovy, grass shrimp, Atlantic menhaden, spot, Atlantic silverside, smallmouth flounder, striped searobin, and blueback herring. The estimated total mortality of these species due to impingement was 30%. Population surveys of fishes and macroinvertebrates indicate that the standing crop lost through impingement was <10% for species in central Barnegat Bay. No evidence exists that losses of organisms through impingement on intake screens have had a discernible effect on invertebrate and fish communities in Barnegat Bay.

To mitigate impingement effects, the conventional vertical traveling screens (0.95 cm mesh) on the intake structure of the station were replaced in 1979 with Ristroph screens composed of a continuously rotating traveling design modified with a low pressure spray wash and fish recovery and return system. The Ristroph screens contain watertight, fish buckets which collect impinged organisms washed from the screens and return them to the discharge canal via a sluiceway (Figure 9). These screens have substantially lowered immediate impingement mortality of some species, such as the bay anchovy (50% decrease), weakfish (40%), and Atlantic silverside (35%).[126]

The potentially greatest impact of the OCNGS on plankton, macroinvertebrates, and fishes is through entrainment of one or more life stages through the cooling-water system. The absolute number of individuals of a given species lost via entrainment can be very high. For example, sampling from September 1976 through August 1977 revealed the following selective entrainment figures: *Acartia* (4.42×10^{12}), rotifers (1.21×10^{12}), barnacle larvae (1.48×10^{12}), arrowworms (3.81×10^8), bay anchovy eggs (4.32×10^8), and goby larvae (1.84×10^8). Even more of these life forms were entrained during an earlier sampling period from September 1975 through August 1976.[75]

Despite the large numbers of eggs, larvae, and other small life forms of Barnegat Bay organisms lost because of in-plant passage at the OCNGS, these losses have not translated into detectable impacts on aquatic communities in the bay. Effects of overall operation of the OCNGS on aquatic communities seem to be restricted to the discharge canal and Oyster Creek. The species composition, abundance, and distribution of phytoplankton, zooplankton, benthic invertebrates, and fishes in these two waterways are substantially different than in Barnegat Bay. Fluctuations within bay communi-

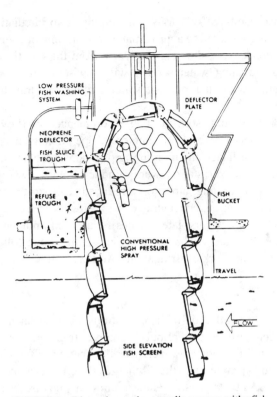

FIGURE 9. Ristroph, rotating traveling screen with a fish
recovery system. (From Kennish, M. J., Roche, M. B., and
Tatham, T. R., in *Ecology of Barnegat Bay, New Jersey*,
Kennish, M. J. and Lutz, R. A., Eds., Springer-Verlag, New
York, 1984, 318. With permission.)

ties appear to be due to the natural population dynamics of constituent populations and not due to operation of the station. Aquatic
communities in Barnegat Bay parallel those of other mid-Atlantic
estuaries that are unaffected by the operation of electric generating
stations.

Pilgrim Nuclear
Power Station

Similar to the OCNGS, the Pilgrim Nuclear Power Station
(PNPS) utilizes a once-through, open-cycle system to cool condens-
ers during operation. The plant, located on the northwestern shore of
Cape Cod Bay at 41°56′N latitude and 70°34′W longitude, has a

design capacity of 655 MWe. It employs two circulating water pumps, each delivering approximately 585 m³/min of cooling water to the condensers, drawing the water from the bay through four vertical traveling water screens with 0.95-cm wire mesh. Each circulating water pump draws water through two traveling water screens.

Ecological studies on the PNPS commenced in 1969, 3 years before the facility went into commercial operation. Intense environmental surveillance and monitoring programs related to the operation of the power station were conducted until 1977, with the programs continuing, albeit on a reduced scale, to the present.[127] Initial investigations concentrated on the effects of the calefaction of Cape Cod Bay waters; later studies focused on impingement and entrainment effects. The summary of findings that follow relies heavily on the work of Bridges and Anderson.[128]

Impact on Aquatic Biota
Thermal Discharges

More than a decade of investigations have revealed no major deleterious effects of thermal discharges from the PNPS on the marine environment. However, near-field benthic effects have been observed in proximity to the discharge canal. For example, surveys by SCUBA in 1980 on the benthic community nearby the discharge canal revealed a 1100 to 1400 m² denuded zone and a more peripheral area of stunted algal growth approximately 1900 to 2900 m². Bridges and Anderson[128] attributed the denuded zone to scour from the thermal discharges and the more distal stunted algal zone to thermal effects of the waste heat from the station.

Aside from these obvious benthic impacts near the discharge canal, significant finfish mortalities ascribable to gas bubble disease have occurred in the discharge canal and thermal plume. In April 1973, for instance, 43,000 adult Atlantic menhaden (*Brevoortia tyrannus*) succumbed to gas bubble disease in these areas. About 5000 more adult Atlantic menhaden died from gas bubble disease in the same locations in April 1975. Water supersaturated with dissolved gases in the discharge canal was responsible for both events.

Impingement

The most aggressive impingement sampling program at the PNPS extended from January 1973 through December 1980. Over this interval, more than 25,000 fish belonging to 56 species were impinged during 10,629 sampling hours. The impingement rate for all fishes from 1973 to 1980 equalled 2.39 fish per hour, with the total weight of all impinged fishes amounting to 600 kg. The numerically dominant species impinged, accounting for about 91% of all impinged fish, included: (1) Atlantic herring (*Clupea harengus harengus*) (49.1%); (2) rainbow smelt (*Osmerus mordex*) (16.8%); (3) Atlantic silverside (*Menidia menidia*) (11.5%); (4) alewife (*Alosa pseudoharengus*) (10.5%); and (5) cunner (*Tautogolabrus adspersus*) (3.0%). Table 7 lists the impingement rates and seasonality of these species at the PNPS.[129]

The rate of fish impingement correlates most strongly with the number of circulating pumps in service. Reduced water pumping lowers total impingement. This relationship is evident in impingement data collected at the PNPS over the 5-year period from 1984 through 1988. Fish impingement rates in 1984, 1987, and 1988 during plant outages were several times less than in 1985 and 1986 when at least one circulating pump operated.[130] Impingement patterns also reflected the seasonal occurrence of the impacted finfish

TABLE 7
Catch Rates and Seasonality of the Five Dominant Finfish Species Collected in Impingement Samples at the Pilgrim Nuclear Generating Station from January 1976 to December 1980

Species	Rate/h	Rate/year[a]	Dominant season of occurrence
Atlantic herring	1.62	14,191	Autumn
Rainbow smelt	0.55	4,818	Autumn
Atlantic silverside	0.38	3,329	Winter/Spring
Alewife	0.35	3,066	Summer
Cunner	0.10	876	Summer
All fish	3.30	28,908	

[a] Assuming operation of both circulating water pumps 100% of the time.

From Lawton, R. P., Anderson, R. D., Brady, P., Sheehan, C., Sides, W., Kouloheras, E., Borgatti, M., and Malkoski, V., in *Observations on the Ecology and Biology of Western Cape Cod Bay, Massachusetts,* Davis, J. D. and Merriman, D., Eds., Springer-Verlag, New York, 1984, 191. With permission.

populations in the vicinity of the PNPS in western Cape Cod Bay.[128] As expected, highest impingement rates for a species developed when it moved into the western bay.

Entrainment

　　Entrainment data collected from 1983 to 1988 strongly suggest that plankton are entrained in direct proportion to through-plant flow rates.[130] Phytoplankton survival ranged from 48 to 98% after in-plant passage; survival decreased at temperatures above 17°C, due to both thermal and biocidal effects. Zooplankton survival after passage through the cooling water system exceeded that of phytoplankton, ranging from 95 to 100% under most operating conditions. However, at temperatures in excess of 29°C, concomitant with chlorination, 100% mortality of zooplankton occurred. As in the case of other large nuclear power plants, the absolute number of ichthyoplankton entrained annually by the PNPS is extremely large. For example, in 1986, the total number of fish eggs entrained amounted to 1.695×10^9. Of the eight numerically dominant species, the entrained eggs of the labrid-*Limanda* group exceeded all others (n = 1.273×10^9), while eggs of the American plaice (*Hippoglossoides platessoides*) were least entrained (n = 3.75×10^5). The total number of fish larvae entrained in 1986 equalled 2.75×10^9. Among the ten numerically dominant forms, the entrained larvae of sculpin (*Myoxocephalus* spp.) surpassed all others (n = 7.4×10^7). Atlantic herring (*C. harengus harengus*) had the least number of larvae entrained (n = 1.83×10^6).[76] Despite the high absolute numbers of ichthyoplankton entrained in the PNPS, no damage has been detected in finfish populations in Cape Cod Bay.[76,128,130]

SUMMARY AND CONCLUSIONS

　　During their operation, electric generating stations may adversely affect aquatic communities in estuaries by the thermal loading or calefaction of receiving waters, the release of chemical substances, the impingement of organisms on intake screens, and the entrainment of various life forms in cooling water systems. In addition, temporary impacts associated with the construction of these facilities (e.g., dredging) can decimate or eliminate habitat areas as well as biotic populations. Much of the impact of large, open-cycle, once-through power plants centers around their demands for large volumes of condenser cooling water which requires that they be built along estuaries and other coastal zones. For nuclear power plants that are thermally less efficient than their fossil-fueled counterparts, a greater amount of waste heat is discharged. Furthermore, the larger quantity of cooling water used at nuclear power plants, often translates into higher impingement and entrainment mortality.

Thermal discharges influence both the water quality and the organisms of impacted estuaries, with the most acute effects found in near-field regions in close proximity to the power plants. Increasing temperature lowers the dissolved oxygen content of receiving waters while raising the metabolic rate of its faunal inhabitants. These changes cause some individuals to become more susceptible to chemical toxins or disease and others to die forthwith. Bacterial decomposition also accelerates at higher temperatures further depressing dissolved oxygen concentrations. Moreover, the density, viscosity, surface tension, and nitrogen solubility of estuarine waters diminish with increasing temperature.

The calefaction of estuarine waters elicits physiological and behavioral responses in organisms, including the alteration of metabolic rates leading to a diminution of growth and behavioral adjustment manifested in avoidance or attraction reactions to the heated effluent. Even though an organism may not succumb directly due to the elevated temperatures, it can experience sublethal physiological effects such as a decrease in reproduction or hatching success of eggs, an inhibition in development of larvae, and variations in respiratory rates that can be detrimental. The avoidance or attraction responses of finfish to thermal plumes have been intensely studied by ichthyologists. Heat and (especially) cold shock mortality of finfishes are related to the attraction of the poikilotherms to the thermal plumes.

The ultimate consequences of artificial warming on estuarine communities are shifts in their structure. Changes in species composition, species diversity, and population density are commonly observed in the near-field regions of power plants. In some cases, opportunistic or nuisance species may replace equilibrium populations at heavily impacted sites.

Several beneficial uses of waste heat have been proposed over the years, principally for industrial and residential applications, agriculture, and aquaculture. Potential industrial users include food processing, textile, furniture, chemical, concrete block manufacturers, pulp and paper, iron and steel foundries, and automobile and truck manufacturers. Residential space heating in winter and absorption-type air conditioning in summer may someday utilize substantial amounts of waste heat from power plants. Prospective uses of waste heat in agriculture are the heating and cooling of greenhouses, soil warming, crop drying, irrigation, frost protection, reclamation of livestock wastes, wastewater treatment, and environmental control of livestock and poultry housing. Commercial aquaculture operations employing thermal effluents from power plants primarily involve the rearing of oysters and salmon, although feasibility studies incorporating waste heat in fin- and shellfish aquaculture have targeted other organisms as well, namely clams, mussels, shrimp, lobsters, prawns, catfish, eels, carp, yellow perch, striped bass, rainbow trout, and tilapia.

Some of the chemicals released by electric generating stations pose a potential hazard to estuarine organisms owing to their toxicity. Most important in this respect are chlorine, which is injected into cooling water to control biofouling on heat exchanger surfaces, trace metals originating from the dissolution of piping within the condenser cooling system, and low-level radioactive wastes produced by nuclear reactors. Among these chemicals, chlorine has received much attention because at times it accounts for much of the mortality of susceptible organisms in discharge waters (e.g., bacteria and phytoplankton).

Perhaps the greatest biological impact of power plants is ascribable to impingement of larger organisms on intake screens and entrainment of plankton, microinvertebrates, and small juvenile fish passively drawn into the cooling water condenser systems. It is not uncommon for millions of macroinvertebrates and finfishes to be impinged and killed over an annual cycle at larger power plants sited on productive estuaries. Seasonal variations in the number of impinged organisms can be substantial due to seasonal migration or emigration of populations and seasonal changes in water temperature that control the ability of the organisms to withstand currents and hence impingement on intake screens. Unequivocally, the swimming performance of finfishes affects the probability of their impingement. Factors influencing swimming performance, in turn, include the type of species, its developmental stage, parasite infestation, nutritional condition, presence of predators or prey, salinity, temperature, oxygen concentration, suspended matter, tides, time of day, and season.

Various modifications of intake systems, incorporating deflection methods and exclusion structures, have been designed to mitigate impingement by attracting or repelling fish from intake areas. The main deflection techniques developed to ameliorate impingement are electrical barriers, air bubble curtains, artificial lights, acoustic barriers, louvers, and velocity caps. Among these methods, louvers and velocity caps appear to offer the most success in directing fish away from intake screens and guiding them safely into areas for removal or bypass. In regard to exclusion devices, fine screens or clinker bunds may be effective, although they clog easily and consequently require much backwashing and general maintenance. The greatest decreases in impingement mortality have been attained at power plants at which intake velocities were substantially reduced. Other improvements in impingement losses have been documented at power plants where conventional vertical traveling screens were replaced with a more advanced technology, such as Ristroph screens composed of a continuously rotating traveling design modified with a low pressure spray wash and fish recovery and return system.

The in-plant mortality of entrained organisms can be enormous in terms of absolute numbers of individuals lost. For example, the annual entrainment estimates for eggs or plankton at large power plants located on estuaries may exceed 1 billion individuals. Entrainment mortality results from thermal stress, mechanical effects (e.g., turbulence, pressure changes, abrasion of suspended organic particles, and impacts against piping), and chemical toxicity (e.g., biocides such as chlorine). Although no consensus exists regarding the relative significance of thermal, mechanical, and chemical stresses to entrainment mortality of biota, when detailed studies have been performed, mechanical damage appears to be of overwhelming importance. Nevertheless, the total entrainment mortality is the cumulative effect of these three types of stress.

Entrainment mortality is also a function of biotic factors. For instance, the seasonal densities of entrainable organisms in the impacted systems are closely linked to observed mortalities. In addition, the sizes, life stages, and relative susceptibility to injury of the entrained species play a critical role in the mortality of estuarine organisms in power plants.

REFERENCES

1. Hall, C. A. S., Howarth, R., Moore, B., III, and Vörösmarty, C. J., Environmental impacts of industrial energy systems in the coastal zone, *Annu. Rev. Energy,* 3, 395, 1978.
2. Clark, R. B., *Marine Pollution,* 2nd ed., Clarendon Press, Oxford, 1989.
3. Marcy, B. C., Jr., Kranz, V. R., and Barr, R. P., Ecological and behavioral characteristics of fish eggs and young influencing their entrainment, in *Power Plants: Effects on Fish and Shellfish Behavior,* Hocutt, C. H., Stauffer, J. R., Jr., Edinger, J. E., Hall, L. W., Jr., and Morgan, R. P., II, Eds., Academic Press, New York, 1980, 29.
4. Cairns, J., Jr., Effects of increased temperature on aquatic organisms, *Ind. Water Wastes,* 1, 150, 1956.
5. Van Vliet, V., Effect of heated condenser discharge upon aquatic life, *Am. Soc. Mech. Eng.,* 57, 1, 1957.
6. Moore, E. W., Thermal pollution of streams, *Ind. Eng. Chem. Ind. Ed.,* 50, 1, 1958.
7. Herry, S., Pollution of rivers by heated discharges, *Bull. Cent. Belge Etude Doc. Eaux,* 46, 226, 1959.
8. Markowski, S., The cooling water of power stations: a new factor in the environment of marine and freshwater invertebrates, *J. Anim. Ecol.,* 28, 243, 1959.
9. Krenkel, P. A. and Parker, F. L., Eds., *Biological Aspects of Thermal Pollution,* Vanderbilt University Press, Nashville, 1969.
10. Marcy, B. C., Jr., Entrainment of organisms at power plants, with emphasis on fishes — an overview, in *Fisheries and Energy Production: A Symposium,* Saila, S. B., Ed., D. C. Heath & Company, Lexington, MA, 1975, 89.
11. Hocutt, C. H., Introduction, in *Power Plants: Effects on Fish and Shellfish Behavior,* Hocutt, C. H., Stauffer, J. R., Jr., Edinger, J. E., Hall, L. W., Jr., and Morgan, R. P., II., Eds., Academic Press, New York, 1980, 1.
12. Jensen, L. D., Ed., *Issues Associated With Impact Assessment, Proc. 5th Workshop on Entrainment and Impingement,* EA Communications, Sparks, MD, 1981.

13. U.S. Environmental Protection Agency, Development Document for Best Technology Available for the Location, Design, Construction, and Capacity of Cooling Water Intake Structures for Minimizing Adverse Environmental Impact, Tech. Rep. EPA 440/1-76/015-a, U.S. Environmental Protection Agency, Washington, D.C., 1976.

14. Eicholz, G. G., *Environmental Aspects of Nuclear Power,* Ann Arbor Science Publishers, Ann Arbor, MI, 1976.

15. U.S. Environmental Protection Agency, Development Document for Effluent Limitations, Guidelines, and New Source Performance Standards for the Steam Electric Power Generating Point Source Category, Tech. Rep. EPA 440/1-74 029a, U.S. Environmental Protection Agency, Washington, D.C., 1974.

16. Tebo, L. B., Jr. and Little, J. A., Permitting procedures for thermal discharges, in *Factors Affecting Power Plant Waste Heat Utilization,* Gross, L. B., Ed., Pergamon Press, New York, 1980, 110.

17. Sorge, E. V., The status of thermal discharges east of the Mississippi River, *Chesapeake Sci.,* 10, 131, 1969.

18. Day, J. W., Jr., Hall, C. A. S., Kemp, W. M., and Yàñez-Arancibia, A., *Estuarine Ecology,* John Wiley & Sons, New York, 1989.

19. Wilson, J. G., *The Biology of Estuarine Management,* Croom Helm, London, 1988.

20. Olla, B. L., Studholme, A. L., and Bejda, A. J., Behavior of juvenile bluefish, *Pomatomus saltatrix,* in vertical thermal gradients: influence of season, temperature acclimation, and food, *Mar. Ecol. Prog. Ser.,* 23, 165, 1985.

21. Tuttle, J. H., Jonas, R. B., and Malone, T. C., Origin, development, and significance of Chesapeake Bay anoxia, in *Contaminant Problems and Management of Living Chesapeake Bay Resources,* Majumdar, S. K., Hall, L. W., Jr., and Austin, H. M., Eds., Pennsylvania Academy of Science, Easton, 1987, 442.

22. Drost-Hansen, W., Thermal aspects of aquatic chemistry, in Effects and Methods of Control of Thermal Discharges, Report to Congress by the U.S. Environmental Protection Agency in Accordance with Section 104(T) of the Federal Water Pollution Control Act Amendments of 1972, Serial No. 93-14, U.S. Environmental Protection Agency, Washington, D.C., 1973, 847.

23. Brett, J. R., Thermal requirements of fish — three decades of study 1940-1970, in Biological Programs in Water Pollution, Taft, R. A., Ed., San. Eng. Center Tech. Rep., unpublished manuscript, 1960, 110.

24. O'Connor, S. G. and McErlean, A. J., The effects of power plants on productivity of the nekton, in *Estuarine Research,* Vol. 1, *Chemistry, Biology, and the Estuarine System,* Cronin, L. E., Ed., Academic Press, New York, 1975, 494.

25. Odum, E. P., *Fundamentals of Ecology,* 3rd ed., W. B. Saunders, Philadelphia, 1971.

26. Woolcott, W. S., The effects of a thermal discharge on algae populations of a Piedmont section of the James River, Virginia, *Assoc. Southeast Biol. Bull.,* 21, 91, 1974.

27. Morgan, R. P., II and Stross, R. G., Destruction of phytoplankton in the cooling water supply of a steam electric station, *Chesapeake Sci.,* 10, 165, 1969.

28. Zieman, J., The Effects of Thermal Effluent Stress on the Seagrasses and Macroalgae in the Vicinity of Turkey Point, Biscayne Bay, Florida, Ph.D. thesis, University of Miami, Coral Gables, FL, 1970.

29. Thorhaug, A., Seger, D., and Roessler, M., Impact of a power plant on a subtropical estuarine environment, *Mar. Pollut. Bull.,* 4, 166, 1973.

30. Zieman, J. and Wood, E., Effects of thermal pollution on tropical type estuaries, with emphasis on Biscayne Bay, Florida, in *Tropical Marine Pollution,* Wood, E. and Johannes, R., Eds., Elsevier, New York, 1975, 75.

31. Heinle, D. R., Temperature and zooplankton, *Chesapeake Sci.,* 10, 186, 1969.

32. Heinle, D. R., Zooplankton, in *Functional Adaptations of Marine Organisms,* Vernberg, F. J. and Vernberg, W. B., Eds., Academic Press, New York, 1981, 85.

33. Kennish, M. J., Roche, M. B., and Tatham, T. R., Anthropogenic effects on aquatic communities, in *Ecology of Barnegat Bay, New Jersey,* Kennish, M. J. and Lutz, R. A., Eds., Springer-Verlag, New York, 1984, 318.

34. Loveland, R. E. and Vouglitois, J. J., Benthic fauna, in *Ecology of Barnegat Bay, New Jersey,* Kennish, M. J. and Lutz, R. A., Eds., Springer-Verlag, New York, 1984, 135.

35. Kennish, M. J. and Olsson, R. K., Effects of thermal discharges on the microstructural growth of *Mercenaria mercenaria, Environ. Geol.,* 1, 41, 1975.

36. Kennish, M. J., Effects of thermal discharges on mortality of *Mercenaria mercenaria* in Barnegat Bay, New Jersey, *Environ. Geol.,* 2, 223, 1978.

37. Abbe, G. R., Thermal and other discharge-related effects on the bay ecosystem, in *Ecological Studies in the Middle Reach of Chesapeake Bay: Calvert Cliffs,* Heck, K. L., Jr., Ed., Springer-Verlag, Berlin, 1987, 270.

38. Jones, R. D., Jensen, L. D., and Koss, R. W., Benthic invertebrates, in Environmental Responses to Thermal Discharges from the Indian River Station, Indian River, Delaware, Jensen, L. D., Ed., Tech. Rep., EPRI Publ. No. 74-049-00-3, Electric Power Research Institute, Palo Alto, CA, 1974, 106.

39. Sullivan, C. M., Temperature reception and responses in fish, *J. Fish. Res. Bd. Can.,* 11, 153, 1954.

40. Stauffer, J. R., Jr., Influence of temperature on fish behavior, in *Power Plants: Effects on Fish and Shellfish Behavior,* Hocutt, C. H., Stauffer, J. R., Jr., Edinger, J. E., Hall, L. W., Jr., and Morgan, R. P., II, Eds., Academic Press, New York, 1980, 103.

41. Meldrim, J. and Gift, J. J., Temperature preference, avoidance, and shock experiments with estuarine fishes, in Ecological Considerations for Ocean Sites Off New Jersey for a Proposed Nuclear Generating Station, Vol. 2, Part 3, Appendix 6, Ichthyological Associates, Ithaca, NY, 1971.

42. Olla, B. L. and Studholme, A. L., The effect of temperature on the activity of bluefish, *Pomatomus saltatrix* L., *Biol. Bull.,* 141, 337, 1971.

43. Olla, B. L. and Studholme, A. L., The effect of temperature on the behavior of young tautog, *Tautoga onitis* (L.), in *Proc. 9th European Marine Biology Symp.,* Barnes, H., Ed., Aberdeen University Press, Aberdeen, 1975, 75.

44. Olla, B. L., Bejda, A. J., Samet, C., and Martin, A. D., The effect of temperature on the behavior of marine fishes: a comparison among Atlantic mackerel, *Scomber scombrus,* bluefish, *Pomatomus saltatrix,* and tautog, *Tautoga onitis,* in *Combined Effects of Radioactive, Chemical, and Thermal Releases to the Environment,* International Atomic Energy Agency, Vienna, 1975.

45. Olla, B. L., Studholme, A. L., Bejda, A. J., Samet, C., and Martin, A. D., Effect of temperature on activity and social behavior of the adult tautog, *Tautoga onitis* under laboratory conditions, *Mar. Biol.,* 45, 369, 1978.

46. Gallaway, B. J. and Strawn, K., Seasonal abundance and distribution of marine fishes at a hot-water discharge in Galveston Bay, Texas, *Cont. Mar. Sci.,* 18, 71, 1974.

47. Sonn, H. W., Electric utilities: the large food production complexes of the future, in *Power Plant Waste Heat Utilization in Aquaculture,* Godfriaux, B. L., Eble, A. F., Farmanfarmaian, A., Guerra, C. R., and Stephens, C. A., Eds., Allanheld, Osmun & Company, Montclair, NJ, 1979, Preface.

48. Kim, B. C., Wilkinson, W. H., Rosenberg, H. S., and Oxley, J. H., Industrial and residential uses of power plant waste heat, in *Factors Affecting Power Plant Waste Heat Utilization,* Goss, L. B., Ed., Pergamon Press, New York, 1980, 32.

49. Stansfield, R. V. and Carroll, B. B., Waste heat projects — state-of-the-art overview, in *Factors Affecting Power Plant Waste Heat Utilization,* Goss, L. B., Ed., Pergamon Press, New York, 1980, 1.

50. Burns, E. R., Behrends, L. L., Maddox, J. J., Madewell, C. E., Mays, D. A., and Pile, R. S., Agricultural uses of power plant waste heat, in *Factors Affecting Power Plant Waste Heat Utilization,* Goss, L. B., Ed., Pergamon Press, New York, 1980, 3.

51. Gaucher, T. A., Thermal enrichment and marine aquaculture, in *Marine Aquaculture,* McNeil, W. J., Ed., Oregon State University Press, Corvallis, 1970, 141.

52. Sylvester, J. R., Biological considerations on the use of thermal effluents for finfish aquaculture, *Aquaculture,* 6, 1, 1975.

53. Godfriaux, B. L., Eble, A. F., Farmanfarmaian, A., Guerra, C. R., and Stephens, C. A., Eds., *Power Plant Waste Heat Utilization in Aquaculture,* Allanheld, Osmun & Company, Montclair, NJ, 1979.

54. Skidmore, D. and Chew, K. K., Mussel Aquaculture in Puget Sound, Tech. Rep., Washington Sea Grant Program, University of Washington, Seattle, 1985.

55. Hubert, W. A., Aquaculture uses of power plant waste heat, in *Factors Affecting Power Plant Waste Heat Utilization,* Goss, L. B., Ed., Pergamon Press, New York, 1980, 18.

56. Chow, W., Condenser biofouling control: the state-of-the-art, in Proceedings: Condenser Biofouling Control — State-of-the-Art Symposium, Chow, W. and Massalli, Y. G., Eds., Tech. Rep., Electric Power Research Institute, Palo Alto, CA, 1985, 1-1.

57. Garey, J. F., Jorden, R. M., Aitken, A. H., Burton, D. T., and Gray, R. H., *Condenser Biofouling Control: Symposium Proceedings,* Ann Arbor Science Publishers, Ann Arbor, MI, 1980.

58. Langford, T. E., *Electricity Generation and the Ecology of Natural Waters,* Liverpool University Press, Liverpool, U.K., 1983.

59. Characklis, W. G., Bryers, J. D., Trulear, M. G., and Zelver, N., Biofouling film development and its effects on energy losses: a laboratory study, in *Condenser Biofouling Control: Symposium Proceedings,* Ann Arbor Science Publishers, Ann Arbor, MI, 1980, 49.

60. Brungs, W. A., Effects of residual chlorine on aquatic life, *J. Water Pollut. Control Fed.,* 45, 2180, 1973.

61. Mattice, J. S. and Zittel, H. E., Site-specific evaluation of power plant chlorination, *J. Water Pollut. Control Fed.,* 48, 2284, 1976.

62. Electric Power Research Institute, Biofouling Detection Monitoring Devices: Status Assessment, Tech. Rep., Electric Power Research Institute, Palo Alto, CA, 1985.

63. Morgan, R. P., II and Carpenter, E. J., Biocides, in *Power Plant Entrainment: A Biological Assessment,* Schubel, J. R. and Marcy, B. C., Jr., Eds., Academic Press, New York, 1978, 95.

64. Morgan, R. P., II, Biocides and fish behavior, in *Power Plants: Effects on Fish and Shellfish Behavior,* Hocutt, C. H., Stauffer, J. R., Jr., Edinger, J. E., Hall, L. W., Jr., and Morgan, R. P., II, Eds., Academic Press, New York, 1980, 75.

65. Seegert, G., Bogardus, R. B., and Horvatk, F., Review of the Mattice and Zittel Paper. Site-Specific Evaluation of Power Plant Chlorination Project, Tech. Rep., Edison Electric Institute, Washington, D.C., 1978.

66. Carpenter, E. J., Peck, B. B., and Anderson, S. J., Cooling water chlorination and productivity of entrained phytoplankton, *Mar. Biol.,* 16, 37, 1972.

67. Eppley, R. W., Ringer, E. H., and Williams, P. M., Chlorine reactions with seawater constituents and the inhibition of photosynthesis of natural marine phytoplankton, *Est. Coastal Mar. Sci.,* 4, 147, 1976.

68. Brooks, A. S. and Seegert, G. L., The effects of intermittent chlorination on rainbow trout and yellow perch, *Trans. Am. Fish. Soc.,* 106, 278, 1977.

69. Capuzzo, J. M., The effects of free chlorine and chloramine on growth and respiration rates of larval lobsters (*Homarus americanus*), *Water Res.,* 11, 1021, 1977.

70. Capuzzo, J. M., Davidson, J. A., Lawrence, S. A., and Libni, M., The differential effects of free and combined chlorine on juvenile marine fish, *Est. Coastal Mar. Sci.,* 5, 733, 1977.

71. Capuzzo, J. M., Lawrence, S. A., and Davidson, J. A., Combined toxicity of free chlorine, chloramine, and temperature to stage I larvae of the American lobster, *Homarus americanus, Water Res.,* 10, 1093, 1976.

72. Thatcher, T. O., The relative sensitivity of Pacific northwest fishes and invertebrates to chlorinated seawater, in *Water Chlorination: Environmental Impacts and Health Effects,* Vol. 2, Jolley, R. L., Gorchev, H., and Hamilton, D. H., Jr., Eds., Ann Arbor Science Publishers, Ann Arbor, MI, 1978, 341.

73. Middaugh, D. P., Couch, J. A., and Crane, A. M., Response of early life history stages of the striped bass, *Morone saxatilis,* to chlorination, *Chesapeake Sci.,* 18, 141, 1977.

74. Morgan, R. P., II and Prince, R. D., Chlorine toxicity to eggs and larvae of five Chesapeake Bay fishes, *Trans. Am. Fish. Soc.,* 106, 380, 1977.

75. Jersey Central Power and Light Company, Oyster Creek and Forked River Nuclear Generating Stations 316(a) and (b) Demonstration, Volume 1, Tech. Rep., Jersey Central Power and Light Company, Morristown, NJ, 1978.

76. Boston Edison Company, Marine Ecology Studies Related to Operation of Pilgrim Station, Semi-Annual Rep. No. 29, Boston Edison Company, Braintree, MA, 1987.

77. Clark, J. R. and Brownell, W., Electric Power Plants in the Coastal Zone: Environmental Issues, American Littoral Society Special Publ. No. 7, Highlands, N.J., 1973.

78. Marcy, B. C., Jr., Beck, A. D., and Ulanowicz, R. E., Effects and impacts of physical stress on entrained organisms, in *Power Plant Entrainment: A Biological Assessment,* Schubel, J. and Marcy, B. C., Jr., Eds., Academic Press, New York, 1978, 135.

79. Van Winkle, W., Ed., *Proc. Conf. Assessing the Effects of Power-Plant-Induced Mortality on Fish Populations,* Pergamon Press, New York, 1977.

80. Murarka, I. P., Validation and Software Documentation of the ANL Fish Impingement Model, Tech. Rep. ANL/ES-62, Argonne National Laboratory, Argonne, IL, 1978.

81. Swartzmann, G. L., Comparison of Simulation Models Used in Assessing the Effects of Power-Plant-Induced-Mortality on Fish Populations, U.S. Nuclear Regulatory Commission Rep. No. NUREG/CR-0474, Washington, D.C., 1978.

82. Ogawa, H., Modelling of power plant impacts on fish populations, *Environ. Manage.,* 3, 321, 1979.

83. Goodyear, C. P., Assessing the impact of power plant mortality on the compensatory reserve of fish populations, in *Proc. Conf. Assessing the Effects of Power-Plant-Induced Mortality on Fish Populations,* Van Winkle, W., Ed., Pergamon Press, New York, 1977, 1985.

84. Stupka, R. C. and Sharma, R. K., Survey of Fish Impingement at Power Plants in the United States, Vol. 3, Estuaries and Coastal Waters, Tech. Rep. ANL/ES-56, Argonne National Laboratory, Argonne, IL, 1977.

85. Uziel, M. S., Entrainment and impingement at cooling-water intakes, *J. Water Pollut. Contr. Fed.*, 6, 1616, 1980.

86. Uziel, M. S. and Hannon, E. H., Impingement: An Annotated Bibliography, Tech. Rep. EA-1050, Electric Power Research Institute, Palo Alto, CA, 1979.

87. Barnthouse, L. W. and Van Winkle, W., The impact of impingement on the Hudson River white perch populations, in *Issues Associated With Impact Assessment, Proc. 5th Natl. Workshop on Entrainment and Impingement,* Jensen, L. D., Ed., EA Communications, Sparks, MD, 1981, 199.

88. Lawler, J. P., Hogarth, W. T., Copeland, B. J., Weinstein, M. P., Hodson, R. G., and Chen, H. Y., Techniques for assessing the impact of entrainment and impingement as applied to the Brunswick Steam Electric Plant, in *Issues Associated with Impact Assessment, Proc. 5th Natl. Workshop on Entrainment and Impingement,* Jensen, L. D., Ed., EA Communications, Sparks, MD, 1981, 159.

89. Holmes, R. H. A., Fish and Weed on Fawley Power Station Screens, C. E. R. L. Lab. Note RD/L/N 129/75, Leatherhead, Surrey, England, 1975.

90. Thomas, D. L. and Miller, G. J., Impingement studies at the Oyster Creek Generating Station, Forked River, New Jersey, From September to December 1975, in *Proc. 3rd Natl. Workshop on Entrainment and Impingement: 316(b) Research and Compliance,* Jensen, L. D., Ed., Ecological Analysts, Melville, NY, 1976, 317.

91. Powers, D. A., Physiology of fish impingement: role of hemoglobin and its environmental modifiers, in *Proc. 3rd Natl. Workshop on Entrainment and Impingement: 316(b) Research and Compliance,* Jensen, L. D., Ed., Ecological Analysts, Melville, NY, 1976, 241.

92. White J. C. and Brehmer, M. L., Eighteen-month evaluation of the Ristroph traveling fish screens, in *Proc. 3rd Natl. Workshop on Entrainment and Impingement: 316(b) Research and Compliance,* Jensen, L. D., Ed., Ecological Analysts, Melville, NY, 1976, 367.

93. Sharma, R. K., Fish Protection at Water Diversions and Intakes: A Bibliography of Published and Unpublished References, Tech. Rep., ANL/ESP-1, Argonne National Laboratory, Argonne, IL, 1973.

94. Hocutt, C. H., Behavioral barriers and guidance systems, in *Power Plants: Effects on Fish and Shellfish Behavior,* Hocutt, C. H., Stauffer, J. R., Jr., Edinger, J. E., Hall, L. W., Jr., and Morgan, R. P., II, Eds., Academic Press, NY, 1980, 183.

95. Hocutt, C. H. and Edinger, J. E., Fish behavior in flow fields, in *Power Plants: Effects on Fish and Shellfish Behavior,* Hocutt, C. H., Stauffer, J. R., Jr., Edinger, J. E., Hall, L. W., Jr., and Morgan, R. P., II, Eds., Academic Press, New York, 1980, 143.

96. Davis, M. H. and Coughlan, J., Response of entrained plankton to low-level chlorination at a coastal power station, in *Water Chlorination: Environmental Impact and Health Effects,* Vol. 2, Jolley, R. L., Gorchev, H., and Hamilton, D. H., Jr., Eds., Ann Arbor Science Publishers, Ann Arbor, MI, 1978, 369.

97. Schubel, J. R. and Marcy, B. C., Jr., Eds., *Power Plant Entrainment: A Biological Assessment,* Academic Press, New York, 1978.

98. Schubel, J. R., Smith, C. F., and Koo, T. S. Y., Thermal effects of power plant entrainment on survival of larval fishes: a laboratory assessment, *Chesapeake Sci.,* 18, 290, 1977.

99. Schubel, J. R., Smith, C. F., and Koo, T. S. Y., Thermal effects of entrainment, in *Power Plant Entrainment: A Biological Assessment,* Schubel, J. R. and Marcy, B. C., Jr., Eds., Academic Press, New York, 1978, 19.

100. Jinks, S. M., Lauer, G. J., and Loftus, M. E., Advances in techniques for assessment of ichthyoplankton entrainment survival, in *Issues Associated With Impact Assessment, Proc. 5th Natl. Workshop on Entrainment and Impingement,* Jensen, L. D., Ed., EA Communications, Sparks, MD, 1981, 91.

101. Hoss, D. E., Hettler, W. F., and Coston, L. C., Effects of thermal shock on larval estuarine fish — ecological implications with respect to entrainment in power plant cooling systems, in *The Early Life History of Fish, Proc. Int. Symp.,* Blaxter, J. H. S., Ed., Springer-Verlag, New York, 1973, 357.

102. Schubel, J. R., Effects of Exposure to Time-Excess Temperature Histories Typically Experienced at Power Plants on the Hatching Success of Fish Eggs, Spec. Rep. No. 32-PPRP-4, Ref. No. 73.11, Chesapeake Bay Institute, Johns Hopkins University, Baltimore, 1973.

103. Marcy, B. D., Jr., Survival of young fish in the discharge canal of a nuclear power plant, *J. Fish. Res. Bd. Can.,* 28, 1057, 1971.

104. Marcy, B. D., Jr., Vulnerability and survival of young Connecticut River fish entrained at a nuclear power plant, *J. Fish. Res. Bd. Can.,* 30, 1195, 1973.

105. Flemer, D. A., Preliminary report on the effects of steam electric station operations on entrained organisms, in Postoperative Assessment of the Effects of Estuarine Power Plants, Ref. No. 71-24a, Mihursky, J. A. and McErlean, A. J., co-principal investigators, Chesapeake Bay Laboratory, Solomons, MD, 1971.

106. Beck, A. D. and Miller, D.C., Analysis of inner plant passage of estuarine biota, Proc. American Society of Civil Engineering, Power Div. Spec. Conf., Boulder, CO, 1974.

107. Carpenter, E. J., Peck, B. B., and Anderson, S. J., Survival of copepods passing through a nuclear power station on northeastern Long Island Sound, U. S. A., *Mar. Biol.,* 24, 49, 1974.

108. Tarzwell, C. M., An argument for open ocean siting of coastal thermal electric plants, *J. Environ. Qual.,* 1, 89, 1972.

109. Schubel, T. R., Effects of exposure to time-excess temperature histories typically experienced at power plants on the hatching success of fish eggs, *Est. Coastal Mar. Sci.,* 2, 105, 1974.

110. Hall, L. W., Jr., Helz, G. R., and Burton, D. T., *Power Plant Chlorination: A Biological and Chemical Assessment,* Ann Arbor Science Publishers, Ann Arbor, MI, 1981.

111. Hirayama, K. and Hirano, R., Influence of high temperature and residual chlorine on marine phytoplankton, *Mar. Biol.,* 7, 205, 1970.

112. Capuzzo, J. M., The effect of temperature on the toxicity of chlorinated cooling waters to marine animals — a preliminary review, *Mar. Pollut. Bull.,* 10, 45, 1979.

113. Davis, M. H. and Coughlan, J., A model for predicting chlorine concentration within marine cooling circuits and its dissipation at outlets, in *Water Chlorination Environmental Impact and Health Effects: Chemistry and Water Treatment,* Jolley, R. L., Brungs, W. A., Cotruva, J. A., Cumming, R. B., Mattice, J. S., and Jacobs, V. A., Eds., Ann Arbor Science Publishers, Ann Arbor, MI, 1981, 347.

114. Dempsey, C. H., The exposure of herring postlarvae to chlorine in coastal power stations, *Mar. Environ. Res.,* 20, 279, 1986.

115. U.S. Atomic Energy Commission, Final Environmental Statement Related to Operation of the Oyster Creek Nuclear Generating Station, Tech. Rep., AEC Docket No. 50-219, U.S. Atomic Energy Commission, Washington, D.C., 1974.

116. Mountford, K., Plankton Studies in Barnegat Bay, Ph.D. thesis, Rutgers University, New Brunswick, NJ, 1971.

117. Loveland, R. E., Moul, E. T., Busch, D. A., Sandine, P. H., Shafto, S. S., and McCarty, J., The Qualitative and Quantitative Analysis of the Benthic Flora and Fauna of Barnegat Bay Before and After the Onset of Thermal Addition, Tech. Rep., Rutgers University, New Brunswick, NJ, 1972.

118. Hein, M. K., Effects of Thermal Discharges on the Structure of Periphytic Diatom Communities, M.S. thesis, Rutgers University, New Brunswick, NJ, 1977.

119. Hoagland, K. E. and Crocket, L., Ecological Studies of Woodboring Bivalves in the Vicinity of the Oyster Creek Nuclear Generating Station, Tech. Rep. NUREG/CR-1855, U.S. Nuclear Regulatory Commission, Washington, D.C., 1981.

120. Hoagland, K. E. and Crocket, L., Ecological Studies of Woodboring Bivalves in the Vicinity of the Oyster Creek Nuclear Generating Station, Tech. Rep. NUREG/CR-1939, Vol. 3, U.S. Nuclear Regulatory Commission, Washington, D.C., 1981.

121. Hoagland, K. E. and Turner, R. D., Range extensions of teredinids (shipworms) and polychaetes in the vicinity of a temperate-zone nuclear generating station, *Mar. Biol.,* 58, 55, 1980.

122. Maciolek-Blake, N., Hillman, R. E., Feder, P. I., and Belmore, C. I., Study of Woodborer Populations in Relation to the Oyster Creek Generating Station, Tech. Rep. No. 15040, Jersey Central Power and Light Company, Morristown, NJ, 1981.

123. Maciolek-Blake, N., Hillman, R. E., Belmore, C. I., and Feder, P. I., Study of Woodborer Populations in Relation to the Oyster Creek Generating Station, Tech. Rep., GPU Nuclear Corporation, Parsippany, NJ, 1982.

124. Shafto, S. S., The Boring and Fouling Community of Barnegat Bay, New Jersey, M.S. thesis, Rutgers University, New Brunswick, NJ, 1974.

125. Young, J. S. and Frame, A. B., Some effects of a power plant effluent on estuarine epibenthic organisms, *Int. Rev. Ges. Hydrobiol.,* 61, 37, 1976.

126. GPU Nuclear Corporation, Summary of Environmental Studies Submitted in Support of the 316(a) and (b) Demonstration for the Oyster Creek Nuclear Generating Station, Tech. Rep., GPU Nuclear Corporation, Parsippany, NJ, 1986.

127. Merriman, D., Preface, in *Observations on the Ecology and Biology of Western Cape Cod Bay, Massachusetts,* Davis, J. D. and Merriman, D., Eds., Springer-Verlag, New York, 1984.

128. Bridges, W. L. and Anderson, R. D., A brief survey of Pilgrim Nuclear Power Plant effects upon the marine aquatic environment, in *Observations on the Ecology and Biology of Western Cape Cod Bay, Massachusetts,* Davis, J. D. and Merriman, D., Eds., Springer-Verlag, New York, 1984, 263.

129. Lawton, R. P., Anderson, R. D., Brady, P., Sheehan, C., Sides, W., Kouloheras, E., Borgatti, M., and Malkoski, V., Fishes of western inshore Cape Cod Bay: studies in the vicinity of the Rocky Point shoreline, in *Observations on the Ecology and Biology of Western Cape Cod Bay, Massachusetts,* Davis, J. D. and Merriman, D., Eds., Springer-Verlag, New York, 1984, 191.

130. Boston Edison Company, Marine Ecology Studies Related to Operation of Pilgrim Station, Semi-Annual Rep. No. 33, Boston Edison Company, Braintree, MA, 1989.

Index

A

Abalone, 419
Absorption, see Sorption
Acartia, 220, 442
Acartia hudsonica, 96
Acartia tonsa, 410, 436
Accumulation, see Bioaccumulation
Acenaphthene, 162
Achromobacter, 82
Acid mine waste, 277
Acinetobacter, 82
Acorn barnacles, 101
Acoustic barriers, 431, 432, 449
Activity, radionuclide, 321
Acute toxicity, 15, see also specific pollutants
Acute toxicity tests, 16–18
Adriatic Sea, 222, 225–227
Adsorption, see Sorption
Aegean Sea, 222, 225–227
Aeromonas, 82
Age, and biotransformation ability, 11,
       167, 214
Agitation dredge, 361, 362, 390
Agricultural activity
       dredged-spoil disposal, 391
       organochlorines, 246, 247, see also
              Pesticides
       PAHs, 133, 134
       power plant waste heat utilization,
              418, 419
       system pollution susceptibility, 50
Agriculture, power plant waste heat
       utilization, 418, 419
Air bubble curtains, 431, 432, 449
Albertini Inlet, British Columbia, 386–388
Alco torda (razorbill), 113
Aldrin, 183, 189, 192–193, 247

Alewife (*Alosa pseudoharengus*)
       power plant, Pilgrim Nuclear
              Power Station, 445
       temperature limits, 413
       thermal stress, 434
Algae
       bioaccumulation in, 8
       heavy metals
              concentration, 287, 288
              homeostasis, 285, 286
       oil spills and, 95, 97
       PAH bioconcentration, 152
       PCBs, 220
       power plant effects
              Pilgrim Nuclear Power
                     Station, 444
              thermal discharge and,
                     407, 408
       radiosensitivity, 341
Aliphatic hydrocarbons, toxicity of, 9
Alkanes
       crude oil composition, 69
       microbial breakdown, 82
       petroleum versus biogenic, 72
Alkenes, 70, 72
Alkyl groups, biotransformation
       processes, 10
Alkynes, 70
Alosa aestivalis (blueback herring), 423
Alosa pseudoharengus (alewife), 413,
       434, 445
Alpha particle emission, 322
Aluminum, 264, 294
Amazon River, PAHs, 145
American Smelting and Refining
       Company, 281
Ammonia
       dredging effects, 369, 380

sewage disposal and, 44
Amoco Cadiz oil spill, 64, 108
Amphipods
    acute toxicity tests, 18
    dredging effects, 385, 386
    PAH toxicity, 161
Amphora coffeaeformis, 274
Amplification of metals, 284, 285
Anachis lafresnayi, 385
Anas acuta (northern pintails), 242
Anas clypeata (northern shovelers), 242
Anas platyrhynchos (mallard duck), 111, 224,
        228, 231
Anas rubripes, see Duck, black
Anchovy, bay (*Anchoa mitchilli*), 441, 442
    impingement losses, 428
    thermal discharges and, 416, 417
Anguilla (eel), 104, 227, 417, 419, 448
Annelids
    bioaccumulation in, 8
    heavy metal homeostasis, 285, 286
    PAHs, 167, 173
Annual cycles, see also Seasonal cycles
    and oil spill recovery, 97
    sedimentation, 290, 291
Annual doses of radiation, 342
Anoxia, see also Oxygen; Oxygen depletion
Anthracene
    Chesapeake Bay, 168
    classification of PAHs, 162, 163
    sewage sludge, 149
    toxicity, 136
    uptake and biotransformation, 159
Antifouling paints, heavy metals, 273, 276
Antimony, see Heavy metals
Apalachicola Bay, 49
Apirolio, 200, see also Polychlorinated
        biphenyls
Aquaculture, 418, 419, 448
Arenicola marina, 91, 103, 222
Argo Merchant oil spill, 64, 80, 96, 108, 109
Argon, 341
Argopecten, 419
Aricidea lopezi, 386
Arochlors, 189, 200
    national surveys, 215
    in oysters, 222
    physical characteristics, 202, 203, 205
Aromatic hydrocarbons, see also Chlorinated

hydrocarbons; Polynuclear
    aromatic hydrocarbons
    crude oil composition, 70, 71
    microbial breakdown, 82
    petroleum versus biogenic, 72, 73
    toxicity of, 9
Arrhenius law, 406
Arrow oil spill, 64, 81, 89, 114, 116, 117, 120
Arrowworms, 442
Arroyo Colorado, Texas, 196, 197
Arsenic, 22, see also Heavy metals
    body burdens in striped bass, 294
    concentration of in marine fauna, 287
    in San Francisco Bay, 298
Arthrobacter, 82
Arthropods
    bioaccumulation in, 8
    PAHs, 167, 173
Ash disposal, 273
Asphyxiation, dredging effects, 359
Assays, see Tests
Assimilation efficiencies, 12, 14
Asterias, 105
Atchafalaya Bay, 49
Atherina presbyter (sand smelt), 429
Atlantic Flyway, PCB contamination,
        228, 231
Atlantic herring, see Herring, Atlantic
Atlantic mackerel, see Mackerel, Atlantic
Atlantic menhaden, see Menhaden, Atlantic
Atlantic Ocean
    PCB contamination, 208, 209
    turbidity, 372
Atmosphere, PCB volatilization, 209
Atmospheric deposition, 9
    heavy metals, 271
    organochlorines
        DDT, 185, 186
        PAHs, 133, 134, 146, 165,
            170, 171
    petroleum hydrocarbons, 66, 67
    pollution sources, 1
    radionuclides, 23, 24
Auks, 113
Autotrophic growth, 43
Avicennia, 92
Aythya affinis (lesser scaups), 242
Aythya valisnenria (canvasbacks), 242
Axinopsida serricata, 386, 387

B

Bacillus, 82
Background radiation, 323–325, 334,
      335, 342
Bacteria, 270
    bioaccumulation in, 8
    oil degradation, 82
    PAH transformation, 166, 167
    PCB degradation, 201
    radiosensitivity, 341
    respiration, temperature and, 405
    sewage sludge
        pathogens, 37, 56
        sorption of toxic pollutants, 36
Bacterium, 82
Balanus gladula, 101
Baltic Sea
    heavy metal pollution, 281, 302
    nitrogen and phosphorus levels in, 45
    PAHs, 145
    PCBs, 208, 224, 233
Baltimore Canyon, PAHs, 156, 157
Baltimore Harbor
    heavy metal contamination, 279,
        290, 293
    PAHs, 168
Bankia gouldii, 441
Barge discharges, 363
Barnacles
    larvae, power plant effects, 410, 442
    oil spills and, 80, 101, 104
Barnegat Bay, New Jersey, 411, 415, see also
        Oyster Creek Nuclear
        Generating Station
Bass, largemouth, 413
Bass, sea, 416
Bass, striped (*Morone saxtilis*)
    aquaculture, 419, 448
    heavy metal contamination, 289, 294
    organochlorines, 240–242, 234
        PCBs, 233
        toxaphene, 197
    PAHs, 151, 152
    power plant effects
        chlorine toxicity, 423
        thermal stress, 413, 416, 434
Bay anchovy, see Anchovy, bay
Beaches, dredged spoil disposal strategies,
        382, 388, 391

Beaufort Sea, 81
Beaufort Sound, North Carolina, 45
Belfast Harbor, Maine, 142
Benthic communities, see also Bivalves;
        Invertebrates; Mollusks;
        Mussels; Oysters
    dredging effects, 359, 363, 364
    measures of ecosystem health, 48
    oil spill effects, 97–105
    organic loading and, see Organic
        loading
    organochlorines
        DDT, 186
        Puget Sound, 243–245
    PAHs, 138, 139
    power plant effects, 410, 440, 441
    radionuclide dispersion, 336
Benthic feeders, see Food chain
Benz(a)anthracene
    Chesapeake Bay, 140, 168
    classification of PAHs, 162, 163
    concentrations of, 146
    fish, concentrations in, 156, 157
    in lobsters from Nova Scotia, 152
    sewage sludge, 148, 149
    in shellfish, 155
    uptake and biotransformation, 159
Benzene, 106
Benzenes, chlorinated, 189, 198, 199, 247
Benzofluoranthenes, 146
    in Chesapeake bay, 168
    in lobsters from Nova Scotia, 152
    in shellfish, 155
Benzo(a)fluorene, Chesapeake Bay, 140, 141
Benzo(g,h,i)perylene
    Chesapeake Bay, 140, 141, 168
    combustion-derived, concentrations
        of, 146
1,2-Benzophenanthrene, crude oil, 71
Benzo(c)phenanthrenes, toxicity, 136
Benzo(a)pyrene, 133, 134, see also
        Polynuclear aromatic
        hydrocarbons, 173
    annual deposition rates, 165
    bioelimination, in fish, 14
    in Chesapeake Bay, 138–141, 168
    classification of PAHs, 162, 163
    combustion-derived, concentrations
        of, 146

in fish, 156–158
in lobsters from Nova Scotia, 152
from petroleum, 71, 106
in Puget Sound, 245
in sewage sludge, 149
in shellfish, 153–155
toxicity, 136, 161
uptake and biotransformation, 159
3,4-Benzo(a)pyrene, carcinogenicity, 173
Benzo(e)pyrene
 Chesapeake Bay, 140, 141
 classification of PAHs, 162, 163
 combustion-derived, concentrations
  of, 146
 in lobsters from Nova Scotia, 152
 sewage sludge, 149
1,2-Benzphenanthrene, crude oil, 71
Bering Sea, organochlorines, 194
Beta radiation, 322
Bhawania heteroseta, 385
Bioaccumulation, 15
 bioconcentration factors (BCFs), 6, 9,
  159, 169
 definitions, 6, 7
 heavy metals, 283–289
 mechanisms of, 6–10
 PAHs 150–160, 171
  Chesapeake Bay, 169
  toxicities, 161, 162
 PCBs, 206
 radioactivity, 339–341
Bioassays, see Tests
Bioavailability
 and heavy metal bioaccumulation, 285
 PAHs, 169
Biochemical oxygen demand (BOD),
 dredging and, 358, 369
Biocides, see also Chlorinated hydrocarbons
 organotin, 275
 phosphorus-containing, 183, 184
 power plants, 437, 438
  effects on biota, 435–437
  Pilgrim Nuclear Power Station,
   446
Bioconcentration
 defined, 6, 7
 PAHs, 152, 159
Bioconcentration factors (BCFs), 6, 9, 159

Biodegradation, 15
 DDT, 186
 PAHs, 137, 173
 PCBs, 201
Biofouling, measures of ecosystem
 health, 48
Biogenous hydrocarbons, petroleum
 hydrocarbons versus, 71–73
Biological Aspects of Thermal Pollution, 401
Biological compensation, 425
Biological effects of waste disposal
 bioaccumulation, 6–10
 biotransformation, 10, 11
 elimination of xenobiotics, 11–14
 toxicity, 14–19
Biological factors
 and nutrient loading effects, 46–48
 oil (petroleum), fate of, 72–85
 PAH transport, 171
 and PCB concentrations, 214
 and toxicity, 15
 waste disposal site selection
  considerations, 5
Biological half-times, 12
 fish, 14
 zooplankton, 13
Biological oxygen demand (BOD), dredging
 effects, 358, 369
Biomagnification
 defined, 6, 7
 heavy metals, 283–285
 organochlorines, 187
Biomass, seaweed, 97
Biomonitors, see Mussel, blue; Sentinel
 species
Biotransformation, 15
 mechanisms of, 10, 11
 PAHs, 166, 167, 173
Bioturbation
 oil spills and, 98, 99
 PAH redistribution, 149
 waste disposal site selection
  considerations, 5
Biphenyls, 133, 162, see also Polynuclear
 aromatic hydrocarbons
Birds
 dredged-spoil effects, beneficial,
  382, 383

heavy metal contamination, 284
oil spills and, 95, 109–114
   Arrow, 117
   hydrocarbon concentration
      in, 104
   organochlorines
      in Chesapeake Bay, 234
      DDT, 183, 185, 186
      PCBs, 224, 228–232
      in Puget Sound, 246
      in San Francisco Bay, 242, 243
Biscayne Bay, Florida, thermal discharges
   in, 408, 409
Bivalves, see also Clams; Mollusks; Oysters
   bioaccumulation by, 8, 9
   depuration rates, 13
   dredging and dredged-spoil disposal
      effects, 385–387
   oil spills, 99–101, 104
   organochlorines
      endrin, 193, 194
      PCBs, 215, 216
      San Francisco Bay, 239–241
   PAHs
      concentrations of, 150–156,
         159, 160
      sentinel function, 174
   as sentinel organisms, 9, 10, 174
   thermal discharges and, 410
Black-crowned night heron (*Nycticorax*
   nycticorax), 232, 242
Black sea bass, 416
Blue crab, see Crab, blue
Bluefish (*Pomatomus saltatrix*)
   oil spills, 94
   thermal discharges and, 412,
      415–417
Bluegill (*Lepomis macrochirus*), 18, 413
Blue-green algae, temperature and, 407, 408
Blue mussel, see Mussel, blue
Blue runner, thermal discharges
   and, 416, 417
Boops boops, 227
Boops salpa, 227
Boston Harbor, Massachusetts
   heavy metal pollution, 279, 302
   PAHs, 142, 143
   PCBs, 216, 218

Brayton Point Power Plant, 424
Brazos River, 49
Brett's temperature tolerance
   polygon, 407, 408
Brevibacterium, 82
Brevoortia patronus (menhaden), 197
Brevoortia tyrannus, see Menhaden, Atlantic
Brown algae, heavy metal concentration,
   287, 288
Brown shrimp, see Shrimp, brown
Brunswick Steam Electric Plant, 428
Bucket dredge, 361, 362, 369
Burning, see Combustion
Butadienes, chorinated, 243
Butterfish, 412
Buzzard's Bay, Massachusetts, 85, 163

C

Cadmium, 22, see also Heavy metals
   body burdens in striped bass, 294
   in Chesapeake Bay, 295
   concentration of in-marine fauna, 287
   dredging and, 359, 378, 379
   in San Francisco Bay, 298, 300
   toxicity, relative, 264
Calanus, 96
Calanus finmarchicus, 81
Calanus hyperboreus, 97
Calcium
   concentration factors for different
      classes of organism, 340
   nutrient requirements, 43
Calefaction, 399, 447, see also Thermal stress
California killifish, 413
Callinectes sapidus, see Crab, blue
Calvert Cliffs Nuclear Power Plant, 411
Cambarus, 18
Camphenes, chlorinated, 189, 190, 196–198
Campylobacter, 37
Cancer irroratus (rock crab), 152
Cancer magister (Dungeness crab), 161,
   162, 280
Cape Cod Bay, see Pilgrim Nuclear
   Power Station
Cape Fear River, North Carolina, 199, 380
Capitella, 115, 363, 387
Capitella capitata, 18, 99, 440
Capping, dredged-spoil dumpsite, 367, 391

Carbon, see also Organic carbon; Organic
    loading; Organic particles
    nutrient requirements, 43
    oxygen utilization, 44, 46
    as pollutant, 1
Carbon-14, natural levels in seawater, 335
Carboxylic acid
    crude oil, 71
    herbicides derived from, 189, 191
Carcinogens
    organochlorines, 185
    PAHs, 136, 173
Cardium edule (cockle), 91, 99, 214
Carp (*Cyprinus*), 419, 448
Casco Bay, Maine, PAHs, 143
Caspian terns, 242, 243
Catfish, channel (*Ictalurus punctatus*), 18,
    413, 419, 448
Cement, 273
Central Bay, see San Francisco Bay
Central Flyway, PCB contamination, 228,
    231
Cerastoderma edule, see Cockle
Cerium, 340
Cesium, 340
Chaetomorpha aerea, 97
Chalk Point Power Plant, 408
Charles River, Massachusetts, 143
Charleston Harbor, South Carolina, 380
Chedabucto Bay, 85, 89
Chelon labrosus (grey mullet), 224
Chemical factors
    nutrient loading, 46–48
    oil (petroleum), fate of, 72–85
Chemical oxygen demand, dredging
    and, 358
Chemical properties
    PAHs, 134, 136
    PCBs, 200–206
    and toxicity, 15
Chemicals
    dredging and disposal of sediments,
        358
    electric power generating station, 448
    organic, see Chlorinated
        hydrocarbons; Polynuclear
        aromatic hydrocarbons
Chemical transformation, PAHs, 166
Chemical uptake, 7, 8

Chernobyl disaster, 326
Chesapeake Bay
    dredging and dredged-material effects,
        376–378, 383–386
    heavy metals, 276, 279, 289–295
        antifouling paints, 275
        sediment composition, 280
    organic loading, 56
        eutrophication, 39–43
        nitrogen and phosphorus levels
            in, 45
    organochlorines, 234–236
        hexachlorobenzene, 198
        PCBs, 211, 215, 219
        toxaphene, 197
    oxygen depletion, temperature
        and, 405
    PAHs, 138–141, 163, 167–169, 174
    pollution susceptibility, 49, 50
    turbidity, 372
Chesapeake-Delaware Canal
    DDT, 191
    PCBs, 217
Chironomus, 18
Chlamydomonas sp., 436
Chlophen, 200, see also Polychlorinated
    biphenyls
Chloramine, 423
Chlorbenside, 239
Chlordane, 183, 189, 192, 194, 195, 247
    Chesapeake Bay, 234
    Laguna Madre system, 197
    San Francisco Bay, 236, 237, 239,
        240
Chlordecone, 195
Chloretol, 200, see also Polychlorinated
    biphenyls
Chlorinated butadienes, Puget Sound, 243
Chlorinated hydrocarbons, 21, 22, see also
    Organochlorines
    biomagnification, 7
    case studies, 234–246
        Chesapeake Bay, 234–236
        Puget Sound, 243–246
        San Francisco Bay biota,
            239–243
        San Francisco Bay sediments,
            236–238
    elimination of, 12

PCBs, 199–233, see also PCBs
    in biota, 211–233
    environmental effects, 203,
      206–233
    physical-chemical properties,
      200–206
    in sediments, 209–211
    in water, 207–209
  pesticides, 185–199, see also
    Pesticides; specific pesticides
    chlorinated benzenes and
    phenols, 198–199
    cyclodienes, 192–195
    DDT, 185–192
  San Francisco Bay, 237, 238
  sewage sludge, 36, 56
Chlorine
  PAH transformation, 166
  power plant biocides, 420–423, 448
    and fish mortality, 416
    Oyster Creek Nuclear
      Generating Station,
      437, 438
Chlorine content, PCBs
  and concentration factors, 220, 221
  and properties, 201–203
Chromium, 22, see also Heavy metals
  Chesapeake Bay, 295
  dredging and, 359, 378
  enrichment factors, 293, 294
  San Francisco Bay, 298
  toxicity, relative, 264
Chronic toxicity, 15
Chronic toxicity tests, 17–19
Chrysene
  Chesapeake Bay, 140, 141, 168
  classification of PAHs, 162, 163
  combustion-derived, concentrations
    of, 146
  in lobsters from Nova Scotia, 152
  sewage sludge, 149
  in shellfish, 155
  toxicity of, 161
Chthamalus fissus, 101
Citharichthys stigmaeus (sand dab), 18
Cladosporium resinae, 82
Clam, Baltic (*Macoma balthica*)
  PAHs
    Chesapeake bay, 167–169

    experimental studies, 174
    PCB monitoring, 214
Clam, brackish water (*Rangia cuneata*), 139,
    159
Clam, hard (*Mercenaria mercenaria*), 101,
    106, 410
  biotransformation ability, 10, 11
  oil spills, hydrocarbon concentration
    in, 104
  PAHs, Chesapeake Bay, 139
  PCB concentrations, 214
  PCB monitoring, 213
  thermal discharges and, 410
Clam, softshell (*Mya arenaria*)
  PAHs, 152, 154–156
  PCBs, 236
Clam, wedge (*Rangia cuneata*)
  dredged-material effects, 378
  PAHs, 139, 159
Clams
  aquaculture, 419, 448
  dredged-material effects, 378, 382
  oil spill effects, 91, 94, 104
  PAHs
    Chesapeake Bay, 167–169
    concentrations, 155
    uptake of, 159
  PCBs, 223
Clamshell dredge, 374, 375
Clam worms (*Nereis succinea*), 167–169,
    174
Clapper rails (*Ralus longirostris*), 232
Clay
    dredged-soil disposal site caps, 367
    oil sorption onto, 80, 81
    PCB sorption, 203
    radionuclide dispersion, 336
Clean Water Act of 1977 (Public Law
    95-217), 401
Clinker bunds, 431, 449
Clupea harengus, see Herring, Baltic
Coal, combustion products, see Combustion
Coal and coal tar products, PAH sources,
    146, 170, 171
Coal-fired power plants, radon
    emissions, 334
Coastal revitalization, dredged-spoil
    disposal, 391
Coastal wetlands, 366

Cobalt, 302, see also Heavy metals
    concentration factors for different
        classes of organism, 340
    nutrient requirements, 43
    toxicity, relative, 264
Cocarcinogens, PAHs as, 137
Cockle (*Cardium edule*), 91, 214
Cod (*Gadus morhua*), 108, 109
Coelenterates, oil spills and, 93
Cold shock, 405, 415, 447
Colloids, PCB affinity, 207, 208
Columbia River, Oregon, 326
    PCBs, 219
    pollution susceptibility, 49
Combustion
    and heavy metal pollution, 273
    PAH sources, 133, 134, 146, 147,
        173, 174
        Chesapeake Bay, 167
        content of effluents, 165
        dredged material levels, 148,
        149
        environmental cycle, 171
        surface sediments, total
        concentrations, 146–148
Commencement Bay, Washington, 198, see
    also Puget Sound
    heavy metal pollution, 279–283, 302
    organochlorines
        hexachlorobenzene, 198
        PCBs, 219, 244
Complexation, heavy metal interactions, 268,
    285
Conger eel, 417
Connecticut Yankee Nuclear Power
    Station, 435
Construction
    dredged-material disposal
        regulation, 388
    Oyster Creek Nuclear Generating
        Station, 438, 439
Consumer products, radioactive waste
    recycling in, 350
Containment
    dredge spoils, 360
    radionuclides, 351
    waste disposal strategies, 3
Convective descent, dredged-material effects,
    376, 377

Cooling systems
    biofouling organisms in, 419, 420
    entrainment and impingement losses,
        see Electric power generating
        stations
Coos Bay, Oregon
    organochlorines, 219, 242
    PAHs
        mollusk concentrations, 156
        in shellfish, 154
Coos River, Oregon, 242
Copepods
    acute toxicity tests, 18
    chlorine sensitivity, 436
    oil spills, 96, 97
    PAH toxicity, 161, 162
    thermal discharges and, 410
Copper, 22, 302, see also Heavy metals
    bioavailability survey, 213
    body burdens in striped bass, 294
    Chesapeake Bay, 295
    concentration factors for different
        classes of organism, 341
    concentration of in-marine fauna, 287
    dredging and dredged-material effects,
        358, 378
    enrichment factors, 293, 294
    nutrient requirements, 43
    San Francisco Bay, 298–301
    toxicity, relative, 264
Copper oxide antifouling paints, 273, 274
Corexit 9527, 93, 110–112
Cormorant, double-crested (*Phalacrocorax
    auritus*), 229
Cormorant, white-crested (*Phalacrocorax
    carbo*), 229
Corophium, 440
Corynebacterium, 82
Cosmic radiation, 323–325, 342
Crab, blue (*Callinectes sapidus*), 18
    acute toxicity tests, 18
    antifouling paints and, 274
    heavy metals, 264, 274
    Oyster Creek Nuclear Generating
        Station impingement mortality,
        441
    PAHs, metabolic processing, 136, 137
    pentachlorophenol, 198
    radiosensitivity, 341

Crab, Dungeness (*Cancer magister*), 161, 162, 280
Crab, fiddler (*Uca*)
    acute toxicity tests, 18
    heavy metals, antifouling paints, 274, 275
    oil spills, 102, 116
Crab, horseshoe (*Limulus polyphemus*), 214
Crab, lady, 104
Crab, rock (*Cancer irroratus*), 152
Crab, spider (*Libinia emarginata*), 274
Crabs
    kepone in, 195
    oil spills and, 94, 101, 102, 104
    PCBs, 223
Crangon crangon (shrimp), 102, see also Shrimp
Crangon septemspinosa, see Shrimp, sand
Crassostrea, see Oysters
Crassostrea angulata, 100
Crassostrea gigas (Pacific oyster), 18, 100, 274
Crassostrea virginica, see Oyster, American; Oysters
Crayfish, 18
Creosote, 171
Crevalle jack, 416, 417
Croaker, Atlantic (*Micropogonias undulatus*), 139
Crustacea, see also Shellfish; specific crustaceans
    bioelimination, 13, 14
    bioconcentration factors for radioactive isotopes, 340, 341
    biotransformation ability, 10
    heavy metals and 285–288
    oil spills and, 93, 104
    PAHs, 156
    radiosensitivity, 341, 342
    toxaphene in, 196
Cumulative effects, 17
Cunner (*Tautogolabrus adspersus*), 445
Cutterhead dredge, 361, 390
Cycloalkanes, 69, 72, 73, 82
Cyclodienes, 189, 247, see also specific pesticides
Cyclostremiscus pentagona, 385
Cynoscion nebulosus (spotted sea trout), 197, 236

Cynoscion regalis, see Weakfish
Cyphonaute larvae, thermal discharges and, 410
Cyprinodon variegatus (sheepshead minnow), 18, 161, 214
Cyprinus (carp), 419, 448
Cytochrome P-450 MFO systems, 10, 11, 105, 136, 137, 150, 167

D

2,4-D, 189, 190
Dacthal, 189, 190, 197
Daphnia, 18
DCPA (dacthal), 190
DDD [1,1-dichloro-2,2-bis(*p*-chlorophenyl) ethane], 187
    Laguna Madre system, 197
    San Francisco Bay, 237
DDE [1,1-dichloro-2,2-bis(*p*-chlorophenyl) ethylene], 187, 189
    Chesapeake Bay, 234
    Laguna Madre system, 197
    San Francisco Bay, 237, 239
    in surface films, 209
DDT, 163, 183, 185–192, 247
    in Arroyo Colorado-Laguna Madre system, 197
    bioaccumulation of, 8
    in bivalves, 191
    in Chesapeake Bay, 234
    in fish, 191, 192
        bioelimination of, 14
        on San Francisco Bay, 240–242
    in Puget Sound, 243–246
    regulation of, 184, 185
    in San Francisco Bay, 236–238, 240–242
    in surface films, 209
Dealkylation, biotransformation processes, 10
Decapods, PAHs, 156
Decommissioning and decontamination wastes, 333, 351
Deep-hole waste disposal, radioactive wastes, 346
Degradation mechanisms
    microbial, see Microbial degradation organochlorines

DDT, 186, 187
    versus organophosphates, 183
Delaware Bay, 294
    DDT, 191
    heavy metal enrichment factors, 293
Delor, 200, see also Polychlorinated
    biphenyls
Density
    power generating station and, 447
    water quality, temperature and, 405,
        406
Dentex macrophthalmus, 227
Depuration, 11, 12
    bivalves, 13
    PAHs, 137, 150–162
    zooplankton, 12, 13
Desorption, heavy metal removal in river
    systems, 269
Detritus feeders, radionuclide distribution,
    337
Developmental abnormalities, oil spill
    effects, 105–107
Developmental stage
    and impingement mortality, 449
    and sensitivity to pollutants, 9
Diagenesis, PAH sources, 173
Diarrhetic shellfish poisoning (DSP), 55
Diatoms
    heavy metals and, 274
    oil spills and, 96
    organic loading from sewage and, 35
    power station effluents and, 407, 408
Dibenz(a,h)anthracene
    Chesapeake bay, 168
    classification of PAHs, 162, 163
    toxicity of, 161
Dibenzo(a,i)pyrene, 136
Dibenzothiophenes, sewage sludge, 148, 149
Dicofol, 189
Dieldrin, 183, 187, 189, 190, 193, 247
    Chesapeake Bay, 234
    fish and shellfish, 188
    Laguna Madre system, 197
    San Francisco Bay, 239, 240
Dielectric constant, PCBs, 201, 203
Diffusion
    dredged-material effects, 376, 377
    radionuclide, 336

Dilution, radioactive wastes, 350, 351
Dimethylnaphthalene, 162
Dinoflagellates, red tides, 53–55
Dinophysis, 55
Dinophysis acuminata, 55
Dinophysis acuta, 55
Dinophysis fortaa, 55
Dinophysis norvegica, 55
Diphenylmethane, 71
Dipper dredge, 361, 390
Direct uptake, 7, 8
Discharge plume, dredged-material effects,
    376, 377
Discovery Bay
    heavy metal contamination, 280
    organochlorines, 245
Dispersal
    optimizing, 5, 6
    radionuclides, 350, 351
    waste disposal strategies, 3
Dispersants, oil
    in mangrove swamps, 92
    and oil biodegradation, 83, 84
    seagrass effect, 93
Dispersion models, 49–52
Disposal conveyance, dredging
    devices, 360–363
Dissolution, see Solubility
Dissolved organic matter, and petroleum
    hydrocarbon solubility, 78
Dissolved oxygen
    dredging and dredged-spoil disposal
        effects, 358, 369, 392
    power generating station and, 447
Dogfish, smooth, 417
Domestic wastes
    heavy metal sources, 264
    pollution sources, 1
Double-crested cormorant (*Phalacrocorax
        auritus*), 229
Downreay, 345
Dredging and dredged-spoil disposal, 24, 25,
    273
    case studies, 383–388
        Albertini Inlet, British
            Columbia, 386–388
        Chesapeake Bay, 383–386
    devices, 359–363

disposal strategies, 4
effects, adverse, 363–380
　　habitat and organisms,
　　　363–371
　　water quality, 369, 372–380
effects, beneficial, 380–383
heavy metals in, 276, 302
organochlorines, PCBs-211, 210
Oyster Creek Nuclear Generating
　　Station construction, 438, 439
PAHs, 148
pollution sources, 1
regulation of, 3, 388, 389
Drum screens, 432
Duck, black (*Anas rubripes*)
　　Chesapeake Bay pollutants and, 234
　　PCBs, 224, 228, 231
Duck, mallard (*Anas platyrhynchos*)
　　oil spills, 111
　　PCBs, 224, 228, 231
Ducks (*Anas; Aythya*)
　　oil spills, 111, 113
　　PCBs, 224, 228, 231
Dunaliella tertiolecta, 274
Dungeness crabs (*Cancer magister*), 161,
　　162, 280
Dustpan dredge, 361, 390
Duwamish River
　　nitrogen and phosphorus levels in, 45
　　PCBs, 219
Duwamish Waterway, 368, 373, 375
Dynamic collapse, dredged-material effects,
　　376, 377
Dynkol, 200, see also Polychlorinated
　　biphenyls

E

Eagle, bald, 186, 234
Eagle, white-tailed sea (*Haliaetus albicilla*),
　　228, 229
Eagles, DDT body burdens, 183
EC_{50}, 16
Echinoderms
　　heavy metal homeostasis, 285, 286
　　oil spills and, 93, 105
　　PAH transformation, 167
Eels (*Anguilla*), 104, 227, 417, 419, 448

Eggs
　　oil effects, 86
　　Oyster Creek Nuclear Generating
　　　Station, 442
　　sensitivity to pollutants, 9
Eggs, bird
　　DDT and, 185, 186
　　PCBs, 228–230, 232
Eggs, bivalve, oil spills and, 100, 101
Eggs, fish, see Ichthyoplankton
Egrets, snowy, 242, 243
Eiders, 113
Elasmopus pectenicrus, 161
Electrical barriers, 431, 432, 449
Electric power generating stations, 25, 26,
　　see also Radioactivity
　　biocides, 419–423, 435–437
　　case studies
　　　Oyster Creek Nuclear
　　　　Generating Station,
　　　　437–443
　　　Pilgrim Nuclear Power Station,
　　　　443–446
　　effects of thermal discharges on
　　　biota, 406–417
　　　fish, 411–417
　　　invertebrates, 410, 411
　　　phytoplankton, 407–409
　　　seagrasses, 408, 409
　　　zooplankton, 409, 410
　　effects of thermal discharges on water
　　　quality, 405, 406
　　entrainment, 433–437
　　heavy metal pollution by, 273
　　historical development, 401, 402
　　impingement, 425–432
　　PAHs, 134
　　mechanical effects, 435
　　nuclear fuel cycle, 327
　　PAHs, 134
　　thermal stresses, 434, 435
　　waste heat utilization, 417–419
Elimination of xenobiotics, see also
　　Depuration
　　mechanisms of, 11–14
　　PAHs, 159
Elizabeth River, 138, 139, 215, 290, 294

Elliott Bay, Washington
  dredged-material behavior, 375
  heavy metal pollution, 279, 302
  organochlorines
    hexachlorobenzene, 198
    PCBs, 217, 219–221, 244, 245
Elutriate test, dredged material, 379
Embryonic development, oil spills and, 106
Endosulfan, 189, 240
Endrin, 183, 187, 189, 192–194, 197, 247
Engraulis encrasicholus, 226
Enrichment, nuclear fuel cycle, 327
Enrichment factors, heavy metals, 293, 294
Enterobacter, 82
Enteromorpha, 104
Enteromorpha intestinalis, 97
Entrainment, 400, 433, 449, 450
    at Oyster Creek Nuclear Generating
      Station, 441
    at Pilgrim Nuclear Power Station, 446
Environmental cycle, PAHs, 170, 172
Epoxidation, 10
Escambia Bay, Florida, PCBs, 211, 219
Esox lucius, 227
Ethyl parathion, 197
Eudorella pacifica, 386
Eurytemora affinis, 161, 162
Euthynnus aletteratus, 226
Eutrophication, 56, see also Organic loading
    Chesapeake Bay, 39–43
    dredged sediments, 380, 383
    sewage sludge and, 36
Evaporation
    oil, crude, 76, 77
    temperature and, 406
Exclusion structures, 431
Excretion kinetics, 12
Exxon Valdez oil spill, 64, 117–120

F

Fallout, see also Atmospheric deposition
    distribution of, 329, 330
    PAHs, 165
Fat solubility, see Lipids
Fawley Generating Station, 428, 430, 431
Federal Water Pollution Control Act
      Amendments (Public Law 92-
      500), 401, 425, 426

Federal Water Pollution Control Act of 1972
      (Clean Water Act), 3
Fenchlor, 200, see also Polychlorinated
      biphenyls
Filters, 432
Finasol OSR-5, 109
Finfish, see Fish
Finland, 364
Fire, see Combustion
Firth of Clyde, Scotland, heavy metal
      pollution, 277, 281, 282
Fish, 106
    bioaccumulation in, 8
    bioelimination rates, 14
    biotransformation ability, 11
    heavy metal contamination, 284–288
    oil spills and, 94, 95, 105
      hydrocarbon concentration
        in, 104
      toxic effects, 106–109
    organic loading and oxygen depletion,
      39
    organochlorines, 187, 188, 190, 248
      aldrin, 192, 193
      chlordane, 194
      dieldrin, 193
      endrin, 193, 194
      hexachlorobenzene, 198
      kepone, 195
      mirex, 195
      PCBs, 206, 207, 216–219, 223,
        233, 235, 236
      Puget Sound, 244, 245
      San Francisco Bay, 240–242
      toxaphene, 196, 197
    PAHs
      bioconcentration and toxicity,
        152, 159
      concentrations of, 150–152,
        156–158, 160
      toxicity and pathology, 138,
        139
    power plant effects
      chlorine, toxicity, 423
      entrainment, 424
      thermal discharges, 411–417
    radioactivity and
      concentration factors, 340, 341

radiation doses, annual, 342
radiosensitivity, 341, 344
Fish eggs, see Ichthyoplankton
Fission product radionuclides, 330
Fjords, nutrient levels in, 45
Flatfish (*Pleuronectes platessa*)
oil spills, 104, 108
PCBs, 217–219
Flocculation, heavy metal removal, 269
Flooding, and sedimentation, 290
Florida oil spill, 64, 88, 89, 94, 108, 115, 116, 120
Flounder (*Pseudopleuronectes americanus*)
aquaculture, 419
impingement mortality, 442
oil spills, 94
thermal shock responses, 434
Flounder, smallmouth, 442
Flounder, starry (*Platichthys stellatus*)
heavy metal contamination, 284
organochlorines, 218, 240–242
Flounder, winter (*Pseudopleuronectes americanus*)
oil pollution, 107
PAHs, 151, 152, 157
temperature limits, 413
Fluoranthene
in Chesapeake Bay, 140, 141, 167–169
classification of PAHs, 162, 163
combustion-derived, concentrations of, 146
in fish, 157
in sewage sludge, 148, 149
in shellfish, 154, 155
toxicity of, 161
Fluorene
classification of PAHs, 162
crude oil, 71
in lobsters from Nova Scotia, 152
sewage sludge, 148, 149
toxicity, 136
uptake and biotransformation, 159
Flying fish, 104
Food chain
bioaccumulation, 9
organochlorines in, 248
DDT, 185–192
PCBs, 184, 207, 208

PAHs, 137, 170–172
radionuclide distribution, 337, 338, 351
Forked River, see Oyster Creek Nuclear Generating Station
Fossil fuels, see also Combustion
and heavy metal pollution, 273
PAH sources, 137, 138, 171, 173
Fratercula artica, see Puffins
Free metal ionic activity, 303
Freshwater inflow, see also Riverine inflow
Chesapeake Bay, 290
and nutrient loading effects, 40, 41, 48, 49
PAH sources, 142
San Francisco Bay, 295–297, 299, 300
Froude models, 49, 51, 56
Fucus, oil spills and, 104
Fucus serratus, 97
Fucus spiralis (rockweed), 117
Fucus vesiculosus, 97
Fundulus, 104
Fundulus heteroclitus (mummichog), 18, 106, 118
Fundulus similis (killifish, longnose), 18
Fungi
oil degradation by, 82
PAH transformation, 167
Fungicides, 246

G

Gadus morhua (cod), 108, 109
Galapagos hydrothermal vent fields, 265
Galveston Bay, Texas, 49
organochlorines
hexachlorobenzene, 198
pentachlorophenol, 198, 199
thermal discharges, 415
Gambusia affinis (mosquitofish), 161
Gamma rays, 323
Gammarus, 18
Gammarus pseudolimnaeus, 18, 161
Gannets, oil effects, 113
Gas bubble disease, 444
Gaseous effluents, nuclear wastes, 333
Gases, nuclear fuel cycle, 328
Gasterosteus aculeatus (stickleback, threespine), 18

Gastropods, 101, 385
   dredging effects, 385
   heavy metal concentration, 287, 288
   thermal discharges and, 410
Genetic effects
   mutagenicity of PAHs, 135, 173
   of radiation, 344
Geochemistry, PAH production, 168
Geology
   and nutrient loading, 46–48
   and radioactive waste disposal, 347,
     348
   San Francisco Bay, 300
German Bight, heavy metal pollution, 281,
     302
Gills
   mixed function oxygenase activity, 11
   petroleum hydrocarbons and, 9
Global distribution, PAHs, 142–146
Gobius paganellus, 227
Goby larvae, 442
Gonyaulaux catenella, 54
Gonyaulaux excavata, 54
Goosefish, 416
Government policy, see Legislation;
     Regulation
Grasses, oil spill effects, 90, see also
     Seagrasses
Gravimetry, 374, 375
Great blue herons, 246
Great Lakes
   dredging in, 374, 375
   mirex in, 196
   PAHs, 144
   PCBs, 224
Grebe, great-crested (*Podiceps cristatus*), 229
Green algae
   heavy metals and, 287–289
   oil spills and, 97
Greenland, 156
Guillemot, common (*Uria aalge*), 113, 229
Gulf of Mexico
   hydrocarbon concentration, 85
   PCBs, 211, 217
   turbidity, 372
Gull, herring (*Larus argentatus*) oil spills,
     104, 112
   PCBs, 228–230

H

Habitat
   dredging effects
     adverse, 363–371
     beneficial, 382, 383
   oil spill effects, 86–94
Haddock (*Melanogrammus aeglefinus*), 109
Hake, red (*Urophycus chuss*), 157, 158
Half-life, radionuclide, 321
Half-time, biological, 12
Haliaetus albicilla (white-tailed sea eagle),
     228, 229
Haliocherus grypus, see Seal, gray
Haliotis (abalone), 419
Halodule wrightii, 93
Hampton Roads, 292, 293
Hanford atomic plant, 326
Harbor seals
Hard clams, see Clams, hard
Hazardous waste disposal leachates, San
     Francisco Bay, 298, 299
Health hazards, 136
Health of ecosystem, measures of, 48
Health status, and impingement losses, 430
Heat exchanger, biofouling, 419, 420
Heat shock, 405, 447
Heat stress, see Electric power generating
     stations; Thermal stresses
Heavy metals, 22, 23
   anthropogenic input, 271–283
   bioaccumulation, 8, 283–289
   bioavailability survey, 213
   bioelimination, 12–14
   defined, 263
   dredging and dredged-material effects,
     358, 359, 369, 378, 379
   excretion kinetics, 12
   metabolism, 264
   NOAA Mussel Watch Program, 160
   nutrient requirements, 43
   Puget Sound, 246
   in sewage sludge, 36, 56
   sources, 1, 265–283
   sources, anthropogenic
       input, 271–283
     antifouling paints, 273–276
     smelting, 282, 283
     waste disposal, 276–282

sources, atmospheric input, 271
sources, river input, 265–271
speciation, 265–267
specific estuarine systems, 289–301
    Chesapeake Bay, 289–295
    San Francisco Bay, 295–301
toxicity, 264
Helminths, 56, 57
Hemigrapsus, oil spills, 104, see also Crabs
Heptachlor, 183, 189
Herbicides, 189, 246, 289, 234
Heron, black-crowned night (*Nycticorax nycticorax*), 232, 242
Heron, great blue, 246
Herring
    oil spills, 109, 118
    temperature limits, 413
Herring, Atlantic (*Clupea harengus harengus*)
    oil spills and, 106, 109
    PCBs, 214, 224
    power plant effects
        impingement losses, 429
        Pilgrim Nuclear Power Station, 445, 446
Herring, Baltic (*Clupea harengus*), 106, 109
Herring, blueback (*Alosa aestivalis*)
    chlorine toxicity, 423
    impingement mortality, 442
    thermal stress, 434
Herring Gull, see Gull, herring
Heterocyclic compounds, 71
Heterophoxus oculatus, 386
Hexachlorobenzenes, 190, 192, 198, 239, 247
    Puget Sound, 245
    San Francisco Bay, 240
Hexachlorocyclohexane, 189
Hexachlorocyclopentadienes, 189, 190
Hexachlorohexane, 239
High-level nuclear wastes, 4, 332, 333, 351
Hippoglossoides platessoides (American plaice), 446
Hogchokers (*Trinectes maculatus*), 138, 139
Homarus americanus, see Lobster
Hopper dredge, 361–363, 390
Hormones
    and biotransformation ability, 11
    and PCB concentrations, 214

Horseshoe crab (*Limulus polyphemus*), 214
Hot spots
    heavy metal contamination, 292
    organochlorines, 247, 248
        DDT, 190, 191
        PCBs, 236
Hudson-Raritan estuary, 2
    organic loading, 34, 35
    PAHs, 148, 163
        concentrations of, 149
        redistribution in sediments, 149, 150
    pentachlorophenol, 199
Hudson River, 2, 49
    PAHs, 151, 152
    PCBs, 208, 210, 211, 219, 233
    power plants, impingement losses, 426–428
Human health hazards
    organochlorines, 203, 206, 248
    PAHs, 136
Humic acids, 268
Hydraulic dredging devices, 361, 362, 390
Hydraulic loading, and accomodative capacity of system, 47–49
Hydraulics, dredging effects, 357, 358
Hydrocarbons
    chlorinated, see Chlorinated hydrocarbons
    PAHs, see Polynuclear aromatic hydrocarbons
    petroleum, composition of, 67–72, see also Oil spills; Petroleum hydrocarbons
Hydrogen sulfide
    crude oil, 71
    dredging and, 358
Hydroids, heavy metal homeostasis, 285, 286
Hydrology
    dredging effects
        beneficial, 380–382
        spoil disposal, 365–368
    and eutrophication, 40, 41
    and nutrient loading effects, 44, 46–48
    and PAH transformation, 167
    radioactive waste disposal criteria, 347, 348

San Francisco Bay, 300
Hydrolysis, 15
Hydrothermal activity, heavy metal
    sources, 265
Hypochlorite, see Chlorine
Hypomesus pretiosus, 106
Hypoxia, see also Oxygen; Oxygen depletion

I

Ichthyoplankton, 410
    chlorine, toxicity, 423
    entrainment effects, 424
    oil sensitivity, 108, 109
    power plant effects
        chlorine toxicity, 423
        entrainment mortality, 435
        Pilgrim Nuclear Power Station,
            446
        thermal shock responses, 434
    radiosensitivity, 344
Ictalurus punctatus, see Catfish, channel
Impingement, 400, 449
    effects of, 425–432
    at Oyster Creek Nuclear Generating
        Station, 441
    Pilgrim Nuclear Power Station, 445,
        446
Impoundments, dredged-material disposal
    regulation, 388
Incineration, see Combustion
Indeno(c,d)pyrene, 146, 168
Indian Point Nuclear Station, 426, 428
Indian River Power Plant, 411
Indicator species, see Mussel, blue; sentinel
    species
Industrial wastes, 1, 2
    estuarine pollution susceptibility, 49
    heavy metal sources, 264
    organochlorines
        pentachlorophenol, 198, 199
        PCBs, 199, 200
    PAH sources, 133, 134, 146
    petroleum hydrocarbons in, 66
    Puget Sound, 243
    regulation of disposal, 3
Inerteem, 200, see also Polychlorinated
    biphenyls
Inflow, see Freshwater inflow; Riverine
    inflow

Insecticides, 246, see also Pesticides
Intake velocities, 432
Intertidal zones, oil spills, 117
Invertebrates, 106, see also Benthic
    communities; Bivalves;
    Mollusks; Musscls; Oysters
    acute toxicity tests, 18
    bioelimination mechanisms, 13
    EC_{50} estimation, 16
    oil spills, 99–105
    PAH transformation, 167
    PCBs, 221–223
    power plants and, 410, 411
    thermal discharges and, 410
Iodine isotopes, 340, 342
Ionizing radiation, see Radioactivity
Iron, 302, see also Heavy metals
    concentration factors for different
        classes of organism, 340
    dredging and dredged-material effects,
        358, 378
    nutrient requirements, 43

J

James River, 42, 43
    kepone in, 195
    power station effluent, 408

K

Kanechlor, 200, see also Polychlorinated
    biphenyls
Kelthane, 189
Kepone, 195, 247
    dredged-material effects, 378
    uses, 189, 190
Killer whales, 245
    organochlorines
        Puget Sound, 245
Killifish, California, 413
Killifish, longnose (*Fundulus similis*), 18
Kingfish, northern, 416, 417
Krypton, 328
Kuwait crude oil, 97, 103

L

Lactic acid, temperature stress and, 407
Ladder dredge, 361, 390
Lady crab, 104

Ladyfish, 417
Lagodon rhomboides (pinfish), 18, 223, 434
Laguna Madre, Texas, 196, 197
Lake Erie, 144, 374
Lake Ontario, 196, 375
Lake Superior, 211
Laminaria saccharina, 97
Larids, 114
Larus argentatus, see Gull, herring
Larval forms
    chlorine and, 422, 423
    fish, see Ichthyoplankton
    oil spills and, 86, 100, 101
    power plant effects, 442
        chlorine biocides, 422, 423
        thermal discharges, 410
 sensitivity to pollutants, 9
LC_{50}, 16, 7
Lead, 22, see also Heavy metals
    body burdens in striped bass, 294
    Chesapeake Bay, 295
    concentration of in-marine fauna, 287
    dredging and dredged-material
        effects, 359, 378, 379
    enrichment factors, 293, 294
    San Francisco Bay, 298, 301
    toxicity, relative, 264
Lead-210, natural levels in sewater, 335
Legislation, 3, 425, 426
    dredged-material disposal, 366, 367,
        388, 389
    PCB ban, 203
    power plants, 401, 402
    radioactive waste policy, 317
Leiostomus xanthurus, see Spot
Lepas (barnacle), 104
Lepomis macrochirus, see Bluegill
Lethal doses (LD_{50}) of radiation, 341, 343,
    344
Lethal toxicity, 15
Levees, 366, 388
Libinia emarginata, 274
Life cycle factors
    molting, see Molting
    and PCB levels, 214
    and sensitivity to pollutants, 9, 11
    and toxicity of petroleum in
        mollusks, 100, 101
Life cycle toxicity tests, 16–18

Ligurian Sea, 222, 225, 226
Limanda, 446
Limnoria lignorum, 386
Limpet, 284
Lindane, 189, 192, 194, 247, see also
        Cyclodienes
    in San Francisco Bay fish, 240
    in surface films, 209
Lipids
    and bioaccumulation, 8
    and biotransformation, 10
    and organochlorines, 183, 247
        DDT levels, 240
        PCBs, 201, 203, 205, 212,
        214,220, 221
    and PAHs, 159, 171
    zooplankton, 220, 221
Littorina, 101, 104
Liver
    organochlorines in
        aldrin, 193
        hexachlorobenzene, 198
        PCBs, 218, 229
    PAH metabolism, 173
Liverpool Bay, England, 34, 55, 56
    heavy metal pollution, 281, 282, 302
    organic loading, 35
Lobster (*Panulirus interruptus*), 101, 102
Lobster, American (*Homarus americanus*),
    106
    aquaculture, 419, 448
    chlorine and, 422, 423
    oil spill effects, 94
    PAHs, 151, 152, 154
London Dumping Convention, 345, 350
Long Island Sound
    dredged-material disposal, 373, 374
    heavy metals, 265
    organic loading, oxygen depletion, 39
    PAHs, 163
        fish concentrations, 157
        in shellfish, 154
        shellfish concentrations, 153
    PCBs, 218, 219
Louver screens, 431, 432, 449
Low-level nuclear wastes, 332, 333, 351
Lugworm, 103
Luidia, 105
Lutra lutra (otter), 232

M

Mackerel, Atlantic (*Scomber scombrus*)
    Oyster Creek Nuclear
        Generating Station
        construction, 438–439
    PAH concentrations, 151, 152
    thermal discharges and, 412
Macoma carlottensis, 386
Macroalgae, 8, see also Seaweed
Macrobrachium, 419
Maena maena, 227
Magnesium
    concentration factors for different
        classes of organism, 340
    nutrient requirements, 43
Mammals
    heavy metal concentration, 287, 288
    oil spill effects, 95, 114
    organochlorines
        PCBs, 232, 233
        Puget Sound, 245, 246
        San Francisco Bay, 243
    radiosensitivity, 341
Manchester Ship Canal, 359
Manganese, 302, see also Heavy metals
    concentration factors for different
        classes of organism, 341
    dredged-material effects, 378
    enrichment factors, 293, 294
Mangroves, oil spills, 88, 91, 92, 98, 120
Marine Protection, Research, and Sanctuaries
        Act (MPRSA) of 1972, 3, 389
Marsh habitat, dredge spoils and, 365–367
Massachusetts Bay, 143
Matagorda Bay, 49
Matagorda Bay, Texas, 49, 163
Mathematical models, 49–52, 56
Mechanical stresses, entrainment and, 435
Mechanical vibration, fish response, 432
Median lethal concentrations, 16
Mediomastus ambiseta, 116, 385
Mediterranean Sea
    oil spills, 85
    PAHs, 144
    PCBs, 208, 211, 221, 222, 224–227
Melanogrammus aeglefinus (haddock), 109
Menhaden (*Brevoortia patronus*), 197
Menhaden, Atlantic (*Brevoortia tyrranus*),
    444

entrainment effects, 424
    impingement losses, 428, 441
    thermal discharges and, 415–417, 434
Menidia (silverside), 18, 214, see also
    Silverside, Atlantic
Menidia beryllina, 107
Menidia menidia, 107
Mercaptans, crude oil, 71
Mercenaria mercenaria, see Clam, hard
Mercury, 22, see also Heavy metals
    bioavailability survey, 213
    Chesapeake Bay, 295
    concentration of in-marine fauna, 287
    dredging and, 359, 379
    Minamata disaster, 2
    San Francisco Bay, 298–301
    toxicity, relative, 264
Merluccius merluccius, 227
Metabolism, see also Bioaccumulation;
        Toxicity
    bioaccumulation, 6–10
    and bioelimination, 13
    biotransformation, 10, 11
    DDT, 187
    elimination of xenobiotics, 11–14
    heavy metals, 264
    heavy metals homeostasis, 285
    microbial, see Microbial degradation
    PAHs, 136, 137, 166, 167, 159,
        171, 172
    of petroleum hydrocarbons, 105
    temperature and, 406, 407
    toxicity, 14–19
Metallothioneins, 264, 285
Metals, see also Heavy metals
    dredging effects, 358
    Puget Sound pollutants, 243
    waste disposal strategies, 5, 6
Methemoglobin, chlorine toxicity, 423
Methoxychlor, 189
Methylated PAHs, uptake and
        biotransformation, 159
3-Methylcholanthrene, 136
Methyl mercury, 12, 14, 284
Methylnaphthalene
    classification of PAHs, 162, 163
    uptake and biotransformation, 159
    toxicity of, 161, 162
Methylparathion, 197

Methylphenanthrene, 162, 163
Microbial degradation
    DDT, 186
    dredged materials, 380
    oil spills, 81–85
    PAH biotransformation, 173
Micrococcus, 82
Micropogonias undulatus (Atlantic
        croaker), 139
Microstomus knitt (lemon sole), 109
Midges, 18
Millstone Nuclear Power Station, 422, 424,
        435
Minamata disaster, 2
Mining
    heavy metal pollution, 277
    nuclear fuel cycle, 327, 332, 351
    San Francisco Bay pollution, 297
Minnow, fathead (*Pimephales promelas*), 18,
        224
Minnow, sheepshead (*Cypinodon
        variegatus*), 18
    PAH toxicity, 161
    PCB concentrations, 214
Mirex, 189, 190, 192, 195, 196, 247, see also
        Cyclodienes
Mississippi River
    hydrocarbon concentration, 85
    PAHs, 163
Mississippi River Flyway, PCB
        contamination, 228, 231
Mitella, 104
Mixed function oxygenase (MFO) system,
        105
    detoxification mechanisms, 10, 11
    PAH metabolism, 136, 137, 150, 167,
        172
Mobile Bay, 49, 190
Modiolus, 104
Mojarra, 417
Molecular structure, 15
Molecular weight
    PAHs, 136
    and toxicity, 9
Mollusks, see also Clams; Oysters
    bioaccumulation by, 8, 9
    biotransformation ability, 10, 11
    heavy metal contamination, 284

bioindicator function, 303
    homeostasis, 285, 286
oil spills, 91, 93
    hydrocarbon concentration
        in, 104
    life stages and, 99–101
organochlorines, 187
    endrin, 193, 194
    PCBs, 214, 215, 221–223
    toxaphene, 196
PAHs
    concentrations of, 150–156,
        159, 160
    sentinel function, 174
    transformation of, 167
radioactivity and
    concentration factors, 340, 341
    radiation doses, annual, 342
    radiosensitivity, 341
Molting
    and biotransformation ability, 11
    PAHs and, 136, 137
    and PAH transformation, 167
    and xenobiotic clearance rate, 12
Molybdenum, 43, 302, 341, see also Heavy
        metals
Monitoring, see also National surveys;
        Sentinel organisms; specific
        pollutants
    organochlorines, 247
    PCBs, 211–213
Morone americana, see Perch, white
Morone saxatilis, see Bass, striped
Mortality
    cold shock, 415
    cumulative percent, 17
    dredging effects, 359
    oil spills and, 86, 93, 94
    time-mortality curves, 15, 16
Mosquitofish (*Gambusia affinis*), 161
Motor vehicle emissions, PAHs, 148
Mudflats, oil spill effects, 88, 91, 121
Mugil auratus, 225
Mugil cephalus, see Mullet, striped
Mulinia lateralis, 410
Mullet, grey (*Chelon labrosus*), 224
Mullet, striped (*Mugil cephalus*)
    oil spills and, 107
    PCB concentrations, 214, 224, 226

Mulus barbatus, 225
Mulus surmuletus, 225
Mummichog (*Fundulus heteroclitus*), 18,
    106, 118
Municipal wastes
    PAHs, 148
    petroleum hydrocarbons in, 66, 67
    pollution sources, 1, 2
    San Francisco Bay, 297
Murre, oil spills, 113
Mussel, bay (*Mytilis edulis*)
    organochlorines, San Francisco Bay,
        239–241
    PAH concentrations, 155
Mussel, blue (*Mytilis edulis*)
    heavy metal contamination
        by antifouling paints, 275
        as sentinel species, 283, 284
    oil spill effects, 99, 106
    organochlorines
        PCBs, 211–213
        San Francisco Bay, 239
    PAHs, 172, 173
        concentrations of, 150, 151
        depuration of, 159
    PCB monitoring, 212, 213
Mussels (*Mytilus*), 289
    aquaculture, 419, 448
    heavy metal contamination, 280–282
        bioindicator function, 303
        concentration of, 287, 288
    oil spills, 99, 100, 101
        hydrocarbon concentration
           in, 104
        long-term effects, 103
    organochlorines, 215–216, 223
        DDT, 191
        PCBs, 221, 223
        San Francisco Bay, 239–241
    PAH concentration, 150–156, 159
Mya, 104, see also Clams
Mya arenaria, see Clam, softshell
Mycobacterium, 82
Mysidopsis bahia (mysid shrimp), 18
Mytilus californianus, 239
Mytilus edulis, see Mussel, blue
Mytilus galloprovincialis, 100, 221
Myxocephalus (sculpin), 446

N

Nanticoke River, 234, 294
Naphthalene
    bioelimination in fish, 14
    classification of PAHs, 162, 163
    and crab population, 102
    from petroleum, develomental effects,
        106
    Puget Sound, 245
    sewage sludge, 148, 149
    toxicity of, 136, 161
    uptake and biotransformation, 159
Naphthenes, petroleum versus biogenic, 72
Narragansett Bay
    hydrocarbon concentration, 85
    nitrogen and phosphorus levels in, 45
    PAHs, 169, 170, 174
    PCBs, 215
National Environmental Policy Act (NEPA)
    of 1969, 401
National surveys, see also specific pollutants
    NOAA Mussel Watch Project, 160–
        163, 174
    organochlorines
        DDT, 189–192
        PCBs, 215–219
Natural sources of pollutants
    PAHs, 167, 168, 173
    petroleum hydrocarbons, 66, 67
    radioactivity, 323–326, 335
Neanthes arenceodentata (sandworm), 18,
    103, 161, 162
Nebalia pugettensis, 386
Needlefish, Atlantic, 417
Nekton, oil spills, 95
Neopanope texana, 102
Nephytes cf. *cryptomma,* 384, 385
Nereis, 167–169, 174
Nereis diversicolor, 222, 223
Nereis succinea, 167
Neuston, 95
Neutron emission, 322
Newark Bay, New Jersey, 263, 302
New Bedford Harbor, Massachusetts
    PAHs, 144
    PCBs, 217, 219
New York Bight, 1–4
    dredged material disposal, 373–375

heavy metal pollution, 302
organic loading, 35, 38, 56
organochlorines
    hexachlorobenzene, 198
    PCBs, 210, 217–219
PAHs, 144, 148, 163
    concentrations of, 149
    fish concentrations, 157
    mollusk concentrations, 155
    redistribution in sediments,
        149, 150
    shellfish and finfish
        concentrations, 151–
        154
New York City, 2, 365
Nickel
    concentration of in-marine fauna, 288
    dredged material effects, 378
    enrichment factors, 293, 294
    San Francisco Bay, 298
    toxicity, relative, 264
Nicotine, 183
Night heron, black-crowned (*Nycticorax nycticorax*), 232, 242
Niobium, 341
Nisqually Reach, Washington, 198, 219
Nitrogen, 33, see also Organic loading
    and chlorine tolerance, 423
    crude oil, 71
    dredged sediments, 380, 383
    nutrient loading
        Chesapeake Bay, 42, 43
        comparison of levels in
        ecosystems, 45
        sewage, 34
    and oil biodegradation, 82, 83
Nitrogen solubility
    power generating station and, 447
    water quality, temperature and, 405,
        406
Nitzochia ovalis, 274
NOAA
    PAH study, 174
NOAA Mussel Watch Project
    PAHs, 160–163
Noflamol, 200, see also Polychlorinated
    biphenyls
Nonachlor, 189, 194

Nonpoint source runoff, 2
    organic loading, 44, 56
    PAHs, 171, 173
    pollution sources, 1
Nonpolar compounds, biotransformation, 10
Northern kingfish, 416, 417
Northern puffer, 416
Northern swellfish, 413
North Sea, 345
North Sea oil, mussel effects, 99
Norwegian Sea, 345
Nuclear explosions, 328–332
Nuclear fuel cycle, 326–328
Nuclear fuel reprocessing, 345
Nuclear power plants, 400
Nuclear weapons testing, 330, 331
Nuisance species, 447
Nutrient enrichment, 56
Nutrients, see also Nitrogen; Organic loading;
    Phosphorus
    dredging effects
        adverse, 380
        beneficial, 383
        spoil disposal and, 391
    measures of ecosystem health, 48
    and oil biodegradation, 82, 83
    as pollutants, 1, see also Organic
        loading
Nutritional status
    and biotransformation ability, 11
    and impingement mortality, 449
    and PAH transformation, 167
Nycticorax nycticorax, 232, 242

O

Ocean Dumping Act, 389
Oceanodroma leucorhoa (petrel), 111, 113
Oil pollution, 2, 9, 20, 21
    case studies, 115–119
        Arrow spill, 116, 117
        Exxon Valdez spill, 117–119
        Florida spill, 115, 116
    composition of oil, 67–72
        alkanes, 69
        alkenes, 70
        alkynes, 70
        aromatic hydrocarbons, 70, 71
        biogenous hydrocarbons, 71

classes or series, 69
cycloalkanes, 69
effects on habitats, 86–94
 mangroves, 91, 92
 mudflats and sandflats, 91
 salt marshes, 88–91
 seagrasses, 93
 subtidal, temperate, 93, 94
effects on organisms, 94–114
 benthos, 97–105
 birds, 109–114
 fish, 105, 108, 109
 mammals, 114
 phytoplankton, 95, 96
 zooplankton, 96, 97
fate of oil, 72–85
 dissolution, 78, 79
 emuylsification, 79
 evaporation, 76, 77
 microbial degradation, 81–85
 photochemical oxidation, 77, 78
 sedimentation, 79–81
 spreading, 74–76
PAHs, 133, 134, 137, 138, 148, 165
sources, 64–67
toxicity of crude oil, 71, 72
Oil refining, PAH sourcces, 273
Oithona nana, 96
Olefins, petroleum versus biogenic, 72
Oligochaetes, dredging effects, 385
Oncorhynchus, 419
Onchyrhynchus gorbuscha (pink salmon), 161
Oncorhynchus kisutch (coho salmon), 161, 223, 224
Open-water habitat, 366
Open-water spoil disposal, 367, 368, 373–375
Opsanus tau (toadfish), 139
Orconectes, 18
Organic carbon
 and PCB sorption, 201, 203
 as pollutant, 1
 waste disposal strategies, 5, 6
Organic chemicals, see also Chlorinated hydrocarbons; Polynuclear aromatic hydrocarbons
 dredging and, 358, 381

Organic loading, 19, 20, see also Eutrophication; Nutrient loading
 dispersion models of release effluent, 49–52
 particle model, 52, 54
 puff model, 51–53
 estuarine susceptibility to, 47–49
 nutrients, 43–47
 organic carbon enrichment, 38–43
 red tides and paralytic shellfish poisoning, 53–55
 sewage, 34–38
 disposal methods, 34–36
 sludge composition, 36–38
Organic particles
 dredging and, 358
 radionuclide dispersion, 336, 337
 sewage sludge, sorption of toxic pollutants, 36
Organochlorines, 183, see also Chlorinated hydrocarbons
 bioaccumulation, 8
 bioelimination, 12–14
 dredged-material effects, 378
 NOAA Mussel Watch Program, 160
 San Francisco Bay, 237, 238
Organometallic compounds, antifouling paints, 273–276
Organophosphates, 183, 184, 197
Organotin biocides, 275, 276
Osmerus mordex (rainbow smelt), 445
Osprey (*Pandion haliaetus*), 183, 186
 DDT body burdens, 183
 DDT effects, 186
Ostrea edulis (mollusk), 172, 173, 419
Otter (*Lutra lutra*), 232
Oxidation
 biotransformation processes, 10
 oil (petroleum), 77, 78
 PAHs, 166, 173
Oxidation-reduction reactions, 15
Oxidation state, and heavy metal distribution, 270
Oxychlordane, 189
Oxygen
 blue-green algae and, 407
 crude oil, 71
 dredging and, 358, 359

and impingement losses, 430, 449
measures of system health, 48
nutrient requirements, 43
and oil biodegradation, 82
organic carbon mineralization
   and, 44, 46
organic loading and, 39, 40, 46
temperature and, 405–407
Oyster, American (*Crassostrea virginica*),
   18, 264
   heavy metals, antifouling paints, 274,
      275
   PAHs, 152, 153
   PCB monitoring, 213
   thermal discharges and, 411
Oyster, Pacific (*Crassostrea gigas*), 18, 100,
   274
Oyster, rock (*Saccostrea commercialis*), 275
Oyster Creek Nuclear Generating Station,
   437–443
   construction effects, 438, 439
   cooling system effects, 439–443
   entrainment losses, 424
   impingement losses, 429–431
   zooplankton mortality, 410
Oysters (*Crassostrea*), 18, 411
   acute toxicity tests, 18
   aquaculture, 419, 448
   dredged-spoil disposal effects, 382
   heavy metal contamination, 274, 287–
      289, 295
   NOAA Mussel Watch Program, 160
   oil spills, 94, 100, 101, 104
   organochlorines
      DDT, 191
      PCBs, 215, 216, 222, 236
      pentachlorophenol, 198, 199
   PAHs
      Chesapeake Bay, 139
      uptake of, 159

P

Pachygrapsus crassipes (crab), 101, 102
Pacifastacus ieniusculus, 18
Pacific Flyway, PCB contamination, 228,
   231
Pagellus acarne, 227
Pagellus erythrinus, 227

PAHs, see Polynuclear aromatic
   hydrocarbons
Paints, heavy metals, 273–276
Palaeomontes pugio, see Shrimp, grass
Palaeomontes vulgaris, see Shrimp, grass
Palos Verdes Peninsula, PCBs, 211, 217–
   219
Pandion haliaetus (osprey), 183, 186
Panulirus interruptus (lobster), 101, 102
Paracentrotus lividus, 106
Paralytic shellfish poisoning, 53–55
Paraprionospio pinnata, 385
Parathion, 197
Parophys vetulus, see Sole, English
Particle models, 52, 54, 56
Particulates, see also Organic particles
   and impingement losses, 430
   and oil sedimentation, 80, 81
   PCB affinity, 207, 208
   radionuclide dispersion, 336, 337
   sewage sludge, sorption of
      pollutants, 36
Partitioning of contaminants, dredged, 379,
   380
Patapsco River, 290, 292
Patapsco subestuary, 292
Patella (limpet), 284
Pathogens, 1, 56
   disposal strategies, 4
   in sewage sludge, 37, 38
Patinopecten (scallops), 419
Pautuxent River, 41–43, 45, 56
PCBs, see Polychlorinated biphenyls
Pecten, see Scallops
Pecten maximus, 275
Pecten opercularis, see Scallops
Pelican, brown (*Pelecanus occidentalis*), 186
   DDT effects, 186
   PCBs, 224, 228–230, 232
Pelican, white, 229
Pelicans, oil effects, 113
Penaeus aztecus, see Shrimp, brown
Penaeus duorarum (pink shrimp), 18, 159
Penaeus setiferus, 18
Penguins, oil effects, 113
Penicillium, 82
Penobscot Bay, Maine
   organic loading, 50
   PAHs, 142, 143

Penobscot River, Maine
  PAHs, 142, 143
  pentachlorophenol, 199
Pentachloroanisole, 199
Pentachlorophenols, 189, 190, 198, 199
Pentachorophenol, 247
Perch, white (*Morone americana*)
  chlorine toxicity, 423
  impingement losses, 426–428
  PCBs, 236
  thermal stress, 413, 434
Perch, yellow (*Perca flavescens*), 414, 419, 448
Persistence, DDT, 186
Perthane, 183
Perylene, 162, 163
Pesticides
  cyclodienes, 192–195
    aldrin, 192, 193
    chlordane, 194, 195
    dieldrin, 193
    endrin, 193, 194
    lindane, 194
  DDT, 185–192
  hexachlorobenzene, 198
  kepone, 195
  mirex, 195–196
  pentachlorophenol, 198, 199
  toxaphene, 196–198
Petrel (*Oceanodroma leucorhoda*), 111, 113
Petroleum hydrocarbons, see also Oil pollution
  bioaccumulation of, 8, 9
  crude oil composition, 67–72, see also Oil spills
  dredging and, 358
  elimination mechanisms, 12–14
  PAHs, 170
  San Francisco Bay, 297
  structure of, 68
pH
  dredging and, 358, 369
  and PCB vaporization, 203
Phaeocystis, 35
Phalacrocorax (cormorant), 229
Phalaropes, 114
Phenanthrene, 133, see also Polynuclear aromatic hydrocarbons
  in Chesapeake Bay, 140, 141, 168

classification of PAHs, 163
  in crude oil, 71
  in lobsters from Nova Scotia, 152
  in Puget Sound, 245
  in sewage sludge, 149
  in shellfish, 154, 155
  toxicity of, 136, 162
  uptake and biotransformation of, 159
Phenochlor, 200, see also Polychlorinated biphenyls
Phenols
  chlorinated, 189, 198, 199, 247
  crude oil, 71, 106
Phoca vitulina (common seal), 232
Phocoena phocoena (porpoise), 232
Phospholipids, 11
Phosphorus, 33, see also Nutrient loading
  dredged sediments, 380, 383
  nutrient loading in Chesapeake Bay, 42, 43
  and oil biodegradation, 82, 83
Phosphorus-containing biocides, see Organophosphates
Photic responses of fish, 432
Photolysis, 15
Photooxidation
  DDT, 187
  oil (petroleum), 77–79
  PAHs, 166, 171, 173
Photosynthesis, petroleum hydrocarbon and, 95, 96
Phthalate esters, 245, 289
Phusa hispida (ringed seal), 232
Phyllophora truncata, 97
Physa integra, 18
Physical factors
  nutrient loading, 46–49
  oil (petroleum), fate of, 72–85
  PAH transformation, 166, 167
  and toxicity, 15
  waste disposal site selection considerations, 5
Physical properties, PCBs, 200–206
Phytoplankton
  bioaccumulation in, 8
  DDT uptake, 187
  dredging effects, 369
  heavy metal concentration, 287, 288
  oil spill effects, 95, 96

organic loading and, 56
eutrophication, 38, 39
from sewage, 35, 36
PCBs, 208, 219, 220
power plant effects
biocides, 422
entrainment, 424, 435
Oyster Creek Nuclear
Generating Station, 440
thermal discharges, 407–409
radioactivity and
radiation doses, annual, 342
radionuclide concentration,
351
radionuclide dispersion, 336
turbidity and, 372
Pilchard (*Sardina pilcardus*), 109, 226
Pilgrim Nuclear Power Station, 424
entrainment, 446
impingement, 445, 446
thermal disharges, 444–446
Pimephales promelas (minnow, fathead), 18,
224
Pinfish (*Lagodon rhomboides*), 18
PCBs, 223
thermal shock responses, 434
Pioneering forms, 363, 364
Pipefish, 104
Pipeline dredge, 362
Placopecten magellanicus (sea scallop), 155
Plaice (*Pleuronectes platessa*), 107, 109, 423
chlorine toxicity, 423
oil spills, 109
Plaice, American (*Hippoglossoides
platessoides*), 446
Planes, 104
Plankton, see also Ichthyoplankton;
Phytoplankton; Zooplankton
entrainment, 433
measures of ecosystem health, 48
oil spills, 95, 104
Oyster Creek Nuclear Generating
Station, 440
radionuclide dispersion, 336
Plants
concentration factors, radionuclides,
340, 341
dredging and dredged-spoil dumping
effects, 364, 365, 391

nutrient requirements, 43
oil spills
Arrow, 117
hydrocarbon concentration in,
104
salt marsh, 89–91
Platichthys stellatus, see Flounder, starry
Pleuronectes flesus, 227
Pleuronectes platessa, see Plaice
Plutonium, 14, 328
Point source control, 3
Polar compounds, biotransformation, 10
Policy, see Legislation; Regulation
Pollutants, types of, 1, 2
Polonium-210, 334, 335, 351
Polychaetes
acute toxicity tests, 18
bioelimination of pollutants, 11, 13,
14
dredging and dredged-spoil disposal
effects, 363, 384–387
mixed function oxygenase
activity, 11
oil spill effects, 91, 93, 103
repopulation, 99
Wild Harbor River recovery,
115, 116
PCBs, 214, 223
thermal discharges and, 410
Polychlorinated biphenyls (PCBs), 163, 184,
185, 247
bioaccumulation, 8
Chesapeake Bay, 234–243, 289
concentrations in biota
factors affecting, 214
national surveys, 215–219
effects on biotic groups, 219–233
benthic invertebrates, 221–223
birds, 224, 228–232
fish, 223–227
mammals, 232, 233
phytoplankton, 219, 220
zooplankton, 220, 221
fish and shellfish, 188, 189
NOAA Mussel Watch Program, 160
physical-chemical properties, 200–
206
Puget Sound, 243–246
San Francisco Bay, 237

sediment concentrations, 209–211
sewage sludge, 36, 56
trade names, 200
water concentrations, 207–209
Polychlorinated terphenyls, 233
Polynuclear aromatic hydrocarbons
    (PAHs), 21
    bioelimination in fish, 14
    biotransformation of, 11
    chemical and physical properties, 134,
      136
    crude oil, 71
    distribution in biota, 150–163
      accumulation, depuration, and
        toxicity, 150–162
      NOAA Mussel Watch Project,
        160–163
    distribution in sediments, 137–150
      Boston Harbor, 142
      Chesapeake Bay, 139–141, 289
      global, 143–145, 148
      New York Bight, 148–150
      pathology, 138, 139
      Puget Sound, 138, 146–148
    distribution in water, 163–165
    environmental cycle, 170–172
    in estuarine systems, 167–170
      Chesapeake Bay, 167–169
      Narragansett bay, 169, 170
    examples of, 135
    metabolism, 136, 137
    from oil spills, 105
    from power plants, 400
    Puget Sound, 246
    sources, 134
    transformation mechanisms, 166, 167
    types and sources, 133–135
Polyvinyl chloride, 276
Pomatomus saltatrix, see Bluefish
Porpoise, common (*Phocoena phocoena*),
    232
Porpoise, harbor, 245
Portunus, 104
Potassium
    concentration factors for different
      classes of organism, 340
    nutrient requirements, 43
Potassium-40, natural levels in seawater, 335

Potomac River, 41–43, 49, 167–169,
    234, 294
Pout (*Trisopterus luscus*), 429
Power plants, see Electric power generating
    stations
Prawns, 419, 448
Precipitation
    DDT transport, 185
    PAH content, 165
Primordial radionuclides, 325, 326
Procambarus, 18
Procellaruformes, 114
Protogonyaulaux catenella, 54
Protogonyaulaux excavata, 54
Protogonyaulaux tamarensis, 54
Protozoa, 56
    radiosensitivity, 341
    in sewage sludge, 37
Prudhoe Bay crude oil, 97, 103, 110–112
Pseudeurythae paucibranchiata, 384, 385
Pseudomonas, 82
Pseudopleuronectes americanus, see
    Flounder, winter
Psychrophilic bacteria, 83
Public Law 92-500, 401, 425, 426
Public Law 95-217, 401
Puffer, northern, 416
Puffin (*Fratercula artica*), 113
Puff model, 51–56
Puget Sound, 113
    heavy metal pollution, 280–283, 302
    organochlorines, 243–246
      hexachlorobenzene, 198
      PCBs, 220–221, 223
      pentachlorophenol, 199
    PAHs, 142, 146, 147
Pyralene, 200, see also Polychlorinated
    biphenyls
Pyranol, 200, see also Polychlorinated
    biphenyls
Pyrene
    Chesapeake Bay, 140, 141, 168
    classification of PAHs, 162,
    concentrations of, 146
    fish, concentrations in, 156, 157
    in lobsters from Nova Scotia, 152
    in shellfish, 154, 155
Pyrethrum, 183

Pyridine, 71
Pyrodinium bahamaense var. *compressa,* 54

Q

Quinolines, 71

R

Radioactivity, 23, 24
    biological effects, 334–338
        bioaccumulation, 339–341
        genetic, 344
        somatic, 339, 341–344
    defined, 318–321
    disposal strategies, 4
    nuclear power plants, 400
    sources of, 323–333
        anthropogenic, 326–331
        natural, 323–326
    types of ionizing radiation, 321–323
    types of wastes, 317, 331–333
    waste disposal, 345–350
        management strategies, 349, 350
        storage and disposal systems, 346–349
Radionuclides
    bioelimination, 12–14
    Chesapeake Bay, 289
Radium, 335
Radon, 334
    natural levels in sewater, 335
    nuclear fuel cycle, 327
    sources of, 334
Ralus longirostris, 232
Rangia cuneata, see Clam, wedge
Rapid filter beds, 432
Rappahannock River, 41–43
Rappahannock Shoals, 384, 385
Raritan Bay, New Jersey, 49, see also Hudson-Raritan estuary
    PAHs, shellfish and finfish concentrations, 151, 152
    PCBs, 210
Razorbill (*Alco torda*), 113
R.C. Storer spill, 108
Recolonization, after dredging, 359, 363
Recycling, 3, 350

Redox conditions, and heavy metal distribution, 270
Redox reactions, 15
Red hake (*Urophycus chuss*), 157, 158
Red tides, 53–55
Regulation, 3
    dredged material disposal, 388, 389
    organochlorines
        DDT, 184, 185
        PCBs, 203, 206
    power plants, 401, 402
    radionuclides, 317, 345
Reprocessing, nuclear fuel cycle, 328
Reproductive cycle
    and PCB concentrations, 214
    temperature and, 407
Respiration
    chloramine and, 423
    temperature and, 405, 407
Rhine River, PAHs, 145
Rhine/Waal/Meuse/Scheldt estuaries
    heavy metal pollution, 281, 302
    PCBs, 207, 208, 211
Rhizophora, 92
Rhone River estuary, PAHs, 142
Ricker's theory of fisheries dynamics, 426
Riprap, 388
Ristroph screens, 431, 442, 443, 449
River and Harbor Act of 1899, 389
Riverine inflow, see also Freshwater inflow
    heavy metals, 265–271
    organic loading, sewage, 34
    PAH sources, 146, 173
    petroleum hydrocarbons in, 66
    pollution sources, 1
Riverine silt, dredged-material behavior, 375
River Mersey, 34
River systems, see also specific rivers
    comparison of nutrient levels in, 45
    PCBs, 208
Road runoff, PAHs, 148
Rockfish, 236
Rock melt waste disposal of radioactive wastes, 347
Rock sole, 284
Rockweed (*Fucus spiralis*), 117
Rotating disc screens, 432
Rotating traveling screens, 449

Rotenon, 183
Rotifers, 442
Rubidium, 335
Runoff
    organochlorines, 184
        DDT, 185
        kepone, 195
    PAHs, 134, 142, 146, 171, 173
    petroleum hydrocarbons in, 66, 67
    pollution sources, 1, 2
    Puget Sound, 243

S

Saccostrea commercialis (rock oyster), 275
Salicornia, 90
Salinity
    dredged-spoil disposal and, 392
    and eutrophication, 40, 41
    and impingement losses, 430, 449
    and oil biodegradation, 82, 83, 95
    and PAH transformation, 167
Salmo gairdneri (rainbow trout), 18
    aquaculture, 419, 448
    oil spill effects, 107
Salmon (*Oncorhynchus; Salmo*)
    aquaculture, 419, 448
    oil spills, 118
    temperature limits, 413
Salmon, coho (*Oncorhynchus kisutch*)
    PAH toxicity, 161
    PCBs, 223, 224
Salmon, pink (*Onchyrhynchus*
        gorbuscha), 161
Salt marshes
    dredged spoil deposition, 382
    oil spill effects, 88–91, 98, 116, 120
Salt marsh plants, 98
Saltwater intrusion, 366
Salvelinus fontinalis (brook trout), 18, 414
San Antonio Bay, 49
Sand dab (*Citharichthys stigmaeus*), 18
Sandflats, 91
San Diego Bay, 219, 275
Sand smelt (*Atherina presbyter*), 429
Sandworm (*Neanthes arenceodentata*), 18,
        103, 161, 162
San Francisco Bay
    dredged material disposal, 373

    heavy metals, 279, 293–301
    nitrogen and phosphorus levels in, 45
    organochlorines, 236–243
        in biota, 239–243
        DDT, 190
        PCBs, 219
        in sediments, 236–238
        toxaphene, 196, 197
    pollution susceptibility, 49
San Onofre nuclear power stations, 422
San Pablo Bay, see San Francisco Bay
San Pedro Harbor, California, 49, 219
Santa Barbara oil spill, 80, 97
Santa Monica Bay, California, 34, 55, 56
    heavy metal pollution, 281
    organochlorines
        hexachlorobenzene, 198
        PCBs, 219
Santee River, 49
Santotherm, 200, see also Polychlorinated
        biphenyls
Sarcina, 82
Sarda sarda, 226
Sardina pilcardus (pilchard), 109, 226
Sargasso Sea, PCBs in, 208
Sargassum, hydrocarbon concentration
        in, 104
Sarotherodon (tilapia), 419
Saurida undosquamis, 227
Saxitoxin, 54
Saybrook (Long Island Sound), 374, 375
Scale models, 56
Scallop, sea (*Placopecten magellanus*), 155
Scallops (*Pecten*), 101
    aquaculture, 419
    heavy metals and, 275
    oil spill effects, 94, 104
Scendesmus, 220
Scheldt estuary, 213, 281
Schistomeringos, 387
Schistomeringos japonica, 387
Schistomeringos rudolphi, 387
Scoloplos pugettensis, 386
Scomber scombrus (Atlantic mackerel), 151,
        152, 412
Scophthalmus aquosus, 157
Scorpaena scrofa, 227
Scoters, 113

Screens, power plant water intake system, 432, 442, 443, 449

Sculpin (*Myxocephalus*), 446

Scup, 416, 417

Sea bass, black, 416

Seabed disposal of radioactive waste, 348, 349

Sea eagle, white-tailed (*Haliaetus albicilla*), 228, 229

Seagrasses, 98, 409
    dredging effects, 364, 365
    oil spill effects, 93, 120
    power plants and, 408, 409

Seal, gray (*Haliocherus grypus*), 117, 232, 233

Seal, harbor (*Phoca vitulina*)
    heavy metal contamination, 284
    oil spills, 114, 117
    organochlorines, 232, 243, 246
        Chesapeake Bay, 243
        Puget Sound, 245

Seal, ringed (*Phusa hispida*), 232

Sea lions, oil spills, 114

Seals
    heavy metal concentration, 287, 288
    PCBs, 232, 233

Sea robin, 416, 442

Seasonal cycles
    and biotransformation ability, 11, 167
    and heavy metal pollution, 299, 300
    impingement losses, 429, 430, 449
    nuclear fallout, 329, 330
    and organic loading, 44
    and PAHs, 153, 154, 167
    and PCB concentrations, 214

Sea trout, spotted (*Cynoscion nebulosus*), 197, 236

Sea urchins (*Strongylocentrotus*), 102

Seaweed, 97, 104

Secondary treatment, sewage sludge, 36

Sediment, 342
    dredged material behavior, 374, 375
    and dredge site repopulation, 363
    dredging devices, 361–363
    dredging effects, 357, 358
        adverse, 369
        beneficial, 382, 383
    heavy metal recycling, 268

oil spills, 98
    biodegradation in, 84
    salt marsh recovery, 89
    sedimentation mechanisms, 80
    Wild Harbor River recovery, 115, 116

organochlorines
    PCBs, 201, 203, 207, 209–211, 234, 235
    Puget Sound, 243–245
    San Francisco Bay, 236–238

PAHs, 138, 146, 163, 165, 173
    Chesapeake Bay, 168
    distribution of, 137–150
    Narragansett Bay, 170
    New York Bight, 148
    Nova Scotian coking facility and, 152
    transformation mechanism, 166

radionuclide dispersion, 336, 337

waste disposal site selection considerations, 5

Sediment, suspended, see Suspended matter

Sedimentation
    heavy metal removal, 269
    oil spills, 79–81

Sediment inflow, 290

Selenium, 22, see also Heavy metals
    body burdens in striped bass, 294
    San Francisco Bay, 298–300

Sentinel organisms, 9, 10, see also Mussel, blue
    PAHs, 174
    PCBs, 211–213

Serratia, 82

Sewage disposal
    and organic loading, 34–38, 43
        methods of disposal, 34–36
        sludge composition, 36–38
    organochlorines
        DDT translocation, 186
        kepone, 195
    PAHs, 133, 165, 170

Sewage sludge, 2, 55
    composition of, 36–38
    heavy metals in, 273, 276, 281, 282, 302

PAHs, 148, 149
PCBs, 210, 211
pollution sources, 1
regulation of disposal, 3
Sex
and biotransformation ability, 11
and PAH transformation, 167
and PCB accumulation in fish, 224
Sexual maturity, and PCB concentrations, 214
Shad, American (*Alosa sapidissima*), 434
Shad, gizzard, 414
Shear-waters, 113
Sheepshead minnow (*Cypinodon variegatus*), 18, 161, 214
Shellfish, see also specific shellfish
dredged-spoil disposal effect, 432
heavy metal contamination, 295
oil spills, 94
organic loading, and oxygen depletion, 39
organochlorines, 187, 188, 190, 248
aldrin, 192, 193
chlordane, 194
dieldrin, 193
lindane, 194
mirex, 195
PCBs, 219, 235, 236
San Francisco Bay, 239–241
PAHs
Chesapeake Bay, 139
concentrations of, 151–156, 159–160
environmental cycles, 172
Shiner, common, 414
Shipworms, 441
Shock, temperature, see Cold shock; Thermal shock
Shrimp, 364
acute toxicity tests, 18
aquaculture, 419, 448
oil spills, 94, 102
Shrimp, brown (*Penaeus aztecus*)
biotransformation ability, 11
PAH toxicity, 161
pentachlorophenol, 198
Shrimp, grass (*Palaeomonetes pugio; Palaeomonetes vulgaris*), 18

acute toxicity tests, 18
impingement mortality, 441
PAH toxicity, 161, 162
radiosensitivity, 341
Shrimp, mysid (*Mysidopsis bahia*), 18
Shrimp, pink (*Penaeus duorarum*), 159
Shrimp, sand (*Crangon septemspinosa*)
acute toxicity tests, 18
impingement mortality, 441
PCB concentrations, 214
Sidecasting dredge, 361, 390
Sigambra tentaculata, 384, 385
Silver, 22
in Chesapeake Bay, 295
concentration of in-marine fauna, 288
in San Francisco Bay, 298–300
toxicity, 264
Silverside, (*Menidia*), 18, 214
Silverside, Atlantic, 441, 442, 445
Sipunculids, 13, 14
Site selection, for waste disposal, 4, 5
Size
and accumulation rates, 8
and PCB concentrations, 214
and xenobiotic clearance rate, 12
Skaggerak, 113
Skeletonema costatum, 96
chlorine sensitivity, 436
heavy metal effects, 274
Skimmers, black, 382
Slimicides, 273
Smelt, hydrocarbon concentration in, 104
Smelt, rainbow (*Osmerus mordex*), 445
Smelt, sand (*Atherina presbyter*), 429
Smelting, metals, 243, 273, 282, 283, 289
Snails, 18, 101, 104
Snapper, gray, 417
Sole (*Solea solea*)
chlorine toxicity, 423
oil spills, 109
Sole, English (*Parophys vetulus*)
heavy metals, 284
oil spills, 107
PCBs, 244, 245
Sole, lemon (*Microstomus knitt*), 109
Solubility, 15
oil (petroleum), 78, 79
PCBs, 201, 203–205

Somatic effects of radiation, 339, 341–344
Sorfjord, Norway, 278, 289
Sorption, 15
    dredged materials, 380
    oil, 80, 81
    PAHs, 163, 165, 171
    PCBs, 201, 203, 207
    radionuclide dispersion, 336
    sewage sludge particulates and, 36
Soval, 200, see also Polychlorinated
        biphenyls
Spartina, 104
Spartina alterniflora, 97, 104, 117
Spartina anglica, 90
Speciation, heavy metals, 265–267, 285
Specific activity, radionuclide, 321
Spoil bank habitat, 365
Spoil disposal, see Dredging and dredged-
        spoil disposal
Spoil-island development, 382
Spot (*Leiostomus anthurus*)
    acute toxicity tests, 18
    heavy metals and, 274
    PAHs, 138, 139
    PCBs, 223
    power plant effects
        impingement mortality, 441
        thermal discharges, 416, 417,
        434
Spotted sea trout (*Cynoscion nebulosus*), 197,
    236
Sprattus sprattus, 227
Squalus acanthias, 227
Starfish, 105
Starry flounder, see Flounder, starry
Stenothermal periods, 407
Stickleback, threespine (*Gasterosteus
    aculeatus*), 18
Storms
    dredged-spoil disposal site
        disturbance, 367
    PAH redistribution, 149
    Puget Sound pollutants, 243
    and sedimentation, 290
Streblospio benedicti, 385
Stress, 301, see also Thermal stress
Striped bass, see Bass, striped
Striped mullet, see Mullet, striped
Strongylocentrotus (sea urchins), 102

Strontium, 302, see also Heavy metals
    concentration factors for different
        classes of organism, 340
    nuclear explosion and, 328
Sub-bottom containment, dredge spoils, 360
Sublethal toxicity, 15
Subtidal habitat, oil spill effects, 93, 94
Subtropical ecosystems, see Tropical and
        subtropical species; Tropical
        and subtropical systems
Succession
    dredging effects, 359
    measures of ecosystem health, 48
Suction dredges, 361, 374, 375, 390
Sueda maritima, 90
Sulfhydryl groups, heavy metal interactions,
    285
Sulfur, crude oil, 71
Surface films, PCBs in, 209
Surface runoff, see Runoff
Surface sediments, PAHs in, 146–148
Surface tension, water, 405, 406, 447
Surface to volume ratio, and toxicity, 8, 214
Surfactants, and oil biodegradation, 83, 84
Surry Power Station, 431
Surveys, see Natioal surveys
Suspended matter, see also Organic particle;
        Particulates
    dredging and dredged-material effects,
        369, 372, 376, 377
    heavy metals, 268
    and impingement losses, 430, 449
    and oil sedimentation, 80
Swellfish, northern, 413
Synergism
    chlorine and temperature, 423
    PAHs, 137
Synthetic organic compounds, see
        Chlorinated hydrocarbons;
        Polynuclear aromatic
        hydrocarbons
Syringodium filiforme, 93, 365

T

2,4,5-T, 190
Tamar estuary
    heavy metals, 269
    PAHs, 142, 144
Tampa Bay, Florida, 163, 364

Tampico Maru oil spill, 101, 102, 108
Tautog (*Tautoga onitis*), 412, 418
Tautogolabrus adspersus (cunner), 445
Tees estuary, 359
Temperature, see also Thermal discharges,
   power plants
 chlorine synergism, 423
 dredged-spoil disposal and, 392
 hydrothermal soures of heavy
   metals, 265
 and impingement mortality, 449
 and oil biodegradation, 81–83, 95
Temperature tolerance polygon, 407, 408
Teredo bartschi, 441
Teredo furcifera, 441
Teredo navalis, 441
Tern, Caspian, 242, 243
Tern, Forster's, 242
Terns
 dredged-spoil effects, 382
 oil effects, 113
 PCBs, 242, 243
Tests
 dredged material, 360, 379
 toxicity, 15–19
Tetramethyl benzenes, 73
Tetramethyl naphthalenes, 73
Tetraselmis suecica, 96
Thalassia testudinum
 dredging effects, 365
 oil pollution effects, 93
 power plant effects, 409
Thames estuary
 heavy metal pollution, 281, 282, 302
 organic loading, 35
Thermal discharges, power plants
 ecosystem effects, 406–417
 Oyster Creek Nuclear Generating
   Station, 439–442
 Pilgrim Nuclear Power Station, 444–
   446
 and water quality, 405, 406
Thermal loading, 399
Thermal shock, 434
Thermal stress, 399, 440
 entrainment and, 434, 435
 and fish mortality, 416, 417
Thorium isotopes, 335

Thunnus thynnus, 226
Thyroid damage, radiation and, 342
Thyrrenian Sea, 364
Tiber River, 211
Tides
 dredged-spoil disposal site
   disturbance, 367
 and impingement mortality, 430, 449
 and nutrient loading effects, 48, 49
 PAH redistribution, 149, 150
Tilapia, 419, 448
Time-mortality curves, 15, 16
Time of day, and impingement
   mortality, 449
Time of year, see Seasonal cycles
Toadfish (*Opsanus tau*), 139
Torrey Canyon oil spill, 2, 63, 79, 91, 109
Toxaphene, 183, 187, 189, 192,
   196–198, 247
 decline of, 190
 San Francisco Bay, 239, 240, 242
Toxicity, 2
 classification of elements, 272
 heavy metals, 264, 265, 303
 mechanisms of, 14
 oil, crude, 71, 72
 organic compounds, molecular weight
   and, 9
 organochlorines, 184, 185
   versus organophosphates, 183
   PCBs, 203, 206
 PAHs, 136, 150–162, 171
Toxins
 microbial, in sewage sludge, 37
 oil effects, 86
 red tides, 53–55
Trace elements, nutrient requirements, 43
Trachurus mediterraneus, 227
Trade names, PCBs, 200
Trailing dredge, 374, 375
Transformation, see Biotransformation
Transition metals, dredging and, 358
Transmissometry, 374, 375
Transport
 DDT, 185, 186
 dredged-material, 372, 373, 376, 377
 heavy metals, 264, 265
 oil, crude, 74–76

PAHs, 146, 149, 150, 170–172
  radionuclides in ecosystems, 336–338
Transuranic wastes, 332, 351
Tridacna maxima, 155
Triggerfish, 104
Trimethylnaphthalene, 162
Trinectes maculatus, 107, 138, 139
Triorganotin compounds, 273, 274
Triphenylene, 140, 141
Trisopterus luscus (pout), 429
Trithion, 197
Tritium, 335
Trochochaeta multisetosa, 387
Tropical and subtropical countries, DDT
    production in, 185
Tropical and subtropical systems
  oil degradation in, 81
  power plant thermal discharges
    and, 409, 441
Trout, brook (*Salvelinus fontinalis*), 18, 414
Trout, rainbow (*Salmo gairdneri*), 18, 107,
    419, 448
Trout, spotted sea (*Cynoscion nebulosus*),
    197, 236
Tubulanus pellucidus, 385
Turbidity, dredging and, 369, 372–380, 385,
    see also Suspended matter
Turbulence
  dredged-spoil disposal site
    disturbance, 367
  and organic loading, 44
  PAH redistribution, 149

U

Uca, see Crab, fiddler
Uca pugilator, 274, 275
Uca pugnax, 102, 116
Ultrasonic vibrations, 432
Ulva angusta, 97
Upeneus moluccensis, 227
Uptake, PCBs, 215–219
Uranium, 327, 332, 335
Urban sources, see also Municipal wastes;
    Sewage sludge
  PAHs, 133, 146–148
  petroleum hydrocarbons in, 66
Uria aalge (guillemot), 113, 229

Urophycus chuss (red hake), 157–158
U.S. Army Corps of Engineers, 388
  dredged material research, 369, 371,
    372
  dredged-spoil disposal site caps, 367
  dredging of Chesapeake Bay, 383
U.S. Department of Energy (DOE), 348
U.S. Environmental Protection Agency
    (EPA), 3
  chlorine emissions standard, 421
  organochlorines
    mirex, 195, 196
    PCB ban, 203
    toxaphene ban, 197, 198
U.S. Food and Drug Administration, 206
U.S. Nuclear Regulatory Commission
    (NRC), 350

V

Vanadium, 43, 302
Van't Hoff law, 406
Vaporization
  DDT, 187
  PCBs, 201
Vapor pressure
  PCBs, 203, 205
  temperature and, 406
Velocity caps, 431, 449
Ventura River, 80
Vermillion Bay, 49
Vertical traveling screens, 442, 449
Viruses, 37, 56
Viscosity
  power generating station and, 447
  water quality, temperature and, 405,
    406
Vitamins, nutrient requirements, 43
Volatility, 15

W

Waste disposal
  approaches to, 3–6
  biological effects, 6–19, see also
    Biological effects of waste
    disposal
  heavy metals, 276–282

petroleum hydrocarbons in, 66
pollution sources, 1, 2
radioactive, 345–350
Waste heat utilization, 417–419, 439, 448
Wastewater discharges, see also specific
    sources and types of wastewater
    organic loading, 55, 56
    PAHs, 134
Wastewater treatment, San Francisco
    Bay, 297
Water-column effects, dredged material, 360
Water flow, 366
Waterfowl, see Birds; specific common
    names
Water Pollution Control Act Amendments
    of 1972, 388
Water quality, 391
    dredging effects, 358, 360, 369, 372–
        380, 391
    power plants and, 405, 406
Water Resources Development Act
    of 1986, 366
Water-soluble compounds, 10
Water use, by power plants, 402–404
Wave action, see Turbulence
Weakfish (*Cynoscion regalis*)
    PAHs, 139
    thermal discharges and, 415–417, 442
Weapons testing, 330, 331
Weathering, heavy metal-sources, 265
Wedge clam, 378
Wedgewire screens, 432
Wetlands
    dredged-material disposal, 365–367,
        388, 391
    oil spills, 88
Whales, killer, 245
White-crested cormorant (*Phalacrocorax
    carbo*), 229
White perch, see Perch, white
White-tailed sea eagle (*Haliaetus albicilla*),
    228, 229
Wild Harbor River, 85, 88, 115
Windowpane (*Scophthalmus aquosus*), 157
Winter flounder, see Flounder, winter

X

Xenobiotics, see Chlorinated hydrocarbons;
    Polynuclear aromatic
    hydrocarbons
Xenon, nuclear fuel cycle, 328
Xiphias gladius, 227
Xylophaga washingtoni, 386

Y

Yellowcake, 327
Yellow-green algae, 408
Yellow perch, see Perch, yellow
York River, 41–43

Z

Zinc, 302, see also Heavy metals
    bioavailability survey, 213
    Chesapeake Bay, 295
    concentration in different classes of
        organism, 288, 340
    dredging and, 358, 359
    enrichment factors, 293, 294
    nutrient requirements, 43
    San Francisco Bay, 298, 301
    toxicity, relative, 264
Zooplankton
    bioaccumulation in, 8
    bioelimination mechanisms, 12, 13
    oil consumption, 81
    oil spill effects, 96, 97
    organic loading from sewage, 36
    PCBs, 208, 220, 221
    power plant effects
        biocides, 422
        entrainment, 424, 435
        thermal stress, 409, 410
    radioactivity and
        annual doses, 342
        Oyster Creek Nuclear
            Generating Station, 440
        radionuclide concentration,
           351
        radionuclide dispersion, 336